1st International Conference on Contemporary Research in Mechanical Engineering with Focus on Materials and Manufacturing (ICCRME-2018)

IOP Conference Series: Materials Science and Engineering
Volume 404

Lucknow, India
6 – 7 April 2018

ISBN: 978-1-5108-7256-1
ISSN: 1757-8981

Printed from e-media with permission by:

Curran Associates, Inc.
57 Morehouse Lane
Red Hook, NY 12571

Some format issues inherent in the e-media version may also appear in this print version.

Printed by Curran Associates, Inc. (2018)

For permission requests, please contact the Institute of Physics
at the address below.

Institute of Physics
Dirac House, Temple Back
Bristol BS1 6BE UK

Phone: 44 1 17 929 7481
Fax: 44 1 17 920 0979

techtracking@iop.org

Additional copies of this publication are available from:

Curran Associates, Inc.
57 Morehouse Lane
Red Hook, NY 12571 USA
Phone: 845-758-0400
Fax: 845-758-2633
Email: curran@proceedings.com
Web: www.proceedings.com

Table of contents

Volume 404

1st International Conference on Contemporary Research in Mechanical Engineering with Focus on Materials and Manufacturing (ICCRME-2018)

6–7 April 2018, Lucknow, India

Accepted papers received: 1 August 2018
Published online: 5 October 2018

Preface

Papers

IOP Conf. Series: Materials Science and Engineering **404** (2018) 011001 doi:10.1088/1757-899X/404/1/011001

ABOUT THE CONFERENCE (ICCRME-2018)

The objective of ICCRME-2018, organized by Department of Mechanical Engineering, Integral University, is to provide a common platform for the academia and industrial experts to exchange their views through interaction, paper presentation and sharing of knowledge and experiences, to tackle projects and problems in challenging areas in their respective fields.

The outcome of the conference will provide a formidable foundation for the building of a superstructure for the future prospects of technological advancements in the country. The organizing committee of conference invited potential authors to submit their paper based on original research findings and case studies suitably fitting the theme of the conference to ICCRME-2018.

All abstracts and papers were peer reviewed by experts from the Review Committee and manuscripts have been checked through plagiarism check software. After a thorough review process the papers were accepted for publication. The conference received over whelming response from the students and academicians within and outside the country.

The two day conference programme has been carefully structured to encouraged mutual inspiration and fruitful debate among researchers. The first day will be centered on a plenary session of five distinguished key note speakers and in-depth discussion of the presentation. We will also maintained focus throughout the day with short presentations by the participants.

All registered and presented papers will be published by IOP, Publishing Ltd. (Bristol, UK) as an open access volume of IOP conference series: Material Science & Engineering,

IOP Conf. Series: Materials Science and Engineering **404** (2018) 011001 doi:10.1088/1757-899X/404/1/011001

Conference photographs

IOP Conf. Series: Materials Science and Engineering **404** (2018) 011001 doi:10.1088/1757-899X/404/1/011001

IOP Conf. Series: Materials Science and Engineering **404** (2018) 011001 doi:10.1088/1757-899X/404/1/011001

IOP Conf. Series: Materials Science and Engineering **404** (2018) 011001 doi:10.1088/1757-899X/404/1/011001

IOP Conf. Series: Materials Science and Engineering **404** (2018) 011001 doi:10.1088/1757-899X/404/1/011001

IOP Conf. Series: Materials Science and Engineering **404** (2018) 011001 doi:10.1088/1757-899X/404/1/011001

IOP Conf. Series: Materials Science and Engineering **404** (2018) 011001 doi:10.1088/1757-899X/404/1/011001

IOP Conf. Series: Materials Science and Engineering **404** (2018) 011002 doi:10.1088/1757-899X/404/1/011002

Peer review statement

All papers published in this volume of *IOP Conference Series: Materials Science and Engineering* have been peer reviewed through processes administered by the proceedings Editors. Reviews were conducted by expert referees to the professional and scientific standards expected of a proceedings journal published by IOP Publishing.

Material Science & Metallurgical Engineering Education in India-Past, Present & Future

Aaditya Ranjan Srivastava[1], Shrish Bajpai[2] and Sushant Khare[3]

[1] UG Student, Mechanical Engineering, Faculty of Engineering, Integral University, Lucknow, India.
[2] Assistant Professor, ECE Department, Faculty of Engineering, Integral University, Lucknow, India.
[3] Executive, Tata Consultancy Services Limited, Bangalore, India.

a) saadityaranjan@gmail.com
b) shrishbajpai@gmail.com
c) khareysushant@gmail.com

Abstract.

Metallurgy as a science discipline has been established from the beginning of stone-age. It is imperative to note as we have progressed we have added facets of material science to it and the field has evolved into a discipline of engineering. From science fiction to reality the inherent inter-disciplinary nature of this branch sets up the need of its education and training. There have been several instances in when the pivot of development has completely changed due to advancement in material science and metallurgy whether it is semiconductor industry or space industry. In this paper, we have reviewed the scope and development of this discipline in India detailing through the course of history to the present status.

Keywords: Material Science, Metallurgical Engineering, Technical Education, India.

1. Introduction:

Material science, also known as material science and engineering is the science of metals and metallurgy is the study of extraction of metals. Its study is important as we deal with metals every day. To sustain life and energy, we are surrounded with metals and materials in various forms. The use of metals was started 5th and 6th millennium BC, even the periods and ages are being named after the major advancement in use of material. First came the Stone Age as during that era equipment and utensils were made of stones, then came the Bronze Age and Iron Age. Earlier, metals found in native state were being used then metals which can be extracted, reformed in simple way by heating were used by masses. These all evidence shows that the metallurgy engineering or material science was needed and used from much earlier period. In India, we can get concepts related to metallurgy in various Vedas as Yajurveda & Atharvaveda. Metals were exchanged in trades between India and nearby regions. Earlier people used metals which can be obtained easily and materials like gold, silver and so on were being extensively used by people. In ancient India, metals were used majority for making weapons for wars and for making utensils. The Iron pillar made in New Delhi, India, is a perfect example of showing that metallurgy or material science was being practiced in earlier India too. To let the pillar stand for generation it was made corrosion resistance and the concept of alloy was used in making of the marvel [1].

In medieval India, with arrival of foreigners, there was more improvement to the use of material in different product making activities. Material science or Metallurgy is the science of material in which every aspect which affects the physical and chemical process of material like nature, structure, bonding, environmental factors are studied. Material science is an interdisciplinary field included in almost every branches of engineering as every field is incomplete without use of materials, so it is important to have proper knowledge to have better understanding of the field. Material science or metallurgy is included as a subject or optional subject in curriculum of mechanical engineering, electrical engineering, chemical engineering, electrical and electronics engineering, aeronautics engineering, and various other branches of technology. Due to rapid development after the world war, the manufacturing field has grown rapidly. These manufacturing units of any country provide a great boon to economy of that country, and developing countries such as India which has biggest population of youth [2].

This paper is divided three parts, first we have discussed about the metallurgy and material science education in India at undergraduate, postgraduate, doctoral and post-doctoral level, then future scope of metallurgy in different industry with the employment of material science engineer and at last conclusion of material science education in India.

2. Basic Structure of Engineering Education in India:

First technical institute in India had been established in year 1847 (Thomson college of civil engineering) after that till now there are around ten thousand technical institutes in India. Technical education in India has been divided into five stages (diploma, under graduate, post graduate, doctoral research, and post doctoral research). There are 60 major polytechnic courses, 45 major under graduate courses, more than 200 post graduate courses, doctoral research in almost all major sub fields of engineering and post doctoral research in emerging areas of engineering are offered by the different technical engineering institute in India. Indian Institute of Sciences (IISc), Indian Institute of Technology (IIT), National Institute of Technology (NIT) and Indian Institute of Information Technology (IIIT) are top technical institutions of India [3, 5].

The duration of diploma polytechnic degree is three years or six semesters, undergraduate technical course is of four year or eight semesters, dual degree program course that is (B.Tech + M.Tech) or (B.Tech + MBA) is of five years or ten semesters. For further course like Doctoral research, the duration is between three years to six years according to the research topic. Post Doctoral program is generally of two years but, can be extended to one more year as per the requirement of research.

First dedicated independent department for Metallurgical Engineering in India was founded at Banaras Hindu University, Varanasi at 1923, awarded first undergraduate degree in 1927 & doctoral degree in 1955 [8]. Next department was open in Bengal Engineering College in 1939 as 3 year degree course in Metallurgy affiliated to Calcutta University. Department of Materials Engineering at Indian Institute of Science was open in year 1945 to offer undergraduate degree in material science. Oldest department related to the field is mentioned in the Table 1.

Material science as a subject is included in various branches of engineering at different period at undergraduate and post graduate level. It was first included with physics, and then was taught with chemistry but later due to its inevitability in various fields it was taken separately as a subject [4]. It also keeps in track the research development and projects related to this field. Then its branches were

Serial	Technical Institute	Institute Establishment Year	Department Establishment Year
1	Indian Institute of Technology (BHU) Varanasi (Old Name : Banaras Engineering College, Varanasi)	1919	1923
2	Indian Institute of Engineering Science and Technology, Shibpur (Old Name : Bengal Engineering College, Shibpur)	1856	1939
3	Indian Institute of Science, Bangalore (Old Name : Tata Institute , Bangalore)	1911	1945
4	College of Engineering, Pune	1856	1948
5	Jadavpur University, Kolkata (Old Name : College of Engineering and Technology, Jadavpur)	1912	1956

Table 1: Oldest departments of material science & metallurgy engineering in India with their establishment year.

Being set up in different IITs and NITs, many private institutions also included it as a separate degree course.

IOP Conf. Series: Materials Science and Engineering **404** (2018) 012001 doi:10.1088/1757-899X/404/1/012001

Both UG and PG degree courses were being offered and doctoral or PhD was done on the specific field one want to pursue. Although it is being taught as a separate degree course but still many branches have one or more of material science course as subject in UG level.

2.1 At Diploma Level:

Diploma courses associated to material science or metallurgy engineering has been offered by only few polytechnic. In first year subjects are common with other linked branches. In second year, specific subjects (Metal Forming, Fuel Furnaces, Metallurgical Analysis etc.) related to the discipline has been offered. Subjects like testing of metals, physical metallurgy etc. are offered as a compulsory subjects in mechanical, civil, industrial, production and manufacturing diploma students. In final semester specific subjects (Steel Making, Corrosion of Metals, Alloy Steel etc.) related to the specialization are offered.

2.2 At Under Graduate Level:

There are about 6-10 core courses which is associated to the Material Science & Metallurgical Engineering and rest of the subjects are related to the mechanical engineering discipline. Couple of courses are inter disciplinary courses (open elective) which are normally offered in final year. First year is common to all branches. In second year there are common subjects with mechanical engineering discipline. From fifth semester specialization in material sciences has been offered. There is one major project in the final year of the course in which student have to choose a specific topic related to material science or metallurgy engineering. Student need to choose the departmental elective associated to the project. Department of Metallurgical Engineering, Indian Institute of Technology Banaras Hindu University (IIT-BHU), Varanasi is the oldest department which laid out a proper format of curriculum for undergraduate studies. Core courses associated to the metallurgical engineering are thermodynamics, structure of material, extractive metallurgy, heat treatment, steel making, foundry metallurgy, mechanical behaviour of materials, iron making etc. Above courses give the in-depth knowledge of physical and chemical behaviour of metallic element, their associated alloys, heat treatment of different elements, plating, shot peen. Production of ores and manufacturing of metal associated with different items.

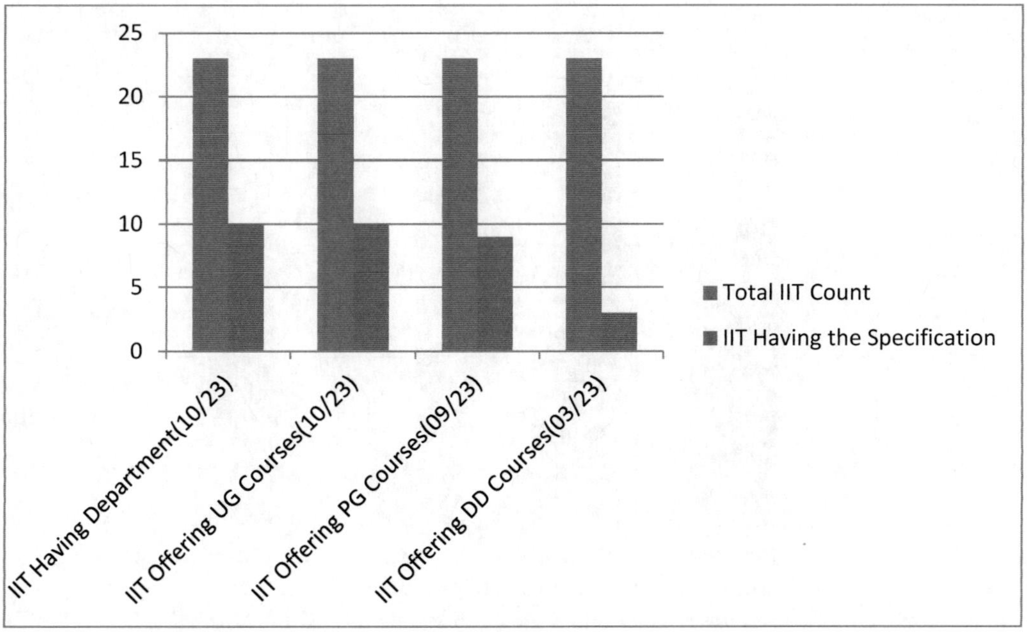

Fig 1: Current scenario of Material Science & Metallurgical Engineering in different Indian Institute of Technology in India.

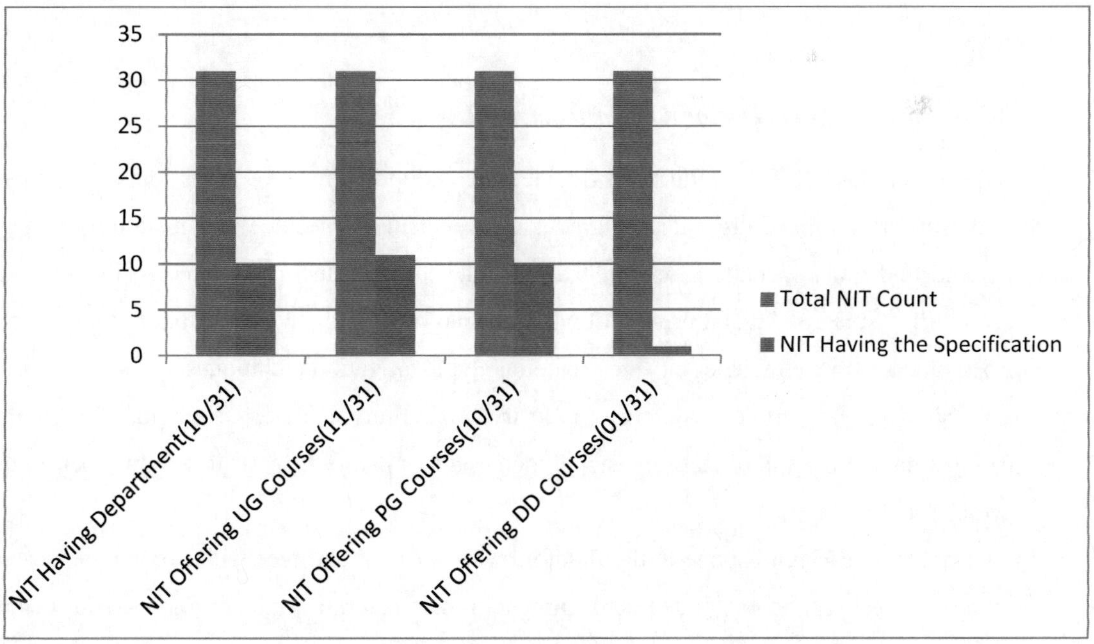

Fig 2: Current scenario of Material Science & Metallurgical Engineering in different National Institute of Technology in India.

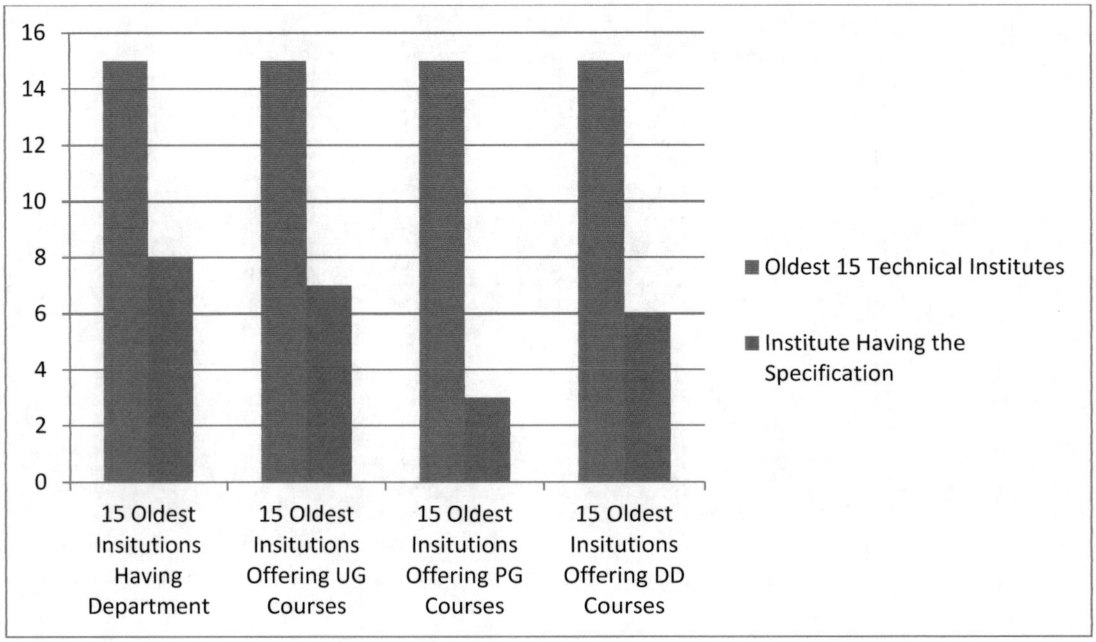

Fig 3: Current scenario of Material Science & Metallurgical Engineering in different Fifteen Oldest Institute in India.

.

2.3 At Post Graduate, Doctoral & Postdoctoral Level:

Material Science & Metallurgical Engineering is divided into two main domains ferrous metallurgy and non-ferrous metallurgy at post graduate level material science and metallurgical engineering is offered under the specialization in Material Science, steel technology, corrosion science, industrial metallurgy, process metallurgy, foundry technology etc. [6, 8]. Major topic covered under postgraduate course includes advance material characterization techniques, ceramics, steel technology, corrosion science, process energy, industrial metallurgy, material engineering, nano material, biomaterials in the end of last semester student have to justify his/her dissertation before the audience, which are the examiner.

Doctoral Research is done in the field of metal forming and mechanical behavior, material joining, iron and steel technology, integrated computational material engineering, material characterization, surface engineering, composite metal (bio-metal, ceramics, nano-metal) are the major fields in which major Indian researcher works.

Post doctoral research has been performed only in few technical premium institutes in the emerging areas of Metallurgical Engineering. At present level post doctoral research has been done in the areas of physical metallurgy and multi component metallic alloys.

2.4 About the Professional Bodies:

The Indian Institute of Metals (IIM) which is started just after independence in 1947 (December 29), is the oldest professional society of Metallurgical & Material Engineers in India. Associate Membership of Indian Institute of Metals (AMIIM) examination is conducted by the IIM which is equivalent to the diploma offered by the Indian polytechnics. Indian Stainless Steel Development Association (ISSDA) is another professional body for the engineers & professional, working with the metal companies which are specific to the steel sector. Aluminium Association of India (AIA) is a professional body of the engineers working towards the aluminium sector. All India Non Ferrous Metal Exim Association (ANMA) is another professional body associated with the working professional in the non-ferrous metal industry of India.

2.5 About Simulation Tools:

There are multiple simulation tools which are used in academics & research associated with the material science. Materials Studio (license software package) is used to create the modelling & simulation environment of different material. Materials Studio Visualizer is associated with the Materials Studio gives a graphical user environment. Avizo is also a license software package for scientific data visualization and analysis of different metals. Ansys, Abaqus and Matlab are the other commercial license software package, which is used in solving the complex mathematical problems as coated with the field. Prism Uncertainty Quantification (PUQ), Atom Touch, Material Simulation Toolkit (MAST) open source software used for uncertainty propagation, statistical model calibration and data analysis and validation of simulations that will eventually enable predictions with quantified confidence.

2.6 About Literature:

There are many fields of engineering (aerospace engineering, ceramic engineering, chemical engineering, civil engineering, electrical engineering, electronics engineering, industrial engineering, manufacturing engineering, mechatronics engineering, mechanical engineering, production engineering, renewable engineering and so on) in which Material Science & Metallurgical engineering subjects are offered as core subjects as well as elective subjects in India.

IOP Conf. Series: Materials Science and Engineering **404** (2018) 012001 doi:10.1088/1757-899X/404/1/012001

R. Balasubramaniam, O.P. Khanna, R.K. Rajput, R.S. Khurmi, U.C. Jindal are the famous authors whose books are adopted as the textbook in the different courses associated with the material science and metallurgy engineering. R.H. Tupkary (Iron and Steel Making), R.S Parmer (Material Joining Technologies), A.V.K.Suryanarayan (Mechanical Behaviour of Materials), R. Shridhar (Non-Ferrous Metal), Raj Narayan (Corrosion Engineering), V.K.Sikka (Physical Metallurgy), P.L.Jain (Foundry Technology), D.F. Sudarshan (Surface Engineering), T. Pradeep (Nanomaterials), R. Raj (High-Temperature Material), S.D. Sehgal (Emerging material)etc. are the famous indian authors who contributed in the field of material science and metallurgy engineering.

2.7 Online Web Resources by Indian Government:

There are twenty-six different courses in NPTEL database and it is offered by different IIT`s. Major courses are Defects in Materials, Material Characterization, Iron Making, Material Science and Engineering, Physics of Materials, Principles of Polymer Synthesis, Surface Engineering of Nanomaterials, Theory and Practice of Non Destructive Testing, Electroceramics, Analysis and Modelling of Welding, Introduction to Crystal Elasticity and Crystal Plasticity, and so on which are offered as four weeks, eight weeks & twelve weeks. Weekly, five to ten videos of duration about forty

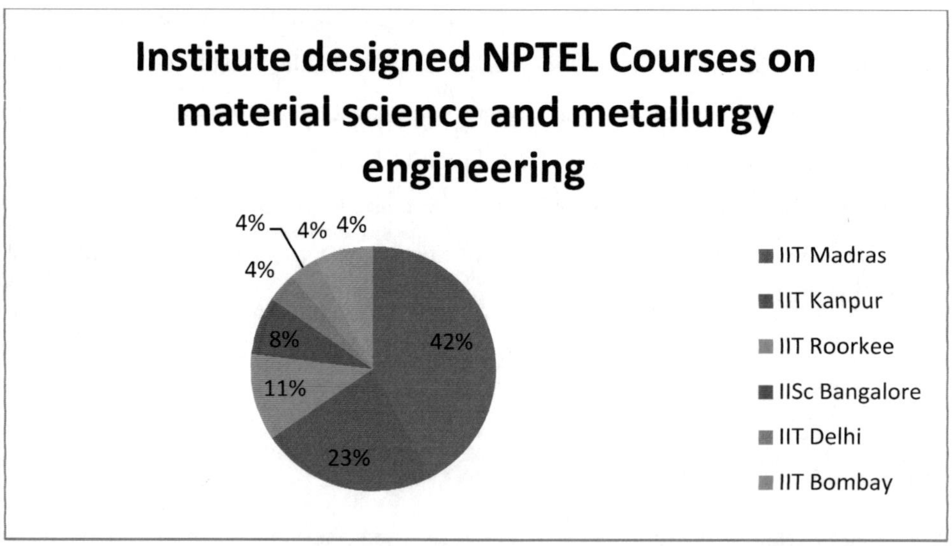

Fig 4: NPTEL lecture designed by the different institution in India.

Minutes uploaded on the course registered candidate profile which he can able to see & download it. Weekly quiz (objective) or assignment related to that week has to be submitted by the candidates. Marks are allocated for each quiz and a final examination is also conducted which covers the whole course. After completing any course successfully, a certificate is issued by the NPTEL with the logo of concerned institute and MHRD [7].

2.8 Job Prospect:

The metallurgical profession is a wide field which offers a variety of job prospects for the students who took its associated courses in technology and engineering. Metallurgical engineers in India take the employment with the companies involved in ore production and metal extraction companies. They can take the profile of plant engineer, metallurgist, welding engineer, quality check engineer, and Metallurgical R&D lab technician and so on. They are also eligible in railways, armed force and government sectors [10].

3. Conclusion

With push coming through programs such as 'Make in India' and other initiatives from government focused on marking India on world as a manufacturing hub, it is quintessential to start from the primal need of machines that is raw material [9]. Through our research we have found that Indian institutes have a good platform through years of educational background. The inherent need is to allocate more budget to research activities in the department to allow them evolve. Industry-academia partnerships are also important as initial set up cost is a big factor in this area. To realize the ambitious goals set up my government agencies material science engineering could be the missing link.

References

[1] Raghavan, V. "Materials Science and Engineering: A First Course, 3rd edn.,(SI unit)." (1993): 269-272.

[2] Singh, Upinder. A history of ancient and early medieval India: from the Stone Age to the 12th century. Pearson Education India, 2008.

[3] Khare, S., Chatterjee, A., Bajpai, S., & Bharati, P. K. (2016). Manufacturing engineering education in India. Management and Production Engineering Review, 7(1), 40-44.

[4] Doyama, Masao. "Materials science education in Japan." Materials Science and Engineering Serving Society. 1998. 6-11.

[5] Bajpai, Shrish, Sushant Khare, and Rishabh Yadav. "Control Education in India: Present & Future." IFAC-PapersOnLine 49.1 (2016): 813-818.

[6] Schuhmann, R. "Integration of Metallurgical Engineering Education." JOM 7.5 (1955): 647-650.

[7] Khare, S., Bajpai, S., & Bharati, P. K. (2015). Production engineering education in India. Management and Production Engineering Review, 6(1), 21-25.

[8] Department of Metallurgical and Materials Engineering, National Institute of Technology, Tiruchirappalli, India "https://www.nitt.edu/home/academics/departments/meta/" , accessed

on January 29, 2018.
[9] Khare, Sushant, Shubham Chowdhry, and Shrish Bajpai. "Control engineering education in India." Power, Control and Embedded Systems (ICPCES), 2014 International Conference on. IEEE, 2014.
[10] Thareja, Priyavrat. "Industry Responsive Metallurgical Engineering Education Quality through TQM." (2014).

IOP Conf. Series: Materials Science and Engineering **404** (2018) 012002 doi:10.1088/1757-899X/404/1/012002

Microstructure evolution of Friction Stir Welded Dissimilar Aerospace Aluminium Alloys

Noor Zaman Khan[1], Arshad Noor Siddiquee[1], Zahid A. Khan[1], Mohammed Ubaid[1,*], Dhruv Bajaj[1], Mohd Atif[2], Adnan Khan[2]

1 Department of Mechanical Engineering, Jamia Millia Islamia, New Delhi, India

2 Department of Mechanical Engineering, Integral University Campus, Shahajanhanpur, U.P., India

*Corresponding author E-mail: mohommad_ubaid@yahoo.in

Abstract. Aluminium alloys exhibiting properties such as high strength to weight ratio, high corrosion resistance etc. has made them a choice for many industrial applications. Therefore, joining of aluminium alloys is of utmost importance, especially for the aerospace industry. Friction stir welding (FSW), a clean novel solid-state joining process, is being used to effectively join dissimilar materials for making different parts of space shuttles, aircrafts etc. 2.5 mm thick dissimilar aluminium alloys, AA2219-O and AA7475-T761, were joined using FSW. Evolution of microstructure for the traverse cross section of welded joint was studied. As a result, significant refinement of grains was observed in stir zone along with dissolution and coagulation of strengthening precipitates due to high rise in temperature and severe plastic deformation. Partial recrystallization was observed in TMAZ along with coarsening of grains and strengthening precipitates. In HAZ, coarser grains and strengthening precipitates was observed with no extent of plastic deformation.

1. Introduction

The use of aluminium alloys has increased significantly in modern industries [1]. Various desirable properties such as high corrosion resistance, high strength to weight ratio, good machinability etc. made the aluminium alloys a choice for various industrial sectors such as aerospace, automobile, railways, marine etc. [2]. Therefore, an effective joining process is required to join these materials. However, joining of aluminum alloys using conventional fusion welding methods is difficult due to the formations of various defects which degrades the joint strength.

W. M. Thomas invented and patented Friction stir welding (FSW) in The Welding Institute (TWI), UK. [3]. As a clean solid state joining process, FSW emerged as a remarkable technique to join high strength aluminium alloys effectively which are difficult to be welded by fusion welding processes [4]. FSW join materials at a temperature below their melting point to avoid defects which arise during fusion welding processes. In FSW, a non-consumable rotating tool is allowed to move along the joint line which results in the formation of weld. Figure 1 shows the schematic diagram of FSW [5]. FSW possess various advantages over conventional welding processes such as higher joint strength, low distortion, no evolution of toxic gases etc.

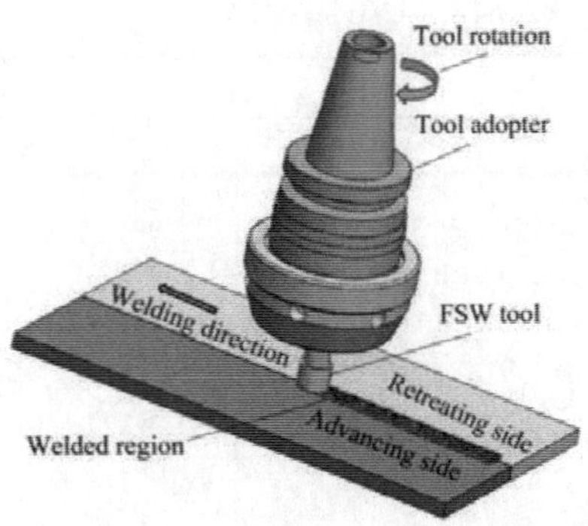

Figure 1. Schematic Diagram of Friction Stir Welding [5]

Joining dissimilar materials allows to produce components having properties of both the materials. Many researchers studied the joining of dissimilar aluminum alloys by conventional welding processes as well as FSW. Conventional welding techniques accompanies various defects such as distortions, solidification cracking, porosity, cracks, chemical reactions between dissimilar materials etc. [6].

Many researchers have studied the evolution of microstructure in FSW of dissimilar materials. Moreira et al. [7] joined dissimilar aluminium alloys (AA6061 and AA6082) using FSW and reported significant changes in the microstructure of the weld region compared to the base material. Lee et al. [8] performed dissimilar FSW of cast A356 and AA6061 and reported that microstructure of SZ was mainly composed of material on RS and joint quality was dependent on location of base materials. Ahmed et al. [9] welded AA7075 and AA5083 using FSW and their results revealed that the difference in base metal properties have strong impact on recrystallization in SZ. Scialpi et al. [10] investigated dissimilar FSW of AA6082 and AA2024 and reported that microstructure of the welded joint significantly affected the micro-hardness.

Aluminium alloys 2xxx and 7xxx series are of utmost importance to the aerospace industry [11]. 2xxx series aluminium alloys are widely used in aircraft components [12] whereas 7xxx series aluminium alloys have applications in aircraft structural components and missile parts etc. [13]. AA2219 and AA7475 are widely used in aircrafts for making cryogenic tanks, upper and lower wing surfaces, body stiffness, surface floors etc. [1].

Joining of dissimilar aerospace grade aluminium alloys by FSW need to be explored for fruitful utilization of various desirable properties of aluminium alloys. Very less literature is available on FSW of dissimilar AA2219 and AA7475. Keeping this in view, an attempt has been made to investigate the microstructural evolution in FSW of dissimilar aluminium alloys (AA2219 and AA7475).

2. Experimental Procedure

2.5 mm thick aluminium alloys 2219-O and 7475-T761 are selected as base materials to be joined by using FSW. Chemical composition of AA2219 and AA7475 is presented in table 1 and table 2 respectively. AA2219 was fixed on advancing side (AS) whereas AA7475 was fixed on retreating side (RS). A high carbon die steel tool with shoulder diameter of 10 mm, pin diameter of 4 mm and pin length of 2.2 mm was employed to perform FSW. Experiments are conducted on robust vertical milling machine which was retrofitted for performing FSW.

Table 1. Chemical Composition of A2219-O (wt. %)

Elements	Al	Cu	Sn	Mn	Fe	Si	Ti	Zn	Ni	Zr	V
AA2219	91.97	6.8	0.02	0.315	0.16	0.06	0.04	0.06	0.023	0.203	0.165

Table 2. Chemical Composition of AA7475-T761 (wt. %)

Elements	Al	Cu	Mg	Mn	Fe	Si	Ti	Zn	Ni	Zr	Cr
AA2219	90.99	1.34	1.93	0.0062	0.101	0.06	0.023	5.36	0.002	0.0076	0.155

After extensive trial experimentation, welding was performed at rotational speed of 1120 rpm and traverse speed of 160 mm/min at a tool tilt angle of 2.5°.

Microstructural specimens were machined from the welded samples using Wire Electric Discharge Machine (Wire-EDM). The acquired samples were polished as per standard metallographic techniques and etched with Keller's reagent. Microstructural examination was done using Optical Microscopy (make: QS Meteorology, Chennai).

3. Results and Discussion

FSW of dissimilar AA2219 and AA7475 was performed successfully without any macro and micro defect. After FSW, three distinct zones viz. the stir zone (SZ), the thermo-mechanically affected zone (TMAZ), and the heat affected zone (HAZ) were observed at AS and RS. All these zones are distinguished on the basis of their microstructures. Mechanical properties of the welded joints depend on the microstructures [14]. Thus, the study of microstructure become important for predicting the quality of the welded joints.

Microstructures of AA7475 (RS) and AA2219 (AS) are shown in figure 2. Grain orientation of material located on RS is in the rolling direction whereas grains are randomly oriented in AS material. Fine black dots visible in BM of AA2219 are non-absorbable precipitates.

FSW resulted in refinement of grains in SZ due to severe plastic deformation [6, 15] as shown in figure 3. The rotating tool moving along the joint line produces frictional heat which softens the base materials and moves them around the tool pin. Tool rotation plastically deforms the base materials due to sticking and slipping action which resulted in grain refinement caused by dynamic recrystallization of grains [16, 17]. SZ experiences high heat due to friction and plastic deformation which causes dissolution of strengthening precipitates [15]. Dissolution of strengthening precipitates reduces the strength of the SZ. However, some amount of re-precipitation might have occurred in SZ depending on the heat input and cooling rate. The temperature in SZ reaches a solid solution temperature which is sufficient for re-precipitation after cooling leading to improved joint strength [8]. The re-precipitated particles visible as white dots in AA7475 (RS) are might be $MgZn_2$ (η) [18]. Whereas in AA2219 some undissolved and coagulated Al_2Cu (θ) precipitates are present. These θ precipitates are due to high copper content in AA2219 which resulted in retention of these precipitates even though the temperature in SZ reaches the solutionizing temperature.

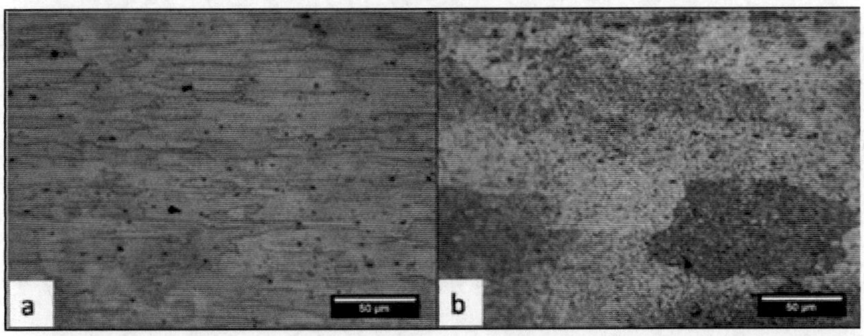

Figure 2. Microstructure of base materials: (a). AA7574 (RS), (b). AA2219 (AS)

Figure 3. Microstructures of SZ: (a). RS (b). AS.

Moving away from the weld center, TMAZ is located on both AS and RS. The interface boundary between SZ and TMAZ of AS is clearly visible [9] due to higher material deformation rate at the AS of the joint as shown in figure 4. However, exceptions may be found [19] due to the difference in the flow-ability of the materials involved in the process. TMAZ experiences plastic deformation as well as thermal cycle due to mechanical stirring of tool shoulder [20]. The extent of plastic deformation and heat experienced by TMAZ is less than that in SZ. TMAZ microstructure is shown in figure 5. Grains in TMAZ are not as fine and recrystallized as observed in SZ because of partial recrystallization and significant amount of grain deformation [21]. Dissolution of GP zones and metastable phases of strengthening precipitates might have occurred in TMAZ of AS and RS due to enough heat input required for their dissolution. Coarse stable θ and η precipitates are observed at AS and RS respectively. Thermal softening causes precipitate dissolution and which lead to the degradation of mechanical properties in TMAZ. However, heat input in SZ is higher than that in TMAZ even though SZ poses better mechanical properties. This might be due to re-precipitation and grain refinement in SZ [5, 6, 9].

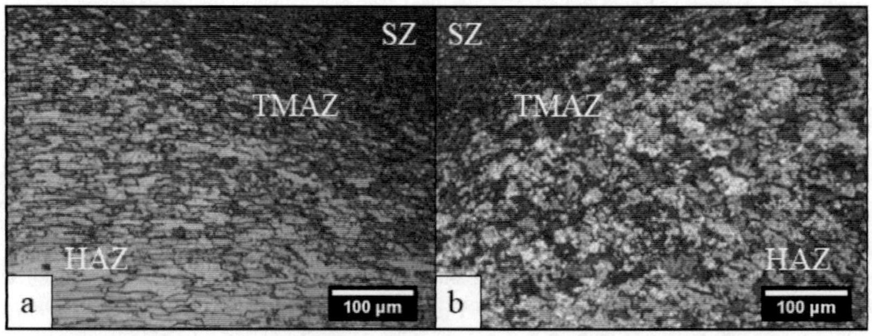

Figure 4. Microstructure of interface for traverse cross section: (a). RS (b). AS.

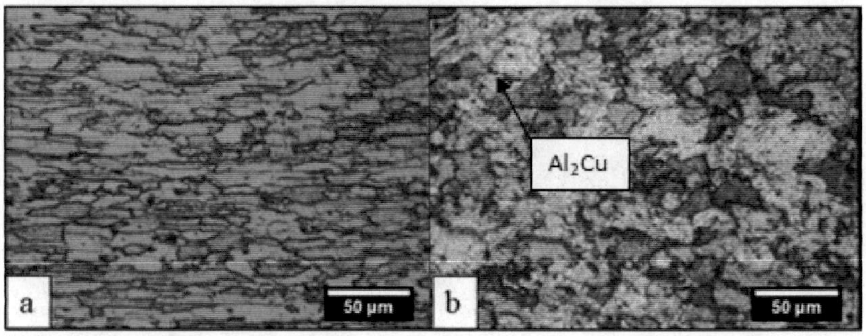

Figure 5. Microstructures of TMAZ: (a). RS (b). AS.

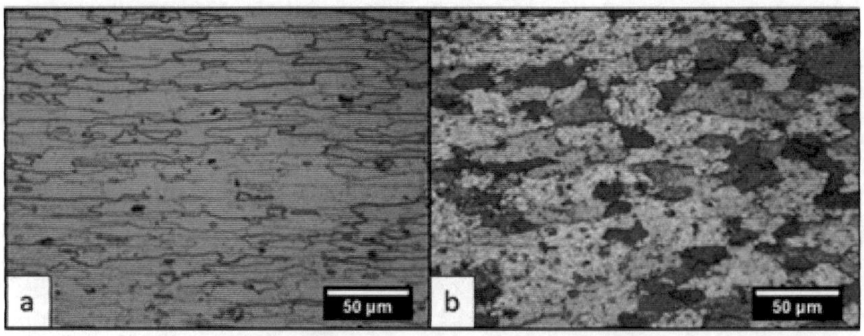

Figure 6. Microstructures of HAZ: (a). RS (b). AS.

HAZ has similar or larger grain size compared with that of base materials [10] as shown in figure 6. This region is observed between the base material and TMAZ. Coarsening of grains and precipitates is observed in HAZ because of experiencing only heat input without any plastic deformation. Also, high volume of coarse strengthening precipitates in HAZ was due to less dissolution of precipitates as shown in figure 6. As less heat is generated in HAZ which might have not sufficient for dissolving the precipitates and also traverse speed in the presence case is higher which has not given the sufficient time required for dissolution of precipitates. Lower mechanical properties of the welded joints are generally observed at either TMAZ or HAZ depending upon the heat input and temper condition of base materials. Coarse precipitates at TMAZ/HAZ interface results in lower mechanical properties [22].

4. Conclusions

Dissimilar FSW of AA2219 and AA7475 was completed successfully. Microstructures of various weld zones was evolved. Following conclusions are drawn based on the above results and discussions:

- Fine recrystallized grains are observed in stir zone.
- Dissolution of strengthening precipitates and grain growth in SZ reduces the joint strength whereas re-precipitation in SZ improves the joint strength.
- Partial recrystallization took place in TMAZ.
- Coarsening and dissolution of precipitates is observed in TMAZ whereas only coarsening of precipitates is observed in HAZ.
- Grain size increases by moving away from the SZ towards the unaffected base material.

References

[1] Starke E and Staley J 1996 Application of modern aluminum alloys to aircraft, Progress in Aerospace Sciences. 32 2-3 pp 131-172

[2] Amini A, Asadi P and Zolghar P 2014 *Friction stir welding applications in industry, in: Advances in Friction Stir Welding and Processing* ed M.K.B. Givi and P. Asadi (Cambridge: Woohead Publishing Limited) pp 671-722.

[3] Thomas W, Nicholas E, Needham J, Murch M, Templesmith P and Dawes C 1991 Friction Stir Butt Welding *G B Patent Application No.* 9125978.8 (USA)

[4] Mishra R and Ma Z 2005 Friction stir welding and processing *Materials Science and Engineering: R: Reports* 50 1-78

[5] Khan N, Siddiquee A, Al-Ahmari A and Abidi M 2017 Analysis of defects in clean fabrication process of friction stir welding Transactions of Nonferrous Metals Society of China 27 1507-1516

[6] Guo J, Chen H, Sun C, Bi G, Sun Z and Wei J 2014 Friction stir welding of dissimilar materials between AA6061 and AA7075 Al alloys effects of process parameters *Materials & Design (1980-2015)* 56 185-192

[7] Moreira P, Santos T, Tavares S, Richter-Trummer V, Vilaça P and de Castro P 2009 Mechanical and metallurgical characterization of friction stir welding joints of AA6061-T6 with AA6082-T6 *Materials & Design* 30 180-187

[8] Lee W, Yeon Y and Jung S 2003 The joint properties of dissimilar formed Al alloys by friction stir welding according to the fixed location of materials *Scripta Materialia* 49 423-428

[9] Ahmed M, Ataya S, El-Sayed Seleman M, Ammar H and Ahmed E 2017 Friction stir welding of similar and dissimilar AA7075 and AA5083 *Journal of Materials Processing Technology* 242 77-91

[10] Scialpi A, De Giorgi M, De Filippis L, Nobile R and Panella F 2008 Mechanical analysis of ultra-thin friction stir welding joined sheets with dissimilar and similar materials *Materials & Design* 29 928-936

[11] Wanhill R, Byrnes R and Smith C 2011 *Stress corrosion cracking (SCC) in aerospace vehicles in: Stress corrosion cracking* ed V Raja, T Shoji (Oxford: Woodhead Publishing Ltd) pp 608-650

[12] Bauccio M 1993 ASM metals reference book *American Society for Metals* (Ohio: Materials Park)

[13] ASM handbook 1990 *ASM International* (OH: Materials Park)

[14] Rajakumar S and Balasubramanian V 2012 Correlation between weld nugget grain size, weld nugget hardness and tensile strength of friction stir welded commercial grade aluminium alloy joints *Materials & Design* 34 242-251

[15] Yan Z, Liu X and Fang H 2016 Effect of Sheet Configuration on Microstructure and Mechanical Behaviors of Dissimilar Al–Mg–Si/Al–Zn–Mg Aluminum Alloys Friction Stir Welding Joints *Journal of Materials Science & Technology* 32 1378-1385

[16] Çam G and Mistikoglu S 2014 Recent Developments in Friction Stir Welding of Al-alloys *Journal of Materials Engineering and Performance* 23 1936-1953

[17] Wan L, Huang Y, Lv Z, Lv S and Feng J 2014 Effect of self-support friction stir welding on microstructure and microhardness of 6082-T6 aluminum alloy joint *Materials & Design* 55 197-203

[18] Jamshidi Aval H 2015 Microstructure and residual stress distributions in friction stir welding of dissimilar aluminium alloys *Materials & Design* 87 405-413

[19] Ilangovan M, Boopathy S and Balasubramanian V 2015 Microstructure and tensile properties of friction stir welded dissimilar AA6061–AA5086 aluminium alloy joints *Transactions of Nonferrous Metals Society of China* 25 1080-1090

IOP Conf. Series: Materials Science and Engineering **404** (2018) 012002 doi:10.1088/1757-899X/404/1/012002

[20] Guerra M, Schmidt C, McClure J, Murr L and Nunes A 2002 Flow patterns during friction stir welding *Materials Characterization* 49 95-101

[21] Su J, Nelson T, Mishra R and Mahoney M 2003 Microstructural investigation of friction stir welded 7050-T651 aluminium *Acta Materialia* 51 713-729

[22] Paglia C, Jata K and Buchheit R 2006 A cast 7050 friction stir weld with scandium: microstructure, corrosion and environmental assisted cracking *Materials Science and Engineering: A* 424 196-204

Grey Based Taguchi Method for Optimization of TIG Process Parameter in Improving Tensile Strength of S30430 Stainless Steel

Asif Ahmad[1], S.Alam[2]

[1]Department of Mechanical Engineering, Integral University, Lucknow, 226021, Uttar Pradesh-India
Email-aauptu2015@gmail.com, Mob +91-7054073718
[2]Department of Mechanical Engineering, Integral University, Lucknow, 226021, Uttar Pradesh-India,
Email-alamshah@iul.ac.in, Mob +91- 9616879753
ORCID-0000-0003-3988-4424

Abstract
Tungsten inert gas welding is popular known welding technique for ferrous & nonferrous. Stainless steel grade 3HQ (S30430) is a specialized wire grade with very wide usage for manufacturer of stainless steel fastener. It has now totally replaced Grade 384 and 305 for heading application. The stable austenitic structure makes 302HQ nonmagnetic, even after substantial cold work, and also results in excellent toughness, even down to cryogenic temperatures. This paper attempts in optimizing the Tungsten Inert Gas (TIG) welding process parameter. The effect of various parameters and their influence is important to determine the strength of welded joint. To obtain a good quality weld, it is therefore, essential to control the input welding parameters. Therefore appropriate selection of input welding parameter is necessary in order to obtain a good quality weld and subsequently increase the productivity of manufacturing industry. This paper present multi objective optimization using grey relation analysis (GRA) for S30430 with TIG process to determine the suitable selection of parameters Experiment were conducted according to Taguchi's design of experiments (DOE) with orthogonal array L9 is used, mathematical model was developed using parameters such as speed (mm/min), current (Amp), voltage (V), depth of penetration (mm). After conducting experiment and collecting data, signal to noise ratio were determined by using Minitab18 and it is used to obtain optimum level for every input parameter.

Keyword- TIG welding, Grey relation analysis (GRA), Taguchi methodology, Signal to Noise ratio, Minitab18.

1. Introduction
In today's scenario welding of thin sheets is being the challenging task in the field of Engineering. The major reason behind this is perfection, which is being the nominal factor considered during the joining process of sheets because, in such joining of plate, two pass and three pass welding cannot be carried out in thin sheets below 4mm, as under welding results in predominant reduction in strength and over weld results in formation of hole. Hence optimization of weld parameter plays a predominant role in joining of thin sheets. This research is to make the thin sheet welding easy and compatible by optimizing the process parameters. Welding is great significant operation in any manufacturing industry [1]. It is absolutely necessary to optimize different welding process parameter, so that industry achieve reliable, productive and good quality product. Nowdays global manufacturing industries are more focused toward R & D. To investigate welding process parameter such as current, voltage, inert gas, pulsed on/off time etc. TIG welding process is commonly used operation for joining of two materials with the application of heat. TIG is also known as Gas Tungsten Arc Welding (GTAW). Arc is maintained between non-consumable electrode and work piece in a protective inert gas atmosphere [2]. The various studies are done to investigate the influence of welding process parameter on tensile strength [3].

2. Taguchi Design of Experiment
Taguchi DOE is a popular statistical technique that provides a proper and efficient methodology for process optimization. Taguchi method allows us improve the consistency of production. Taguchi design recognizes that not all factors that cause variability can be controlled. These uncontrollable factors are called noise factor. Taguchi design tries to identify controllable factor that minimize the effect of noise factor. During experimentation, you manipulate control factor to evaluate variability that occurs and then determine optimal control factor setting, that minimize the process variability. Process designed with this goal produce more consistent output and performance regardless of the environment in which it is used [4]. It is world widely used for product design and process optimization [5-10]. As a result, time as well as cost is reduced considerably.

IOP Conf. Series: Materials Science and Engineering **404** (2018) 012003 doi:10.1088/1757-899X/404/1/012003

Taguchi DOE methodology uses orthogonal array that gives different combinations of parameters and their levels for each experiment. According to this methodology, every parameter is studied with minimal number of necessary experiments only [11].

3. Layout of Experiment

In order to perform experiment for data collection following sequence is followed.

- Selecting the base and filler material.
- Selecting pulsed TIG welding process parameters.
- Finding the upper and lower limits of the identified process parameters.
- Select the appropriate orthogonal array.
- Conduction of the experiments as per the selected orthogonal array.
- Find the optimum condition.

3.1 Selecting base material and there mechanical properties.

S30430 stainless steel sheets of dimension 100 × 150 × 3 mm are welded autogenously with butt joint without edge preparation [12]. The chemical composition of S30430 stainless steel sheet is given in Table1

Table 1: Chemical Composition of base material (wt %)

US DESIGNATION	% Cr	% Ni	% C	% Mn	% Si	% P	% S
S30430	18	8	0.03	2	0.75	0.045	0.03

Table 2: Mechanical Properties of S30430

Tensile strength	Yield Strength	Hardness	Melting Point	Density
564 MPA	241 MPA	B80 HBN	1400-1450 ^{0}C	8 g/cm^3

3.2 Identify the pulsed welding process parameter.

From the literature survey [13-14] and previous work done the most important process parameters which are having greater influence on the weld bead geometry. They are welding speed, welding current, welding voltage and gas flow rate.

3.3 Level of working range

Numbers of experiment has been conducted by varying one of the process parameters and keeping the others constant [13-14]. The working range of the process parameters are shown in Tables 3.

Table 3: Process parameters working range at standoff distance 2mm.

Factor/Process Parameter	Code	Level 1	Level 2	Level 3
Welding current (Amp)	A	140	150	160
Welding voltage (V)	B	24	25	26
Welding speed (mm/min)	C	165	179	193

3.4 Orthogonal array is selected

Process parameters selected for this experiment is three, and the level of each parameter is three. Taguchi orthogonal design of experiment uses a special set of predefined arrays called orthogonal arrays (OAs) to design the plan of experiment [15]. These standard arrays provide the way to full information of all three factors that affects the process performance. For the present experimental work, three factors with their three levels are used for which the corresponding orthogonal array is L9 as shown in below table.

Table 4: Orthogonal Array L_9 (Minitab18)

Experiment No.	Process Parameter		
	A	B	C
1	1	1	1
2	1	2	2
3	1	3	3
4	2	1	2
5	2	2	3
6	2	3	1
7	3	1	3
8	3	2	1

IOP Conf. Series: Materials Science and Engineering **404** (2018) 012003 doi:10.1088/1757-899X/404/1/012003

9	3	3	2

3.5 Conduction of experiments

By putting the values of three levels of three parameters in L9 Orthogonal array, the nine set of experiments with different values of parameters and results are obtained as fallows. The S/N ratio help in measuring the sensitivity of quality characteristic to external noise factor which is not under control. Taguchi techniques are experimental design optimization techniques which use standard Orthogonal Arrays (OA) for forming a matrix of experiments. Using an OA to design the experiment helps the researcher to understand the influence of different controllable factors on the quality characteristics and the variations in a fast and economic way. The highest S/N ratio indicates more impact of the process parameter on the performance characteristics. On the basis of characteristic three S/N ratios are available namely lower the better, higher the better and nominal the better. In this paper higher the better is used for maximizing depth of penetration [16].

$$S/N\ Ratio = -10\log 1/n\ [\textstyle\sum_{i=0} 1/y_i^2]$$

Signal-to Noise ratio (S/N ratio) was first introduced by Sir Michael A Choma. The purpose of the Signal-to-Noise ratio (S/N ratio) is to find which design parameters significantly affect the quality characteristic. In the Taguchi Method the term Signal represents the desirable value (mean) for the output characteristic and the term Noise represents the undesirable value (standard deviation) for the output characteristic. The S/N ratio is defined as n=10log (M.S.D.) Where, M.S.D is the mean square deviation for the output characteristic. To obtain optimal welding performance, higher the better quality characteristic can be taken and S/N ratio is calculated for each experiment.

Where n is the repetition of output response in the same trial and is the response.

Table 5: Experimental Results

Experimental No.	A	B	C	Penetration	S/N Ratio
1	140	24	165	5.4	14.6479
2	140	25	179	5.2	14.3201
3	140	26	193	5.0	13.9794
4	150	24	179	5.3	14.4855
5	150	25	193	5.1	14.1514
6	150	26	165	5.4	14.6479
7	160	24	193	5.3	14.4855
8	**160**	**25**	**165**	**6.9**	**16.7770**
9	160	26	179	6.4	16.1236

Table 6: Response Table for Signal to Noise Ratios (Larger is better)

Level	Welding current (amp)	Welding voltage (v)	Welding speed (mm/min)
1	14.32	14.54	15.36
2	14.43	15.08	14.98
3	15.80	14.92	14.21
Delta	1.48	0.54	1.15
Rank	1	3	2

Table 7: Response Table for Mean

Level	Welding current (amp)	Welding voltage (v)	Welding speed (mm/min)
1	5.2	5.333	5.9
2	5.267	5.733	5.633
3	6.2	5.600	5.133
Delta	1.0	0.400	0.767
Rank	1	3	2

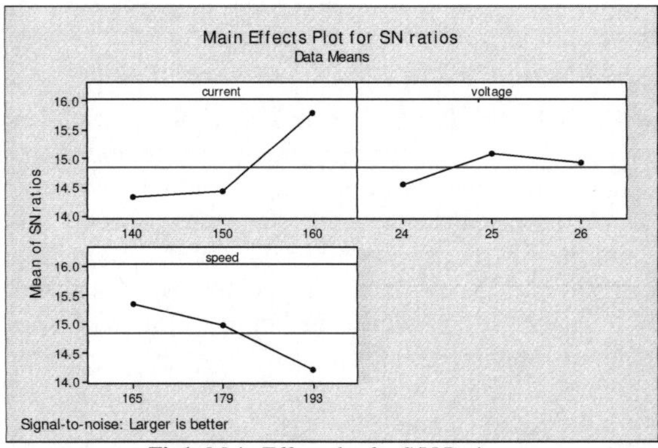

Fig1: Main Effect plot for S/N Ratio

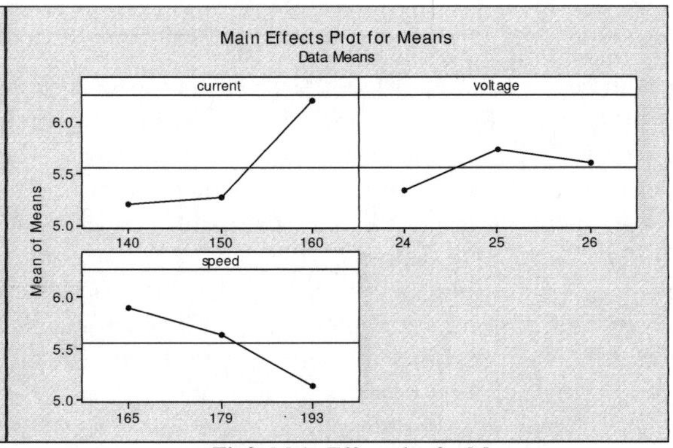

Fig2: Main Effect plot for Mean

4. ANOVA Approach

Analysis of variance (ANOVA) is a method for testing the hypothesis that there is no difference between two or more population means. ANOVA is a hypothesis testing technique used to test the equality of two or more population (or treatment) means by examining the variances of samples that are taken.

Table 8: Analysis of Variance for SNRA1, using Adjusted SS for Tests

Source	DF	Seq SS	Adjss	AdjMS	F	P	% contribution
Current	2	4.0708	4.0708	2.0354	6.71	0.130	56.46
Voltage	2	0.4649	0.4649	0.2325	0.77	0.566	6.45
Speed	2	2.0671	2.0671	1.0335	3.41	3.41	28.67
Error	2	0.6066	0.6066	0.3033			
Total	8	7.2094					

5. Integration of Taguchi method Grey Relation Analysis

In the grey relation analysis experimental results were first normalized and then grey relation coefficient was calculated from normalized experimental data to express the relationship between desired and actual experimental data. Then the grey relation grade was computed by averaging the grey relation coefficient corresponding to each process response [17].

5.1. Grey relation generation(GRA)

In Grey Relation Analysis when the range of the sequence is large or the standard value is enormous, the function of the factor is neglected [18]. However, if the factor goal and direction are different, grey relation might produce incorrect results. There preprocess of the data which are related to a group of sequence which is called grey relational generation. Experimental results are normalized in the range between zero to one. The normalization is done three different ways:-

Larger the better $x_i^*(K) = \frac{x_i^0(k) - min\, x_i^0(k)}{max\ x_i^0(k) - min\, x_i^0(k)}$ .... (1)

Smaller the better $x_i^*(K) = \frac{max\ x_i^0(k) - \ x_i^0(k)}{max\ x_i^0(k) - min\, x_i^0(k)}$ .... (2)

Where $x_i^*(K)$ = value after data preprocessing, $max\ x_i^0(k)$ & $min\, x_i^0(k)$ are max. & min value of $x_i^0(k)$ SNRA$x_i^0(k)$ = 14.6479, $max\ x_i^0(k)$ = 16.7770, $min\, x_i^0(k)$= 13.9794, i = 1,2,3..............m
k =1,2,3..............n, m = no. experimental data

Table 9: Normalization of experimental data

Sample No.	Penetration (mm)
Reference Sequence	1
1	0.2389
2	0.12178
3	0
4	0.1809
5	0.06148

6	0.6685
7	0.06848
8	1
9	0.7664

5.2. Grey Relation coefficient
After pre-processing, a grey relation coefficient can be calculated with the preprocessed sequence. It expresses the relationship between ideal and actual normalized experimental results. Grey relation coefficient can be expressed as

$$\xi_i\ (k) = \frac{\Delta \min + \xi\Delta \max}{\Delta_{oi}\ (k) + \xi\Delta \max}$$

Taking $\xi = 0.5$, $\Delta \max = 1$, $\Delta \min = 0$, $\Delta 1 = 1 - 0.23891 = 0.76109$

Table 10: Grey Relation coefficient of each performance characteristic

Sample No.	Grey Relation Coefficient	Grey Relation Grade	Rank
1	0.3964	0.3964	4
2	0.362786	0.362786	6
3	0.3333	0.3333	9
4	0.3790	0.3790	5
5	0.34757	0.34757	8
6	0.601322	0.601322	3
7	0.349279	0.349279	7
8	1	1	1
9	0.68157	0.68157	2

Table 11: Analysis of Variance for Grey Relation Grade

Source	DF	Seq SS	Adjss	AdjMS	F	P	% contribution
Current	2	0.15890	0.15890	0.07945	5.63	0.151	38.670
Voltage	2	0.06594	0.06594	0.03297	2.34	0.300	16.04
Speed	2	0.15786	0.15786	0.07893	5.59	0.152	38.417
Error	2	0.02822	0.02822	0.01411			
Total	8	0.41091					

6. Result & Discussion
In this research article, L_9 orthogonal array is used to investigate the influence of welding process parameter, welding current found to be the most crucial factor that effect the process characteristic. By using statistical technique ANOVA, that will assesses the contribution of each parameter. The statistical software Minitab 13 is used to investigate the significance contribution of each parameter. It has been found that experiment no. 8 has the optimum parameter setting for better control of the process characteristics. By using Grey Relation Analysis integrated with Taguchi Method, higher Grey Relation Grade (GRG) will have better multiple responses; Rank 1 is given to higher GRG. Comparing the results obtained from both the method *i.e* taguchi & GRA is has been found that welding current is most crucial parameter which affects the performance characteristics followed by welding speed and welding voltage. In Taguchi experiments, we always want to maximize the S/N ratios and the means were maximized when the Current was 160Amp, voltage 25 Volt and welding speed was 165 mm/min based on these results, we should set the factor at the calculated value.

Table 12: Response Table for levels

Welding Current	Level 1	160 Amp
Welding Voltage	Level 3	25 Volt
Welding speed	Level 2	165 mm/min

So, on the basis of these results we can say that Tensile Strength of stainless steel will be higher when we will use Current at 160A, 25 Voltage and Welding speed 165 mm/min. So these are optimum welding parameters on which we can attain the higher tensile strength of Stainless Steel S30430 welds.

7. Future Scope
Different techniques are used for investigation of processes and process parameter. The most common techniques is trial and error method based experiment, it is time consuming and uneconomical. The results

obtained is verified by various other optimization techniques is used to get optimum results *i.e* Genetic Algorithm (GAs), fuzzy logic, Artificial Neural Network (ANN), Finite Element Method (FEM).

8. Acknowledgement

I would gratefully acknowledge to, Dr.S.Alam, Integral University, Lucknow, Uttar Pradesh, India, for providing research facility & guidance, I also thankful to Dr. Ishtiyaq Ahmad, NIT-Raipur, Chhattisgarh, India, for their consistent help in writing this research paper.

References

[1] Dinesh Kumar R, Elangovan S, Siva Shanmugam N, June, 2014 "Parametric optimization of Pulsed TIG Welding Process in Butt Joining of 304L Austenitic Stainless Steel Sheet" IJRET: International Journal of Research in Engineering and Technology, eISSN: 2319-1163, pISSN: 2321-7308

[2] S Elangovan, S Venkateshwaran and K Prakasan, 2012 'International Journal of Advanced Engineering Research and Studies. Vol. I/ Issue III/April-June, E-ISSN2249–8974.

[3] Sanap Deepali, Prof. Galhe D.S, Prof. Burkul R.M, May, 2015 "A Review Paper on Effect of welding speed and groove angle on strength of butt joint using TIG welding" International Journal Of Engineering Science & Research Technology ISSN:22779655

[4] Zulquernain Mallick 'Optimization of operating parameters for a back-pack type grass trimmer' 2007, Elsevier

[5] Wang W.H, and Tarng Y.S, 1998, "Design Optimization of cutting parameters for turning operations based on the Taguchi method". Journal of Materials Processing Technology, 84, pp. 122-129.

[6] Syrcos G.P, 2003, "Die casting process Optimization using Taguchi methods". Journal of Materials Processing Technology, 135, pp. 68-74.

[7] Ghani, J.A, Choudhury I.A, and Hassan, H.H, 2004, "Application of Taguchi method in the Optimization of end milling parameters". Journal of Materials Processing Technology, 145, pp. 84-92.

[8] George, P.M, PillaiN, and Shah, N, 2004, "Optimization of shot peening parameters using Taguchi technique". Journal of Materials Processing Technology, 153, pp. 925-930.

[9] Yang H.J, Hwang P.J, and Lee S.H, 2002, "A study on shrinkage compensation of the SLS process by using the Taguchi method". International Journal of Machine Tools and Manufacture, 42, pp. 1203-1212.

[10] Lee B.H, Abdullah J, and Khan Z.A, 2004, "Optimization of rapid prototyping parameters for production of flexible ABS object". Journal of Materials Processing Technology, Accepted (In Press).

[11] Phadke M.S, 1989, "Quality Engineering Using Robust Design". Prentice Hall International Inc., New York.

[12] P.K. Giridharan, N.Murugan 2009 'Optimization of pulsed GTA welding process parameters for the welding of AISI 304L stainless steel sheets'The International Journal of Advanced Manufacturing TechnologyJanuary 2009, Volume 40, Issue 5–6, pp 478–489

[13] Sunadaresan S, Janaki Ram G.D, "Use of magnetic arc oscillation for grain refinement of gas tungsten arc welds in titanium alloys". SciTechnol Weld Joining 1999;4 (3):151–60.

[14] Kumar A, Sundarrajan S. "Selection of welding process parameters for the optimum but joint strength of an aluminum alloy". Mater Manuf Process 2006;21(8):789–93.

[15] Amitava Mitra, 2012 'Experimental Design and the Taguchi Method' Fundamentals of Quality Control and Improvement: Solutions Manual to Accompany, Third Edition. Print ISBN: 9780470256978 Online ISBN: 9780470409831

[16] Tarng YS, Juang SC, Chang CH (2002) The use of grey-based Taguchi methods to determine submerged arc welding process parameters in hardfacing. J Mater Process Technol 128:1–6

[17] HosseinHasani, SomayehAkhavanTabatabaei, GhafourAmiri "Grey Relational Analysis to Determine the Optimum Process Parameters for Open-End Spinning Yarns" Journal of Engineered Fibers and Fabrics Volume 7, Issue 2 – 2012

[18] Ching-kaochang and H.S.Lu, 2006 'Design optimization of cutting parameters for side milling operations with multiple performance characteristics' The International Journal of Advanced Manufacturing Technology February 2007, Volume 32, Issue 1–2, pp 18–26

Correlation between pulse 300 series SS wire parameters

R Saluja[1] and K M Moeed[2]
[1] Department of Mechanical Engineering, Integral University, Lucknow, India
[2] Principal, Integral Polytechnic, Integral University, Lucknow, India

E-mail: ratisaluja@gmail.com

Abstract. A centralized and extensive relationship linking arc characteristics is presented to explain the metal transfer phenomenon, electrode melting rate, metal transport and weld-solidification during the Pulse gas metal arc welding. The study details interpretation of numerous arc, heat and mass transfer correlations during molten stage of the electrode, initiation of droplet, growth, detachment, transfer and impingement into the weld-pool. An interactive coupling between pulse process parameters, temperature distribution and weld pool dynamics has been discussed. This review elucidates conditions of the pulse metal transfer in the arc column and into the weld pool on the basis of current waveform, magnitude of peak, base and pulse current for 300stainless steel wire.

Keywords: Droplet, filler-wire, metal transfer, pulse gas metal arc welding, wire feed rate

1. Introduction

A high-temperature plasma arc between a continuous, consumable filler wire and the weld pool is built up in pulse-gas metal arc welding (p-GMAW) and initiates periodical impingement of one droplet per pulse at the end of wire tip. P-GMAW is a welding technique, in which the high arc current is upheld in contrast to conventional GMAW to initiate pulse transfer [1].

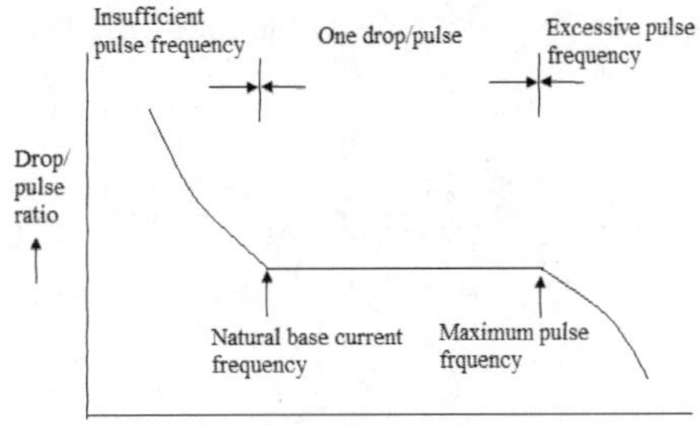

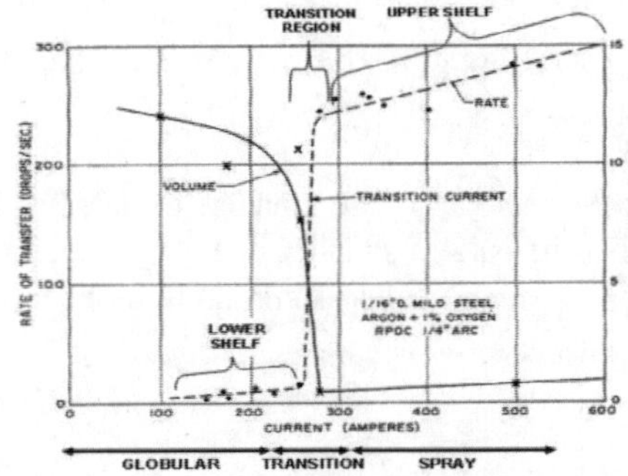

Figure 1(a). Illustration of pulsed wire current as a function of drop/pulse ratio and pulse frequency [3].
Figure 1(b). Influence of arc-current on drop transfer rate and droplet volume in an argon-rich shielding [4].

Where traditional GMAW uses continuous AC or DCEP/ DCRP current supply, unlikely p-GMAW requires intermittent DC. Thus current is giving 'cooling off' during the welding process in pursuance to meet precise and quality weld requirements. The droplet formation, growth and detachment in arc region, its transfer in weld pool, and its periodical impingement are governing by the equilibrium of forces and heat transfer in the arc plasma, within the droplet in conjunction to the weld pool [2]. Essential factors, in connection with static characteristics of the power supply, qualitative characteristics that affect the productivity of p-GMAW are arc stability, wire melting and feed rate [5].

The transition phenomenon of the filler wire is escorted by a chronological alteration in welding current during pulse welding as displayed in figure 1 (a). The transition is recognized with an increase in droplet detachment frequency and an inverse reduction in the droplet volume as shown in figure 1(b) [4]. The shortcoming of short circuit and globular metal transfer are prevailed opting pulse mode transfer for stainless steel. With solid SS wires, another advantage of pulsed power welding is that larger diameter wires upto 1.6mm can be used. In this research, we will pursue aspects and methodologies espoused for selection of pulse parameters, heat retained by the droplet, effect of the pulse waveform and resistance heating of the electrode together with index of the wire melting and feeding rate along with pulse transfer phenomenon.

2. Pulse Metal Transfer Phenomena

As arc force and deposition rate are exponentially dependent on current, the pulse welding process above the transition current crafts the arc forces irrepressible in the vertical and overhead positions. Decrease in the magnitude of average and pulse current, lessen the arc forces and deposition rates, therefore enhancing feasibility for all positions welding for 300 series stainless steel [7]. The metal transfer and the curved weld pool surface have significant effects on the transitory distributions of amperage, arc-region temperature and pressure [2].

Conversely, droplets-impingement significantly influences the surface contour and the convective heat transfer in the weld-pool [8]. The buckled weld pool controls the current distribution within the arc-region. These circumstances instantly affect the transport of heat and momentum flux from the arc plasma to the molten weld-puddle [9].

3. Governing equations for pulse parameters

Suitable combinations of pulse parameters provide steady and controlled metal transfer [10]. Typical degree and amount, fairly accurate properties of welding parameters for steel, can be quantified by detailed relations described in this study [9].

3.1 Formulation of pulse, average and base current

Peak current duration (T_p) is the time span between commencements of increase of current to the beginning of decrease of current during a pulse and base current duration (T_b) is the time duration spent at lesser current as mentioned in figure 2[11]. The primary parameters of pulsed GMAW pulse cycle time (T_{pul}) is expressed as

$$T_{\text{pul}} = (T_{\text{p}} + T_{\text{b}}) \tag{1}$$

Load duty cycle (D) is ratio of the pulse width to the cycle time, can be expressed by the following equation

$$D = T_{\text{p}}/(T_{\text{p}} + T_{\text{b}})\text{X}100\% \tag{2}$$

Although above parameters are entailed to quantify magnitude of average current in pulse waveform, still attention has to be paid to simplify operating conditions by keeping the average current below the transition current [12,13]. Average and peak current for a square wave p-GMAW can be resolved as

$$(I_{\text{avg}}) = \frac{(I_{\text{p}}T_{\text{p}} + I_{\text{b}}T_{\text{b}})}{(T_{\text{pul}})} \tag{3}$$

$$I_{\text{p}} = \left(\frac{T_{\text{pul}}}{T_{\text{p}}}\right).I_{\text{avg}} - \left(\frac{T_{\text{pul}}}{T_{\text{p}}} - 1\right).I_{\text{b}} \tag{4}$$

For a known wire feed rate (V_w), droplet volume (V_d), slope (m) and intercept (C) a stable parametric region over a range of possible combinations of pulse parameters could be evaluated under the following constraints.

$$I_{\text{avg}} = \text{m}V_{\text{w}} + \text{C} \tag{5}$$

The theoretical condition of entailed droplet volume (V_d), can be approximated as

$$V_d = \frac{A_w.V_w}{f} \tag{6}$$

Pulse frequency (f) is number of droplets transfer per second. Assuming one drop per pulse is triggered in p-GMAW [9].

$$f = 1/(T_p + T_b) \tag{7}$$

During the p-GMAW, deviation in the temperature field and melting of the electrode wire speed influence size of droplet. The transition type is an important factor that dominates the arc stability, depending on the current waveform. Therefore, consider the effect of pulse waveform along with the assumption that wire-tip melts continuously. The role of pulse frequency is to command overall heat input, wire melting rate and arc characteristics [14]. Thus frequency (*f*) can also be quantified as a function filler wire rate (V_w), weld-area (A_w) and volume of drop transfer (V_d)

$$f = \frac{A_w.V_w}{V_d} \tag{8}$$

The optimum condition of drop detachment parameter (D_n) symbolizing frequency of droplets can be achieved for all possible combinations of I_p and T_p, and can be symbolized as

$$I_p^{\ 2}\, T_p = D_n \tag{9}$$

Various researches have been proposed using this power law relationship for evaluation of I_p and T_p by assuming 'n=2' and $I_p^2 T_p$= constant, presuming other process parameters fixed. The condition for minimum base current (I_b) below which arc extinguishes and minimum base current limit (β) for stable arc, is stated as

$$I_b \geq \beta \tag{10}$$

Investigators have opted minimum base current, for minimizing overall heat input. If base current is not maintained up to a level large contoured bead and lack of fusion results as a consequence. Its level can diverge significantly, however, for Grade 304 SS its range will be 50A. A linear association subsists between the peak (I_p) and base currents (I_b) for known combinations of variables as wire feed rate (V_w) pulse frequency(*f*) and peak current duration (T_p) for a fixed arc length[14]. Relation between a known peak current excess over the base current (I_e), variation in wire feed rate (V_w) will be

$$I_p \propto I_b \tag{11}$$

$$I_e = I_p - I_b \tag{12}$$

$$(I_b, f) \propto V_w \tag{13}$$

From above equations, it may be understood that at a given I_e, both the base current and either pulse frequency or peak current duration vary directly with wire feed rate.

$$I_{avg} = I_b + I_e I_b f \tag{14}$$

Substituting the value of average current from equation 14 in equation 5, a combined result for a synergic p-GMAW can be obtained. It signifies that deviation in base current, peak current excess over the base current, the peak current duration and pulse frequency varies with WFR for synergy p-GMAW [10].

$$m.V_w + C = I_b + I_e T_p f \tag{15}$$

3.2 Computation of peak current time and droplet detachment time

Considerations are also given to the heating time span (t_1) required for droplet necking preparation, necking and droplet growth time (t_2), and duration during which part of the neck reaches to the melting point is known as detachment time (t_3) [15]. Accordingly, droplet detachment time (t_d), can be calculated by

$$t_d = t_1 + t_2 + t_3 \tag{16}$$

T_p should be greater than t_1, otherwise no droplet will be formed, and as droplet necking would not be triggered. Consequently, initiation of pulse spray transfer occurs, if magnitude of peak current (T_p) is lesser than sum of heating time and drop growth time. A. Thus condition for current duration (T_p) can be expressed as:

$$t_1 < T_p < t_1 + t_2 \tag{17}$$

Equation 18 correlates the peak and background conditions by a combination of exponential and Lorentzian function for calculating correct value of peak current. This equation can be applied to characterize the minimum time essential for droplet detachment at a preferred peak current value [16].

$$Tp = \left[496.1\,X\,(1 - e^{-0.003 I_b T_b})) + \frac{270.1}{(\frac{(I_b T_b - 188.2)2}{8423.5})}\right] / I_p \tag{18}$$

3.3 Quantification of temperature distribution of the electrode wire

Steady motion of the solid wire dependent variables are enthalpy (H) and specific internal thermal energy (I) of system. If radial velocity (v) for steady wire disappears while keeping axial velocity (u) uniform than the volumetric thermal energy generation rate is a consequence of resistive/Joulean heating (qG'''). The thermal energy equation for the electrode-wire and droplet can also be stated in cylindrical coordinates in equation 19.

$$\frac{\partial \rho l}{\partial t} + \rho u \frac{\partial H}{\partial x} + \rho v \frac{\partial H}{\partial r} = \frac{\partial}{\partial x}\left(kr\frac{\partial t}{\partial x}\right) + \frac{\partial}{r \partial x}\left(kr\frac{\partial T}{\partial x}\right) + qG''' \tag{19}$$

If T (x, t) is the temperature of the electrode stick-out, and dependent variable wire feeding speed (V_w) is kept constant, the next one dimensional thermal transmission can be expressed by governing equation 20 [17].

$$\partial T/\partial t + V_w\,\partial T/\partial x = (K/c\rho)\,\partial^2 T/\partial x^2 + (j^2/4.2c\rho)R(T) \tag{20}$$

Where ρ is filler wire density, j is current density ($j = 4I_w/\pi W d_e{}^2$), K is thermal conductivity, C is Bulk specific heat, d_e is wire diameter and R (T) is specific resistance at temperature T. If φ is the equivalent voltage of molten-wire due to heat generation at arc region, H_o is the equivalent voltage of molten droplet temperature holding per unit mass, T_m is melting point of wire. At stable GMAW arc with constant wire extension (E_x) with respect to time (dE_x / dt = 0), the temperature distribution of wires can also be expressed as

$$dE_x/dt = -(\emptyset j - 4.2K\,\partial T/dx|_{x=Bx})/4.2(\rho H_0 - \rho c T_m) + V_w \tag{21}$$

Arc stability in p-GMAW is mainly due to variation in arc length, means the extent of the temporal change of the arc length or the wire extension length E_x. Ignoring the influence of heat conduction in the filler wire, the temperature distribution of the wire gradually varies with time, the wire melting rate can be expressed by the formula. Therefore, the temporal change of E_x is

$$\frac{dE_x}{dt} = V_w - v(t) = V_w - \{\emptyset j(t) + a\,E_x j(t)^2)\}/4.2\rho\,(H_0 + b) \tag{22}$$

As shown in figure 2(a), the melting speed of the wire is proportional to E_x, the effect of anti-heat generation is large for stainless steel at high current value and high wire melting rate. It gets worse during pulse welding, as the current flows intermittently. Assume the case of steady state, under the completely constant current value, the temperature of the electrode wire (∂T / ∂t = 0) is constant. It is also assumed that the resistivity of the drop does not depend on the temperature ($R(T) = R_o$). To determine T_m, solution of equation (20) becomes as follows.

$$T\,(x) = T_m - \left(\frac{I^2 R_0}{4.2c\rho}\cdot\frac{E_x}{v_w}\right)\cdot\frac{e^{\left(V_w \frac{x}{k}\right)-1}}{e^{\left(V_w \frac{Ex}{k}\right)-1}} + \frac{j^2 R_0}{4.2c\rho}\cdot\frac{x}{V_w} \tag{23}$$

However, k = K / cρ, use small diameter wire MIG. Since it can be thought that $\left(V_w \frac{Ex}{k}\right) - 1$ in welding, and formula of wire melting rate is obtained from heat input state of the wire stick out and resistance of the wire extension, and it is understood that it receives the amount of heat retained by the droplet.

$$V_w = (\emptyset j + R_0\,E_x j^2)/4.2\rho R_0 \tag{24}$$

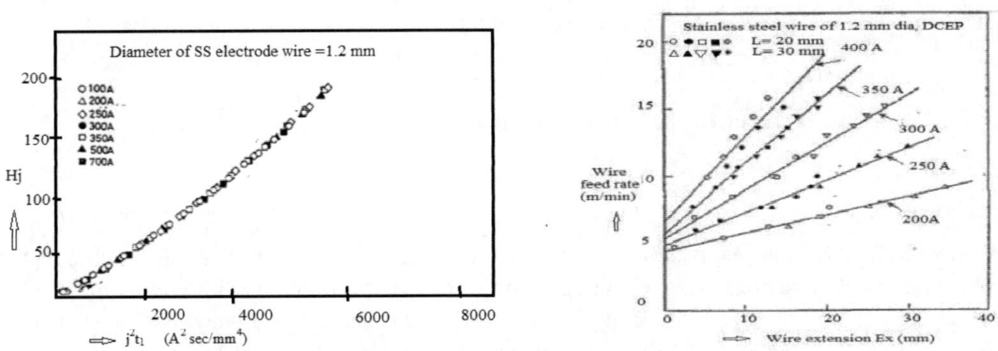

Figure 2 (a) Heat content per unit mass of GMA electrode wire by joule's heating [17]
Figure 2 (b) Alliance between WFS and electrode stick out for a 1.6 mm diameter steel wire with argon as shielding gas [5].

3.4 Calculation of heat retained by droplets

There are various theoretical and empirical techniques for calculation of wire melting rate in p-GMAW [5]. For wire melting coefficient (α_m), wire density (ρ), welding current (I_w), wire cross-sectional area ($F = \pi r d_e^{\,2}/4$), where d_e is diameter of filler wire, WMR can be expressed as

$$V_w = \alpha_m I_w / 3600 \rho F \tag{25}$$

The amount of heat retained by the droplet (H_o) is heat quantity given from arc (H_A) and resistance obtained by heat generation (H_J). For stable MIG arc, the heat holding amount H_o of the droplet is

$$H_o = H_A + H_j \tag{26}$$

$$H_A = \emptyset j / 4.2 \rho v_w \tag{27}$$

$$H_J = R_o E_x j^2 / 4.2 \rho v_w \tag{28}$$

Resistance calorific value (H_J) can be derived from the direct energization of the wire without generating it. During the energization time t_1, the wire reaches the arc generation point. During initiation of the current flow, t_1 is time ($t_1=E_x/V_w$) for second pass than the amount of heat held per unit mass of wire at the time point. However, V (t) is the voltage between the gauge points and l_o is the gauge distance at room temperature. H_j can also be calculated as

$$H_J = j \int_0^{t1} V(t) dt / 4.2 \rho l_0 \tag{29}$$

However, V (t) is the voltage between the gauge points and l_o is the gauge distance at room temperature.

$$V_w = (\emptyset j + a\, E_x j^2) / 4.2 \rho\, (H_0 + \mathrm{b}) \tag{30}$$

H_o is obtained by using the resistance characteristic values 'a' and 'b' of the wire from figure 2(a) with maximum error 5%. The values of a' is 1.16 X 10^{-3} 'Ω.mm and 'b' is 29 cal/g for Grade 304 SS 1.2 diameter filler wire. Results conclude that experiment is prominent to measure long E_x to dissolve, hence it is practicable in the current range where droplets migrate with relatively small diameter. If, V_A = anode drop, V_{WA} is anode work function, V_T is an equivalent voltage of the plasma temperature in the case of steel. The equivalent voltage (φ) required for melting of wire can be estimated as

$$\emptyset = V_A + V_{WA} + V_T \tag{31}$$

3.5 Quantification of Wire melting rate

Wire melting or burn off rate of pulse GMAW is kept comparatively high, intended for droplet formation and transition of metal transfer mode. If overall wire melting rate is V_{wo} and V_{wp} and V_{wb} are the wire burn off rates for the period of peak current duration (T_p) and base current duration (T_b) respectively [18,19]. In consideration V_{wo} may be expressed as

$$V_{wo} = \int_0^{t_{pul}} V_w T_{pul} dt_{pul} \tag{32}$$

For a square pulse-GMAW equation 32 can be determined by

$$V_{wo} = (V_{wp}T_p + V_{wp}T_b)T_{pul}^{-1} \quad (33)$$

Here V_{wp} and V_{wb} can be evaluated by the following formula

$$V_{wp} = AI_p + BE_wI_p^2 \quad (34)$$

$$V_{wb} = AI_b + BE_wI_b^2 \quad (35)$$

Here first term (AI_b) represents the contribution from arc melting, and the second term of ($BE_wI_b^2$) represents the contribution from electrode extension resistance heating as displayed in figure 2 (b). The arc melting term is dependent on the polarity, electrode type and size. For DCEP constant voltage pulse transfer magnitude of coefficient 'A' is approximated as 0.28 and value of 'B' lies within 0.9–1.0 [1].Therefore equation can be rewritten as

$$V_{wo} = [A(I_pT_p + I_bT_b) + BE_w(I_p^2T_p + I_b^2T_b)T_{pul}^{-1} \quad (36)$$

Overall wire melting rate (V_{wo}) can be articulated as function of as a function of average current (I_{avg}) and pulse frequency (f) assuming $I_p^2T_p \gg I_b^2T_b$ and ignoring the ohmic heating droplet detachment. Thus equation 37 will provide constant wire feed rate (V_w) for given wire parameters by fixing the fraction of average current (I_{avg}) to pulse frequency (f) [1, 20].

$$V_{wo} = AI_{avg} + B\,E_w\,D_n\,f \quad (37)$$

Assuming wire feed rate (V_w) as equivalent linear function of overall wire melting rate (V_{wo}), equation 36 can be stated in terms of the droplet volume (v_d).

$$v_d = AA_w\frac{I_{avg}}{f} + A_w\,E_w\,B\,D_n \quad (38)$$

If droplet volume (v_d) is not influenced with average current by previous assumptions, and equation 36 can modified as.

$$V_{wp} = AI_m + BE_w(I_{eff})^2 \quad (39)$$

Where

$$I_{eff}^2 = \{DI_p^2 + (1\text{-}D)\,I_b^2\} \text{ or} \quad (40)$$

$$I_{eff}^2 = \{I_{avg}^2 + D\,(1\text{-}D)\,I_e^2\} \quad (41)$$

Pulse wire feed rate V_p can also mention in terms of wire feed rate of conventional GMAW V_{gmaw}.

$$V_{wp} = V_{gmaw} + BE_wD\,(1 - D)\,I_e^2 \quad (42)$$

Expression states that wire melting rate of p-GMAW is relatively higher than of conventional GMAW.

4. Conclusion

Various researchers investigated the intricate association of SS pulse parameters and detailed a critical and selective application using multiple techniques. Though, there is a squabble between researchers to consider base current for the formulation of practicable parameters for 300 SS series or not. It is mandatory to control arc current as is directly proportional to wire feed rate for steady GMAW process. The exact match between wire feed rate and wire melting rate is essential, as lesser wire feed rate grounds melt back the filler wire, whether higher feed rate extinguishes the arc. It is found that for a fixed peak current, its average value is also invariable at a particular pulse frequency. It is invariably influenced by the wire feed rate, even if the average current is constant, still.

The summarized review state that the wire melting rate of MIG welding is determined by arc melting equivalent voltage of fusion, length, and the resistance value of projecting part, heat retained by droplet, the magnitude of arc current and its types. In pulse welding, in addition to the pulse current waveform, the average current and the effective current are the determining factors of the wire melting rate. It is anticipated that quantification of wire and its governing parameters, investigational interpretations and numerical analysis for the p-GMAW process have directed towards innovative results enterprising the modeling and simulation feasibility of the process for welding of 300 series stainless steel.

Acknowledgement

Authors are grateful to Mr. S. A. Prabhu for sharing his knowledge regarding p-GMAW process parameters. Authors express gratitude to Dr. Inderjeet Singh and Dr. P. K. Bharti (HOD-ME dpt.) for their continuous support and motivation.

References

[1] Ghosh P K and Gupta P C 1996 Use of pulse current MIG welding improves the weld characteristics of Al–Zn–Mg Alloy *Indian Weld. J.* **29** pp 24–32

[2] Hu J and Tsai H L 2007 Heat and mass transfer in gas metal arc welding Part I: The arc *International Journal of Heat and Mass Transfer* **50** p 833

[3] Essers W G and Gompal V 1984 Arc control with pulsed GMA welding, *Weld. J.* **64** pp 26s–32s

[4] Dennis J H et al 2001 Model for prediction of fume formation rate in gas metal arc welding (GMAW), globular and spray modes, dc electrode positive *Ann. occup. Hyg.* **45** pp 105–113

[5] Jacobsen N 1992 Monopulse investigation of drop detachment in pulsed gas metal arc welding *J. Phys. D: Appl. Phys.* **25** pp 783–797.

[6] Ferraresi V A et al 2003 Metal Transfer in the Aluminum Gas Metal Arc Welding *J. of the Braz. Soc. of Mech. Sci. & Eng.* **XXV** pp 229-234

[7] Agarwal B A et al 2016 Challenges in application of pulse current gas metal arc welding process for preparation of weld joint with superior quality *International J. of Eng. Research & Technology* **5** pp 319-27

[8] Fan H G et al 1999 Droplet formation, detachment, and impingement on the molten pool in gas metal arc welding *Metallurgical and Materials Trans. B* **30** pp 791–801l

[9] Palani P K et al 2006 Selection of parameters of pulsed current gas metal arc welding *Jourinal of Materials Processing Technology* **172** pp 1–10

[10] Allum 1983 MIG welding-time for a reassessment *Met. Construct.* Pp 347–353.

[11] Ueguri S et al 1985 Study of metal transfer in pulsed GMA welding *Weld. J.* **64** pp 242s–250s.

[12] Lambert J A 1989 Assessment of the pulsed GMA technique for tube attachment welding *Weld. J.* **68** pp 35–43

[13] Choi SK et al 1998 Dynamic simulation of metal transfer. Part II: short circuit transfer modes *Weld. J.* **77** pp 45s–51s

[14] Zhu P et al 1997 Theoretical predictions of the start up phase in GMA welding *Weld. J.* **78** pp 269s-74s

[15] Boughton and J.A. Lucey J A 1965 The use of pulsed current to control metal transfer in welding *Brit. Weld. J.* **4** pp 159–166

[16] Harvey R C 1995 Gas metal arc welding fume generation using pulsed current *Weld. J.* **74** pp 59s-68s

[17] S. Subramaniam et al 1999 Experimental approach to selection of pulsing parameters in pulsed GMAW *AWS Weld. J.* **78** pp 166s–172s

[18] Hirata Y 1995 Physics of welding (III) - Melting rate and temperature distribution of electrode wire, *Welding International* **9** pp 348-51

[19] Cornu 1988 Advanced welding system. IFS Publication Ltd. UK 2, 127–165

[20] Smati Z 1986 Automatic pulsed MIG welding. *Weld. J.* **65** pp 38s–44s

Desiccant materials for air conditioning applications - A revie

R P Singh[a], V K Mishra, R K Das
Indian Institute of Technology (ISM) Dhanbad, Jharkhand
[a]talkrpsingh@gmail.com

ABSTRACT

Desiccants are a group of adsorbent materials that have a great affinity for water vapour. Due to this merit, desiccant material can produce hot and dry air which can be used for the drying process as well as in air-conditioning applications to minimize the latent heat load. A good desiccant must have better moisture absorption capability as well as lower regeneration temperature. The moisture absorption capability basically depends upon the desiccant characteristics like pore volume, apertures size and void fraction. Depending upon normal physical state, desiccants can be either solids or liquids. Solid desiccant materials such as silica gel, natural zeolites, molecular sieves, activated alumina, synthetic polymers are highly porous in nature and adsorb water by using mechanisms of chemical adsorption. Liquid desiccants such as Calcium chloride, tri-ethylene glycol, and lithium chloride are generally very strong solutions of ionic salts and their behaviour is controlled by changing its temperature and concentration. This paper presents a detailed study on different solid, liquid and composite desiccant which combine two or more desiccant materials for better properties and performance. This review has a great significance for the research and technical development in the field of desiccant dehumidification and air-conditioning technology.

Keywords: solid desiccant, liquid desiccant, regeneration, dehumidification, air-conditioning

Contents

IOP Conf. Series: Materials Science and Engineering **404** (2018) 012005 doi:10.1088/1757-899X/404/1/012005

Nomenclature

COP	– Coefficient of Performance
DDAC	– Desiccant Dehumidification and Air Conditioning
FAM-Z	– Functional Adsorbent Material Zeolite
FSM	– Folded Sheet Mesoporous Material
HKUST	– Hong Kong University of Science and Technology
IUPAC	– International Union of Pure and Applied Chemistry
KIT	– Korea Advanced Institute of Science and Technology
MCM	– Mobil Composition of Matter Number
MIL	– Materials of Institute Lavoisier
RD	– Regular Density
SBA	– Santa Barbara Amorphous
VC	– Vapour Compression

1. Introduction

With the amplified thermal conditions, standard of living and comfort demands of occupants, the demand for air-conditioning has increased and the recent statistics estimate that the energy consumed for air-conditioning, both commercial and domestic, is about 45% of the total energy consumption [1]. In the meantime, because of the increasing global warming and extinction of fossil fuels, challenge for development of new air conditioning technologies has also been increased, for the assistance or replacement of conventional Vapour Compression (VC) systems and one of the best alternative is Desiccant Dehumidification based Air Conditioning (DDAC) [2]. These DDAC systems were initially adopted and were being used for a long time in industrial and agricultural field applications, like humidity control in fabric mills and low temperature post-harvest crop drying in stores. Later, they became more protuberant in air conditioning field as they make complete use of surface vapour pressure difference for humidity transfer between the process air and desiccant material [3]. The DDAC systems are most efficient as they are advantageous in many ways, as [2,4-8]:

1. The desiccant materials are capable of removing impurities from the air stream to improve supply air quality.
2. In DDAC system, the regeneration process can be powered by renewable energy sources like solar energy, geothermal energy etc. and also low grade energy like waste heat which makes the system more economical.
3. The desiccant material in the DDAC system will dry the process air stream and this reduces the corrosion of piping and clogging of valves and capillaries by ice crystals which occurs in case of VC system.
4. The DDAC system uses zeolites and silica gels as desiccants which are environment friendly unlike CFC's and other refrigerants.
5. DDAC systems are capable of handling air below dew point temperatures like -40^{0}C.
6. Desiccant dehumidification completely eliminates processes of evaporation and condensation, which are prevalent in VC system and reduces the growth of disease causing microbes.
7. In DDAC systems, increase in humidity of process air is not a problem, as it is handled by the desiccant wheel before being transferred to the conditioned space.
8. Environment cooling can be used for handling of sensible load, as ambient air is mostly used as the process, which reduces the energy consumption of the system.

For the past few decades, experimental researches in development of simulation models [9-14], desiccant materials [15-20] and improvement of systems and cycles [21-24], etc. were carried out in the field of DDAC systems. The DDAC systems mainly depend on the desiccant materials opted, as its characteristics (regeneration temperature, adsorption isotherms, etc.) significantly affect the performance of the system [25]. With advancement in material science, various desiccant materials have been explored and technologically advanced. Researches are also being carried out in the fields of development of advanced desiccant materials with improved long-term stability, adsorption characteristics and regeneration capacity.

2. Desiccants

Generally, the materials which have the ability to attract and hold other gases or liquids are termed as sorbents. These sorbents are mainly used in chemical separation processes and are used for absorbing gases or liquids other than water vapour. Desiccants are a subset to the sorbents, which specifically have affinity for water. This process of absorbing and holding water vapour by the desiccants may be defined as either absorption or adsorption depending on whether the desiccant material undergoing chemical change on attracting moisture (absorption) or not (adsorption). Materials like wood, natural fibres, clays and various synthetic materials can also attract and hold water vapour, but they have less holding capacity. Generally, a desiccant attracts between 10 to 1100% of its dry mass of water vapour depending on the moisture content available in the surroundings and the type of desiccant material used. Desiccants continuously attract moisture, even from the dry air, until it reaches equilibrium with the environment. Moisture is removed by heating the desiccant material to temperatures from 50° to 260°C and exposing to a regeneration airstream. After the desiccant is completely dried, it should be cooled so that it can attract water vapour again [4]. The desiccants are available either as solids or liquids, namely, solid desiccants and liquid desiccants. Mostly adsorbents are solids and absorbents are liquids.

3. Solid Desiccants

Solid desiccants are highly porous materials which adsorb water by various mechanisms like chemical adsorption onto the walls of pores or consecutive layered physical adsorption of water molecules or capillary condensation into the pores. These have large internal surface area per unit mass, as large as 4600 m^2 and the surface area which attracts water is always in the crystalline structure of the material. These materials attract moisture because of electrical field present at the desiccant surface, which utilizes the atomic and electrostatic forces to attract water molecules into the microscopic pores of the desiccant surface [4, 26]. Solid desiccants are generally classified into following classes.

3.1. Silica Gels

Silica Gel is a naturally available mineral which is non-toxic, odourless and generally processed into beaded form. Its porosity is greater than 70% of its surface area and is as high as 650m^2/g with pore size of 2-3 nm (type A) to 0.7 nm (type B) and it has a heat absorption capacity of about 2800 kJ/kg [27]. Silica gels are mainly of two kinds namely, macro-porous and micro- porous. Macro porous silica gel saturates readily with the surroundings whereas micro porous silica gel absorbs water for a long time. Silica gel generally requires temperature between 90° to 150°C as regeneration temperature to remove the moisture from it [28,29].

3.2.Zeolites

Zeolites are aluminosilicate minerals of alkali and alkaline earth metals like sodium, potassium and calcium. They have crystalline lattice which are wide open that makes ease of holding water vapour like in a cage and have porous structure to accommodate the alkali and alkaline earth metal ions. Water vapour can be removed by heating the material, leaving unchanged aluminosilicate skeleton with a void fraction ranging from 0.2 to 0.5. After heating, the size of apertures of the skeleton ranges from 3 to 8 A^0 for further adsorption of water vapour [30]. Generally, zeolites will be contacted with aqueous solutions of appropriate cations (practically 0.1M) at a temperature of 60-70°C at solid to liquid (S/L) weight ratios ranging from 1/20 to 1/50, for cation exchange [31]. Analcime, Clinoptilolite, Heulandite, Natrolite, Phillipsite and Stilbite are some of the common zeolite minerals. These zeolites are also used in some of the industrial chemical processes like gas separations, ion exchanges, water treatment and catalysis [32].

3.3. Activated Alumina

Activated alumina are hydrides and oxides of aluminium, generally prepared by thermal dehydration or activation of aluminium tri-hydrate or gibbsite. The structural characteristics of alumina can be controlled either by temperature and duration of the thermal process or by the gases used for producing them. Activated alumina, like silica gel, have greater capacity for water vapour than zeolites and surface acidity is the important property for adsorption as well as catalysis. Because of this surface acidity, Lewis acid sites (sites that accept electrons) are abundant on alumina and for fully hydrated alumina, Bronsted acid sites (-OH groups that donat proton) are present. These activated aluminas have surface area ranging from 150 - 500 m^2/g and heat adsorption capacity as high as 3000kJ/kg. The pore configuration of the activated alumina mainly depends on the conditions of heat treatment and will be of size ranging from 1.5 – 6 nm [27,30]. Because of its high sorbent properties, it has been prominently used in desiccant dehumidification and cooling applications [33,34] and also in removal of oxygenates, Lewis bases, polar organic compounds and hydrogen fluorides from gases and liquids. Depending on the degree of hydration, the high temperature forms of alumina are θ-alumina and δ-alumina having only Lewis acid sites and γ-alumina and η-alumina having both type of acid sites (both Lewis and Bronsted).

3.4. Molecular Sieves

Molecular Sieves, which can also be termed as synthetic zeolites, are crystalline aluminosilicates which are obtained by isomorphous substitution of aluminium or silicon from zeolite framework with other elements in thermal process. Compared to zeolites, molecular sieves are advantageous for their large pores or channels and catalytic sites in place of acid sites. The structure and surface features are mainly controlled by the temperature of the process and composition of the ingredient materials. The largest windows of these synthetic zeolites are of order 8.1 A^0 for a 12-membered oxygen ring pore system. Some of the commercial synthetic zeolites are VPI-5, a crystalline alumino-phosphate with 18-membered oxygen rings; $AlPO_4$-8, contains 14-membered oxygen rings and Cloverite, a gallo-phosphate consisting of 20-membered oxygen rings as well as 8-membered oxygen dual rings as pore systems [30]. Because of its pore system, it has great affinity in adsorbing water vapour resulting in its application in dehumidification and air-conditioning systems [35-38].

3.5. Activated Carbon

It is one of the most widely used adsorbent because of its huge micropore and mesopore volumes and higher surface area of 300-4000m^2/g. Activated carbon is generally prepared by low temperature carbonization (400-500°C) to remove the volatile matter, followed by activation process, either gas or chemical, where partial gasification (800-1000°C) to develop surface area and porosity. Commercial activated carbon is generally available in both liquid and gas phase. The liquid phase activated carbon has a pore volume with pores having diameter larger than 30 A^0 and gas phase activated carbon has pores of size 10-25 A^0, so based on the application the phase of activated carbon is selected [30]. Apart from the activate carbon available as powder or granules, activated carbon fibres have better adsorption characteristics. These activated carbon fibres are prepared by taking synthetic fibres like poly-acrylonitrile, coal tar, viscose rayon and phenolic resin as precursors. Similar to activated carbons, these precursors are also carbonized at high temperatures in inert atmosphere and activated by steam at temperatures of 200-300°C followed by the calcination at 400-1000°C in inert atmosphere. Stabilized fibres with desired properties are activated at 800-1000°C. The pore volume of the activated carbon fibres is of range 0.5-1.4 ml/g with average pore diameter of 1.2-3.0 nm and specific surface area of 700-2500 m^2/g. Due to these properties, the adsorption characteristics of activated carbon fibres is better than the activated carbon powder or granules [39].

3.6. Polymer Desiccants

A polymer can be termed as a substance which is made of macromolecule [40,41]. Desiccants based on polymers are mainly investigated because of their ability to adsorb water vapour upto 80% of its own weight and potential to reach the desired water adsorption characteristics. Following are some of the polymer desiccant materials on which experimental investigations are largely carried out.

3.6.1. Polyelectrolytes

A polyelectrolyte is a polymeric material which consists of large portion of its repeating units having ionic or ionisable groups, or both [41]. These polyelectrolytes have been employed as desiccant materials for a long time because of their ease of regeneration, high adsorption capacity and adjustable functionality [42-44]. The hydrophilicity of the polyelectrolytes can be varied by introducing hydroxyl and carboxyl groups into them but the exchange of cations affects the water adsorption characteristics [45,46]. The tremendous development in material synthesis made possible the development of polyelectrolytes with higher adsorption capacities and durability and employed in dehumidification [47-50]. Joohyun Lee and Dae-Young Lee [51] developed a polymeric desiccant by ion modification of sodium salt of polyacrylic acid and fabricated it into desiccant wheel whose experimental results showed 2-3 times higher adsorption capacity than silica gel. White et al. [52] conducted experimental investigation on desiccant wheel made using a polymeric desiccant developed by ion modification of polyacrylic acid and results were compared with two desiccant wheels made of silica gel and FAM-Z01. The results indicated 10-20% more dehumidification than the other two desiccant wheels at low regeneration temperature of 50°C.

3.6.2. Metal-Organic frameworks

According to IUPAC, metal-organic frameworks are coordination polymers or networks that have potential voids in the open framework [53]. They have large surface area, high pore volume and exceptional physiochemical variability leading to its variable composition [54-56]. These properties of metal-organic frameworks lead their application to catalysis [57-59], gas storage [60-63] and gas separation processes [64-67]. Most of the metal-organic frameworks used in desiccant dehumidification processes have a specific pore volume of 2 cm^3/g and surface area as high as 6000 m^2/g [68-71] and some of the MIL type compounds like MIL-100(Cr) and MIL-101(Cr) have water adsorption capacities as high as 1.43g/g and inimitable hydrothermal stability [72-75]. Seo et al. [76] conducted experiments on MIL-101(Cr) and MIL-100(Fe) and the performance results were compared with that of silico-alumino-phophate molecular sieve and alumino-silicate zeolite as reference materials. The MIL-101(Cr) and MIL-100(Fe) exhibited more than 1.5 g/g and 0.84 g/g water capacity respectively, which were 6.0-11.5 times that of the reference materials. Akiyama et al. [77] investigated on a MIL material consisting of substituents like –H, $-NO_2$, $-NH_2$ and $-SO_3H$ as ligands and observed that they exhibit higher water capacities of 0.8-1.2 g/g and water adsorption trapping pressure could be controlled by changing substituents or their hydrophobicity. Henninger et al. [75] investigated on the properties of three materials namely, HKUST-1(Cu), MIL-53(Al) and MIL-100(Fe). He found that of all the three materials, MIL-100(Fe) showed higher water adsorption capacity of 0.6 g/g. Similarly, Rezk et al. [78] investigated and reported that MIL-100(Fe) had 78.6% more adsorption capacity than silica gel RD-2060 at 20°C and P/P_0 of 0.9.

3.7. Composite Desiccants

Composite materials are generally formed by the impregnation of hygroscopic salts into the pores of the host, i.e., a porous desiccant material in this case. The hygroscopic salts (nitrates, sulphates and haloids etc.) possess high water adsorption characteristics but they are unstable at higher humidity ratios due to lyolysis, so porous desiccant materials with stable characteristics like silica gels, activated carbon, mesoporous silicates and natural rocks are used as host material. Aristov [79], Gordeeva and Aristov [80] investigated on some composite desiccants for cooling applications and reported that the composite desiccants achieved high COP as well as water adsorption characteristics. Some of the composite desiccants used in dehumidification and cooling applications are discussed below.

3.7.1. Silica gel-host composites

Silica gel is one of the host material which is widely used in composite desiccants because of its low regeneration temperature [81] and composite desiccants in which hygroscopic salts are impregnated into silica gel, have enhanced water adsorption characteristics [82-84]. Many investigators developed various silica gel-host composites and studied their properties. Jia et al. [85] synthesized a silica gel host-composite desiccant where saturated LiCl solution was impregnated into the pores and results showed that the adsorption capacity was 2-3 times higher than that of pure silica gel. The desiccant wheel of the above composite desiccant resulted in 20-40% more dehumidification than the silica gel desiccant wheel [86]. Simonova et al. [87] developed a composite desiccant by impregnating saturated aqueous $Ca(NO_3)_2$ solution into silica gel and experimentally determined that the material had an adsorption capacity ranging from 0.15-0.17 g/g at a desorption temperature of 75-80°C. Aristov et al. [88] experimentally determined that a source temperature of 65-75°C was sufficient for regeneration of $LiNO_3$-Silica gel composite desiccant. Bu et al. [89] prepared three $CaCl_2$-Silica gel composite desiccants with different pore sizes of 2-3 nm, 4-7 nm and 8-10 nm and found that the composite desiccants of 4-7 nm and 8-10 nm exhibited enhanced water adsorption capacities with the concentration of $CaCl_2$. Similarly, Gordeeva et al. [90] also proved that the adsorption capacity of $MgSO_4$-Silica gel composite desiccant could be increased with increasing the concentration of $MgSO_4$. Cortes et al. [91] investigated on three composite desiccants, namely, $CaCl_2$-Silica gel, $MgCl_2$-Silica gel and LiBr-Silica gel and found that the $CaCl_2$-Silica gel had the highest water adsorption capacity and $MgCl_2$-Silica gel had the lowest. Mrowiec-Bialion et al. [92] synthesized a $CaCl_2$-Silica gel composite desiccant using sol-gel method and found experimentally that the water adsorption capacity of the composite desiccant was higher than 1 g/g. Similarly, using sol-gel method, Sukhyy et al. [93] developed Na_2SO_4-Silica gel composite desiccant whose water adsorption capacity was reported to be 0.85 g/g.

3.7.2. Mesoporous silicate-host composites

After the discovery of MCM-41 [94,95] and FSM-16 [96], a family of mesoporous silicates emerged, which are prepared by hydrothermal synthesis of silica gel in the presence of long chained surfactant templates. Later, this lead to the discovery of more of these kind like MCM-48, MCM-50, KIT-1, SBA-15 and many others [97,98]. They have microporous distribution, high surface area (> $500 m^2/g$), regular structures and abundant silanol groups resulting in higher water adsorption characteristics than original silica gel [99]. These mesoporous silicates exhibit type V adsorption isotherm characteristics which makes them a better host material for composite desiccants [100,101]. Many researchers developed these mesoporous silicate desiccants with enhanced adsorption properties. Ponomarenko et al. [102] developed a composite mesoporous silicate by impregnating $CaCl_2$ in SBA-15 and observed experimentally that the adsorption capacity as high as 0.47 g/g could be achieved. Glaznev et al. [103] synthesized two composite desiccants of $CaCl_2$-SBA-15 with 8.1 nm and 11.8 nm pore sizes and found that the desorption was achieved at a low temperature of 88.5°C for larger pore size and was 6°C lower than that of the smaller pore size desiccant. Zhang and Pei [104] developed three mesoporous silicate desiccants with sodium polystyrene sulfonate, sodium polyacrylate and sodium malate as impregnating salts and found that the desiccant with sodium malate had exhibited higher water adsorption characteristics and the water intake was double than that of the pure mesoporous silica at 25^0 C.

3.7.3. Carbon-host composites

Activated carbon is widely used adsorbent due to its high pore volume and large surface area and also, it is a good host material for composite material as it exhibits type V adsorption isotherm characteristics with S-shaped curve [105]. Combining activated carbon with hygroscopic salts is one of the paramount ways for enhancing water adsorption characteristics of activated carbon. Huang et al. [106] synthesized a composite desiccant by impregnating activated carbon into $NaSiO_3$ solution and found that the water adsorption characteristics were higher than the original activated carbon. Tso and Chao [107] developed a composite desiccant by impregnating both $CaCl_2$ and $NaSiO_3$ into activated carbon. The water adsorption capacity was found to be 0.805 g/g at atmospheric pressure and at temperatures of 25-115°C. They also observed that utilizing waste heat at temperature of 60°C as source heat for regeneration resulted in almost 80% of adsorption of water vapour.

3.7.4. Natural rock-host composites

The investigations on composites of chlorine salts and natural porous materials brought attention towards natural mineral clay hosted composites [108-111]. Chen et al. [112] synthesized a composite of LiCl and attapulgite clay and investigated on its properties. The results indicated that at 30°C temperature and 1.5 kPa pressure, the water adsorption capacity was 0.44 g/g, which was 6-7 times that of the original attapulgite clay, but with a regeneration temperature of 170°C. Butut et al. [113] investigated on a composite desiccant of Turkish bentonite and $CaCl_2$ and found that it had water adsorption characteristics but it required regeneration temperature of 105-150°C. Sapienza et al. [114] synthesized a composite desiccant of vermiculite and $LiNO_3$ and found that the water adsorption capacity was about 0.4 g/g where adsorption took place at $33\text{-}36^0$ C and desorption at $62\text{-}65^0$ C. Nakabayashi et al. [115] fabricated three desiccant composites by impregnating $CaCl_2$, NaCl and LiCl into the pores of wakkanai siliceous shale. The investigation results indicated that the adsorption characteristics of composite desiccant with NaCl was 5-7 times that of the pure wakkanai siliceous shale at 25°C.

3.7.5. Binary salt impregnated composites

The research works mentioned above were based on the composite desiccants with single salt impregnations. Recently composite desiccants with double hygroscopic salts impregnation became more prominent in researches. Gordeeva et al. [116,117] synthesized two composite desiccants by impregnating $CaCl_2$+$CaBr_2$ and LiCl+LiBr into silica pores and found that the water vapour can be desorbed at 70°C by increasing the chlorine salt concentration in the binary salts. These binary salt impregnated composite desiccants were found to be having higher water adsorption characteristics when compared to single salt composite desiccants and low regeneration temperatures less than 100°C. The water adsorption characteristics of these composite desiccants mainly depends on the concentration and categories the of salts [88,90,107,118], fabrication processes [83,84,91,92] and pore distribution of host material [86,101,106,110]. By varying these above characteristics, composite desiccants with desired and required adsorption characteristics can be synthesized.

4. Liquid Desiccants

Liquid Desiccants are generally strong ionic salt solutions whose behaviour can be controlled through varying their concentration, temperature or may be both. The control of temperature is handled by coolers or heaters, whereas concentration is handled by heating the desiccant to remove water vapour into the atmosphere or a scavenger airstream. In air washer, when air is passed through it, the dew point of the air supplied approaches the water temperature with which the machine is supplied. In this process, more humid air gets dehumidified and less humid air humidified. In the same way, the air is brought into contact with the liquid desiccant solution. Generally, the vapour pressure of liquid desiccant solution is lower than that of water at same temperature and the air gets dehumidified due to this vapour pressure difference, [4]. The selection of liquid desiccants depends on many parameters like density of energy storage, thermo-physical properties, vapour pressure, availability, regeneration temperature, boiling point temperature, cost, etc. and of all the above parameters, vapour pressure of the surface is of main concern. Some of the commonly used liquid desiccants, because of their low surface vapour pressure at low temperature and high concentration, are calcium chloride, lithium chloride, lithium bromide and triethylene glycol [3]. Thermodynamic properties of individual liquid desiccants are provided by many researchers like Uemura [119], Patil [120] and Ahmed [121]. Conde et al. [122] developed the formulae for determining the thermodynamic properties of calcium and lithium chlorides and Sun et al. [123] analysed and calculated vapour pressures, based on thermodynamic properties, of various liquid desiccants. Thermodynamic properties of aqueous lithium bromide solution are provided by McNeely [124]. Kaita [125] developed equations for determining thermodynamic properties of lithium bromide-water solutions for temperatures of 40-210°C and for concentrations of 40-60% weight. Ertas and Kiris [126] investigated and provided the properties of calcium chloride and lithium chloride at temperatures of 26.6-65.5°C. Recently, researchers like Lucas et al. [127] and Park et al. [128], started studying the properties of mixed liquid desiccants at appropriate proportions with lower surface vapour pressure and enhanced thermodynamic properties. Below is a brief discussion on commonly used liquid desiccants and their properties.

4.1. Calcium Chloride

Calcium chloride is a typical ionic halide which serves calcium ion in aqueous solution and at room temperature it is a solid. It is generally produced by direct reaction of limestone with hydrochloric acid but in large quantities it is produced as a byproduct of Solvey process. It has a boiling point temperature of 1395°C with 2.15 g/cc density. In $CaCl_2$-H_2O solutions, the crystallization line is complex because of the formation of α and β tetrahydrates [129].

4.2. Lithium Chloride

Lithium chloride is an ionic salt that is widely used in air-conditioning systems. It has better hygroscopic properties and amazing solubility of about 83g/100 ml at 20°C in polar solvents. It has a boiling point temperature of 1382°C with 2.068 g/cc density. The crystallization line of LiCl-H_2O solution is an increasing mass fraction of LiCl and reducing water content [129].

4.3. Lithium Bromide

Lithium bromide is another lithium salt that is widely used as desiccant in air-conditioning applications. It is generally produced by treating lithium carbonate with hydrobromic acid and it can also be produced as a precipitate in water by treating lithium hydroxide with hydrobromic acid. It generally forms many crystalline hydrates comparative to other bromides of alkali metals. It is quite soluble in water, methanol, ethanol, ether and also slightly in pyridine. It has a boling point temperature of 1265°C with a density of 3.464 g/cc [129].

5. Conclusion

The ability of a desiccant to attract water vapour depends on the difference in vapour pressure between its surface and the process air. All desiccants attract moisture until they touch equilibrium point with the surrounding air. The absorbed water vapour is usually removed from the desiccant by regeneration in which desiccant is exposed to a regeneration airstream having temperatures between 50 and 260°C.

Generally, the capacity of solid desiccant is less than the capacity of liquid desiccant but the use of composite desiccant materials may improve the moisture adsorption capacity. In case solid desiccant for air dehumidification with low grade waste heat as driving energy silica gel always performs better than other solid desiccant. Where as in case of liquid desiccant the most suitable is LiCl because of its lower vapour pressure but the major problem with LiCl is its high cost, this issue can be solve by combining it with cheapest desiccant $CaCl_2$. At the same time the composite desiccant of silica gel and LiCl has adsorption capacity 2-3 times higher than that of pure silica gel. The limitation faced by desiccant system is availability of regenerating heat to regenerate desiccant material. But the use of solar energy and waste heat for regeneration of desiccant material will make the system more economical. The use of desiccant system can solve lot of environmental problems well, as it can also minimise the high demand of electrical energy for conventional air-conditioning system and poor indoor air quality.

References

1. Henning H-M. Solar assisted air conditioning of buildings - an overview. Appl Therm Eng 2007;27(10):1734-49.
2. X. Zheng, T.S. Ge, R.Z. Wang. Recent progress on desiccant materials for solid desiccant cooling systems. Energy 74 (2014) 280-294.
3. L. Mei, Y.J. Dai. A technical review on use of liquid-desiccant dehumidification for air-conditioning application. Renewable and Sustainable Energy Reviews 12 (2008) 662–689
4. 2009 ASHRAE Handbook - Fundamentals (SI Edition) - American Society of Heating, Refrigerating and Air-Conditioning Engineers, Inc. (2009). Pg. 32.1-32.6.
5. K. Daou, R.Z.Wang*, Z.Z.Xia, Desiccant cooling air conditioning: a review, Renewable and Sustainable Energy Reviews 10 (2006) 55-77.
6. http://www.gasairconditioning.org/desiccant_advantages_applications.htm
7. G. Panaras*, E. Mathioulakis, V. Belessiotis, Solid desiccant air-conditioning systems - Design parameters, Energy 36 (2011) 2399-2406.
8. Dennis Schlepp, Kenneth Schult, Analysis of Advanced Solar Hybrid Desiccant Cooling Systems for Buildings, SERI/TR-252-2527. Pg. 01-26.
9. F.E. Nia, D. Van Paassen, M.H. Saidi, Modeling and simulation of desiccant wheel for air conditioning. Energy Build, 38 (10) (2006), pp. 1230–1239.
10. De Antonellis S, Joppolo CM, Molinaroli L. Simulation, performance analysis and optimization of desiccant wheels. Energy Build 2010;42(9):1386-93.
11. M. Ali Mandegari, H. Pahlavanzadeh, Introduction of a new definition for effectiveness of desiccant wheels. Energy, 34 (6) (2009), pp. 797–80
12. Fong KF, Lee CK, Chow TT, Fong AML. Investigation on solar hybrid desiccant cooling system for commercial premises with high latent cooling load in subtropical Hong Kong. Appl Therm Eng 2011; 31:3393–401.
13. Hürdoğan E, Büyükalaca O, Yılmaz T, Hepbasli A, Uçkan R. Investigation of solar energy utilization in an novel desiccant based air conditioning system. Energy Build 2012; 55:757–64.

14. José MCL, Francisco FH, Fernando DM, Antonio CA. The optimization of the operation of a solar desiccant air handling unit coupled with a radiant floor. Energy Build 2013; 62:427–35.
15. C.X. Jia, Y.J. Dai, J.Y. Wu, R.Z. Wang. Use of compound desiccant to develop high performance desiccant cooling system. International Journal of Refrigeration 30 (2007) 345-353.
16. Ramzy K A, Kadoli R, Ashok Babu T. Improved utilization of desiccant material in packed bed dehumidifier using composite particles. Renew Energy 2011;36(2):732-42.
17. Ali Al-Alili, YunhoHwang,ReinhardRadermacher. Performance of a desiccant wheel cycle utilizing new zeolite material: Experimental investigation. Energy 81 (2015) 137-145.
18. Cui Q, Chen H, Tao G, Yao H. Performance study of new adsorbent for solid desiccant cooling. Energy 2005;30(2):273-9.
19. Tao Cao, Hoseong Lee, Yunho Hwang, ReinhardRadermacher, Ho-Hwan Chun. Experimental investigations on thin polymer desiccant wheel performance. International Journal of Refrigeration 44 (2014) 1-11.
20. Zhang X, Qiu L. Moisture transport and adsorption on silica gel-calcium chloride composite adsorbents. Energy Convers Manage 2007;48(1):320-6.
21. M. Fatouh, Talaat A. Ibrahim, A. Mostafa. Experimental investigation on a solid desiccant system integrated with a R407C compression air conditioner. Energy Conversion and Management 50 (2009) 2670–2679.
22. R. Narayanan, W.Y. Saman, S.D. White, M. Goldsworthy. Comparative study of different desiccant wheel designs. Applied Thermal Engineering 31 (2011) 1613-1620.
23. FatemehEsfandiari Nia, Dolf van Paassen, Mohamad Hassan Saidi. Modeling and simulation of desiccant wheel for air conditioning. Energy and Buildings 38 (2006) 1230–1239.
24. WeiweiLiua, Zhiwei Liana, ReinhardRadermacherb, Ye Yao. Energy consumption analysis on a dedicated outdoor air system with rotary desiccant wheel. Energy 32 (2007) 1749–1760.
25. Springer. Desiccant-Assisted Cooling Fundamentals and Applications. Springer-Verlag London 2014. Pg 109-141.
26. ASHRAE Air-Conditioning System Design Manual. Refrigerating and Air-Conditioning Engineers, Inc. (2007). Pg. 213-52.
27. Srivastava, N. C., and I. W. Eames. "A review of adsorbents and adsorbates in solid–vapour adsorption heat pump systems." Applied Thermal Engineering, 18.9 (1998): 707-714.
28. Pramuang, Surajitr, and R. H. B. Exell. "The regeneration of silica gel desiccant by air from a solar heater with a compound parabolic concentrator." Renewable energy, 32.1 (2007): 173-182.
29. Chung, Tsair-Wang, and Chin-Chun Chung. "Increase in the amount of adsorption on modified silica gel by using neutron flux irradiation." Chemical engineering science, 53.16 (1998): 2967-2972.
30. Ralph T Yang. Adsorbents: Fundamentals and Applications. Wiley Interscience. John Wiley & Sons. Inc. (2003). Pg. 79-190.
31. Serena Esposito, Antonello Marocco, Gianfranco Dell'Agli, Bruno De Gennaro, Michele Pansini.Relationships between the water content of zeolites and their cation population. Microporous and Mesoporous Materials 202 (2015) 36–43.
32. F. Benaliouche, N. Hidous, M. Guerza, Y. Zouad, Y. Boucheffa. Characterization and water adsorption properties of Ag and Zn-exchanged A zeolites. Microporous and Mesoporous Materials 209 (2015) 184–188.
33. Abd-Elrahman, W. R., et al. "Experimental investigation on the performance of radial flow desiccant bed using activated alumina." Applied Thermal Engineering 31.14 (2011): 2709-2715.
34. Hamed, A. M., et al. "Theoretical and experimental investigation on the transient coupled heat and mass transfer in a radial flow desiccant packed bed." Energy Conversion and Management 65 (2013): 262-271.
35. Karamanis, D., and E. Vardoulakis. "Application of zeolitic materials prepared from fly ash to water vapor adsorption for solar cooling." Applied energy 97 (2012): 334-339.
36. Rao, Zhonghao, Shuangfeng Wang, and Zhengguo Zhang. "Energy saving latent heat storage and environmental friendly humidity-controlled materials for indoor climate." Renewable and Sustainable Energy Reviews 16.5 (2012): 3136-3145.
37. Kubota, Mitsuhiro, et al. "Regeneration characteristics of desiccant rotor with microwave and hot-air heating." Applied Thermal Engineering 50.2 (2013): 1576-1581.
38. Kubota, Mitsuhiro, et al. "Water desorption behavior of desiccant rotor under microwave irradiation." Applied Thermal Engineering 31.8 (2011): 1482-1486.
39. Y. Hamamotoa, K.C.A. Alam, B.B. Saha, S. Koyama, A. Akisawa, T. Kashiwagi. Study on adsorption refrigeration cycle utilizing activated carbonfibers. Part 1. Adsorption characteristics. International Journal of Refrigeration 29 (2006) 305–314.
40. Sawyer L, Grubb D, Meyers GF. Polymer microscopy. Springer; 2008.
41. Kratochvfl P, Jenkins AD, Stepto RFT, Suter UW. Glossary of basic terms inpolymer science (IUPAC recommendations 1996). Pure Appl Chem 68(12) 1996:2287-311.
42. Cote R, Hosette S, Amazouz M. Polymeric desiccant articles and process fortheir manufacture. Google Patents; 2000.
43. Collier R. Advanced desiccant materials assessment, phase 2, final report. GasResearch Institute Report GRI-88/0125; 1988.
44. Czanderna AW, Tillman NT, Herdt G. Polymers as advanced materials fordesiccant applicationsepart 3: alkali salts of PSSA and polyAMPSA and copolymers of polyAMPSASS. ASHRAE Trans 1995;101(1):697-712.

45. Zhao J-H, Wang J, Tu M, Luo B-H, Zhou C-R. Improving the cell affinity of apoly (D, L-lactide) film modified by grafting collagen via a plasma technique. Biomedical Materials 2006;1(4):247.
46. Langowski BA, Uhrich KE. Microscale plasma-initiated patterning (μpip). Langmuir 2005;21(23):10509-14.
47. Shim J-Y, Song G-E, Lee D-Y. Dehumidification and regeneration of a polymericdesiccant. In: Proceedings of the International Sorption Heat Pump Conference, Seoul, Korea; 2008.
48. Inaba H, Komatsu F, Horibe A, Haruki N, Machida A. Sorption Characteristicsof Sorption Material Coated on Heat Transfer surface of a Heat Exchanger. Thermal Science and Engineering 2008; 15:141-50.
49. Lee D-Y. Desiccant, dehumidifying element and methods for manufacturingthe same, WO Patent 2,010,011,016; 2010.
50. Mathiowitz E, Jacob JS, Jong YS, Hekal TM, Spano W, Guemonprez R, et al.Novel desiccants based on designed polymeric blends. Journal of applied polymer science 2001;80(3):317-27.
51. Lee J, Lee D. Sorption characteristics of a novel polymeric desiccant. Int JRefrig 2012;35(7):1940-9.
52. White SD, Goldsworthy M, Reece R, Spillmann T, Gorur A, Lee DY. Characterizationof desiccant wheels with alternative materials at low regeneration temperatures. International Journal of Refrigeration 2011;34(8):1786-91.
53. Batten SR, Champness NR, Chen XM, Garcia-Martinez J, Kitagawa S, Ohrstrom L, et al. Terminology of metal-organic frameworks and coordination polymers (IUPAC recommendations 2013). Pure Appl Chem 2013;85(8):1715-24.
54. Meek ST, Greathouse JA, Allendorf MD. Metal-organic frameworks: a rapidlygrowing class of versatile nanoporous materials. Adv Mater 2011;23(2):249-67.
55. Rowsell JLC, Yaghi OM. Metaleorganic frameworks: a new class of porousmaterials. Microporous Mesoporous Mater 2004;73(1):3-14.
56. Liang X, Zhang F, Feng W, Zou X, Zhao C, Na H, et al. From metal-organic framework (MOF) to MOF-polymer composite membrane: enhancement of low-humidity proton conductivity. Chemical Science 2013;4(3):983-92.
57. Dang D, Wu P, He C, Xie Z, Duan C. Homochiral Metal Organic Frameworksfor Heterogeneous Asymmetric Catalysis. Journal of the American Chemical Society 2010;132(41):14321-3.
58. Lee J, Farha OK, Roberts J, Scheidt KA, Nguyen ST, Hupp JT. Metal-organicframework materials as catalysts. Chemical Society Reviews 2009;38(5):1450-9.
59. Ma L, Abney C, Lin W. Enantioselective catalysis with homochiral metal-organicframeworks. Chemical Society Reviews 2009;38(5):1248-56.
60. Li Y, Yang RT. Gas adsorption and storage in metal-organic framework MOF-177. Langmuir 2007;23(26):12937-44.
61. Ma S, Zhou H-C. Gas storage in porous metaleorganic frameworks for cleanenergy applications. ChemCommun 2010;46(1):44-53.
62. Millward AR, Yaghi OM. Metal-organic frameworks with exceptionally highcapacity for storage of carbon dioxide at room temperature. J Am ChemSoc 2005;127(51):17998-9.
63. Li J-R, Ma Y, McCarthy MC, Sculley J, Yu J, Jeong H-K, et al. Carbon dioxidecapture-related gas adsorption and separation in metal-organic frameworks. Coordination Chemistry Reviews 2011;255(15):1791-823.
64. Britt D, Furukawa H, Wang B, Glover TG, Yaghi OM. Highly efficient separationof carbon dioxide by a metal-organic framework replete with open metal sites. Proceedings of the National Academy of Sciences 2009;106(49):20637-40.
65. Pan L, Parker B, Huang X, Olson DH, Lee J, Li J. Zn (tbip)(H2tbip¼ 5-tert-butylisophthalic acid): a highly stable guest-free microporous metal organic framework with unique gas separation capability. Journal of the American Chemical Society 2006;128(13):4180-1.
66. Li J-R, Kuppler RJ, Zhou H-C. Selective gas adsorption and separation inmetaleorganic frameworks. ChemSoc Rev 2009;38(5):1477-504.
67. Comotti A, Bracco S, Sozzani P, Horike S, Matsuda R, Chen J, et al. Nano-channelsof two distinct cross-sections in a porous Al-based coordination polymer. Journal of the American Chemical Society 2008;130(41):13664-72.
68. Furukawa H, Ko N, Go YB, Aratani N, Choi SB, Choi E, et al. Ultrahigh porosityin metal-organic frameworks. Science 2010;329(5990):424-8.
69. Furukawa H, Go YB, Ko N, Park YK, Uribe-Romo FJ, Kim J, et al. IsoreticularExpansion of MetaleOrganic Frameworks with Triangular and Square Building Units and the Lowest Calculated Density for Porous Crystals. Inorganic Chemistry 2011;50(18):9147-52.
70. Farha OK, Eryazici I, Jeong NC, Hauser BG, Wilmer CE, Sarjeant AA, et al. Metal-Organic Framework Materials with Ultrahigh Surface Areas: Is the Sky the Limit? Journal of the American Chemical Society 2012;134(36):15016-21.
71. Ferey G, Mellot-Draznieks C, Serre C, Millange F, Dutour J, Surble S, et al. Achromium terephthalate-based solid with unusually large pore volumes and surface area. Science 2005;309(5743):2040-2.
72. Küsgens P, Rose M, Senkovska I, Frode H, Henschel A, Siegle S, et al. Characterizationof metal-organic frameworks by water adsorption. Microporous and Mesoporous Materials 2009;120(3):325-30.

73. Akiyama G, Matsuda R, Kitagawa S. Highly porous and stable coordinationpolymers as water sorption materials. Chem Lett 2010;39(4):360-1.
74. Ehrenmann J, Henninger SK, Janiak C. Water adsorption characteristics ofMIL-101 for heat-transformation applications of MOFs. Eur J InorgChem 2011;2011(4):471-4.
75. Henninger SK, Jeremias F, Kummer H, Janiak C. MOFs for Use in AdsorptionHeat Pump Processes. European Journal of Inorganic Chemistry 2011;2012(16):2625-34.
76. Seo YK, Yoon JW, Lee JS, Hwang YK, Jun CH, Chang JS, et al. Energy-EfficientDehumidification over Hierachically Porous MetaleOrganic Frameworks as Advanced Water Adsorbents. Advanced Materials 2012;24(6):806-10.
77. Akiyama G, Matsuda R, Sato H, Hori A, Takata M, Kitagawa S. Effect offunctional groups in MIL-101 on water sorption behavior. Microporous and Mesoporous Materials 2012.
78. Rezk A, Al-Dadah R, Mahmoud S, Elsayed A. Characterisation of metalorganic frameworks for adsorption cooling. International journal of heat and mass transfer 2012;55(25-26):7366-74.
79. Aristov YI. New family of solid sorbents for adsorptive cooling: materialscientist approach. J Eng Thermophys 2007;16(2):63-72.
80. Gordeeva L, Aristov YI. Composites 'salt inside porous matrix' for adsorptionheat transformation: a current state-of-the-art and new trends. Int J Low-Carbon Technol 2012;7(4):288-302.
81. Wang LW, Wang RZ, Oliveira RG. A review on adsorption working pairs forrefrigeration. Renew Sustain Energy Rev 2009;13(3):518-34.
82. Aristov YI, Tokarev M, Cacciola G, Restuccia G. Selective water sorbents formultiple applications, 1. CaCl2 confined in mesopores of silica gel: sorption properties. Reaction Kinetics and Catalysis Letters 1996;59(2):325-33.
83. Gordeeva L, Restuccia G, Cacciola G, Aristov YI. Selective water sorbents formultiple applications, 5. LiBr confined in mesopores of silica gel: Sorption properties. Reaction Kinetics and Catalysis Letters 1998;63(1):81-8.
84. Yu N, Wang R, Lu Z, Wang L. Development and characterization of silica gel-LiCl composite sorbents for thermal energy storage. Chemical Engineering Science 2014.
85. Jia C, Dai Y, Wu J, Wang R. Experimental comparison of two honeycombeddesiccant wheels fabricated with silica gel and composite desiccant material. Energy Conversion and Management 2006;47(15):2523-34.
86. Jia C, Dai Y, Wu J, Wang R. Use of compound desiccant to develop highperformance desiccant cooling system. International Journal of Refrigeration 2007;30(2):345-53.
87. Simonova IA, Freni A, Restuccia G, Aristov YI. Water sorption on composite"silica modified by calcium nitrate". Microporous and Mesoporous Materials 2009;122(1):223-8.
88. Aristov YI, Sapienza A, Ovoshchnikov D, Freni A, Restuccia G. Reallocation ofadsorption and desorption times for optimisation of cooling cycles. International journal of refrigeration 2012;35(3):525-31.
89. Bu X, Wang L, Huang Y. Effect of pore size on the performance of compositeadsorbent. Adsorption 2013;19(5):929-35.
90. Gordeeva L, Glaznev I, Aristov YI. Sorption of water by sodium, copper, andmagnesium sulfates dispersed into mesopores of silica gel and alumina. Russ J PhysChem 2003;77(10):1715-20.
91. Corte's FB, Chejne F, Carrasco-Marín F, Pe'rez-Cadenas AF, Moreno-Castilla C.Water sorption on silica-and zeolite-supported hygroscopic salts for cooling system applications. Energy Conversion and Management 2012;53(1): 219-23.
92. Mrowiec-Białon J, Jarze¸bski AB, Lachowski AI, Malinowski JJ, Aristov YI.Effective inorganic hybrid adsorbents of water vapor by the sol-gel method. Chemistry of materials 1997;9(11):2486-90.
93. Sukhyy KM, Belyanovskaya EA, Kozlov YN, Kolomiyets EV, Sukhyy MP.Structure and adsorption properties of the composites 'silica gelesodiumsulphate', obtained by solegel method. Applied Thermal Engineering 2014;64(1):408-12.
94. Beck J, Vartuli J, Roth W, Leonowicz M, Kresge C, Schmitt K, et al. A newfamily of mesoporous molecular sieves prepared with liquid crystal templates. Journal of the American Chemical Society 1992;114(27):10834-43.
95. Kresge C, Leonowicz M, Roth W, Vartuli J, Beck J. Ordered mesoporous molecularsieves synthesized by a liquid-crystal template mechanism. Nature 1992;359(6397):710-2.
96. Inagaki S, Fukushima Y, Kuroda K. Synthesis of highly ordered mesoporousmaterials from a layered polysilicate. J ChemSocChemCommun 1993;(8):680-2.
97. Llewellyn P, Schüth F, Grillet Y, Rouquerol F, Rouquerol J, Unger K. Watersorption on mesoporous aluminosilicate MCM-41. Langmuir 1995;11(2):574-7.
98. Katiyar A, Yadav S, Smirniotis PG, Pinto NG. Synthesis of ordered large poreSBA-15 spherical particles for adsorption of biomolecules. Journal of Chromatography A 2006;1122(1):13-20.
99. Collier R, Cale T, Lavan Z. Advanced desiccant materials assessment. Finalreport, February 1985-May 1986. Glendale, AZ (USA): Enerscope, Inc.; 1986.
100. Ji JG, Wang RZ, Li LX. New composite adsorbent for solar-driven fresh waterproduction from the atmosphere. Desalination 2007;212(1):176-82.

101.Tokarev M, Gordeeva L, Romannikov V, Glaznev I, Aristov Y. New compositesorbent CaCl2 in mesoporesfor sorption cooling/heating. International Journal of Thermal Sciences 2002;41(5):470-4.
102.Ponomarenko I, Glaznev I, Gubar A, Aristov YI, Kirik S. Synthesis and watersorption properties of a new composite "CaCl2 confined into SBA-15 pores". Microporous and Mesoporous Materials 2010;129(1):243-50.
103.Glaznev I, Ponomarenko I, Kirik S, Aristov Y. Composites CaCl2/SBA-15 foradsorptive transformation of low temperature heat: Pore size effect. International Journal of Refrigeration 2011;34(5):1244-50.
104.Pei L, Zhang L. Preparation and selective adsorption of core-shell desiccantfor heat and moisture recovery. Colloids Surf APhysicochem Eng Asp 2012;406(20):68-74.
105.Li X, Chen X, Li Z. Adsorption equilibrium and desorption activation energy ofwater vapor on activated carbon modified by an oxidation and reduction treatment. J Chem Eng Data 2010;55(9):3164-9.
106.Huang H, Oike T, Watanabe F, Osaka Y, Kobayashi N, Hasatani M. Developmentresearch on composite adsorbents applied in adsorption heat pump. Applied Thermal Engineering 2010;30(10):1193-8.
107.Tso CY, Chao CYH. Activated carbon, silica-gel and calcium chloride compositeadsorbents for energy efficient solar adsorption cooling and dehumidification systems. Int J Refrig 2012;35(6):1626-38.
108.Montes-H G, Geraud Y. Sorption kinetic of water vapour of MX80 bentonitesubmitted to different physicalechemical and mechanical conditions. Colloids Surf APhysicochem Eng Asp 2004;235(1):17-23.
109.Gonzalez J, Molina-Sabio M, Rodrıguez-Reinoso F. Sepiolite-based adsorbentsas humidity controller. Appl Clay Sci 2001;20(3):111-8.
110.Aristov YI, Restuccia G, Tokarev M, Buerger H-D, Freni A. Selective watersorbents for multiple applications. 11. CaCl2 confined to expanded vermiculite. Reaction Kinetics and Catalysis Letters 2000;71(2):377-84.
111.Thoruwa T, Johnstone C, Grant A, Smith J. Novel, low cost CaCl2 based desiccantsfor solar crop drying applications. Renewable energy 2000;19(4): 513-20.
112.Chen H-j, Cui Q, Tang Y, Chen X-j, Yao H-q. Attapulgite based LiCl compositeadsorbents for cooling and air conditioning applications. Applied Thermal Engineering 2008;28(17):2187-93.
113.Bulut G, Chimeddorj M, Esenli F, Çelik M. Production of desiccants fromTurkish bentonites. Applied Clay Science 2009;46(2):141-7.
114.Sapienza A, Glaznev IS, Santamaria S, Freni A, Aristov YI. Adsorption chillingdriven by low temperature heat: new adsorbent and cycle optimization. Applied Thermal Engineering 2012; 32:141-6.
115.Nakabayashi S, Nagano K, Nakamura M, Togawa J, Kurokawa A. Improvementof water vapor adsorption ability of natural mesoporous material by impregnating with chloride salts for development of a new desiccant filter. Adsorption 2011;17(4):675-86.
116.Gordeeva LG, Grekova AD, Krieger TA, Aristov YI. Adsorption properties ofcomposite materials (LiCl+ LiBr)/silica. Microporous and Mesoporous Materials 2009;126(3):262-7.
117.Gordeeva L, Grekova A, Krieger T, Aristov Y. Composites "binary salts inporous matrix" for adsorption heat transformation. Applied Thermal Engineering 2013; 50:1633-8.
118.Tretiak C, Abdallah NB. Sorption and desorption characteristics of apacked bed of clayeCaCl2 desiccant particles. Sol Energy 2009;83(10): 1861-70.
119.Uemura T. Studies on the lithium chloride-water absorption refrigeration machines. Technol Rep KansaiUniv 1967; 9:71–88.
120.Patil KR. Thermodynamic properties of aqueous electrolyte solutions. Vapour pressure aqueous solutions ofLiCI, LiBr, and LiI. J Chem Eng Data 1990; 35:166–8.
121.Ahmed SY, Asan PIGS, AL-Farayedhi AA. Thermodynamic analysis of liquid desiccants. Sol Energy1998;62(1):11–8.
122.Conde MR. Properties of aqueous solutions of lithium and calcium chlorides: formulations for use in airconditioning equipment design. Int J Therm Sci 2004; 43:367–82.
123.Sun J, Gong XL, Shi MH. Study on vapor pressure of liquid desiccants solution. J Refrig 2004;25(1):28–30.
124.McNelly L. Thermodynamic properties of aqueous solutions of lithium bromide. ASHRAE Trans1979; 85:412–34.
125.Kaita Y. Thermodynamic properties of lithium bromide-water solutions at high temperatures. Int J Refrig2001; 24:371–90.
126.Ertas AA, Kiris I. Properties of a new liquid desiccant solution-lithium chloride and calcium chloridemixture. Sol Energy 1992;49(3):205–12.
127.de Lucas A, Donate M, Rodriguez JF. Vapour pressures, densities, and viscosities of the (water+lithiumbromide+potassium acetate) system and (water+lithiumbromide+sodium lactate) system. J Chem Eng Data 2003;48(1):18–22.
128.Park Y, Kim J-S, Lee H. Physical properties of the lithium bromide+1, 3-propanediol+water system. Int JRefrig 1997;20(5):319–25.
129.Ye Yao, Shinqing Liu. Ultrasonic Technology for Desiccant regerneration. WILEY, Shanghai Jiao Tong University Press. Pg. 293-305.

Performance prediction of solid desiccant rotary system using artificial neural network

V K Mishra[a], R P Singh, R K Das
Indian Institute of Technology (ISM) Dhanbad
[a]mishravivekkumar0303@gmail.com

Abstract

This paper presents an artificial neural network model for solid desiccant rotary system to predict its performance in terms of temperature and relative humidity of process air leaving the desiccant wheel after losing latent heat. Present paper also explains the experimental test setup that is used for taking reading. The experimental readings are all taken at steady state by varying the input conditions such as process air inlet velocity, regeneration air inlet velocity and regeneration temperature and process air inlet temperature and relative humidity. Majority of data taken from experiments is used to train the model (85%) and rest (15%) is used for testing of the model. The performance output predicted by the ANN model have high correlation factor($R>0.98336$). The results predicted by the ANN model shows that ANN model can be successfully applied to predict the performance of solid desiccant wheel with sufficient accuracy and reliability.

1. Introduction

Addition of solid desiccant wheel as desiccant dehumidification system in the traditional air conditioning system gives efficient cooling system, since this type of system handles latent and sensible heat load separately. Now a day's this kind of hybrid systems are widely used in several applications such as in pharmaceutical plants, supermarkets, theatres, office buildings, health clubs, and swimming pools [1]. Desiccant air conditioning involves the technology of desiccant dehumidification and evaporative cooling. While the first involves the water as refrigerant and can be activated by low grade thermal energy such as solar energy, waste heat, bioenergy, and district heating and the second one is near zero cost energy [2].desiccant air conditioning system is not only energy efficient and environment friendly but it is also cost competitive, particularly for hot humid and hot dry conditions. Also desiccants remove moisture in vapor form without using any liquid condensate, so desiccant dehumidification can work when dew point temperature of air is below freezing, while dehumidification based on cooling is restricted by freezing point i.e. 0^0C. While desiccant systems can work up to temperature as low as -40^0C [3], traditional vapor compression systems can work up to 4^0C [4]. Rotary desiccant air condition systems are compact and less prone to corrosion as compared to other systems and can continuously work for sufficient time this may be one of the reasons that they attract more attention. Extensive studies have been carried out for rotary desiccant air conditioning based on mathematical simulation [5-7], thermodynamic analysis [8-10], experimental investigation [11-

13], and practical application [14-16].Many efforts have been made to predict and enhance the performance of rotary desiccant dehumidifier incorporated in desiccant cooling systems. Some of the important ones are such as analogy theory by Banks [17], pseudo-state model given by Barlow [18], method by Maclaine-Cross of finite difference [19] finite difference method for cross-cooled dehumidifiers [20] and method by Jurinak of combined potential [21].

From the literature review one can observe that mathematical models have been developed by some of the researchers for performance evaluation of air conditioning system using rotary desiccant systems, and some of them have also developed model for the performance evaluation of desiccant system but they involve large no of mathematical equation and engineering calculation on the other hand model developed by the author's using ANN involve least mathematical and calculation effort and works satisfactorily for the given normal working conditions. To the best of the author's knowledge none of the previous authors have used ANN in predicting the performance of rotary desiccant system although ANN is used by some authors in predicting the performance of solid desiccant- vapor compression hybrid air conditioning system [22]. In the present study, artificial neural network (ANN) model has been developed using MATLAB software using feed forward back propagation method which uses previous experimental data to evaluate the performance of solid desiccant rotary system in terms of output process air temperature and relative humidity. Model uses inlet process air velocity, inlet regeneration air velocity, regeneration temperature, and inlet temperature and relative humidity of process air as well as regeneration air as input parameters. These experimental readings are used for training and prediction of ANN model results and also used for validation of results obtained using ANN model. Designed model of ANN works satisfactorily within the range of given data.

2. Experimental setup

Experimental setup for solid desiccant wheel system is at mechanical engineering department of Indian Institute of Technology (ISM) Dhanbad. The setup provides the arrangement for determining the performance of solid desiccant wheel for range of regeneration temperature i.e. 60-100^{0}C. Arrangement has also been made to vary the inlet flow velocity of process air as well as regeneration air by varying the inlet area. Heat energy used for heating the regeneration air is obtained from electric heater. In the desiccant wheel silica gel is used as desiccant material since it can be easily regenerated even at low temperature of about 60-100^{0}C. The matrix of wheel is made in honeycomb structure with alternate layers of metal silicate sheets and silica gel. Honeycomb structure is chosen for matrix construction since it provides large surface area with relative low pressure drop and has high structure durability. Figure below shows the schematic of experimental setup.

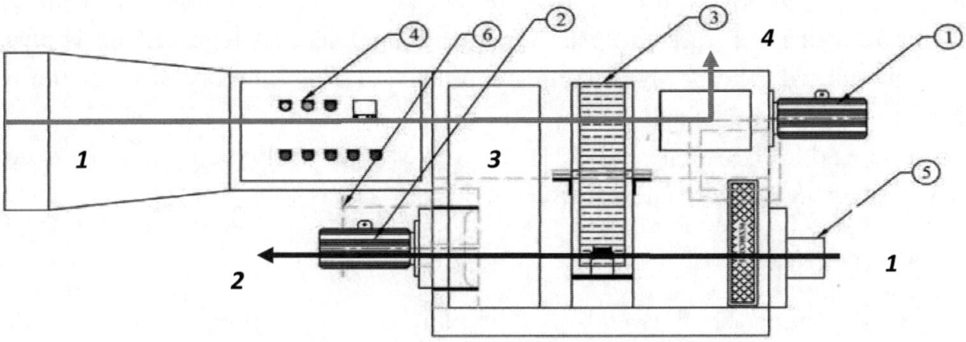

Figure-1 Schematic of experimental setup

Desiccant Wheel can handle 600 m^3/h of process air and 200 m^3/h of regeneration air. Wheel area is divided into two parts from $1/4^{th}$ part regeneration air passes while from $3/4^{th}$ part process air passes the diameter and thickness of desiccant wheel is 460mm and 100mm respectively, and the wheel is rotating at 27 RPH(revolutions per hour).

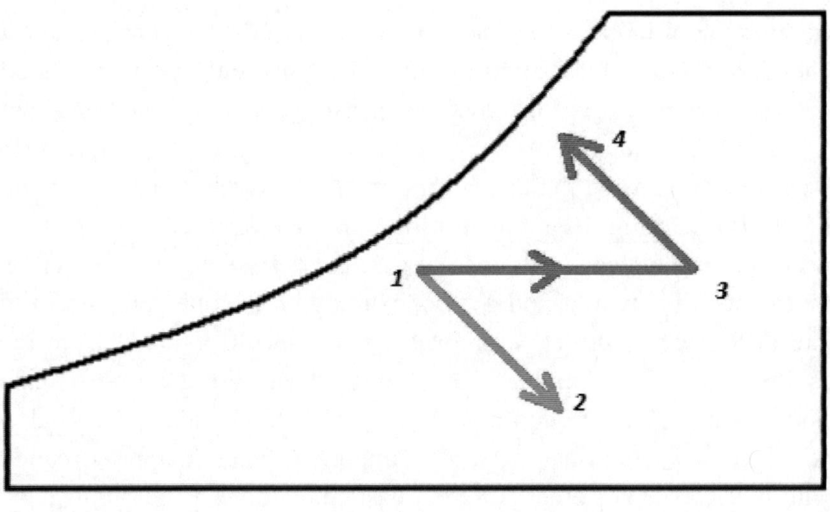

Figure-2 Process diagram on psychrometric chart

3. Measurement

Process marked 1-2 corresponds to path followed by process air while process marked 1-3-4 corresponds to flow path of regeneration air. All the readings for testing are taken at steady state condition. Two air streams, of regeneration air and process air are taken at temperature T_1, and relative humidity RH_1.process air passes through the desiccant wheel and gets dehumidified, and

IOP Conf. Series: Materials Science and Engineering **404** (2018) 012006 doi:10.1088/1757-899X/404/1/012006

regeneration air first passes through the electric heater at stage 3 and then passes through the desiccant wheel to absorb the moisture and gets humidified final state of regeneration is air is that of heated and humidified. The dry bulb temperature and relative humidity of regeneration and process air streams at inlet and outlet are measured with the use of 'K' type thermocouple(±0.70^{0}C) and RH sensors(±2.5%RH) respectively. The air flow rate of process and regeneration air is taken using hot wire anemometer (±03%F.S).

4. Artificial neural network model

A neural network is an interconnected assembly of simple processing elements, units or nodes whose functionality is loosely based on animal neuron. The processing ability of the network is stored in the inter unit connection strengths or weights, obtained by a process of adaptation to, or learning from, a set of patterns. Neural network implementation involves loading of data source, setting of attributes required ,decision on training, validation, and testing method, manipulation of data and target generation, selection of network architecture and initialization, network training and testing and evaluation of performance. Neural network is consisted of large number of neurons which acts as processing elements, and are connected by weights to each other which acts as communication links. A basic neural network model usually consisted of an input layer and an output layer and at least one hidden layer can be more also. The introduction of hidden layers makes it possible for the network to exhibit non linear behavior. The optimal number of hidden units could easily be smaller than the number of inputs, there is no rule or relation between numbers of input and number of hidden layer. Sometimes just 2 hidden layer works best with little data, but researches have shown for difficult object many hidden layers give good results. A simplified neural network consists of inputs, synapses which are nothing but connection between input and activating function, bias function and output. Bias function help neural network to learn the patterns by shifting the activating function to the left or to the right, which may be critical for successful learning. Bias functions are separate for each layer as they are not connected to previous layer. The network is trained to give the desired output from given input. The given output from the neural network model is compared with the one obtained from experiments, if there is some difference in the experimental and the result obtained from the model then weights and bias are again adjusted and network is trained and results are then again compared till the error minimizes and comes within a satisfactory limit. The above method is back propagation theorem which comes under one of the most important function approximation in neural network namely Multilayer preceptor other is Radial basis function. The output of node i of ANN network is given by

$$y_i = g_i = g\left[\sum_{j=1}^{k} w_{ji}x_j + \theta_i\right]$$

Where $y_i$output of ith node, g is activation function, w is weight, x is input variable, and θ is the bias term.

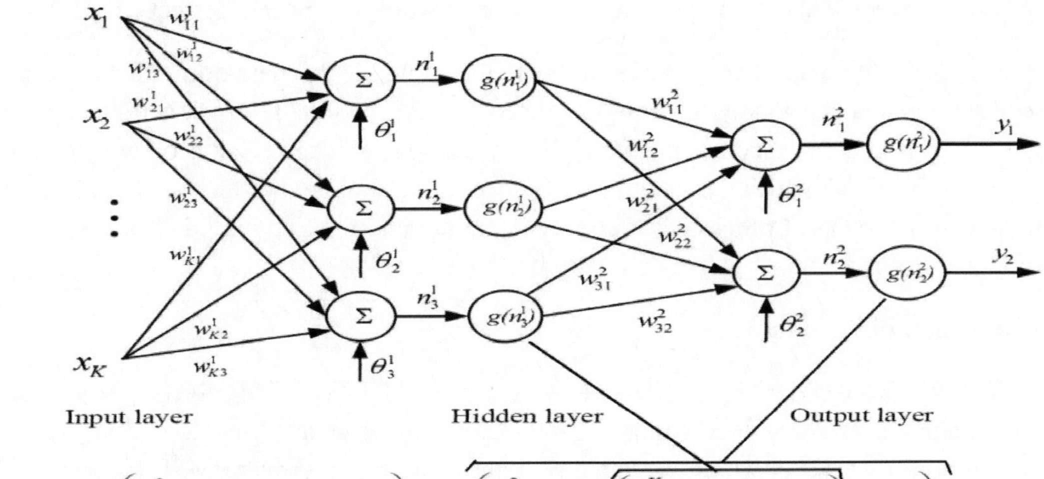

$$y_i = g\left(\sum_{j=1}^{3} w_{ji}^2 g(n_j^1) + \theta_j^2\right) = g\left(\sum_{j=1}^{3} w_{ji}^2 g\left(\sum_{k=1}^{K} w_{kj}^1 x_k + \theta_j^1\right) + \theta_j^2\right)$$

Figure-3 Representation of neural network model

Artificial neural network structure is made by feeding weight to each input and sending it to a summation function and there after adding a bias term and the result is fed to the activating function as input which results in a desired output(in simple ANN model) or acts as input to other layer present(in multiple hidden layer model as shown in figure.3). The values of weight are calculated using feed forward back propagation method. Error evaluation is done during training the model by calculating the RMS (root mean square) error and procedure is done until the RMS error value comes under tolerance level. The prediction of performance of neural network is done by calculating MSE (mean square error) between the predicted value and experimental value using expression no.

$$MSE = \frac{\sum_{i=1}^{N}\left(x_{calculated(model)} - x_{measured(experiment)}\right)^2}{N}$$

Lesser value of MSE means better the result is. RMS error is given by expression no.

$$RMSE = \sqrt{\frac{1}{N}\sum_{i=1}^{N}\left(x_{calculated(model)} - x_{measured(experiment)}\right)^2}$$

Value of absolute regression factor(R) is given by equation no.

$$R^2 = \frac{\sum_{i=1}^{N}\left(x_{calculated(model)} - x_{measured(experiment)}\right)^2}{\sum_{i=1}^{N}\left(x_{calculated(model)}\right)^2}$$

Value of R^2 lies between 0 and 1 where 0 indicates poor fit while 1 indicates perfect fit.

Working of neural network: experimental data is fed as input, neural network understands the relation between input data and store them as corrected weights, and the network is trained using a set of given input and corresponding output data, training includes revision of weights and desired result is obtained which lies within tolerance limit. In the present model of ANN calculation and comparison of process air outlet temperature and relative humidity is done with the experimental results.

5. ANN results and discussion

Feed forward back propagation technique is used with TRIANLM, LEARNGDM, MSE, and TANSIG for training, learning, performance, and activating (transfer) function. The size of network structure is 12-12-10-1 which gives the best performance in terms of MSE (MSE <1e^{-04}) for both temperature and relative humidity. Five parameters namely process air input velocity, regeneration air input velocity, regeneration temperature, and inlet air temperature and relative humidity (same for process air and regeneration air (varies only with season)) are at the input layer and two parameters which are output temperature and relative humidity of process air are added at the output layer. Range of input parameters given for training the model is given the Table.1

Table-1 Operating parameters for the model

Sr. No.	Operating parameters	Range
1	Process air inlet velocity	5.4 - 12.1 m/s
2	Regeneration air inlet velocity	13.6 – 18.39 m/s
3	Regeneration temperature	60 – 100 ^{o}c
4	Ambient temperature	24.1 – 29.5 ^{o}c
5	Ambient relative humidity	21 – 91

Temperature: Figure-4 shows the training performance of case of temperature. The graph is a plot between MSE and number of epoch (a run through all training input – output data set) (iterations) as can be seen from the figure best performance is achieved at 6th iteration corresponding to which mean squared error is 5.1418e^{-005}. It is also clear from the graph that as the number of iteration increases mean squared error decreases.

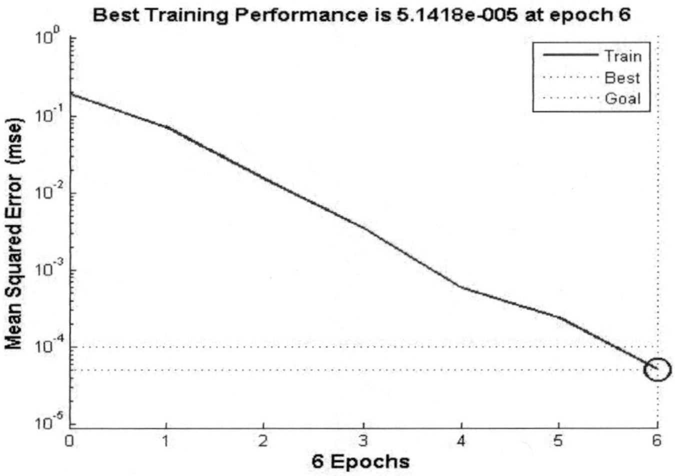

Figure-4 Variation of mean squared error with Epochs

Figure-5 below shows regression plot for training state with target set on x- axis and output of model on y- axis figure shows majority of values coincide with target data resulting in fairly high value coefficient of regression(R=0.98336), which shows closeness of calculated results and experimental results.

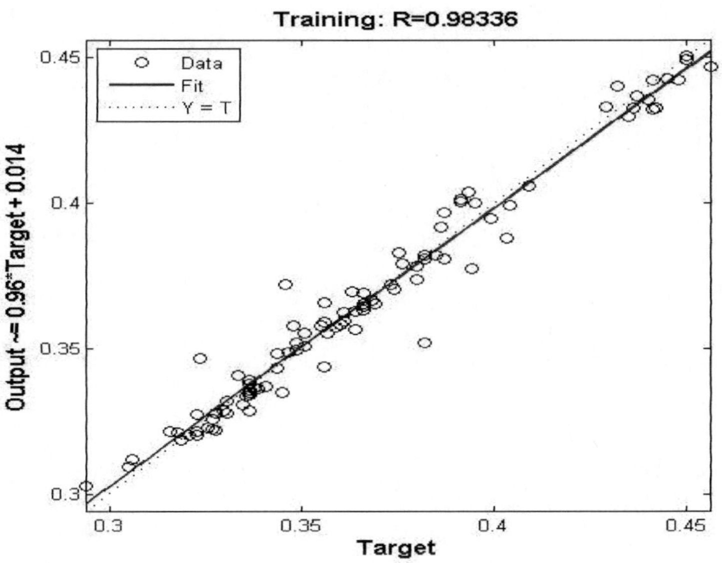

Figure-5 Regression plot for training state between experimental (target) and output

From table 2 one can observe the closeness of results obtained during testing stage from ANN model to the results obtained experimentally maximum percentage difference that occurs between the experimental value and calculated value is nearly 8%(7.89%) and minimum

IOP Conf. Series: Materials Science and Engineering **404** (2018) 012006 doi:10.1088/1757-899X/404/1/012006

percentage difference that occurs is 0.14% which are well under the acceptable range of error and can be considered as correct.

Table 2. Comparison of result of testing of ANN with the results obtained from the experiment

Experimental value(^{0}C)	Calculated value(^{0}C)	Percentage difference (%)
45	44.9354	0.143556
32.4	34.6674	-6.99815
34.6	37.2269	-7.5922
38.7	39.6537	-2.46434
42.9	43.2851	-0.89767
43.2	44.0278	-1.9162
38.2	35.1832	7.897382
39.4	37.7413	4.209898
40.9	40.6059	0.719071
44.1	44.2339	-0.30363
45.6	44.6948	1.985088

Figure-6 below represents the regression plot between the values calculated by ANN during testing and experimentally and these results also show good match with the experimental value and therefore having high regression coefficient (R^2=0.855). Such good fit represents how well the model has learned during the training of data.

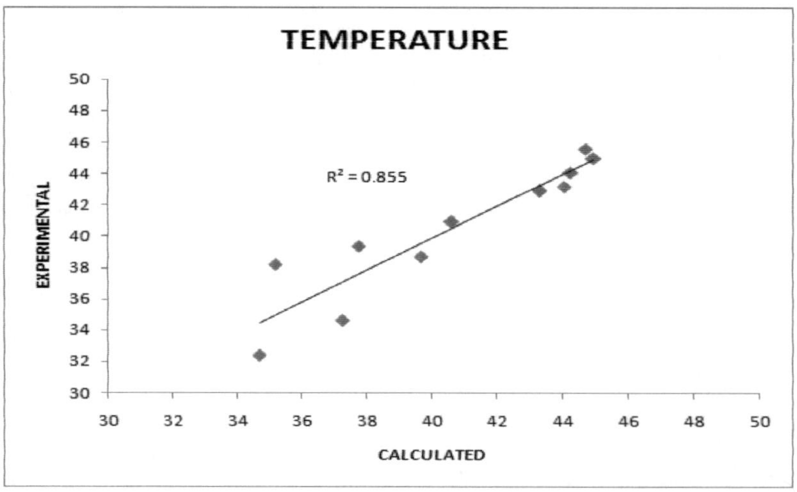

Figure-6 Regression plot for testing stage between experimental and calculated (ANN)

IOP Conf. Series: Materials Science and Engineering **404** (2018) 012006 doi:10.1088/1757-899X/404/1/012006

Relative humidity: Figure-7 below shows plot between mean squared error and number of iteration(epoch),it is evident from graph that as number of iteration are increasing mean squared error is also decreasing and it attains its minimum value 9th iteration which is $2.3739e^{-005}$.

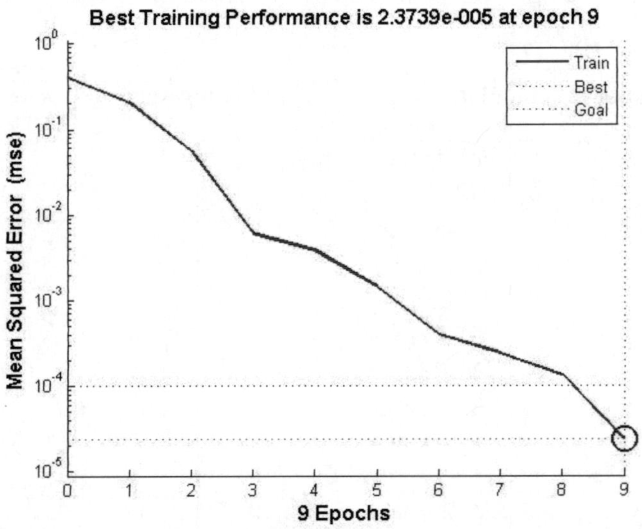

Figure-7 Variation of mean squared error with Epochs

Figure-8 below shows training results for relative humidity with target data, model output plotted on x and y axis respectively. Since majority of data lies along straight fit line, it also shows closeness of actual experimental data with the output data calculated during training period of data and achieves a high correlation factor value R=0.99408

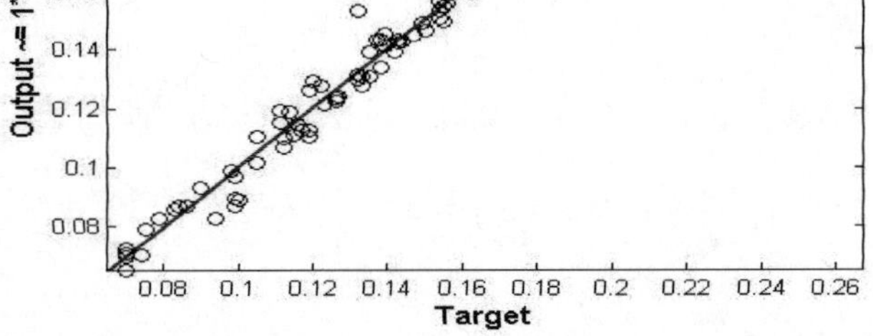

Figure-8 Regression plot for training stage between experimental (target) and output

Table-3 below shows testing stage results of ANN and experimental results and percentage difference between them for relative humidity. Analyzing the table gives a clear idea how much close the calculated values are to the experimental values and have minimum percentage difference of -0.214% and maximum percentage difference of 15% which are well within acceptable range

Table.3 Comparison of experimental values and calculated values during testing stage

Experimental value (%)	Calculated value (%)	Percentage difference (%)
7	7.2418	3.454286
20.4	21.0451	3.162255
17.1	17.0634	-0.21404
11.9	12.584	5.747899
8.4	8.6779	3.308333
7	7.2299	3.284286
13.2	15.2931	15.85682
12.2	12.7627	4.612295
10.5	11.0207	4.959048
7.9	8.2685	4.664557
7	7.1063	1.518571

Figure-9 below shows the regression plot between experimental and calculated values.

It can be deduced from the figure that the testing performance of ANN model is quite good since calculated values are in good agreement with the experimental values and are having good fit with regression coefficient R^2=0.985. Due to this reason one can successfully use the ANN model for predicting the values of relative humidity.

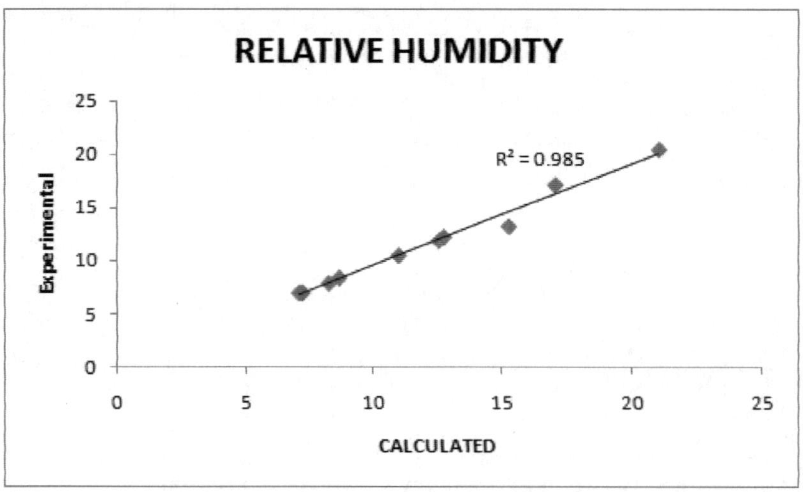

Figure-9 Regression plot for testing stage for relative humidity between experimental and calculated (ANN).

6. Conclusions

The ANN model of 12-12-10-1 network structure (number of neurons in input-hidden-output layer) has been devised to predict the performance of solid desiccant wheel. Relative humidity and temperature of outgoing process air has been considered as output parameter and are compared with the experimental results. Experimental results are used to train and test the model. ANN model represents good results in statistical terms through correlation coefficient(R) and mean squared error, following conclusions are drawn:

- The maximum percentage difference between the experimental and the ANN model result for temperature and relative humidity are 8% and 15% respectively.
- Results show that ANN model values are quite close to the experimental results and are satisfactory.
- ANN model can be successfully used to predict the performance of solid desiccant wheel in terms of temperature and relative humidity of outgoing process air.
- Accuracy of model also depends on type of model having number of hidden layer and amount of data used to train the network, and also learning method of model.

Due to high accuracy and low computational time above model can be effectively used to predict the performance of solid desiccant wheel rather than conducting expensive experiments which consume time as well as money or using other complex mathematical method requiring much more engineering effort.

References

1. Wurm, Jerry, Douglas Kosar, and Tom Clemens. "Solid desiccant technology review." *Bulletin of the International Institute of Refrigeration* 82.3 (2002): 2-31.
2. Wang, R. Z. "Advances in refrigeration and HVAC." (2007).
3. Zhang, X. J. *Study on dehumidification performance of silica gel-haloid composite desiccant wheel.* Diss. PhD Thesis. Shanghai, China: Shanghai Jiao Tong University, 2003.
4. Mazzei, Pietro, Francesco Minichiello, and Daniele Palma. "HVAC dehumidification systems for thermal comfort: a critical review." *Applied Thermal Engineering* 25.5 (2005): 677-707.
5. Nia, Fatemeh Esfandiari, Dolf Van Paassen, and Mohamad Hassan Saidi. "Modeling and simulation of desiccant wheel for air conditioning." *Energy and buildings* 38.10 (2006): 1230-1239.

IOP Conf. Series: Materials Science and Engineering **404** (2018) 012006 doi:10.1088/1757-899X/404/1/012006

6. Zhang, He-Fei, Jin-Di Yu, and Zu-She Liu. "The research and development of the key components for desiccant cooling system." *Renewable energy* 9.1-4 (1996): 653-656.
7. Ge, T. S., et al. "A review of the mathematical models for predicting rotary desiccant wheel." *Renewable and Sustainable Energy Reviews* 12.6 (2008): 1485-1528.
8. Shen CM, Worek WM. The second-law analysis of a recirculation cycle desiccant cooling system: cosorption of water vapor and carbon dioxide.Atmospheric Environment 1996; 30(9):1429–35.
9. Kanoğlu, Mehmet, Ali Bolattürk, and Necdet Altuntop. "Effect of ambient conditions on the first and second law performance of an open desiccant cooling process." *Renewable energy* 32.6 (2007): 931-946.
10. Kanoğlu, Mehmet, Melda Özdinç Çarpınlıoğlu, and Murtaza Yıldırım. "Energy and exergy analyses of an experimental open-cycle desiccant cooling system." *Applied Thermal Engineering*24.5 (2004): 919-932.
11. Kabeel, A. E. "Solar powered air conditioning system using rotary honeycomb desiccant wheel." *Renewable Energy* 32.11 (2007): 1842-1857.
12. Ge, T. S., et al. "Experimental study on a two-stage rotary desiccant cooling system." *International Journal of Refrigeration*32.3 (2009): 498-508.
13. Ge, T. S., et al. "Experimental investigation on a one-rotor two-stage rotary desiccant cooling system." *Energy* 33.12 (2008): 1807-1815.
14. Sand, James R., and John C. Fischer. "Active desiccant integration with packaged rooftop HVAC equipment." *Applied Thermal Engineering* 25.17 (2005): 3138-3148.
15. Casas, W., and G. Schmitz. "Experiences with a gas driven, desiccant assisted air conditioning system with geothermal energy for an office building." *Energy and buildings* 37.5 (2005): 493-501.
16. Henning, H. M., et al. "The potential of solar energy use in desiccant cooling cycles." *International journal of refrigeration* 24.3 (2001): 220-229.
17. Banks, P. J. "Coupled equilibrium heat and single adsorbate transfer in fluid flow through porous medium— I Characteristic potential and specific capacity ratios." *Chemical Engineering Science* 27.5 (1972): 1143-1155.
18. Barlow, Robert S. "Analysis of the adsorption process and of desiccant cooling systems: A pseudo-steady-state model for coupled heat and mass transfer." *NASA STI/Recon Technical Report N* 83 (1982).
19. Maclaine-Cross, I. L. "Proposal for a hybrid desiccant air-conditioning system." *ASHRAE transactions* 94 (1988): 1997-2009.
20. Worek, W. M., and Z. Lavan. "Performance of a cross-cooled desiccant dehumidifier prototype." *Journal of Solar Energy Engineering* 104.3 (1982): 187-196.
21. Jurinak, Jeff Jerome. "OPEN CYCLE SOLID DESICCANT COOLING--COMPONENT MODELS AND SYSTEM SIMULATIONS." (1983): 2665-2665.

Parameter study and optimization of WEDM process: A Review

Md Ehsan Asgar[1] and Ajay Kumar Singh Singholi[2]
[1]PhD Scholar, GGSIPU, New Delhi, India
[2]Associate Professor, GBPGEC, New Delhi, India
E-mail: asgarehsan@gmail.com
ajay.igit@gmail.com

Abstract. Wire electrical discharge machining (WEDM) is a non traditional electro thermal machining process used for the manufacturing of 3D complex shape and geometry of the alloys or composites. Wire EDM capable of precisely machining parts with different Sizes or hardness which have keen edges that are very tough to be machined by the traditional manufacturing processes. There are many numbers of process parameters and performance measures. Lots of researchers have studied and investigated this machining process. This paper shows the reviews in the current research trends in wire electric discharge machining. There are many process parameters, such as pulse on time (TON), pulse off time (TOFF), servo voltage, peak current (IP), wire tension, wire speed for the improvement of different performance parameter such as material removal rate (MRR), surface roughness (SR), surface integrity factors, Kerf width and Tool wear ratio (TWR). The different optimization technique include Taguchi technique, GRA (Grey Relational Analysis), RSM (Response Surface Methodology) and analysed by ANOVA (Analysis of Variance), on different material like Alloys, Superalloys and MMC (Metal Matrix Composites). This literature survey gives the acceptable process parameters and its effect on different performance measure under normal or cryogenic treated tool electrode condition and their ranges in machining of different materials

1. Introduction

Electric discharge machining (EDM), sometimes also called as spark machining, wire erosion, die sinking or spark eroding, is a process of manufacturing through which a required shape is achieved by using sparks.

1.1. WEDM Process

Wire electric discharge machining (WEDM) is an electro-discharge erosion process where material is removed by a chain of sparks from the workpiece. A film of dielectric fluid is used to isolate the workpiece and electrode wire which is regularly supplied to the machining zone to wash out the eroded particles and heat. In this process the spark occurs around more than lakhs times per second and temperature of electrical spark, estimated at around 8,000° to 11,700° C which melt the work piece.

IOP Conf. Series: Materials Science and Engineering **404** (2018) 012007 doi:10.1088/1757-899X/404/1/012007

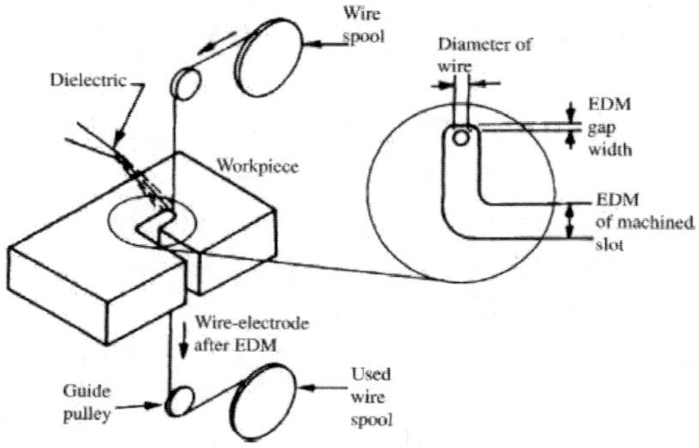

Fig 1 Wire EDM Process (Source: *International Journal of Engineering Development and Research*)

The movement of wire electrode is CNC controlled to obtain the required 2 or 3D shape of the work piece precisely [1]. Wire EDM utilizes a continuously travelling wire electrode made of thin copper, brass or tungsten [2]. The wire is kept in tension for reducing the tendency of producing inaccurate parts [4]. In production manufacturing aspect, Wire EDM contributes a prime share in the production of complicated profile dies, moulds, and critical parts used in automotive, aerospace, nuclear, and other industrial applications.
WEDM process can machine conductive (electrically) material namely tool steel, aluminium, copper, graphite, Inconel, titanium, tungsten carbide, polycrystalline diamond compacts, Cr-Ni based alloys and ceramics.

1.2 Component of WEDM

A Wire EDM system is comprised of following major components.
Work-piece-all the conductive material can be worked by WEDM. *Tool Electrode*-The WEDM electrode is the tool that determines the shape of the cavity is to be produce. The tool in WEDM process is simply a wire. *Dielectric fluid*-The WEDM setup consists of tank in which the dielectric fluid is filled. Wire Electrode and wokpiece submersed into the dielectric fluid.

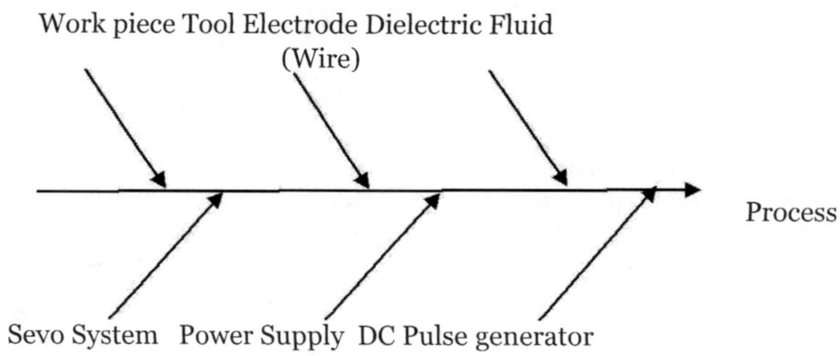

Fig 2 Components of WEDM

Servo system-The servo system is commanded by signals from gap voltage sensor system in the power supply and control the feed of electrode and workpiece to precisely match the rate of material removal. *Power supply*- It transform the AC (alternating current) from the main utility supply into the pulse DC (direct current) required to produce the spark discharge at the machining gap. *The DC pulse generator* is responsible for supplying pulses at a certain voltage and current for specific amount of time.
The focus of this paper is to survey the process parameters of WEDM (such as pulse on time, pulse off time, wire tension, wire feed, peak current) on different materials with respect to various process responses such as surface

quality, electrode wear(EW), kerf width and material removal rate(MRR)

There are many researches done for optimization of different process parameter on different materials like alloys and metal matrix composite for better performance measures. Taha et. al. [6] investigated and evaluated the parameters on machining of CK45 steel through wire cut EDM and measure the process performance such as MRR, TWR and SR. Durairaj et al. [7] examines the consequences of parameter on machining of SS304 through wire cut EDM. the optimal value is achieved for surface roughness and kerf width by using multi objective optimization technique grey relational theory. Alias et al. [8] examines the consequences of parameters on machining of Titanium Ti-Al -4V using wire EDM with Taguchi method. In this study Wire speed, wire tension and open voltage is at constant while the machine feed rate changes. Smallest kerf width is achieved by minimum value of machine feed rate.

2. WEDM Process Parameters

The process parameters in EDM are used to control the performance measures of the machining process. Process parameters are generally controllable machining input factors that determine the conditions in which machining is carried out. These machining conditions will affect the process performance result, which are gauged using various performance measures [2].

- Pulse on Time (TON):- During this time actual Electrical discharge occurs between the wire (gap) and workpiece and electric potential is also applied along. For getting long discharge, large value of ON time must be selected. It is expressed as TON. Wire breakage may be caused by higher value of electric discharge.

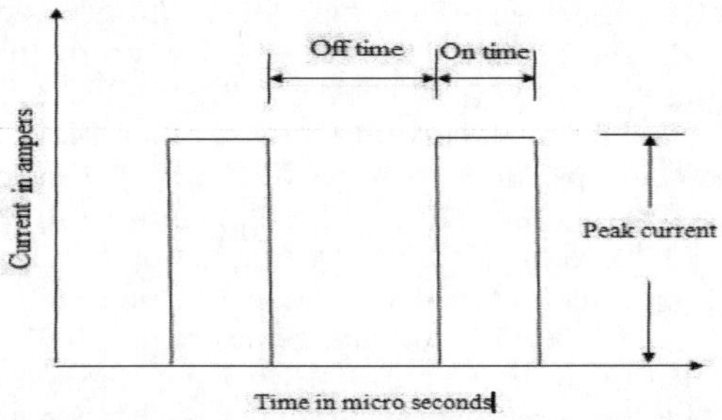

Fig 3 shows TON, TOFF and IP (Source: *Indian Journal of Science and Technology*)

- Pulse off time (TOFF):- No voltage is supplied during this time therefore no electric discharge between workpiece and wire electrode takes place. It is expressed as TOFF. Using low value of discharge may cause wire breakage and due to this cutting efficiency also reduces [3].
- Peak Current (IP):- The peak current plays a vital role in WEDM. It is describe in unit of amperage. During TOFF time, the current increases until it reaches a preset level, which denoted as the IP. In roughing operations higher amperage is used [2].
- Wire Speed: - It is another significant parameter in Wire EDM. If speed of wire increases, the wire dissipation and machining cost will increase while wire breakage occurs at low wire speed [2].
- Wire Tension: - This parameter reveals about the straightness of the electrode wire. That means if wire tension is high then wire will remain straight otherwise wire might drag behind or even bent [3].

3. WEDM Process Response (Output)

Lot of researchers and scholars experimentally tried to use the above different combinations of WEDM process parameters while machining different material such as (alloys, MMC etc) to improve the performance measures of WEDM like (Cutting speed, Material removal rate, Surface roughness, Kerf width, electrode wear ratio,) so that machining rate, productivity and accuracy of the WEDM can be improved. Hence selection of optimal parameters, play important role for achieving better cutting speed, better removal rate, and higher surface finish.

3.1 Review and Discussion of WEDM Research Work based on Performance Measures

We categorize the EDM Process literature into three broad categories based on the problem context in EDM Machining: literature on the MRR; Survey on TWR; and Survey on Kerf Width. The classification is for the purpose of easier understanding of different problem contexts of EDM Process Parameters, interactions and their relationships –for further study and research.

3.1.1 Material Removal Rate.

Material removal rate is an important parameter which influences productivity of any process. As MRR increases, the economic benefits of using WEDM will be worthwhile for any firm [5]. MRR depends on number of input parameters which are associated with WEDM process. Kapgate et al. [9] examines and to determine MRR for triangular, circular, rectangular shape cut on Al/Sic10%. Experimental work was done on SODICK 350W CNC WEDM with brass as wire electrode of size 0.25mm, and mathematical models were developed for different shapes to help the manufactures to maximize the use of AL/Sic10% MMC for higher performance. Kumar D. et. al. [10] examines the performance of machining parameter while machining Al6063/Sicp composite. In this work 4 parameters (TON, TOFF, gap voltage, wire speed) are chosen. For this, Sic was mixed as 5%, 10% & 15% in Al using stir casting method and then machining of pure Al 6063 and Al MMC is done. Different results were compared for MRR after ANOVA with selected combinations. Result shows increase in volume percentage of Sic, as MRR starts decreasing. Kozak. et al. [12] give experimental study on that material whose conductivity is low, and it was concluded that electrical resistance between material and electrode changes with change in clamping position, resulting in change in MRR. For reducing electrical resistance, silver coating technique is done on material surface and as a result, variation in resistance decreases and MRR increases, so productivity is improved. Kuang and Chiang [13] had chosen four machining parameters to evaluate their effect on MRR using RSM technique. Investigation demonstrates that TON and Peak current increases then MRR also increases. Rajmohan T. et al. [11] examines the effect of process parameter on machining of 304 stainless steel using WEDM. By using Taguchi technique it was observed that better material removal rate can be achieved by different combinations of parameters. On the basis of the study by calculating S/N ratio and ANOVA the current and pulse off time are the most significant process parameters for material removal rate. Goyal [21] investigated the MRR on inconel 625 superalloy by cryogenic treated tool provides good machining in terms of MRR. Also shows the effect of parameters on MRR using ANOVA analysis and noticed that at 95% confidence level ($P<0.05$) when P is less than 0.05 then tool electrode and T_{on} have significant effect on MRR.

3.1.2 Surface Roughness and Surface integrity.

It is also the very important response parameter whose value is required to be minimized. It is important to the finish cut of WEDM. Lots of researchers tried to optimize the surface roughness by different methods. Rao P.S. et al. [14] studies and optimize surface roughness(SR) using various process parameters and it was concluded that Peak current and Pulse on time are main & important factors affecting SR. Kulkarni et al. [15] surveys the effect of different process parameters on SR using AISI D3 steel. Coated brass wire (CuZn50) used as electrode material of diameter 0.25 mm. Experimental results were drawn and empirical models were developed for validation and shows good alignment with experimental results. It was observe from the result that, as the Pulse on time and peak current increases, Surface roughness increases and as Pulse off time and servo voltage increases then Surface roughness decreases and surface finish becomes good. Dabade [16] investigates to enhance surface integrity on surface of Al/Sicp Composite using GRA. In this experiment Feed rate, Cutting speed, Tool noise radius and Depth of cut were considered as process parameters. Three level of each process parameter were selected and experiments were performed using L27 orthogonal array as per taguchi optimization technique. After calculating GRC (Grey relation coefficient) and GRG (Grey relational grade), GRG is analysed using ANOVA to identify the significance of individual parameter on particular response. It was concluded that best optimized condition for improving surface integrity on machined surface of composite are; a)Tool radius =0.8mm b) Feed rate= 40mm/min c) Cutting speed = 40m/min d) Depth of cut=0.2mm. SR increases with discharge current and TON [21].This occurs when spark energy generates high current density due to which materials are eroded from the surface. D'Addona et al. [22] observation that, at highest speed (255 m/min) roughness value increases with every cut.

3.1.3 Kerf Width or Cutting Width

Kerf width is mainly depends on parameters Pulse on time, pulse of time, Spark voltage, Wire feed [17] and Flushing pressure of dielectric. Gupta et al. [18] investigates and study the effect of process parameters on kerf width by using HSLA as material. It was concluded that decrease in kerf width causes increase in TON, TOFF, IP and spark gap voltage. But kerf width increased with increase in wire tension. Basil et al. [19] examines the effect of voltage, dielectric pressure, pulse on time and pulse off time on spark gap while machining a super alloy (Ti6Al4V). It was seen that TON and TOFF has more effect on spark gap. It was also noted that wrong setting of TON and TOFF can cause wire breakage. Shivanaga et al.[20] chosen the best process parameter of WEDM while machining 18-4-1 grade speed steel (HSS) of different thickness from 5mm to 80mm. Experiments were done on material on different power input and condition of best power value with stable machining and high cutting speed is recognized. Experiments were repeated with different thickness of work piece material. Then mathematical correlations was developed and are studied to choose best process parameter.

4. Conclusion

After a detailed analysis of the literature, the following conclusions can be drawn. The survey clearly shows that the Wire EDM on super alloy and composites are widely studied in literature. Nickel based alloy, SiC and Aluminium oxide reinforced composites widely reviewed in the literatures. Investigation in Wire-EDM on particulate reinforced Metal Matrix Composites is still increasing through as it has made superior evolution in few areas.

It is observed by experimental results that on increasing Pulse off time & Servo-voltage surface finish become better. Wire Feed & Wire Tension has not major effect on MRR and SR.

Various researchers have made attempts in the machining of Al/SiC in WEDM. While machining of Aluminium Matrix Composites, Wire breakages are initiate to pose limitations on the cutting speed. Work regarding different wire material and coated wire has not been fully explored.

It is also found that major of the research has been carried out on improvement and optimization of same performance measures like MRR, Surface roughness for different materials, but some performance measures like Gap current, dimensional deviation, hardness, etc. are either not much focused or not focused yet. So, this area is yet to be explored more.

References

[1] Bose G K and Pain P 2015, Parametric Optimization of WEDM Process on OHNS Tool Steel Using Statistical Techniques, *International Journal of Emerging Technology and Advanced Engineering* ISSN 2250-2459, **5(12)**, pp 227-238

[2] Kapoor J, Singh S and Khamba J 2010, Recent Developments in Wire Electrodes for High Performance WEDM, *Proceedings of the World Congress on Engineering*, **2**, pp.1-4.

[3] Kumar N, Malhotra N and Sharma S K 2015, Review on Current Research Work in Wire Electrical Discharge Machining *International Journal of Advance Research and Innovation*, ISSN 2347 - 3258, **3(1)**, pp 68-76

[4] Bhavsar P R, Patel S M and Prajapati N K 2015, A Review of Parametric Optimization of Wire Electric Discharge Machining *Indian Journal of Applied Research*. ISSN 2249-555X, pp 60-62

[5] Ghodsiyeh D, Golshan A and Shirvanehdeh J A 2013, Review on Current Research Trends in Wire Electrical Discharge Machining (WEDM), *Indian Journal of Science and Technology* ISSN: 0974-6846, **6(2)**, pp 4128-4140

[6] Taha A. El. Taweel, Ahmed M. Hewidy 2013, Parametric Study and Optimization of WEDM Parameters for CK45 Steel, *International Journal of Engineering Practical Research (IJEPR)*, **2**, pp156- 169.

[7] Durairaj M, Sudharsun D and Swamynathan N 2013, Analysis of Process Parameters in Wire EDM with Stainless Steel using Single Objective Taguchi Method and Multi Objective Grey Relational Grade. *Procedia Engineering 64*, pp 868- 877.

[8] Aliasa A, Abdullaha B and Abbas N M 2012, WEDM: In_uence of machine feed rate in machining titanium ti-6al-4v using brass wire and constant current (4a), *Procedia Engineering 41*, pp 1812-1817.

[9] Kapgate R A, Tatwawadi V H and Modak J P 2013. Process Parameter modeling of Wire Electrical discharge Machining on Al/Sic10% MMC using dimensional analysis. *International Journal of Scientific & Engineering research*, **4(4)**, pp 948-951

[10] Kumar D S, Kanthababu M, Vajjiravelu, Anburaj R, Arul H 2011, Investigation of Wire electrical discharge machining characteristics of Al6063/SiCp composites. *Int J Adv Manuf Technol*, **56**, pp975- 986
[11] Rajmohan T, Prabhu R, Subba Rao G and Palanikumar K 2012, Optimization of machining parameters in electric discharge machining (EDM) of 304 stainless steel. *Procedia Engineering 38*, pp. 1030 – 1036
[12] Kozak J, Rajurkar K. P and Chandarana N 2004, Machining of low electrical conductive materials by WEDM. *Journal of Material Processing Technology*, **149**, pp266-271
[13] Yuan K, Chiang K T 2008, Modelling and Analysis of Machinability Evaluation in WEDM Process of Al oxide based Ceramic. *Material and Manufacturing processes*, **23(3)**, pp241-250
[14] Rao P.S. Ramji K and Satyanarayana B 2011, Effect of WEDM conditions on surface roughness. A Parametric Optimization using Taguchi Method. *International Journal of Advanced Engineering Sciences and Technologies*, **6**, pp 041-048
[15] Kulkarni M L, Sonawane S A 2013, Effect of WEDM Machining Parameters on Output Characteristics. *International Journal of Mechanical & Production Engineering Research and Development*, **3(2)**, pp 57-62
[16] Dabade U A 2013, Multi-objective Process Optimization to improve Surface Integrity on Turned Surface of Al/SiCp Metal Matrix Composite using Grey Relational Analysis, Procedia CIRP, **7**, pp299-304.
[17] Seernivasa Rao M, Venikaiah N 2013, Review on Wire- Cut EDM Process. *International Journal of Advanced Trend in Computer Science & Engineering*, **2(2)**, pp12-17
[18] Gupta P, Khanna R, Gupta R D and Sharma N 2012, Effect of Process Parameters on Kerf Width in WEDM for HSLA Using Response Surface Methodology, *Journal of Engineering & Technology*, **2(1)**, pp1-6
[19] Basil K, Paul J, Issac M J 2013, Spark Gap optimization of WEDM process on Ti6A14V, *International Journal of engineering Science and Innovative Technology*, **2(1)**, pp364-369
[20] Malleswara Routhor S S, Parameswara Rao V S 2014, Optimization and Influence of Process Parameters for Machining with WEDM. International *Journal of Innovative Research in Science Engineering and Technology*, **3(1)**, pp8667-8672
[21] Goyal A 2107, Investigation of material removal rate and surface roughness during wire electrical discharge machining (WEDM) of Inconel 625 super alloy by cryogenic treated tool electrode, *Journal of King Saud University-Science*,**29(4)**, pp528-535
[22] D'Addona D M, Raykar S J and Narke M M 2017, High speed machining of Inconel 718: tool wear and surface roughness analysis, *Procedia CIRP 62*, pp269 – 274

IOP Conf. Series: Materials Science and Engineering **404** (2018) 012008 doi:10.1088/1757-899X/404/1/012008

Multi attribute outranking approach for supplier selection

Nitesh Kumar[1], Tarun Soota[1,2], Neetesh Gupta[1] and Sunil Kumar Rajput[1]
1. Department of Mechanical Engineering, Bundelkhand Institute Of Engineering & Technology, Jhansi
2. Corresponding Author
Email-niteshkumar1127@gmail.com, tarunsoota@rediffmail.com

Abstract

Multi attribute outranking approach has been widely used in supply chain management. Compromise ranking and Outranking method has been used extensively for solving the multi criteria decision problems. Elimination and Choice Translating Reality (Electre) and Preference Ranking Organization Method for Enrichment Evaluations (Promethe) are the most widely used outranking methods to solve multi-criteria decision problems. The above methods may be used to solve problems involving both conflicting qualitative and quantitative criteria. This work focuses on the application of improved outranking based method electre, for solving a real problem of supplier selection. A case study is done for selection of supplier for the spherical roller bearing for finding the rank of the alternatives. The need of this case study is to select the best possible supplier out of the available alternative suppliers.

1. Introduction

One major task of the purchasing department is supplier selection which includes the acquisition of required materials, services and equipment for all type of business organization. The increasing importance of supplier selection decision is forcing the organization to rethink their purchasing and evaluation strategies because a successful purchasing decision directly depends on selecting the best supplier to fulfill the strategies goals apart from the operational requirements of the organization. One of the important areas in purchasing research that has significant practical implication is supplier evaluation and selection. The decision of selecting the best supplier from a wide supplier base is an unstructured, complicated and time consuming task. The decision-making process involves evaluation of different alternative based on various criteria. [Chatterjee et al. 2011]. A general decision making process can be divided into the following steps : Define the problem, Determine requirements, Establish goals, Identify alternatives, Define criteria, Select a decision making tool, Evaluate alternatives against criteria and Validate solutions against problem statement.

Roy (1991) introduced the concept of *Elimination and Choice Expressing Reality (ELECTRE)* which evolved as ELECTRE I, ELECTRE II, ELECTRE III, ELECTRE IV, ELECTRE IS and ELECTRE TRI (ELECTRE Tree). This method consists of two sets of parameters: importance coefficient and the veto thresholds. Effect of Normalization Norms in Flexible manufacturing system selection using multi criteria decision making tool Promethe has been investigated by Chatterjee et al 2014.

Various research works have been done in the past on Multi criteria decision making, using outranking methods like Electre, Promethee, etc. In Electre method a pair wise comparison of alternatives in each attribute is done in order to determine partial binary relations denoting the strength of preference of one alternative over the other. Roodhooft et al. (1996) proposed an activity based costing approach for supplier selection and evaluation which allow computation of total cost caused by a supplier in an organization's production process, thereby increasing the objectivity in the selection process. Savadogo et al. (2006) applied Electre (Elimination and Choice Expressing the Reality) method which would provide a more precise selection of material for a particular application while producing a material selection decision matrix and criteria sensitivity analysis. Almeida et al. (2007) proposed a multi-criteria decision model for outsourcing supplier selection using contribution from utility theory associated with Electre method.
Chatterjee et al. (2011) attempted to solve the supplier selection problem using two most potential multi-criterion decision making approach and compare the relative performance for a giving organization environment. Electre and Vikor method are used to rank the alternative supplier for whom several requirement are consider simultaneously. Chatterjee et al. (2014) study on the application of a very popular Multi-Criteria Decision-Making (MCDM) tool, i.e. Elimination and Et Choice Translating Reality (ELECTRE) for solving an automated inspection device selection problem in a discrete manufacturing environment. Kumar et al. (2015) compared the different multi criteria decision making methods (MCDM) such as TOPSIS & VIKOR for the selection of alternative industrial welding machine. Both the method are aggregating function that represent closeness to the ideal solution.

2. ELECTRE METHOD

The ELECTRE (Elimination and Choice Translating Reality) method was first introduced in 1966 to overcome some deficiencies of popularly used MCDM tools to deal with ordinal attributes without the need for transforming them into cardinal values. ELECTRE is a well known MCDM method that has a history of successful real world applications for its robust ranking technique. It has been applied in various types of decision-making situations. The basic concept of the ELECTRE method is to deal with "outranking relations" by using pairwise comparisons among alternatives under each one of the criteria separately. In the ELECTRE approach, the concordance and dis-concordance indexes are used for outranking the supplier in choosing one alternative over the other alternative. In the first stage the opinion of experts and the study of outranking literature is used to recognize variables and effective criteria in supplier selection. The dominating criteria are extracted, which will be used in the evaluation. Thereafter,

IOP Conf. Series: Materials Science and Engineering **404** (2018) 012008 doi:10.1088/1757-899X/404/1/012008

the lists of qualified suppliers are identified. The decision criteria identified are then approved by decision-making team, who assigns the weights on them. Finally the calculated weights of the criteria are approved by decision making team and the ranks are determined, using ELECTRE method. Based on the average ranking, then that alternative is selected which has best average rank. (Chatterjee et al 2014)

The basic concept of the ELECTRE method is to deal with "outranking relations" by using pairwise comparisons among alternatives under each one of the criteria separately.

- Concordance-C (i, j): for any two alternative i and j, this is a weighted measure of the number of criteria for which i is preferred over alternative j.
- Discordance-D (I, j): It handles the set of criteria for which i is not preferred over j and gives a measure of the degree of "discomfort or discontent" as a result of preferring i to j.

The ELECTRE II method is basically devoted to the ranking problems and the obtained results are in the form of a total ranking preorder among the non-dominated alternatives. The procedural steps as involved in ELECTRE II method are presented below

- In the first step determine the objective and to identify the attribute value for each alternative.
- Then develop the initial decision matrix, X

$$X = [X_{ij}]_{m\times n} = \begin{bmatrix} X_{11} & X_{12} & \cdots & X_{1n} \\ X_{21} & X_{22} & \cdots & X_{2n} \\ \cdots & \cdots & \cdots & \cdots \\ X_{m1} & X_{m2} & \cdots & X_{mn} \end{bmatrix} \quad (1)$$

where x_{ij} is the performance value of ith alternative on j^{th} criterion, *m* is the number of alternatives compared and *n* is the number of criteria.

- Then using the above matrix to develop the normalized decision matrix using the equation (2) The purpose of normalization is to calculation of weight of different attribute and obtain dimensionless values of different criteria to make them comparable with each other. Several normalization techniques have been proposed by the past researchers to transform the different units into dimensionless values. In ELECTRE based methods, vector normalization is generally adopted in which each element of the quantified decision-making matrix is divided by its own Euclidian norm. The norm represents the square root of the addition of element value squares, according to each criterion.[10]

$$G=[g_{ij}]_{mxn} = X_{ij}/[\sum_{i=1}^{m} x_{ij}^{2}]^{1/2} \quad 1\leq i\leq m, \quad 1\leq j\leq n$$

(2)

Where $x_{ij} = \begin{cases} x_{ij} & \text{for beneficial attribute} \\ \frac{1}{x_{ij}} & \text{for non} - \text{beneficial attribute} \end{cases}$

- Depending upon the relative importance of different attributes obtain weight for each attribute which can be calculated given as

$$w_j = v_j / \sum_{i=1}^{m} v_j \ \& \ \sum_{j=1}^{m} w_j = 1$$

(3)

Where v_j is the variance of each attribute which can be calculated as

$$V_j = (1/n) \sum_{i=1}^{n} (g_{ij} - g_{ijmean})^2 \quad (4)$$

- Determine the weighted normalization decision matrix Y

$$Y = [y_{ij}]_{m \times n} = w_j \times g_{ij} \quad (i=1,2....m;,j=1,2....n) \quad (5)$$

Where g_{ij} is the normalized performance value of ith alternative on j^{th} criterion and w_j is the weight of jth criterion.

- Determine the concordance index c(i,j) for every pair of the alternatives Ai and A_j

$$c(i,j) = \sum w_k \quad i,j=1,2..m, i \neq j \quad (6)$$

$$g_k(i) \geq g_k(j)$$

where gk(i) and gk(j) are the normalized measures of performance of the ith and j^{th} alternative respectively with respect to the jth criterion in the decision matrix. Thus, for an ordered pair of alternatives (A_i, A_j), the concordance index c(i,j) is the sum of all the weights for those criteria where the performance score of Ai is at least as that of Aj. Clearly, the concordance index lies between 0 and 1.

- The discordance matrix D expresses the degree that a certain alternative A_i is worse than a competing alternative A_j. The elements d_{ij} of the discordance matrix are defined as follows:

$$d(i,j) = 0 \text{ if } y_k(A_i) \geq y_k(A_j) \quad \text{for all k}$$

$$d(i,j) = \frac{(\max (y_k(j) - y_k(i))}{(\max |(y_k(j) - y_k(i)|)} \quad \text{otherwise} \quad (7)$$

- Compute pure concordance and pure disconcordance indices as follow.

Pure Concordance index $C_j = \sum_{k=1}^{n} c(j,k) - \sum_{j=1}^{n} c(k,j) \quad (j \neq k)$ (8)

Pure Disconcordance index $D_j = \sum_{k=1}^{n} d(j,k) - \sum_{j=1}^{n} d(k,j) \quad (j \neq k)$ (9)

Once two indices are estimated, two ranking are obtained on the basis of these two indices and an average ranking is determined from these two ranking. Based on the average ranking, then that alternative is selected which has best average rank.

3. Case Study

In order to show the application of ELECTRE method, the supplier selection problem for the Spherical Roller Bearing is considered. The Spherical Roller Bearing (Bearing no-22326) considered has a Bore diameter, Outside diameter, width and radius of 130mm,280mm,93mm and 4mm respectively. The data used for supplier selection for spherical roller bearing shown in Table 6.1 has been taken from government workshop located at Jhansi. For this spherical roller bearing there are four alternative suppliers with different criteria. The suppliers are NBC, TIMKEN, FAG and SKF indicated by S1, S2, S3, S4 as per academic interest only. The assumptions for supplier selection are that the delivery time, performance, service and price are similar in nature. Hence the criteria considered for selection of supplier are dynamic load (D), static load (S), mass (M) and reference speed (R). Among these four criteria, D, S, R are beneficial attribute where higher value is desirable and M is non-beneficial attribute requiring smaller value.

Table 1. Quantitative data for Supplier selection

Suppliers	Basic Dynamic Loading Rating(KN)	Basic Static Load Rating(KN)	Mass (Kg)	Reference Speed (rpm)
S1	1043	1343.6	26.8	1800
S2	1270	952	28.2	1900
S3	1250	1370	28	2400
S4	1176	1320	29	2400

The normalized decision matrix for calculating weight criteria shown in Table 2 calculated by using equation (2) according to beneficial and non beneficial criteria

$$G=[g_{ij}]_{mxn} = X_{ij}/[\sum_{i=1}^{m} x_{ij}^{2}]^{½} \qquad 1\leq i\leq m \qquad 1\leq j\leq n$$

(2)

$$\text{Where } x_{ij}=\begin{cases} x_{ij} \text{ for beneficial attribute} \\ \frac{1}{x_{ij}} \text{ for non} - \text{beneficial attribute} \end{cases}$$

For Supplier S1 beneficial attribute D, S & R

$X_{ij}^2=1043^2+1270^2+1250^2+1176^2=5646225$ $\qquad g_{11}=1043/[5646225]^{½}=0.4389$

Similarly for X_{ij}^2= 6330864.96, g_{21}= 0.5339

IOP Conf. Series: Materials Science and Engineering **404** (2018) 012008 doi:10.1088/1757-899X/404/1/012008

For X_{ij}^2=1837000 , g_{41}=0.4199

For Supplier S1 non-beneficial attribute M

$1/X_{ij}^2 = 5.1144\times10^{-3}$ $g_{31} = 0.5224$

Table 2. Normalized decision matrix for selection of bearing

Supplier	D	S	M	R
S1	0.4389	0.5339	0.5224	0.4199
S2	0.5344	0.3783	0.4957	0.4433
S3	0.5260	0.5444	0.5001	0.5599
S4	0.4949	0.5246	0.4817	0.5599

The variance of different attribute shown in Table 3 calculated by using equation (4)

$$V_j=(1/n)[\sum_{i=1}^{n}(g_j - g_{ijmean})^2] \quad (4)$$

(4)

For Dynamic Loading

V_1= [(0.4389-0.49855)^2+ (0.5344-0.49855)^2+ (0.5260-0.49855)^2+ (0.4949-0.49855)^2]/4=0.0014

Table 3. Variance of different attribute

Variance	D	S	M	R
V_j	0.0014	0.0046	0.0002	0.0042

The weight of different attribute shown in Table 4 calculated by using equation (3)

$$w_j=v_j/\sum_{i=1}^{m} v_j \& \sum_{j=1}^{m} w_j = 1 \quad (3)$$

w_1=0.0014/0.0104=0.14

Similarly $w_2 = 0.44$, $w_3 = 0.02$, $w_4 = 0.40$

Table 4. Weight of different attributes

Weight	D	S	M	R
W_j	0.14	0.44	0.02	0.40

The supplier selection problem is solved by using ELECTRE- II

IOP Conf. Series: Materials Science and Engineering **404** (2018) 012008 doi:10.1088/1757-899X/404/1/012008

The normalized decision matrix of ELECTRE Method shown in Table 5 calculated by using equation 2 according to beneficial and non beneficial criteria

Table 5. Normalized decision matrix for selection of bearing

Suppliers	D	S	M	R
S1	0.4389	0.5339	0.5224	0.4199
S2	0.5344	0.3783	0.4957	0.4433
S3	0.5260	0.5444	0.5001	0.5599
S4	0.4949	0.5246	0.4817	0.5599

Table 6. Priority Weights

Weights	0.14	0.44	0.02	0.40

Weighted normalized matrix shown in Table 7 calculated by using equation (5)

$$Y=[y_{ij}]_{m\times n}=w_j\times g_{ij} \qquad (i=1,2....m;,j=1,2....n)$$

(5)

Weight criteria for Supplier S1

$y_{11}=0.14\times0.4389=0.0614$

Similarly $y_{12} = 0.2349$, $y_{13} = 0.0104$, $y_{14} = 0.1679$

Table 7. Weighted normalized matrix

Supplier	D	S	M	R
S1	0.0614	0.2349	0.0104	0.1679
S2	0.0748	0.1664	0.0099	0.1773
S3	0.0736	0.2395	0.0100	0.2239
S4	0.0692	0.2308	0.0096	0.2239

Concordance matrix shown in Table 8 calculated by using equation (6) While calculating the concordance index, if there are ties between the alternatives, they would receive one half of the criteria weights

$$c(i,j)=\sum w_k \quad i,j=1,2..m,i\neq j \qquad (6)$$

$$g_k(i) \geq g_k(j)$$

comparing normalized matrix of S1 with S2 then $g_{12}>g_{21}$ & $g_{13}>g_{23}$

the corresponding weight is 0.44 and 0.02

c(1,2)=0.44+0.02=0.46

comparing normalized matrix of S1 with S3 then $g_{13}>g_{33}$

the corresponding weight is 0.02

c(1,3)=0.02

comparing normalized matrix of S1 with S4 then $g_{12}>g_{42}$ & $g_{13}>g_{43}$

the corresponding weight is 0.44 & 0.02

c(1,4)=0.44+0.02=0.46

Table 8. Concordance matrix

Supplier	S1	S2	S3	S4
S1	-	0.46	0.02	0.46
S2	0.54	-	0.14	0.16
S3	0.98	0.86	-	0.80
S4	0.54	0.84	0.20	-

Discordance matrix shown in Table 9 calculated by using equation (7)

$$d(i,j)=0 \text{ if } y_k(A_i)\geq y_k(A_j) \quad \text{for all k}$$

$$d(i,j)=\frac{(\max\,(y_k(j)-y_k(i))}{(\max|(y_k(j)-y_k(i)|)} \tag{7}$$

d(1,2) = (0.0748-0.0614)/|0.1664-0.2349|=0.1956

Similarly d(1,3) = 1, d(1,4) = 1

Table 9. Dis-concordance matrix

Supplier	S1	S2	S3	S4
S1	-	0.1956	1	1
S2	1	-	1	1
S3	0.0071	0.0164	-	0
S4	0.0737	0.0869	1	-

Pure Concordance index and Pure Disconcordance index shown in Table 10 calculated by using equation (8) and (9)

Pure concordance index $C_j=\sum_{k=1}^{n} c(j,k)-\sum_{j=1}^{n} c(k,j)$

(8)

IOP Conf. Series: Materials Science and Engineering **404** (2018) 012008 doi:10.1088/1757-899X/404/1/012008

Pure dis-concordance index $D_j=\sum_{k=1}^{n} d(j,k)-\sum_{j=1}^{n} d(k,j)$

(9)

$C_1 = (0.46+0.02+0.46)-(0.54+0.98+0.54) = -1.12$

$D_1 = (0.1895+1+1)-(1+0.0071+0.0737) = 1.1087$

4. RESULT

Table 10. Ranking of supplier

Supplier	Pure Concordance index	Intial Ranking	Pure Disconcordance index	Intial ranking	Average ranking	Final rank
S1	-1.12	3	1.1087	4	3.5	3
S2	-1.32	4	2.7104	3	3.5	4
S3	2.28	1	-2.9765	1	1	1
S4	0.16	2	-0.8394	2	2	2

The pure concordance and pure discordance indices for the supplier are computed using Equations (8) and (9) respectively, as exhibited in Table 10. From this table, the ranking of the supplier is observed as S3>S4>S1>S2. S3 is the best supplier, S4 emerges out as the second best choice and S2 is the worst device.

5. Conclusion

The present study explores the use of Electre methods in solving a supplier selection problem and the results obtained can be valuable to the decision maker in framing the supplier selection strategies. MCDM methodology applied for selection of supplier for spherical roller bearing. With the help of normalized decision matrix we estimate criteria weight so that human judgment can be avoided by assigning weights to different attributes. The results show that supplier 3 is best in term of ranking index.

6. References

[1] Almeida, A.T., 2007, Multi-criteria decision model for outsourcing contract selection based on utility function and Electre method, *Computer & Operation Research*, **34**, 3569-3574.

[2] Chatterjee P., Mukherjee P., & Chakraborty S., 2011, Supplier selection using compromise ranking and outranking methods, *Journal of Industrial Engineering International*, **7**, 61-73.

[3] Chatterjee P., Mondal S., & Chakraborty S., 2014, A Comparative Study of Preference Dominance-based Approaches for Selection of Industrial Robots, *Advances in Production Engineering & Management*, **9**, 5-20.

[4] Chatterjee, P. and Chakraborty, S., 2014. Investigated the effect of Normalization Norms in Flexible manufacturing system selection using Multi criteria decision making methods, *Journal of Engineering Science and Technology review*, **7(3),** 141-150.

[5] Govindan K., and Jepsen M.B., 2014, Electre: A comprehensive literature review on methodologies and applications, *Eur. K. Oper. Res.*, **250**, 1-29.

[6] Kumar M., Vrat P., & Shankar R., 2006, A fuzzy programming approach for vendor selection problem in a supply chain. *International Journal of production Economics,* 101, 273-285.

[7] Kumar J., Soota T., & Sagar G., 2015, Compare the different multi criteria decision making method (MCDM) for the selection of alternative industrial welding machine, *Journal of scientific research*, **2**, 125-130.

[8] Roy, B., 1991, The outranking approach and the foundations of Electre Methods, Theory and Decision, **31**, 49-73.

[9] Rao R.V., 2010, A subjective and objective integrated multiple attribute decision making method for material selection, Materials and Design, **31**, 4738-4747.

[10] Rao, R.V., 2007, Decision making in the manufacturing environment: Using graph theory and fuzzy multiple attribute decision making methods, *Springer Science & Business Media.*

IOP Conf. Series: Materials Science and Engineering **404** (2018) 012009 doi:10.1088/1757-899X/404/1/012009

Facilitating Fair Trade Practices as a Development Strategy

Mr Mohd Imran Khan[1*], Mr Shahbaz Khan[1], Dr Abid Haleem[2]
[1]Research Scholar, Department of Mechanical Engineering, Faculty of Engineering and Technology, Jamia Millia Islamia, New Delhi-110025, India
[2]Professor of Mechanical Engineering, Faculty of Engineering and Technology, Jamia Millia Islamia, New Delhi-110025, India.

imrankhan@st.jmi.ac.in

Abstract: Fair Trade emerged as a response to the unjust and inequitable nature of contemporary globalisation It aims at providing greater economic justice to the disadvantaged producers and workers by redesigning the pattern of international trade. This paper gives a brief overview of Fair Trade and critically examines the barriers to the adoption of Fair Trade practices in the conventional international business model. The identified barriers are evaluated using Decision Making Trial and Evaluation Laboratory (DEMATEL) approach. DEMATEL categorised these barriers into influential and influenced group which helps to understand their inter-relationship. The findings suggest that lack of awareness about Fair Trade and absence of well-defined criteria to decide fairness of trade practices are the most influential barrier in mainstreaming the Fair-Trade practices in the business operations. This work empowers policymakers, professionals, consumers, and practitioners to design strategy for adoption of Fair Trade practices to ensure minimum decent living conditions and access to health and education to economically disadvantaged producers, especially women and indigenous people and to protect children from exploitation in the production process.
Keywords: Fair Trade; Globalisation; Poor; Consumer Awareness; Capitalist Economy; DEMATEL.

1. Introduction

Fair Trade has emerged as a global social initiative in response to unjust and inequitable nature of contemporary global system of production and trade. This movement aims at ensuring equity and justice in trade to marginalized producers associated with the production of a commodity by reshaping the patterns of international trade and the very processes of corporate expansion in the global economy that have historically undermined ecological and social conditions around the world [1]. Fair Trade has five primary constituents namely "a minimum price", "a price premium", "pre-financing to allow artisans to purchase raw materials", "training in quality control and marketing", and "use of cooperatives" [2]. Fair Trade was found with a target to achieve greater social control over the commodity with a trans-national supply chain. Fair Trade certification is used as a tool to realise this objective. Some economist views Fair Trade as an annexe of traditional trade while others interpret this as an effort to redesign the neoliberal economic policy. Despite India's position as the third largest economy Fair Trade is not so admired in the business domain as well as by the consumers as compared to the developed countries, although the interest of academicians is increasing in this emerging area. Many academicians have shown their interest to explore Fair Trade's ability to assimilate into the traditional market, to assess consumer experience and to tackle the criticism associated with different Fair-Trade practices [3]. Barriers to mainstreaming the Fair Trade practices is rarely reported in scholarly research.

1.1 Research Objectives

The following are the primary aims of the research undertaken and communicated through this paper:

- To identify barriers to adoption of Fair Trade practices
- To evaluate the identified barriers to understand their interrelationships using DEMATEL approach

The present work seeks to identify and evaluate the causal relationship among the barriers of integrating Fair Trade practices within the conventional trade practices. The DEMATEL methodology can help to structure the complex causal relations through digraphs which depict relationships between barriers.

2. Major Barriers to adoption of Fair Trade practices

The adoption of Fair Trade practices depends upon various key factors. An understanding of the barriers may help management, as well as policymakers to evaluate their Fair-Trade initiatives for overall development of marginalised groups. Barriers are identified through systematic literature review and later verified with expert's opinion are shown in Table 1.

Table 1: Barriers to integrate Fair Trade practices in prevalent production and trade practices

S. No.	Barriers	References
Br1.	Lack of awareness and promotion	[4,5]
Br2.	Lack of consensus on definition and criteria	[2,6]
Br3.	Issue for impact assessment	[6,9]
Br4.	Lack of proper labelling	[7,8]
Br5.	Lack of knowledge among stakeholders	[10,11,16]
Br6.	Sensitivity of commercially valuable information	[12]
Br7.	Lack of regulatory framework	[13]
Br8.	Lack of attractive Fair-Trade brand	[6,12]

3. Research Methodology

The objective of this work is an evaluation of Fair Trade barriers through causal relationship. Appropriate barriers are explored through an extensive literature review and further identified as barriers to Fair Trade. These barriers are finalised through experts' opinions, and a causal relationship is developed through a DEMATEL approach.

3.1 DEMATEL Method

The DEMATEL technique is adopted as a solution methodology in this article. DEMATEL is a comprehensive method to provide the causal relationship among the complex factors through graphs. Some other MCDM techniques such as TOPSIS and AHP are suitable selecting and prioritising the factors rather than providing causal relationships. Therefore, we adopt the DEMATEL approach for the analysing the complex causal relationship among the barrier of adopting Fair Trade practices. The DEMATEL method is summarised step by step as follows.

Step 1: The scale may be designated four levels (i.e. 0,1,2,3,4) to construct direct influence matrix (See Table 2). Each expert was asked to evaluate the direct influence between each two-factor combination through a score 0,1,2,3,4. The notation of x_{ij} represents that the factor *i* influence the factor *j*. The diagonal element (i.e. *i=j*) of the direct inflectional matrix is zero. A non-negative n×n matrix is obtained for each expert as $X^k = [x_{ij}^k]$ where k is the no of experts $1 \le k \le H$. Thereafter obtain X^1, X^2, X^3.... X^H from H experts.

Table 2: Scale and their interpretation

Scale	0	1	2	3	4
Interpretation	No influence	Very Low influence	Medium influence	High influence	Very high influence

Step 2: Establish an overall direct-relation matrix from *H* respondents, the average matrix *A* = [*aij*] can be obtained from the equation:

$$a_{ij}=\frac{\sum_{k=1}^{H} x_{ij}^{k}}{H} \tag{1}$$

Step 3: Obtain normalized initial direct-relation matrix, *D* using Equation (2) & (3).

$$D=A.S \tag{2}$$

$$\text{Where } S=\frac{1}{max_{1\le i\le n}\sum_{j=1}^{n} a_{ij}} \tag{3}$$

Each element in matrix *D* falls between zero and one.

Step 4: Calculate Total relation matrix "T" using Equation (4)

$$T=D.(I-D)^{-1} \quad \text{Where "I" is the identity matrix.} \tag{4}$$

Step 5: Determine the causal parameters through Equations. (4) and (5):

$$r_i = \sum_{i=1}^{n} t_{ij} \ \forall_j \quad (5) \qquad c_i = \sum_{j=1}^{n} t_{ij} \ \forall_i \tag{6}$$

Where R_i indicate the sum of rows and C_i represent the sum of columns.

Step 6: Develop a causal and effect diagram using the dataset consisting of prominence (Pi) and net effect(Ei) using expressions (6) and (7):

$$P_i = R_i + C_j \mid i=j \quad (7) \qquad E_i = R_i - C_j \mid i=j \tag{8}$$

The difference ($R_i - C_i$) shows the net effect that factor *i* contributes to the system. Moreover, if ($Ri - Ci$) is positive, factor *i* is a net cause, while factor *i* is a net receiver or result if ($Ri - Ci$) is negative.

4. Results

The barriers to mainstreaming fair-trade practices are identified through extensive literature review. These barriers are discussed with a five members expert's group to obtaining the deeper insight of these barriers and finalised eight barriers (See Table 1). After finalising the barriers, the experts were asked to evaluate the direct influence of the barriers to mainstream the fair-trade practices on the scale of 0-4 (See Table 2). The evaluation is taken in form overall direct relation matrix is obtained through expert input and shown in table 3 using equation 1.

Table 3: Overall Direct-Relationship Matrix (Z)

Barrier	Br1	Br2	Br3	Br4	Br5	Br6	Br7	Br8
Br1	0	1	2	3	2.2	3	1.8	3
Br2	3	0	3	1	2.8	0.8	1	2
Br3	0	1	0	2.8	1.8	1	3	4
Br4	0	1.4	2	0	1	1.2	0	2
Br5	2	1	2	2	0	1	3	2.2
Br6	1	1	3	3	1	0	2	2
Br7	1	2	3	3.6	2.2	1	0	1
Br8	1	0	1	2	2	1	0	0

Further, the Normalized direct-relation matrix (N) is developed using Equation (2) and (3). Thereafter, this matrix is transformed into Total-Relation Matrix (T) using Equation (4).

In matrix *T*, the summation of rows and column are represented by R and C respectively using equation (5) and (6) (Please see table 4). Ri represents the total effect of barrier i to the other barrier, and Ci represents net influence received by j barrier from other barriers. After the determination of R and C for every row and column, the prominence (Pi) and net effect (Ei) is calculated using equation (7) and (8). The net cause/effect of each barrier are decided through "Ei (i.e. R-C)". If Ei is positive, then the barrier is considered to produce the net cause and if negative then the barrier is the net effect. These cause and effect are shown in Table 4.

Table 4: Cause and Effect of barriers

Barriers	R	C	R+C	R-C	Cause/Effect
Br1	3.43985	1.665882	5.105732	1.773968	Cause
Br2	3.27989	1.612547	4.892437	1.667343	Cause
Br3	2.89017	3.42421	6.31438	-0.53404	Effect
Br4	1.562412	3.97542	5.537832	-2.41301	Effect
Br5	2.95502	2.88148	5.8365	0.07354	Cause
Br6	2.83358	1.99677	4.83035	0.83681	Cause
Br7	3.07738	2.42844	5.50582	0.64894	Cause
Br8	1.574907	3.62846	5.203367	-2.05355	Effect

The R+C and R-C are plotted in Figure 1, which shows the causal relationship among the barriers of mainstreaming Fair-Trade practices. These results are discussed with the experts for further insights.

5. Discussion of the Results

The importance order of the identified barriers based upon the R + C dataset is given as Br3-Br5-Br4-Br7-Br8-Br1-Br2-Br6. Based on influential score (R-C dataset), the barriers are arranged in decreasing order of their influence on the other barriers is represented as Br1, Br2, Br6, Br5, Br3, Br8, Br4. Also, based on R-C dataset values, five barriers, namely Br1, Br2, Br6, Br7, Br5, have been categorised as cause group barriers. Three barriers, namely Br3, Br8, Br4 have been categorised into the effect group barriers by using their respective R-C score. The research findings were discussed with the experts to understand the contemporary barriers of Fair Trade practices better; the details are given as follows:

5.1 Influencing Barriers

Fair Trade movement aims at protecting human rights by promoting social justice, sound environmental practices, and economic security. It encourages trading that are not principally or entirely profit-driven. The findings of this study reflect that lack of consumer awareness is hampering the Fair-Trade campaign. Business players are sceptical that whether there is fundamentally any sizable market of Fair Trade exists. Consumers unwillingness to pay a premium is impeded the mainstreaming of Fair Trade practices. The second most influential barrier is the lack of consensus on definition and criteria of

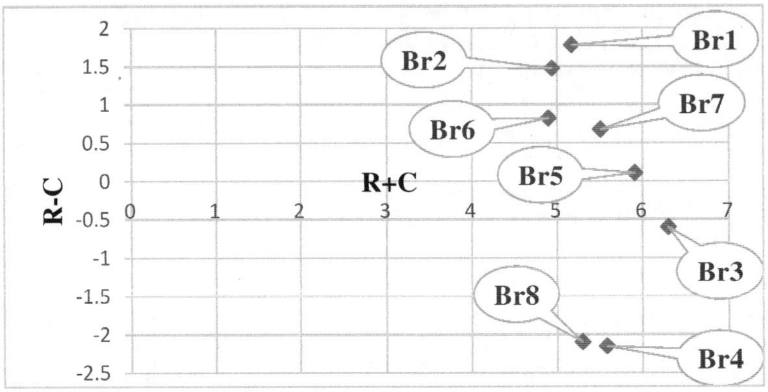

Figure 1. The causal relationship among the barriers towards adopting practices of Fair Trade

deciding what is fair in a trade and has contributed to confusion among the major stakeholders in assessing the fairness of their trade. This variable is the most influential among all the barriers with (R-C value). This movement will gain momentum only when a Fair-Trade organisation will agree upon a broad-based definition considering all aspects of modern trade policies [14]. Also, consensus definition will help in tackling misleading claim about Fair Trade. Sharing of sensitive commercial information in another significant barrier in the assimilation of Fair Trade practices. Even though organisations are willing to work together, they are reluctant to share strategic information about Fair Trade practices. The absence of a framework for enforcing fair practices in trade is also and discouragement to the people involved in Fair Trade movement [15]. Due to lack of consensus about the boundaries of Fair Trade in scholarly literature, executives of mainstream companies are unaware of Fair Trade [2]. Mitigating these barriers may help in maintaining and advancing the Fair-Trade movement.

5.2 Influenced Barriers

Barriers in the effect group make a significant contribution to the system. Among them, Issue for impact assessment is highly influenced barrier. Statements about the benefits of Fair Trade are often made without a rigorous assessment of its impact. Moreover, it is also difficult to demonstrate how outcome of Fair Trade practices is related to the premium paid for that. The other influenced barriers, following the sequence of the priority list, include 'Lack of attractive Fair-Trade brand' and 'Lack of proper Labelling'. A well-established brand of Fair Trade may help in propagating Fair Trade practices to larger masses. Further lack of proper labelling confuses the customers while making a purchasing decision.

6. Managerial Implication

IOP Conf. Series: Materials Science and Engineering **404** (2018) 012009 doi:10.1088/1757-899X/404/1/012009

This paper empowers policymakers, professionals, consumers, and practitioners to identify and evaluate barriers related to mainstreaming the Fair-Trade practices. This work may help the policy planners to design strategy by the modern economic challenges. Through the causal relationship (shown in figure 1) the barriers to embracing Fair Trade practices by industry players can be evaluated, which will ultimately help in raising the living standards of marginalised sections. Fair Trade practices with well-defined boundary may help stakeholders in making an informed decision.

7. Conclusion and Future Scope

The barriers to adoption of Fair Trade practices are identified, and their causal relationships are evaluated using DEMATEL approach. The research conducted opens up direction in conducting data-driven research to measure the impact of Fair Trade practices on modern business style and assessing the condition of marginalised sections in a capitalist economy after adopting Fair Trade practices. The identified barriers can further be evaluated using tools like ANP and DEMATEL under fuzzy environment. A case-based validation will further improve the study. We hope, once these barriers are ironed out, Fair Trade principles will become a minimal standard in international production and not just a niche market.

References

1. James, A. (2014). A Theory of Fairness in Trade. *Moral Philosophy and Politics* 1 (2) 177–200.
2. Miller, D. (2016). Fair Trade: What Does It Mean and Why Does It Matter? *Journal of Moral Philosophy*
3. Smith, S. and Barrientos, S. (2005). Fair Trade and ethical trade: are there moves towards convergence? *Sustainable Development.* 13. 190–198.
4. Irving, S., R. Harrison and M. Rayner: (2002), Ethical Consumerism: Democracy Through the Wallet. *Journal of Research on Consumers,* 3.
5. Catherine Sinclair McClure & Yoo-Kyoung Seock (2017). Does Fair Trade matter to consumers? A study of African cultural products. *International Journal of Fashion Design, Technology and Education* 10(1).
6. Hira Anil and Ferrie Jared, (2006). Fair Trade: Three Key Challenges for Reaching the *Mainstream Journal of Business Ethics* 63. 107–118
7. Moore, G. (2004). The Fair-Trade Movement: Parameters, Issues, and Future Research. *Journal of Business Ethics* 53. 73–86.
8. Khan, M. I., Khan, U., & Haleem, A. (2014). Corporate Social Responsibility: Modelling of Critical Factors using an Integrated ISM and Fuzzy-Micmac approach. *International Journal of Global Business and Competitiveness,* 9(1).
9. Barrientos, S. (2000). Globalization and ethical trade: assessing the implications for development. *Journal of International Development* 12. 559–570.
10 Moore, G., Gibbon, J., and Slack, R. (2006). The mainstreaming of Fair Trade: a micromarketing Perspective. *Journal of Strategic Marketing*. 14. 329–352.
11.Khan, S., Haleem, A., Khan, M., Abidi, M., & Al-Ahmari, A. (2018). Implementing Traceability Systems in Specific Supply Chain Management (SCM) through Critical Success Factors (CSFs). *Sustainability,* 10(2), 204.
12.Murray, D. L. and L. T. Raynolds: (2000). Alternative Trade in Bananas: Obstacles and Opportunities for Progressive Social Change in the Global Economy. *Agriculture and Human Values*
13.Tallontire, A. (2002). Challenges facing Fair Trade: which way now? *Small Enterprise Development*
14.Raynolds, L.T., Murray, D., and Taylor, P.L. (2004). Fair Trade coffee: building producer capacity via global networks. *Journal of International Development.* 16 1109–1121.
15 Haleem, A., & Khan, M. I. (2017). Towards successful adoption of Halal Logistics and its' implications for the stakeholders. *British Food Journal,* 119 (7).
16.Oberoi, J. S., Khamba, J. S., Sushil and Kiran, R. (2008). An empirical examination of advanced manufacturing technology and sourcing practices in developing manufacturing flexibilities. *International Journal of Services and Operations Management*, 4(6).

Selection of Traceable Technology in Food Supply Chain

Mr Shahbaz Khan[1], Mr Mohd Imran Khan[1*], Dr Abid Haleem[2], Mr Mohd Shuaib[3]

[1]Research Scholar, Department of Mechanical Engineering Faculty of Engineering and Technology, Jamia Millia Islamia, New Delhi-110025, India

[2]Professor of Mechanical Engineering, Faculty of Engineering and Technology, Jamia Millia Islamia, New Delhi-110025, India,

[3] Assistant Professor, Mechanical Engineering, Delhi Technological University, Delhi

imrankhan@st.jmi.ac.in

Abstract Traceability is an essential tool to provide the product information throughout the Food Supply Chain (FSC), this help protecting the consumer welfare and in developing global competencies. Several technologies are available to implement the traceability system in FSC; these are barcoding, QR code, edible label, DNA- barcoding and paper-based systems. In this paper, we have tried to identify the significant factors towards the selection of the traceable technology within FSC through the literature review and supported by the expert opinion. These factor and sub-factors are prioritised using Analytic Hierarchy Process (AHP). The findings of this work suggest that product-related factors have the highest weight, and needs attention while selecting the traceable technology in FSC. This work may assist the practitioners in the selection of traceable technology for their FSC in an efficient manner.

Keywords: Analytic Hierarchy Process (AHP); Factors; Food supply chain (FSC); Traceability; Traceable technology

1. Introduction

Consumer concerns have increased towards the food safety, hygiene, and environment which has made the role of traceability important. Traceability is implemented not only to assure the consumer expectation but also to provide the effective management of product flow, increase the speed throughout the FSC [1]. Traceability systems are considered as the prerequisite for the food industry to explore the global market [2] as well as protecting the public health.

Olsen and Borit [3] define the traceability in a generic way as "The ability to access any or all information relating to that which is under consideration, throughout its entire lifecycle, using recorded identifications". In the context of food "all information" refer to the information related to ingredient origin, processing, logistics and retailing of the food products. The recoded information is stored and carried with the help of the "traceability information carriers" such as a barcode, RFID, edible label, DNA barcode and other advanced technologies. The term "tracing" and "tracking" is widely used in the literature of traceability. Tracing is a backward process where the origin is identified by history or records in the supply chain, and tracking is the forward process where consumers and supply chain partners are identified by location in the supply chain [4]. The traceability is the combination of the "tracing" and "tracking" of the product information in a backward and forward direction in the supply chain.

The implementation of traceability in the FSC gets strengthened with the adoption of technologies having the ability to track the product related information. Effective monitoring is the focus of the food industries which want low cost and easily applicable traceability technology. Thus, the selection of these traceable technologies is based on the several factors to provide product-related information to consumers. In this paper, the selection criteria of traceable technology are identified and ranked through the Multi-Criteria Decision Making (MCDM) techniques.

1.1 Need for the Study

The selection of the traceable technology is a difficult and complex task because of the several factors are involved. In the context of the food industry, it becomes more complicated because of the huge variation in the perception of the food consumers. Thus, in such situation, the selection of the traceable technology for the food product is quite important for practitioners. Through this paper, we have tried to identify the factors and subfactors which are important in the selection of traceability technology is identified and prioritise to assist the practitioners.

1.2 Objective of the Research

The notable objectives of this research are:

- To identify the factors and sub-factors for the selection of traceable technology FSC
- To prioritise the factors and sub-factors for the selection of traceable technology in FSC
- To provide the managerial implications

1.3 Factors related to the selection of traceable technology in FSC

The factors which influence the selection of traceable technology are categorised and grouped namely: "product-related factor", "process related factor", "social factor" and "other factor". These factors are further divided into sub-factors and shown in figure 1.

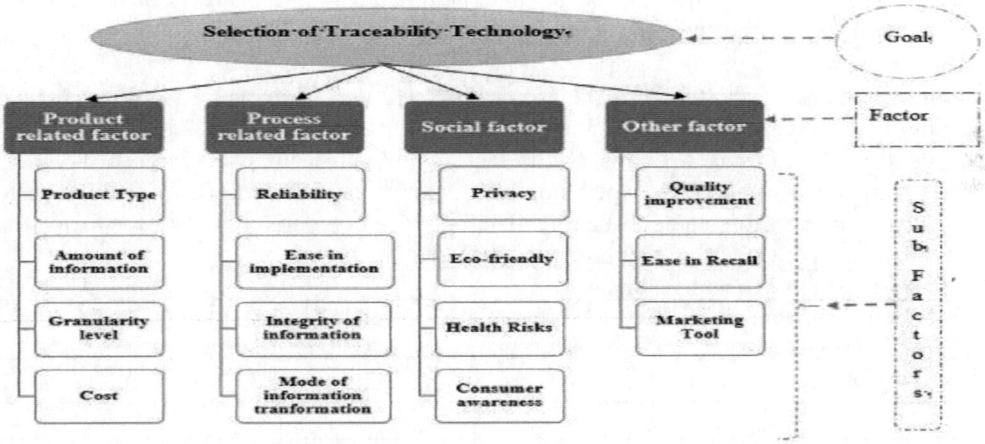

Figure 1: AHP hierarchical framework for the selection of traceable technology

2. Methodology

To accomplish the research objective, we have identified to use a suitable multi-criteria decision-making (MCDM) technique. Previous studies suggest that for prioritising the criteria/factors or factors some MCDM tools such as AHP [5], TOPSIS [6] and DEMATEL [7]. AHP seems to be a proper MCDM technique to provide the relative measurement or priorities by relative importance in this study. A group of five experts from academician and professionals gave their opinion through the AHP decision matrix. A brief about the AHP and process adopted are given below:

2.1Analytical Heretical Process (AHP)

AHP is a technique from the MCDM family and is extensively used in different applications such as supply chain, supplier selection, technology selection [8,9]. MCDM is used to rank or select a set of alternatives under usually independent, incommensurate or conflicting attributes [10]. In brief, the steps followed are given below:

Step 1: MCDM is structured as a hierarchy and decomposed into a hierarchy of interrelated decision elements. The hierarchical structure is used to arrange criteria/factors and sub-criteria/sub-factors.

Step 2: Here the pairwise comparisons between criteria and sub-criteria undertook using a nine-point ratio scale by experts (please see Table 1).

Table 1: Scales in pairwise comparisons

Definition	Equally important	Moderately important	Strongly important	Extremely important	Extremely more important	Intermediate values
Value	1	3	5	7	9	2,4,6,8

Step 3: The criteria are compared pair-wise according to their influence and based on the specified criteria in the higher level [11]. The weights are calculated from equation (1) as given below:

$$AW=\lambda_{max}W \quad (1)$$

In Equation 1; A is the priority matrix; W is weight, and λ_{max} is the maximum Eigenvalue of Matrix A

Step 4: The matrix A is checked for the consistency ratio (CR), this to be less than 0.10.

The consistency ratio is the ratio of consistency index (CI) and the random index number (RI). The CR and CI are defined as follows:

$$CI = \frac{\lambda max-1}{n-1} \quad (2) \quad CR=\frac{CI}{RI} \quad (3)$$

For the different count of criteria, it has a different value as is shown in Table 2.

Table 2: Random Index (RI)

N	1	2	3	4	5	6	7	8	9
RI	0	0	0.58	.90	1.12	1.24	1.32	1.41	1.45

If CR is less than 0.10, the result can be acceptable, and matrix A is sufficient consistent. Otherwise, one has to return to step 1 and repeat the steps.

3. Results

In evaluating the relative importance of the factors to select the traceable technology AHP is used, here, a group of five experts from food manufacturing company and academician is formed. Experts inputs were collected and used in finalising the factors and sub-factors (Please see figure 1). After finalising the factors and sub-factors, expert's opinion is again used through Saaty scale to perform the pairwise comparison of factors and sub-factors. The pairwise comparison matrix for factors has been framed, and their relative weights are summarised in Table 3.

Table 3: Pair-wise comparison matrix for factors

	Product-related factor	Process-related factor	Social factor	Other factor	Weights	Rank
Product-related factor	1	2	2	4	0.435	1
Process-related factor	0.5	1	2	3	0.286	2
Social factor	0.5	0.5	1	2	0.182	3
Other factor	0.25	0.33.	0.50	1	0.097	4
Consistency Ratio (CR)=0.017						

The "product-related factor" having weight 0.435 is the most important factors for the selection of traceable technology for the FSC, followed by "process related factor" (0.286); "social factor" (0.182), and "other factor" (0.097). The pairwise comparison is performed for the subfactors and relative weights are calculated. These relative weights of sub-factors along with the global weight are shown in Table 4.

It is observed that the consistency ratio of pairwise comparison matrix of each factor and sub-factors is less than 0.1 in all the cases. Hence these matrices are highly consistent and acceptable. Table 4 also shows the relative rank and global rank of the sub-factors for the selection of traceable technology.

Table 4: Global Ranking of the factor and sub-factors to select the traceable technology in FSC

Factors	**Relative weight**	**Sub Factors**	**Relative weight**	**Relative rank**	**Global weight**	**Global rank**
Product-related factor	0.435	Product Type	0.274	2	0.11919	**3**
		Amount of information	0.086	4	0.03741	**10**
		Granularity level	0.178	3	0.07743	**5**
		Cost	0.462	1	0.20097	**1**
Process-related factor	0.286	Reliability	0.073	4	0.020878	**13**
		Ease in implementation	0.498	1	0.142428	**2**
		Integrity of information	0.172	3	0.049192	**9**
		Mode of information	0.258	2	0.073788	**6**
Social factor	0.182	Privacy	0.467	1	0.084994	**4**
		Eco-friendly	0.160	3	0.02912	**11**
		Health Risks	0.277	2	0.050414	**8**

		Consumer awareness	0.095	4	0.01729	**14**
Other factor		Quality improvement	0.297	2	0.028809	**12**
	0.097	Ease of Recall	0.54	1	0.05238	**7**
		Marketing Tool	0.163	3	0.015811	**15**

4. Discussion

Based on Table 4, the order of relative importance of the factors for selection of traceable technology for FSC as: 'Product-related factors' > 'Process related factors' > 'Social factors' > 'Other factors'. The order of relative importance of sub-factors is given in Table 4.

The 'product-related factors' are the most important factors for the selection of traceable technology in the FSC. This factor has four sub-factors namely: "product type", "amount of information", "granularity level" and "cost". The major sub-factors for the selection of traceable technology is the cost of the product having the highest weight. The second highest weight is the "product type" and the second major consideration for traceability technology selection. The product type refers to the perishability level, freshness, physical state of the product and their size. These aspects affect the selection of the traceable technology. Next important factors are the "Granularity level" after the "product type". The "granularity level" is defined as the *"...reflects the levels and size of the Identifiable unit, that are handled by the particular system"* [12]. The fine granularity level requires high cost and detailed information than the coarse granularity level. The last important sub-factors are "amount of information". The traceable information carrier limitations such as QR code having the capacity to contain more information than the barcode. Thus, these factors play an important role in the selection of traceable technology for the FSC.

The second important factor is the "Process related factor" which are having the weight 0.286. This factor has four sub-factors. In the context of this factor, most important sub-factor is the "Ease in implementation". Before implementing any technology, the organisation assesses the implementation issue. Similar to other technology, 'ease in implementation' is an important factor in the traceable technology selection. Next important sub-factor is the "mode of information" which refer to the "how information is processed (paper-based or digital)?". After the mode of information, the next important sub-factor is the "integrity of the information". The information integrity is an aspect of the food integrity which is protected through the traceability. Reliability of the traceable technology is the lowest relative weight to select the traceable technology for FSC.

The social factor is the next most important factor for the selection of traceable technology in FSC. This factor also has four sub-factors. "Privacy" is a most important factor for the traceability technology selection for the food product. The consumer's concern about their privacy and some traceable technology traces the product up to their fridge, and so the consumer loses their privacy [13]. Second important sub-factor is the "Health risk of the consumers". Some traceability carriers such as a barcode on fruits and fruits edible label are placed on the surface of the food product which is harmful to the health of the consumers [14]. Next important sub-factor is the "Eco-friendly". The traceable technology should be an eco-friendly and having the less effect on the environment. The advanced traceable technology is eco-friendly such as an elidable label. Selection of traceable technology also depends on the consumer awareness about the traceability of food product because the industries do not want to engage their resources in an environment where demand is very low [15].

Finally, the fourth important factor is "Other factor" which have the three sub-factors namely: 'Quality improvement'; 'ease in recall' and 'marketing tool'. The suitable traceable technology is utilised for the effective recall system especially in case of the food crisis. So, this sub-factor is most important sub-factor in this context. Second important sub-factor is the "quality improvement". Traceability technology is used for the purpose to improve the quality of the food product by effective monitoring the food throughout the supply chain. The next important sub-factor is the "marketing tool". The food companies provide the information about the food product through the traceability, and these systems are used as a marking tool.

5. Implications

This study provides the understanding of the factors and sub-factors for the selection of traceable technology in FSC. The significant implication of this research suggests for facilitating the managers and professionals of food industries; to select a suitable traceable technology for their respective supply chain. The finding of this work provides a direction to the decision maker of the specific FSC such as fruit, vegetables, and dairy supply chain in selecting the traceable technology.

6. Conclusions

The factor and sub-factor for the selection of traceable technology in FSC are identified through the literature survey and are supplicated by the expert's opinion. These factors played a significant role in the selection of the traceable technology for FSC. After achieving the four main factors and fifteen sub-factors, AHP is used for prioritising these factors and their associated sub-factors as per their relative importance and global importance. Finally, the finding of this work is presented, and the implications are provided.

References

1. Leat, P., Marr, P., & Ritchie, C. (1998). Quality assurance and traceability—the Scottish agri-food industry's quest for competitive advantage. Supply Chain Management, 3,115–117.
2. Khan, S., Haleem, A., Khan, M., Abidi, M., & Al-Ahmari, A. (2018). Implementing Traceability Systems in Specific Supply Chain Management (SCM) through Critical Success Factors (CSFs).*Sustainability,* 10(2), 204.
3. Olsen, P., & Borit, M. (2013). How to define traceability. Trends in Food Science & Technology
4. Van Dorp, K. J. (2002). Tracking and tracing: A structure for development and contemporary practices. Logistics Information Management, 15(1), 24-33.
5. Luthra S, Garg D, Haleem, A (2013) Identifying and ranking of strategies to implement green supply chain management in Indian manufacturing industry using Analytical Hierarchy Process. Journal of Industrial Engineering and Management 6(4):
6. Kumar S, Singh B, Qadri MA, Kumar YVS, Haleem A (2013) A framework for comparative evaluation of lean performance of firms using fuzzy TOPSIS. International Journal of Productivity and Quality Management 11(4): 371-392
7. Kumar S, Luthra S, Haleem A, Mangla S.K, Garg D. (2015) Identification and Evaluation of Critical Factors to Technology Transfer Using AHP Approach. International Strategic Management Review 3 (1): 24–42
8. Subramanian N, Ramanathan R (2012) A review of applications of Analytic Hierarchy Process in operations management. International Journal of Production Economics 138 (2): 215–241
9. Kumar S, Luthra S, Haleem A (2015) Benchmarking supply chains by analyzing technology transfer critical barriers using AHP approach. Benchmarking: An International Journal 22(4)
10. Hwang CL, Yoon K (1981) Multiple attribute decision making: Methods and applications. A state of the art survey. New York: Springer-Verlag
11. Chrysochou, P., Chryssochoidis, G., & Kehagia, O. (2009). Traceability information carriers. The technology backgrounds and consumers' perceptions of the technological solutions. *Appetite,53*(3), 322-331. doi:10.1016/j.appet.2009.07.011
12. Bollen, A.F., Riden, C.P. & Cox, N.R. (2007). Agricultural supply system traceability, part I: Role of packing procedures and effects of fruit mixing. Biosystems Engineering, 98(4): 391-400
13. von Locquenghien, K. (2006). On the potential social impact of RFID-containing everyday objects. Science, Technology and Innovation Studies, 2, 57–78.
14. Khan, M. I., Khan, U., & Haleem, A. (2014). Corporate Social Responsibility: Modelling of Critical Factors using an Integrated ISM and Fuzzy-Micmac approach. International Journal of Global Business and Competitiveness, 9(1).
15. Haleem, A., & Khan, M. I. (2017). Towards successful adoption of Halal Logistics and its' implications for the stakeholders. British Food Journal, 119 (7).
16. Khan, S., Asjad, M., and Ahmad, A. (2015), "Review of Modern Optimization Techniques", International Journal of Engineering Research and Technology, 4(4).

Prioritising Barriers towards Adoption of Sustainable Consumption and Production Practices using TOPSIS

Mr Mohd Imran Khan[1], Mr Shahbaz Khan[1*], Dr Abid Haleem[2], Mr Mohd Javaid[3]

[1]Research Scholar, Department of Mechanical Engineering Faculty of Engineering and Technology, Jamia Millia Islamia, New Delhi-110025, India

[2]Professor of Mechanical Engineering, Faculty of Engineering and Technology, Jamia Millia Islamia, New Delhi-110025, India

[2]Assistanat Professor of Mechanical Engineering, Faculty of Engineering and Technology, Jamia Millia Islamia, New Delhi-110025, India

shahbaz.me12@gmail.com

Abstract: Unsustainable form of consumption and production in current business practices and lifestyle of the consumers seems inviting problems for the planet, such as scarcity of natural resources, environmental issues and increase in waste generation. Therefore, industries and consumers need to focus to integrate Sustainable Consumption and Production (SCP) practices in their daily operations to resolve these emerging problems and ensure sustainable development. However, the adoption of SCP practices faces many challenges due to the presence of various barriers. This paper identifies ten major barriers towards the adoption of SCP practices through extensive literature review and supported by experts. Further, these barriers are prioritised using the Technique for Order Preference by Similarity to Ideal Solution (TOPSIS). The finding of this study suggests that the lack of proper awareness consumers is the most significant barriers among the identified barriers towards the adoption of SCP practices. The prioritising of these barriers can be beneficial for the policy makers, managers and government in developing a framework to mainstream the SCP practices.

Keywords: Sustainable Consumption and Production (SCP); Sustainable Development; TOPSIS; Natural Resources

1. Introduction

During any economic activity, Sustainable Consumption and Production (SCP) aims at reducing the use of resources along the whole lifecycle, increasing the net welfare gains as well as improving the quality of life. This requires a systemic approach and cooperation among the different actors of the supply chain, from producer to final consumer. SCP is one of the '2030 Agenda for Sustainable Development' among its 17 Sustainable Development Goals of United Nations.

Oslo symposium in 1994 [1], defines SCP as, "the use of services and related products, which respond to basic needs and bring a better quality of life while minimising the use of natural resources and toxic materials as well as the emissions of waste and pollutants over the life cycle of the service or product so as not to jeopardize the needs of further generations". Thus, SCP practices are to develop sustainable infrastructure, efficient use of resources and better quality of life for all.

Due to globalisation, it is expected, in next two decades, the disposable income of masses may increase, this will lead to increase in demand for already constrained natural resources. If consumption and production pattern is not adjusted, irreversible damage to the environment may happen [2]. Making consumption and production sustainable is a pressing challenge to the modern world, and this needs to be addressed in business sector by instigating a responsible business model. Moreover, consumers need to be made aware of sustainable consumption & lifestyles, by providing information through proper labelling/standards and engaging them in sustainable procurement.

The trend of scholarly literature on SCP practices shows overwhelming respond of the researcher to tackle this challenge [3,4]; the area still needs a rigorous ranking of the barriers in making the SCP practices more relevant. This paper has two research objectives:

- to identify the barriers to the adoption of SCP practices in every supply chain operations and
- to prioritise the identified barriers to uncover their relative importance using TOPSIS technique.

Hwang developed TOPSIS, and Yoon [5] and this MCDM technique selects the best alternative to the shortest Euclidean distance from the ideal solution and the farthest from the negative ideal solution [5]. TOPSIS has a minimum rank changes reversal if any alternative is added or removed.

2. Barriers to adopt SCP practices in supply chain

Adoption of SCP practices requires 'lifecycle thinking', to extend sustainable development in production as well as consumption phases. The adoption of SCP practices inhibited by several barriers. The mitigation of these barriers may help the policymakers to evaluate their existing practices of SCP practices. Table 1, shows ten barriers to adoption of SCP practices as identified by discussions with the experts and by review of the literature. However, different organisations might have different views regarding barriers to adopting SCP initiatives in supply chains.

Table 1: Ten barriers to adoption of SCP practices

S.No	**Barriers**	**References**
B1.	Lack of policy for the sustainable consumption	[4]
B2.	Reluctance to change	[2]
B3.	Lack of resources	[6]
B4.	Lack of government support	[7,8]
B5.	Lack of regulatory framework	[9]
B6.	Lack of awareness program to promote the sustainable consumption	[7,10]
B7.	Lack of skilled workforce	[11]
B8.	Lack of knowledge among stakeholder	[3]
B9.	Lack of effective communication	[12]
B10.	Lack of consumer awareness	[13,14,16,17]

3. Solution Methodology

The TOPSIS approach is used as a research tool in this work. TOPSIS is purposed by Hwang and Yoon [5] widely used to solve MCDM problems. TOPSIS consists of the following steps [15]:

Step1: Construct a decision matrix

The m barriers called $B = \{B1, B2 \ldots Bm\}$ are to be evaluated against n criteria, $C = \{C1, C2 \ldots Cn\}$. The decision matrix is constructed with m row and n column (please see table 3). Table 2 shows the five-point scale (1 to 5) is selected for the rating of the importance. The element of the decision matrix is shown as $x_{ij,}$ which represent the importance of each barriers *ith* concerning each criterion *jth.*

Table 2: Linguistic scale for the importance

Linguistic scale	**Importance intensity**
Very low	1
Low	2
Medium	3
High	4
Very high	5

Step2 Determine the normalised decision matrix

The normalised value of x_{ij} is calculated using equation 1

$$r_{ij}=\frac{x_{ij}}{\sqrt{\sum_{j=1}^{n} x_{ij}^2}} \qquad i = 1, 2, \ldots, m;\ j = 1, 2, \ldots\ldots, n \qquad (1)$$

Step 3: Calculate the weighted normalised decision matrix.

IOP Conf. Series: Materials Science and Engineering **404** (2018) 012011 doi:10.1088/1757-899X/404/1/012011

The weighted normalised decision matrix is calculated as the multiplying the normalised decision matrix by its associated weights:

$$v_{ij} = r_{ij} * w_j \qquad (2)$$

where w_j is the weight of the jth attribute or criterion, and $\sum_{j=1}^{n} wj=1$.

Step 4: Determine the positive-ideal and negative-ideal solutions

$$A^{+} = \{v_1^{+}, v_2^{+}, \dots\dots, v_n^{+}\} \qquad i = 1, 2, \dots, m; j = 1, 2, \dots, n \qquad (3)$$

$$A^{-} = \{v_1^{-}, v_2^{-}, \dots\dots, v_n^{-}\} \qquad i = 1, 2, \dots, m; j = 1, 2, \dots, n \qquad (4)$$

In our case, we are prioritising the importance of the barrier. Thus, positive ideal solution is taken as 5 (highly importance) whereas negative ideal solution is taken as 0.

Step 5: Calculate the separation measures through Euclidean distance

$$D^{+} = \sqrt{\sum_{j=1}^{n}(v_{ij} - v_j^{+})} \qquad i = 1, 2, \dots, m; \qquad (5)$$

$$D^{-} = \sqrt{\sum_{j=1}^{n}(v_{ij} - v_j^{-})} \qquad i = 1, 2, \dots, m; \qquad (6)$$

Step 6: Calculate the *relative closeness,*

The relative closeness of the alternative ith is calculated as:

$$CC_i = \frac{D^{-}}{D^{+}+D^{-}} \qquad (7)$$

Here, $0 \leq CC_i \leq 1$, $i = 1, 2, \dots, m$.

Step 7: Arrange the *relative closeness* in a descending order to obtain the ranking of the alternative.

4. Result

The barriers to adoption of SCP practices are identified through extensive literature review and expert opinion. These barriers are further discussed with a five members expert's group to obtain a deeper insight these barriers and finalised the barriers (please see Table 1). After finalising, the experts were again asked to rate the importance of the barriers on a 5-point scale (please see Table 3).

Table 3: Linguistic assessment for Barriers

S No	Barriers	DM1	DM2	DM3	DM4	DM5
B1.	Lack of policy for the sustainable consumption	M	H	H	H	M
B2.	Reluctance to change	M	M	L	M	L
B3.	Lack of resources	M	M	M	L	M
B4.	Lack of government support	H	M	H	M	M
B5.	Lack of regulatory framework	M	H	L	M	L
B6.	Lack of awareness program to promote the sustainable consumption	VH	H	H	H	H
B7.	Lack of skilled workforce	L	M	L	L	M
B8.	Lack of knowledge among stakeholder	H	H	H	H	H
B9.	Lack of effective communication	H	M	M	H	L
B10.	Lack of consumer awareness	VH	VH	H	VH	H

By steps of TOPSIS provided in the previous section, the identified barriers are prioritised and shown in Table 4.

Table 4: Ranking of Barriers

Barrier	D^{+}	D-	CC	Ranking
B1.	3.896197	7.166524	0.352192	4
B2.	2.352759	9.041691	0.206483	9
B3.	2.495766	8.721722	0.222489	8
B4.	3.43729	7.596113	0.311535	6
B5.	2.983431	8.683911	0.255708	7
B6.	5.385841	5.769811	0.48279	2

B7.	2.070138	9.350718	0.181259	10
B8.	4.680932	6.218949	0.429448	3
B9.	3.599948	7.96145	0.311377	5
B10.	6.57264	4.745685	0.580708	1

Through the analysis, it was found that the 'Lack of consumer awareness' is ranked on the top followed by 'Lack of awareness program to promote the sustainable consumption' and 'Lack of skilled workforce' is the least prioritised barrier.

5. Discussion based on the Result

The order of relative importance of the barrier obtained is B10-B6-B8-B1-B9-B4-B5-B3-B2-B7. To begin with, 'Lack of consumer awareness' gains the top rank. In spite of the increasing levels of awareness, it is reported that consumers do not widely practice sustainable consumption. But, consumers are said to consume sustainably for it's symbolic, rather than, economic value. The next barrier in the rank of relative importance is 'Lack of awareness program to promote the sustainable consumption.' Modern business players adhere to the SCP practices only for using it as a marketing strategy. They usually do not educate the consumers through proper labelling/standards about the advantage of integrating sustainable consumption practices into their lifestyle. A proper marketing strategy for propagating SCP practices may give better results. 'Lack of knowledge among stakeholder' is the third barrier in prioritised scale. The possible solution to this barrier is related to the previous strategy. The organisation/individual involved in making SCP practices needs to educate the stakeholder of SCP such as business, researchers, scientists, retailers, media, and development cooperation agencies, among others. 'Lack of policy for the sustainable consumption' is also a major barrier to the adoption of SCP practices in the supply chain. Due to this barrier business organisation are sceptical about the ways of achieving sustainability in their business operations. The absence of a framework for regulation of SCP practices in case violation is making the supply chain vulnerable to the unsustainable production and consumption. Government support is critical in developing, implementing, monitoring and enforcing strong regulations any strategy regarding SCP practices. Many business organisations are implementing cleaner initiatives in their supply chain to comply with governmental regulations. Lack of sustainable resources is also a significant barrier. Financial constraints have been identified as a challenge for business organisations to make the essential changes required in the production and consumption methods. Reluctance to change and lack skilled labour are prioritised last in the relative importance scale. Capacity building of the workforce should be done to mainstream the SCP practices. The mitigation of these barriers may facilitate the adoption of SCP practices in supply chain and make these practices self-sustaining in future.

6. Managerial Implication

The implication of the research has several implications for the managers and may facilitate the professionals to mitigate the barrier. These identified barriers may help to formulate strategies to coordinate and collaborate to enable SCP practices in SC effectively. The order of relative importance of barrier obtained through TOPSIS may help the policymakers to formulate effective policy. The rank presented may act as a benchmark to mitigate the SCP adoption-related issues.

7. Limitation and Future Scope

The limitation of this study is that it is based on the opinion of the experts, there may be judgmental bias. Also, a wider rating of the 7 or 9-point linguistic scale could be used instead of using a 5-point linguistics scale. The identified barriers can further be evaluated using tools like ANP and DEMATEL under fuzzy environment. A case-based validation will also improve the study.

8. Conclusion

The objective of the SCP practices is reducing the use of resources and provide the less impact on the environment along the whole life cycle to increase the quality of life. This study facilitates the adoption of SCP practices through identifying the major barriers to SCP adoption. Ten major barriers

IOP Conf. Series: Materials Science and Engineering **404** (2018) 012011 doi:10.1088/1757-899X/404/1/012011

to adoption of SPC practices were identified through literature review and supported by experts. These barriers are prioritised through TOPSIS method using the expert inputs. The "lack of consumer awareness" having the highest priority among all identified barriers and which implies that it requires the attention of the organisation as well as government bodies. The ranking of the barriers is analysed with the experts, and this provided useful insight as discussed in the above section. Finally, the managerial implication, limitation and future scope of this research are presented

References

1. Norwegian Ministry of the Environment, 1994. Oslo Roundtable on Sustainable Production and Consumption.
2. Luthra, S., Mangla, S.K., Xu, L., Diabat, A., 2016. Using AHP to evaluate barriers in adopting sustainable consumption and production initiatives in supply chain. *Int. J. Prod. Econ.* 181 (PB), 342–349.
3. Govindan, K., Seuring, S., Zhu, Q., Azevedo, S.G., 2016b. Accelerating the transition towards sustainability dynamics into supply chain relationship management and governance structures. *J. Clean. Prod.* 112, 1813–1823.
4. Tseng, M., Lim, M., Wong, W.P., 2015. Sustainable supply chain management: a closed-loop network hierarchical approach. *Ind. Manag. Data Syst.* 115 (3), 436–461.
5. Hwang C. L. and Yoon K. P. (1981) *Multiple attribute decision making: methods and applications. Business and Economics*. Springer, New York.
6. Luthra, S., Mangla, S.K., Kharb, R.K., 2015. Sustainable assessment in energy planning and management in Indian perspective. *Renew. Sustain. Energy* Rev. 47, 58–73.
7. Blok, V., Long, T.B., Gaziulusoy, A.I., Ciliz, N., Lozano, R., Huisingh, D., Csutora, M., Boks, C., 2015. From best practices to bridges for a more sustainable future: advances and challenges in the transition to global sustainable production and consumption: introduction to the ERSCP stream of the special volume. *J. Clean. Prod.* 108 (A), 19–30.
8. Khan, M. I., Khan, U., & Haleem, A. (2014). Corporate Social Responsibility: Modelling of Critical Factors using an Integrated ISM and Fuzzy-Micmac approach. *International Journal of Global Business and Competitiveness*, 9(1).
9. Mangla, S., Madaan, J., Sarma, P.R.S., Gupta, M.P., (2014). Multi-objective decision modelling using interpretive structural modelling for green supply chains. *Int. J. Logist. Syst. Manag.* 17 (2), 125–142
10. Khan, S., Haleem, A., Khan, M.I., Abidi., M. H., & Al-Ahmari, A., (2018). Implementing traceability systems in specific SCM through CSFs. *Sustainability* 10 (1)
11. Merli, R., Preziosi, M., Massa, I., 2015. Social values and sustainability: a survey on drivers, barriers and benefits of SA8000 certification in Italian firms. *Sustainability* 7 (4), 4120–4130.
12. Dubey, R., Gunasekaran, A., Ali, S.S., (2015). Exploring the relationship between leadership, operational practices, institutional pressures and environmental performance: a framework for green supply chain. *Int. J. Prod. Econ.* 160, 120–132.
13. Lorek, S., Spangenberg, J.H., (2014). Sustainable consumption within a sustainable economy–beyond green growth and green economies. *J. Clean. Prod.* 63, 33–44.
14. Haleem, A., & Khan, M. I. (2017). Towards successful adoption of Halal Logistics and its' implications for the stakeholders. *British Food Journal*, 119 (7).
15. Shyur, H. J., & Shih, H. S. (2006). A hybrid MCDM model for strategic vendor selection. *Mathematical and Computer Modeling*, 44, 749–761
16. Khan, M. I. & Haleem, A. (2016). Understanding "Halal" and "Halal Certification & Accreditation System"- A Brief Review". *Saudi Journal of Business and Management Studies*, 1(1).
17. Saurikhia, A., Ahmed, S., Haleem, A., Gangopadhyay, S., Khan, M. I., (2017). Evaluating technology management factors for fly-ash utilization in the road sector using an ISM approach. *International Journal of Management Science and Engineering Management.*

Assessing Barriers to Adopting and Implementing Halal Practices in Logistics Operations

Dr Abid Haleem[1*], Mr Mohd Imran Khan[2], Mr Shahbaz Khan[2], Dr Abdul Hafaz Ngah[3]
[1]Professor of Mechanical Engineering, Faculty of Engineering and Technology, Jamia Millia Islamia, New Delhi-110025, India
[2]Research Scholar, Department of Mechanical Engineering Faculty of Engineering and Technology, Jamia Millia Islamia, New Delhi-110025, India
[3]Senior Lecturer Faculty of Maritime and Business Management, Universiti Malaysia Terengganu Kuala Terengganu, Terengganu, Malaysia

ahaleem@jmi.ac.in

Abstract: Due to the increasing demand for Halal products, a need of adopting and implementing Halal Logistics (HL) practices are observed. This paper presents a structural model for barriers to adoption of HL practices with conventional logistics. Halal goods and services are becoming popular in public domain because of the improving disposable income of the masses. Another reason for this emergence as Halal appeals to the consumers who care about the origin of the product, fair trade, sustainability, cleanliness and humane animal husbandry, as Halal products are based on ideas of cleanliness, purity, safety and wholesomeness and set of ethical values. Through this paper, significant barriers to the adoption of HL practices in conventional logistics are identified which are instrumental in delivering consumables with Halal attributes. A structural and hierarchical model of the identified barriers is developed using Interpretive Structural Modelling (ISM) technique. Then, MICMAC analysis has been done to categorise these barriers against their driving and dependence power. We identified that the HL is not mandatory for the issuance of a Halal certificate for Halal manufacturers and this is the primary reason for the lack of demand. The structural model obtained, is in harmony with expert's opinion and literature, this model can help in mitigating these barriers towards effective adoption and implementation of HL practices and ultimately guarantee Halal integrity till consumption. The model so generated can be quite helpful in adopting and implementing HL practices. Through structural and hierarchical model this paper provides a basis for defining the problem faced by practitioner and policy planner involved in incorporating Halal practices in a logistics operation.

Keywords: Halal; Halal Logistics; Halal logistics practices; Halal Supply Chain (HSC); Halal Supply Chain Management (HSCM); Interpretive Structural Modelling (ISM); MICMAC.

1. Introduction

Products with Halal attributes are no more only a religious matter and not only consumed by Muslims, but it is becoming a symbol of hygiene, safety, wholesome and ethical practices. Thus, Halal is now in the domain of modern business & exchange and is emerging as a new paradigm for assuring quality and affecting way of living by changing people's attitude, tastes and values. Halal word is a 'Quranic' term that deciphers legal, allowed, accepted or permitted whose essence lies in safe consumption. Halal is turning into a worldwide substance in light of improved disposable income of Muslims [1,2] the recognition of Halal products by non-Muslims who are health conscious and care about the environment and the increase in acknowledgement that products labelled as Halal are, safer, healthier, cleaner and tastier.

The concept of Halal does not only consider at the time of manufacturing and but also matters during the point of procuring and consumption of the products. This infers that managing Halal necessitates a supply chain management (SCM) like approach which takes care of product right from the origin till final consumption or disposition. Thus, a strong Halal logistics (HL) system lays down a robust Halal Supply Chain Management policy, which substantiates the commitment and intention of the corporates in establishing an HL practices/HSCM system.

1.1 Objectives of the Research

The major objectives of the research undertaken and communicated throughout this paper may be outlined as follows:

- To identify the important barriers to adopting and implementing Halal practices in logistic activities
- To establish contextual relationships among these barriers with the help of expert opinions, group discussion & brainstorming and ultimately develop a hierarchical & structural model using ISM methodology
- To classify the identified barriers using MICMAC and make recommendations

2. The concept of Halal Logistics practices

The term logistics can be interpreted as the developing a mechanism control the movement of goods, their handling and storage, and allied information from the source to the point of disposition/consumption with an objective to fulfil the requirement of a customer. Logistics Service Providers (LSPs) act as a medium to maintain the product quality and associated information, till the product reach the designated consumers. Incorporating Halal practices in conventional logistics increases the responsibility of LSPs as products with Halal attributes requires dedicated equipment [3]. All Halal products must be per 'Shariah' which states that products must be 'Halal' as well as 'Toiyyab'; this suggests that product is safe to consume as well possess wholesome characteristics right from the origin till final consumption. Thus, while integrating Halal into logistics operations, the LSPs must assure the consumers that the products during any logistics activity, such as transportation storage or handling, has not breached any Shariah principle [4]. Zailani et al., [5] concluded that HL practices do not degrade the integrity of the Halal product from the source of origin to final consumption during any of its activities. Lodhi [6] maintains that to ensure an efficient & effective practice of HL, that all Halal products must be segregated from other products or substances which can degrade its integrity during key stages in logistics and supply chain activities until it arrives for direct consumption at the final destination.

3. Major Barriers to adopting Halal logistics practices

Identifying and overcoming the barriers to adopting HL practices may assure the successful adoption and implementation of HL practices and may help to overcome the major food threats in the modern world. In expert's opinion, idea engineering workshop, extensive literature reviews the important barriers to adoption and implementation of HL practices were identified and enlisted in Table 1.

Table 1: Identification of Major Barriers to adopting Halal Logistics Practices

Barrier No.	Barrier in Adopting Halal Logistic Practices	References
B-1	Lack of Standardization, Codification and Proper Guidelines for Adopting HL Practices	[3,7]
B-2	Economic Viability of Logistics Services with Halal Practices	[8,9]
	Lack of Government Policies and Support for HL; Weak Enforcement and Lack of a Proper Regulatory Framework for Halal LSPs	[10,11]
B-4	Reluctance of Logistics Service Providers to adopt Halal Practices in their Operations	[1,12]
B-5	Lack of Organizational Culture and Change Management to adopt for HL practices	[13,14]
B-6	Lack of Compatibility and Upgradeability to HL of the existing Logistics	[11]
B-7	Lack of Global Branding of Halal Practices	[15,16]
B-8	Lack of Trained Halal Logisticians and Lack of Capacity Building Programs	[17,18]
B-9	Lack of Robust Information and Communication Technologies Support	[19,2]

B-10	Lack of Agility due to Ineffective Coordination and Collaboration Among Different Halal LSPs	[19,20]
B-11	Dissimilar Practices and Lack of Dedicated Halal Assets & Facilities	[18,1]
B-12	Lack of Customer Demand and Reluctance to pay for HL	[7,21]
B-13	Lack of Ethical Practices and Distrust in Halal Integrity of Products	[17,22]

4. Solution methodology

As mentioned in above section the objective of this study is to analyse the barriers to adopting and implementing HL practices. In this study, we are developing a hierarchical structural model using ISM. The input to ISM is expert's opinion and the available literature.

4.1 Interpretive Structural Modelling (ISM) Methodology

ISM as a research tool was developed by Warfield [23] and is a well-defined methodology to identify and outline various components categorising a problem. In this research, ISM methodology is adapted from Sushil, [24], Luthra [25], Khan et al., [9] Saurikihia et al., [26]and a structural model is obtained using this methodology which is shown in Figure 1.

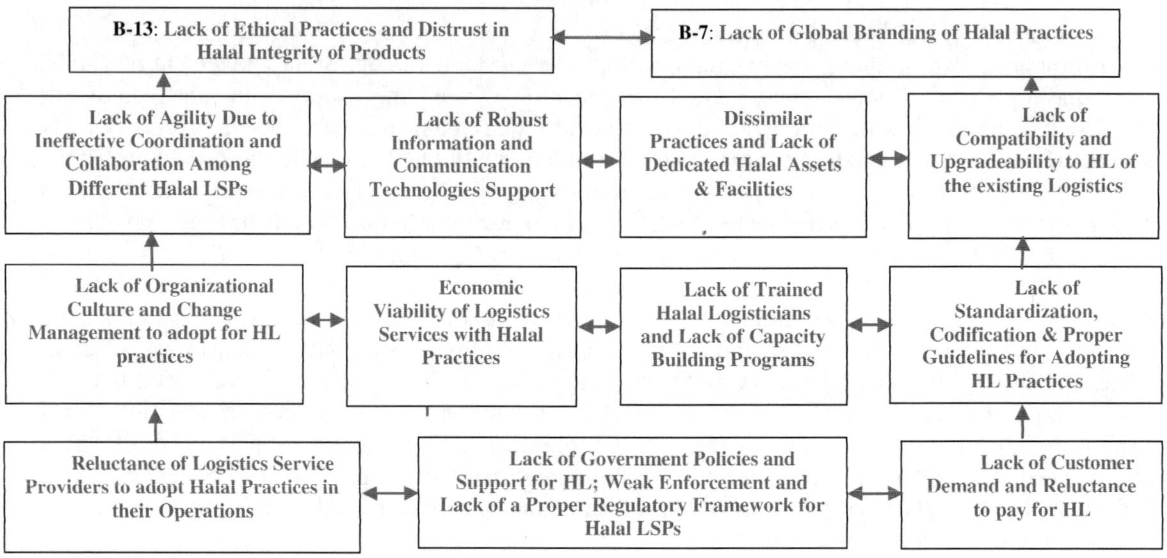

4.2 MICMAC Analysis of Barriers to Adoption and Implementation of Halal Logistics Practices

After developing a structural and hierarchical model, MICMAC analysis has been undertaken to classify the barriers by their driving and dependence power (Shown in Figure-2). Here the sum of "1's" in the row (taken from final reachability matrix-Annexure III) for all the corresponding variables give the driving power. Similarly, the sum of "1's" in the columns provide dependence power. The classification as mentioned above is discussed in the coming sub-sections.

5. Discussion on Results and Findings

Despite increasing market share of the Halal products, LSPs are somewhat reluctant to adopt and implement of HL practices, and specifically in preserving the Halal integrity of a food product.

5.1 Discussion on ISM based model

The structural model developed for barriers in adopting HL practices gave the hierarchical relationship between them. It is observed that 'Reluctance of Logistics Service Providers to adopt Halal Practices in their Operations', 'Lack of Customer Demand & Reluctance to pay for HL' and 'Lack of Government Policies & Support for HL'; 'Weak Enforcement & Lack of a Proper Regulatory Framework for Halal LSPs' are the major barriers to adopting and implementing HL practices in the conventional logistics. It was found that the HL is not mandatory for the issuance of a Halal certificate for Halal manufacturers is the primary reason for the lack of demand. This conclusion seems to be coherent with other recent studies like [5].

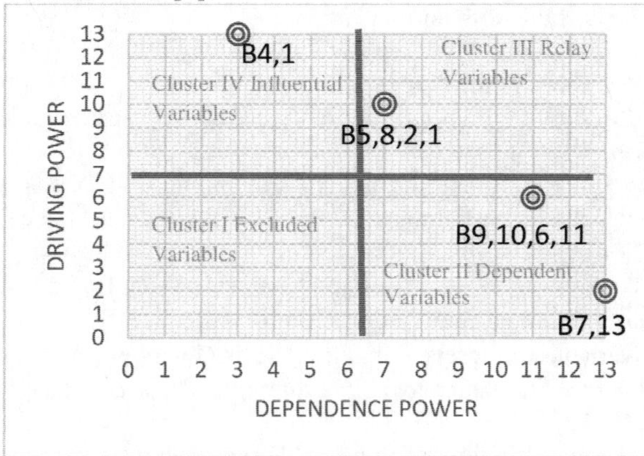

Figure 2: Driving and Dependence graph of Barriers to Adopting and Implementing Halal Logistics Practices.

Tieman [19] mentioned that concept of Halal evolved from based on trust towards supply chain approach. In product approach only, origin (whether it is ingredient or slaughtering method) was main concern but in supply chain approach the complete chain, starting from farming to the end product needs to comply Halal standards. Next level of the developed model comprises of 'Lack of Organizational Culture & Change Management to adopt for HL practices', 'Economic Viability of Logistics Services with Halal Practices', 'Lack of Trained Halal Logisticians and Lack of Capacity Building Programs', 'Lack of Trained Halal Logisticians & Lack of Capacity Building Programs' and 'Lack of Standardization, Codification & Proper Guidelines for Adopting HL Practices'. Due to lack of top management and government support, standardisation and codification of HL practices are not done which hinder the mainstreaming of Halal. Leadership provides impetus to the organisational willingness for adopting HL (Talib et al., 2015). This readiness becomes more prominent when it is backed by organisational culture and change management [4]. Also, unavailability of trained logisticians related to the change management in such a way that practitioner doesn't want to change their behaviour and habits towards Halal practices. 'Lack of Agility Due to Ineffective Coordination and Collaboration Among Different Halal LSPs', 'Lack of Robust Information and Communication Technologies Support', 'Lack of Compatibility and Upgradeability to HL of the existing Logistics', 'Dissimilar Practices and Lack of Dedicated Halal Assets & Facilities' falls in to below topmost levels of the model obtained. Packaged Halal goods transported across nations before it consumption, improper collaboration among different LSPs results in loss of integrity of Halal products. Further, ICT enablement of HL will result in better coordination & collaboration and compliance of different Halal LSPs. Another challenge to integrate Halal practices into logistics is that there is no operational consensus in handling, storage and transportation of Halal products. Different LSPs, follow practices as per their convenience. This could be the result of lack of dedicated assets and facilities. Improper coordination and collaboration lead to poor agility in Halal practices, which often caused breaching of Halal integrity. 'Lack of Ethical Practices and Distrust in Halal Integrity of Products' and 'Lack of Global Branding of Halal Practices' falls into the topmost level of the model developed. A poorly designed Halal logistics with poor ethical consideration will lead to distrust in Halal integrity and ultimately will make Halal brand questionable.

5.2 Discussion based on MICMAC analysis

The result obtained through MICMAC analysis are represented and discussed in Table: 2

Table 2: Representing the findings based on MICMAC analysis

Quadrant	Findings

Excluded Variables	• Autonomous variables are those variables which have a low level of dependence and low level of driving power they are shown in the cluster of the excluded variable • In this case, no such variables fall into this category. This also shows that model so developed is built with correct variables.
Dependent Variables	• Dependent variables are sensitive to the evolution of influent variables and relay variables. They are the output of the system. • B-10: Lack of Agility Due to Ineffective Coordination and Collaboration Among Different Halal LSPs • B-9: Lack of Robust Information and Communication Technologies Support • B-6: Lack of Compatibility and Upgradeability to HL of the existing Logistics • B-11: Dissimilar Practices and Lack of Dedicated Halal Assets & Facilities • B-13: Lack of Ethical Practices and Distrust in Halal Integrity of Products • B-7: Lack of Global Branding of Halal Practices
Relay Variables	• Relay Variables represent a high level of driving power and high level of dependence. • B-5: Lack of Organizational Culture and Change Management to adopt for HL practices • B-2: Economic Viability of Logistics Services with Halal Practices • B-8: Lack of Trained Halal Logisticians and Lack of Capacity Building Programs • B-1: Lack of Standardization, Codification and Proper Guidelines for Adopting HL Practices
Influential Variables	• Variable with high driving and low dependence power are termed as influential variables. • B-4: Reluctance of Logistics Service Providers to adopt Halal Practices in their Operations • B-12: Lack of Customer Demand and Reluctance to pay for HL • B-3: Lack of Government Policies and Support for HL; Weak Enforcement and Lack of a Proper Regulatory Framework for Halal LSPs

6. Managerial Implications of the Research

The findings of the study provide insight to the professionals involved in this area and add value to the body of knowledge. To commercialise and to make logistics with Halal practices economically viable customer needs to be more sensitised towards the requirement of Halal. Other stakeholders, such as Government and top management needs to develop a standardised and codified guideline for Halal practices in the logistics industry, according to which logisticians should be trained, and a robust ICT be implemented as the backbone of the Halal supply chain. However, enforcement of Halal LSPs is a shared responsibility of government and corporate. Anyone not supporting the other may create a situation, where HL practices will not be effectively and responsibly developed. Dedicated warehouse facilities and separate use of handling equipment and operating dedicated transport fleets will eventually affect the overall logistics costs and charges. This hinders more industry players to venture into Halal logistics and eventually poses more risks in the broken Halal supply chain. Additionally, it is negatively impacting the overall development of Halal and Halal logistics business. The identified barriers are inter-related and need to be addressed collectively. Lack of collaboration among various Halal LSPs is due to the reluctance of top management and absence of a codified framework for implementing Halal practices. Mitigation of the identified barriers will lead to the development of a trustable Halal brand.

7. Conclusions

Extending Halal practices in logistics is an emerging phenomenon, which is driven by industries interest to assure the Halal integrity of the product till fork. In this study, the concept of HL practices is elaborated, and the major barriers for the adoption of HL practices were identified. Further, a structural and hierarchical model for adoption of these barriers were presented using ISM. In addition, MICMAC analysis is done to classify these barriers against their driving and dependence power.

References

1. Soon, J.M., M. Chandia, and J.M. Regenstein. (2017). "Halal integrity in the food supply chain" British Food Journal. 119(1).
2. Khan, S., Haleem, A., Khan, M.I., Abidi., M. H., & Al-Ahmari, A., (2018). "Implementing traceability systems in specific SCM through CSFs." Sustainability. 10 (1).
3. Haleem, A., & Khan, M. I. (2017). "Towards successful adoption of Halal Logistics and its' implications for the stakeholders". British Food Journal, 119 (7).
4. Tieman, M. (2011), "The application of Halal in supply chain management: In-depth interviews", Journal of Islamic Marketing, Vol. 2 No. 2, pp. 186-195.
5. Zailani, S., Azmin, I., Aziz, A. A., & Kanapathy, K., (2017). "Halal logistics opportunities and challenges", Journal of Islamic Marketing, 8(1).
6. Lodhi, A.U. (2009), "Understanding Halal Food Supply Chain", HFRC UK Limited, Chesham.
7. Fathi, E., Zailani, S., Iranmanesh, M., & Kanapathy, K. (2016). "Drivers of consumers' willingness to pay for Halal logistics." British Food Journal, 118(2).
8. Ali, M. H., & Suleiman, N. (2016). "Sustainable Food Production: Insights of Malaysian Halal Small and Medium Sized Enterprises". International Journal of Production Economics, 1–12.
9. Khan, M. I., Khan, U., & Haleem, A. (2014). "Corporate Social Responsibility: Modelling of Critical Factors using an Integrated ISM and Fuzzy-Micmac approach." International Journal of Global Business and Competitiveness, 9(1).
10. Khan, M. I. & Haleem, A. (2016). "Understanding "Halal" and "Halal Certification & Accreditation System"- A Brief Review". Saudi Journal of Business and Management Studies, 1(1).
11. Ngah, A. H., Zainuddin, Y., & Thurasamy, R. (2015). "Barriers and enablers in adopting of Halal warehousing". Journal of Islamic Marketing, 6(3).
12. Ngah, A. H., Zainuddin, Y., & Thurasamy, R. (2014). "Barriers and enablers in adopting Halal transportation services: A study of Malaysian Halal Manufacturers". International Journal of Business and Management, 2(2).
13. Hoque, N., Khan, M. A., & Mowla, M. M. (2013). "Organisational culture: features and framework from Islamic the perspective." Humanomics, 29(3).
14. Alserhan, A.B. (2010). "Islamic branding: a conceptualisation of related terms", Brand Management, 18 (1).
15. Salleh, N. M. S. A. S. A. (2016). "Does Halal image strengthen consumer intention to patronize Halal stores? Some insights from Malaysia". Journal of Islamic Marketing, 7(1).
16. Mukhtar, A., & Butt, M. M. (2012). "Intention to choose Halal products: the role of religiosity". Journal of Islamic Marketing, 3(2).
17. Karia, N., & Asaari, M. H. A. H. (2016). "Halal value creation: its role in adding value and enabling logistics service." Production Planning & Control, 27(9).
18. Ngah, A. H., Zainuddin, Y., & Thurasamy, R. (2017) "Applying the TOE framework in the Halal warehouse adoption study", Journal of Islamic Accounting and Business Research, 8(2).
19. Tieman, M. (2012). "Principles in Halal supply chain management", Journal of Islamic Marketing, 3(3).
20. Tieman, M. (2015). "Halal Clusters." Journal of Islamic Marketing 6 (1)
21. Tieman, M. (2013). "Establishing the Principles in Halal Logistics." Journal of Emerging Economies and Islamic Research, 1(1).
22. Bakar, A., Hamid, A., Syazwan, M., & Talib, A. (2014). "Halal Logistics: A Marketing Mix Perspective". Intellectual Discourse, 22(2), 191–214.
23. Warfield, J.N. (1974), "Developing interconnection matrices in structural modeling", IEEE Transactions on Systems, Man and Cybernetics, 4 (1).
24. Sushil. (2012). "Interpreting the interpretive structural model." Global Journal of Flexible Systems Management, 13(2).

IOP Conf. Series: Materials Science and Engineering **404** (2018) 012012 doi:10.1088/1757-899X/404/1/012012

25. Luthra, S., D. Garg, and A. Haleem. 2015. "An Analysis of Interactions among Critical Success Factors to Implement Green Supply Chain Management towards Sustainability: An Indian Perspective." Resources Policy 46 (1).
26. Saurikhia, A., Ahmed, S., Haleem, A., Gangopadhyay, S., Khan, M. I., (2018). "Evaluating technology management factors for fly-ash utilization in the road sector using an ISM approach." International Journal of Management Science and Engineering Management 13(2).

An Overview on Recent progresses and future perspective of biomaterials

Sukriti Yadav[1] and Swati Gangwar[2]
[1]Research scholar, Department of mechanical engineering, Madan Mohan Malaviya University of Technology, Gorakhpur, UP, India
[2]Assistant Professor, Department of mechanical engineering, Madan Mohan Malaviya University of Technology, Gorakhpur, UP, India

E-mail: sukritiyadav09@gmail.com

Abstract. Over the past many decades, tremendous advances in the field of bio materials have been made. Many of the human tissues, such as bones, teeth, tendons, ligaments and other load bearing implants replacement in the human body are carried out successfully with the help of biomaterials. At present, essential concerns in the field of biomaterials are still to be met. The present demand for lifelong implants and bone replacement is characterized by biocompatibility, bioactivity, and mechanical properties of biomaterials, without the immune rejection shows a great challenge. Presently stainless steel, titanium alloys, Co-Cr alloys, zirconia, alumina, ultra-high molecular weight polyethylene, Poly methyl methacrylate etc. are some of the biomaterials that satisfy the requirements of physical and chemical properties, mechanical strength, and biocompatibility for biomedical applications. Significant progress has been made to improve the function of artificial joints, but currently the major efforts has been devoted to minimizing the wear and to increase the lifespan of the implants or prosthesis during its functionality in the human body. In this review, the evolution of different metals/ceramics and polymers that are most commonly used in biomedical applications has been discussed along with the different approaches used to fulfil the challenges faced by medical field.
Key words: Biomaterials; Biocompatibility; Metals/Ceramics; Polymers; Composites etc.

1. INTRODUCTION

An important and challenging use of metallic materials and their composites can be traced back to 19th century. Over the centuries, considerable advancement in the field of synthetic materials has been made in understanding the interaction between materials and living body for effective use of biomaterials in many ways [1,2]. Biomaterials are the synthetic materials that widely used to substituting/recondition the biological functions of body tissue and constantly or periodically remain in contact with body fluid and thus enhance the quality of life of human being [3]. In engineering design, materials designed by matching the properties required for specific application. Biocompatibility is the one of most important parameter that need to be considered while designing the materials for biomedical applications along with the mechanical, physical and chemical properties, whether the material is bioactive, bio inert or biodegradable [4].

In ancient time many natural materials like wood, rubber and glue and other synthesized materials such as iron, gold, zinc and glasses were used for biomedical applications. Metals and their alloys, ceramics and polymers are extensively used as biomaterials. All these materials have different atomic arrangement that provides distinct atomic structure and physical, mechanical and chemical properties. These properties of metallic biomaterials are greatly governed by their composition, microstructure and phases and on the basis of different properties existing in the materials they provide alternative applications for different body parts. Due to attractive mechanical properties of metals such as strength, stiffness, fatigue life these are efficiently used for most of the load bearing applications of the biomedical system [5]. Although Many metals and its alloys have been manufactured in the industry, only some of them are biocompatible which provide good bio functionality and capable to use as distant future implantation materials such as stainless steel, cobalt chromium alloys, Nickel, pure titanium and its alloys (Ti-6Al-4V and Ti-6Al-7Nb), magnesium etc.

[1,3]. To endorse prosthetics stabilization and bone ingrowth highly porous ceramics has been developed. `However, due to its porous nature (low mechanical strength) they are used for low loaded or .unloaded bearing applications. Recently, Ceramics have become important for biomedical application considering three basic types bioactive, bioinert and bioresorbable ceramics [2].
According to Hench's classification the first generation (1960s and 1970s) ceramics biomaterials were 'inert' (alumina and zirconia), they minimize the foreign body response and also immune response. The replacement of conventional metallic femoral heads, acetabular cups of hip joint by high density and highly pure alumina (α- Al_2O_3) was one of the outstanding uses of ceramics because of its excellent wear and corrosion resistance along with high strength and good biocompatibility [4,6,7]. Afterward, second generation (1980 and 2000) ceramics were 'bio-active' (Hydroxyapatite, glass ceramics) as well as bioresorbable (calcium phosphate ceramics). Bioactive materials form chemical bond with soft tissues or promote them to particular responses and behavior. These ceramics are effectively used for bone repair and fixation applications. Bioresorbable materials undergo a progressive degradation in the body while regenerating new tissues and degrading materials are ingest and liberated by metabolic action of body. And for new generation ceramics both the concept of bio activity and biodegradability are united and material properties are combined with their capability to signal and activate specific cellular activity and behavior [2,6].
Natural and synthetic polymers are also the choice for the biomedical application because of its outstanding properties. Collagen and fibronectin, silk fibroin, fibrin, chitosan etc. are the natural polymers which possess good bioactivity and cytocompatibility, whereas synthetic polymers are those which allow precise manipulation of the physicochemical properties i.e. microstructure, degradation rate, mechanical properties, porosity etc. [6,8] and widely used in various tissue engineering applications. Many of the synthetic polyester have been developed that possess structural tunability and good mechanical strength such as polycaprolactone (PCL), poly lactic-co-glycolic acid (PLGA), poly-lactic acid (PLA), polyvinyl acetate (PVA) and these are the polymeric implant biomaterials that are expected to degrade over a period of time [9,10]. This paper briefly reviews the various metals/ ceramics and polymeric biomaterials that have potential to face the challenges of the biomedical engineering.

2. ESSENTIAL CONSIDERATION IN DESIGN OF ARTIFICIAL BIOMATERIALS

The specific medical application decides the design and selection parameters for biomaterials. A material designed for biomedical application should acquire some conspicuous properties (shown in figure 1) for successful and distant future use of biomaterial inside/outside usage in the body without immune rejection, that described as; [11,12].

1. Excellent biocompatibility
2. Adequate mechanical properties
3. Good physical and chemical properties
4. High wear resistance
5. High corrosion resistance
6. Osseo-integration (For bone implants)

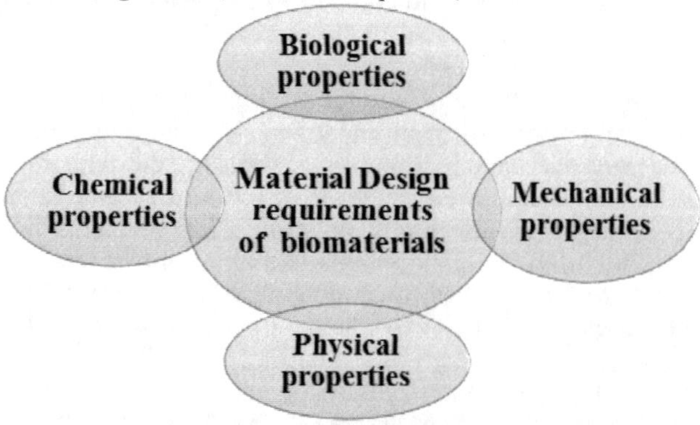

Figure 1. Schematic representation of material design requirements of biomaterials

3. CLASSES OF BIOMATERIALS

Although Many of the materials its alloys are available in the industry, only some of them are biocompatible which provide good bio functionality and capable to use as distant future implantation materials [13]. The important classes of the biomaterials that is widely used for biomedical application are, metals, ceramics, polymers and composites Few of the important properties and applications of different materials are summarized in table1 and table 2.

Table 1. Important properties for different classes of biomaterials [14,15,16]

S.No.	Materials	Advantages	Limitations	Applications
1.	Metals (Ti and its alloy, Ag, Au, Stainless steel etc.)	• Good strength • High Toughness, • Excellent fatigue resistance • High Ductility and malleability etc.	• High modulus • May corrode, • High density, • Difficult to prepare etc.	Load bearing implants such as Joint replacements, bone plates and pins, wires, screws, dental root implants etc.
2.	Ceramics (Alumina, Zirconia, Hydroxyapatite etc.)	• Biocompatible • Great strength and stiffness • Good corrosion and wear resistance • Low density	• Brittle • Not resilience • Low fatigue resistance • Anisotropic mechanical properties under different loading conditions	Dental and orthopedic implants, coating for load bearing implants, medical sensors etc.
3.	Polymers (Nylon, silicon, polyester etc.)	• Biocompatible • Good resilience • Light weight • Easy to fabricate • Good corrosion resistance	• Low strength • deform with time • May degrade	Hip joint socket, blood vessels , contact lenses, heart valves, artificial hearts, etc.
4.	Composites / Bio composites	• Strong • Tailor made • Good corrosion and wear resistance • High Biocompatibility	• Lack of consistency/ homogeneity • Difficult to prepare	Bone cement, joint replacements, Dentistry, bone plates, rods, screws etc.

Table 2. Specific Applications of monolithic materials for biomedical application [17-22]

S.No.	Biomedical Applications	Materials		
		Metals	Ceramics	Polymers
1.	Orthopedic: Joint substitutes, Screws, Pins and wires, hip	Stainless steels, Cobalt chromium	Alumina, Zirconia, porous and dense Hydroxyapatite ,	Poly methyl methacrylate (PMMA), ultra-high molecular

	prosthesis, rods and fracture plates for bone fixation, etc.	alloys, magnesium and its alloys, Titanium and its alloys	Tri-calcium Phosphate, Calcium Phosphate and its' salts etc.	weight polyethylene (UHMWPE), Poly lactic acid (PLA), Poly hydroxyl butyrate (PHB), Poly-D,L-lactide-co-glycolic acid (PLGA), Poly-L-lactate (PLA), etc.
2.	Orthodontic: Dental implants (filling and posts), screws, staples, Orthodontic arch wires,	Cobalt chromium alloys, Titanium and Ti-based alloys (Ni-Ti and Ti6Al4V), stainless steel, platinum, silver etc.	Alumina, Zirconia, Glass, Hydroxyapatite , Tri-calcium Phosphate, Calcium Phosphate and its' salts etc.	Poly methyl methacrylate (PMMA), ultra-high molecular weight polyethylene (UHMWPE), poly tetra fluro ethylene (PTFE), poly-sulfone(PS), polyethylene terepthalate (PET) etc.
3.	Cardiovascular: Stents, valves, Aneurysm clip, Metal coils etc.	Stainless steel, cobalt and its alloys, chromium and its alloys, Nitinol, Titanium alloy (NiTi), magnesium-based alloys	Hydroxyapatite , Tri-calcium Phosphate, Calcium Phosphate etc.	Polyamides, poly tetra fluro ethylene (PTFE), polyolefin, polyester, polyurethane etc.
4.	Instrumental and Supplies: Needles, Scissors, Clamps, Equipment etc.	Stainless steel	-	-

Ti-6Al-4V alloy has been borrowed from aerospace industry and very effectively used in biomedical application due to its superior mechanical properties (high strength weight ratio, relatively low elastic modulus, corrosion resistance and high fatigue strength etc.) along with good biocompatibility. However, the main drawback of using the titanium alloy is its incapability to bond with neighboring hard tissues in in-vivo (inside the body) environment [23]. These metallic materials are widely used for orthopedics surgery and other load bearing implants ranges from simple wires, pins, fracture fixation plates, to artificial joints for knees, ankles, hips, shoulders etc. Simultaneously, metals are also used for dental and orthodontic practice as tooth filling and roots. Metallic materials have tendency to degrade by the reaction with environment (corrosion process) and these corrosion reactions releases some toxic byproducts from biomaterials such as insoluble chemical compounds and ions that may leads to adverse effect (reactions, irritations, toxicity etc.) in the body [24].
The dominant reason in failure of metallic prosthetics is friction and wear of surfaces due to relative motion between loosened parts that causes inflammatory response to the body and makes the patient uncomfortable. It can be avoided by ceramics coating/ bioactive materials, that provides biocompatibility, wear and corrosion resistance toughness etc. [25]. Both natural and synthetic polymers are widely utilized for biomedical application. Polymers are relevant materials as compared to metals and ceramics for biomedical applications because of its outstanding properties such as low density, controlled degradability, environmental friendly and ease of fabrication [26].

4. CONCLUSION

It has been observed that, only one material (i.e. biomaterial) can not suffice all the requirement of a bio implant. For the specific application we need to choose the specific material to fulfil all the requirements of particular biomedical application. These requirements of biomaterials can be fulfilled by having mixture of materials, i.e. composite material. metal/ polymer based composites are extensively used in biomedical application because of its superior combination of properties such as high strength and stiffness, better wear and corrosion resistance compared to unreinforced metal (i.e. also a requirement for the successful use of biomaterial). Alumina, zirconia, Hydroxyapatite, niobium pentoxide etc. are the biocompatible ceramics. The proper selection of matrix system and reinforcing materials along with volume fraction shape and size of reinforcing phases play crucial role for getting the combination of properties such as high strength, wear and corrosion resistance and biocompatibility.

References

[1] Chen Q. , Thouas G. A., 2015, Metallic implant biomaterials, *Materials Science and Engineering* **R 87**, 1–57.
[2] Patel N.R., Gohil P.P., 2012, A Review on Biomaterials: Scope, Applications & Human Anatomy Significance, *International Journal of Emerging Technology and Advanced Engineering*, **2,** 2250-2459.
[3] Overview of Biomaterials and Their Use in Medical Devices, 2003, ASM International**, #06974G.**
[4] Navarro M., Michiardi A. , Castaño O. , and Planell J.A., 2008, Biomaterials in orthopaedics, J R Soc Interface, **27,** 1137–1158.
[5] Yoruc A.B.H., Şener B.C., 2012, Biomaterials, A Roadmap of Biomedical Engineers and Milestones, 67-114.
[6] Nie H., Kim J., Fu S.Y. Mao J.J., 2011, Biomaterials Selection for Dental Pulp Regeneration. *In Comprehensive Biomaterials,* Elsevier, Oxford, **6**, 245–254.
[7] Triplett R.G., Budinskaya O., 2017, New Frontiers in Biomaterials, Oral Maxillofacial Surg Clin N Am, **29**, 105–115.
[8] Wu W., Cheng R., Neves J.D., Tang J., Xiao J., Ni Q. ,Liu X. , Pan G., Li D., Cui W. , Sarmento B. , 2017, Advances in biomaterials for preventing tissue adhesion, *Journal of Controlled Release,* doi: 10.1016/j.jconrel.2017.06.020.
[9] Gajendiran M., Choi J., Kim S.J., Kim K., Shin H., Koo H.J., Kim K., 2017, Conductive biomaterials for tissue engineering applications, *Journal of Industrial and Engineering Chemistry*, **51,** 12–26.
[10] PA Lalor P.A., Jones L.C., 2017, The Innate Response to Biomaterials, *In Comprehensive Biomaterials,* doi:10.1016/B978-0-12-803581-8.10237-1, 30-40.
[11] Kadam A.G., Pawar S.A., Abhang S.A., 2017, A Review on Finite Element Analysis of Different Biomaterials used in Orthopedic Implantation, *International Research Journal of Engineering and Technology* , **4**, 2192-2195.
[12] Vesque J.L., Hermawan H., Dube D., Mantovani D., 2008, Design of a pseudo-physiological test bench specific to the development of biodegradable metallic biomaterials, *Acta Biomaterialia* **4,** 284–295.
[13] Jenkoa M., Sekb M.G., Godeca M., Hodnikc M., Batica B.S., Donika C., Grantd J.T., Dolinar D., 2018, Surface chemistry and microstructure of metallic biomaterials for hip and knee endoprostheses, *Applied Surface Science,* **427**, 584–593.
[14] Konofaos P., Szpalski C., Rogers G.F., Rae M.M., Bumgardner J.D., Warren S.M., 2017, Biomaterials and Their Application in Craniomaxillofacial Surgery, *Comprehensive Biomaterials II*, **7**, 407-428.
[15] Bose S., Banerjee D., Bandyopadhyay A., 2017, Introduction to Biomaterials and Devices for Bone Disorders, *Materials and Devices for Bone Disorders*, 1-27.
[16] Introduction to biomaterials, NPTEL lecture by B.basu , K.Balani.
[17] Allain J.P., Rendón M.E., 2018, Surface treatment of metallic biomaterials in contact with blood to enhance hemocompatibility, Hemocompatibility of Biomaterials for Clinical Applications. https://doi.org/10.1016/B978-0-08-100497-5.00008-2, 279-326.
[18] Malekani J., Schmutz B., Gu Y., Schuetz M., Yarlagadda P., 2012, Orthopedic bone plates:Evolution in Structure, Implementation technique and biomaterial, GSTF International Journal of Engineering Technology, **1,** 135-140.
[19] Sensharma P., Madhumathi G., Jayant R.D., Jaiswal A.K., 2017, Biomaterials and cells for neural tissue engineering: Current Choices, *Materials Science & Engineering C,* , **MSC 7776,** doi: 10.1016/j.msec.2017.03.264.
[20] Dormont F., Varna M., Couvreur P., 2017, Nanoplumbers: Biomaterials to fight cardiovascular disease, **MATTOD-971**, *Elsevier Ltd..,* http://dx.doi.org/10.1016/j.mattod.2017.07.008, 1369-7021.
[21] Banoriya D., Purohit R., Dwivedi R.K., 2017, Advanced Application of Polymer based Biomaterials, *Materials Today: Proceedings* **4,** 3534–3541.
[22] Jaganathan S.K., Suprio E., Murugesan S., Balaji A., Asokan M.K., 2014, Biomaterials in cardiovascular research: applications and clinical implications, *BioMed reaearch International,* **Article ID 459465**, http:/dx.doi.org/10.1155/2014/459465.
[23] Ghosh S.,Dandapat N., Balla V.K. 2015, Preparation and *in vitro* characterization of fluroapatite based bioactive glass-ceramics for biomedical applications, *Materials Today: Proceedings* , **2** , 1326 – 1331.
[24] Rivera V.E., Ramírez M.U. , Gaón M.E., Rodriguez L. , Betancur A.L.G., Villegas L.G., Mendoza L.A., Ruiz A.D.V., 2016, Biomimetic titania/hydroxyapatite coating of CrCoMo micro implants enhances biocompatibility and reduces metal-associated toxicity, *Abstracts / Toxicology Letters* **259S** , S73–S247.
[25] Ghaemi M.H.,Reichert S., Krupa A., Sawczak M., Zykova A., Lobach K., Sayenko S., Svitlychnyi Y., 2017, Zirconia Ceramics with Additions of Alumina for Advanced Tribological and Biomedical Applications, *Ceramics International,*http://dx.doi.org/10.1016/j.ceramint.2017.04.150.
[26] Vedadghavami A., Minooei F., Mohammadi M.H., Khetani S., Rezaei A., Mashayekhan S., Nezhad A.S., 2017, Manufacturing of hydrogel biomaterials with controlled mechanical properties for tissue engineering applications, *Acta Biomaterialia,* http://dx.doi.org/10.1016/j.actbio.2017.07.028, 1-21.

IOP Conf. Series: Materials Science and Engineering **404** (2018) 012014 doi:10.1088/1757-899X/404/1/012014

Russian industrial sector in the conditions of the Fourth Industrial Revolution

E Balashova[1] and E Gromova[2]

[1]Doctor of economics, Peter the Great St. Petersburg Polytechnic University, St. Petersburg, Russia
[2]PhD student, Peter the Great St. Petersburg Polytechnic University, St. Petersburg, Russia

E-mail: lizaveta-90@yandex.ru

Abstract. The Fourth Industrial Revolution is taking on an ever-increasing scope in the world. The construction of the digital economy is the core of the revolution. Russia's potential in digitalizing the economy is promising. The potential effect on Russia's GDP from digitalization of the economy by 2025 is estimated at 4.1-8.9 trillion rubles, which is 19-34% of the total increase in GDP. The purpose of the study is to describe and to analyze the current state of the Russian industry under the conditions of the Fourth Industrial Revolution with a focus on the agile management methodology in industry. It is a review-based work. Theoretical and practical aspects of agile management are presented. It is concluded that the agile approach in industry corresponds to the goals of the Russian economy at the present stage of the development of the world community.

1. Introduction

The Fourth Industrial Revolution is actively developing in the world. Since 2011 it has become widespread. As part of the previous industrial revolution, the model of the economy is based on the idea of improving production efficiency through automation without changing the processes themselves. It is noteworthy that these processes can exist without information technologies. Within the framework of the Fourth Industrial Revolution, information technologies will become the core of business and will lead to irreversible transformation in production and business. All decisions in the digital economy will be based on the capabilities and resources of information and communication technologies. The technological basis for the transition to a new paradigm is the industrial Internet of things. Thanks to it, an intelligent system through interaction with the global environment will manage production in real time. All technical devices will give feedback and make technological solutions with the subsequent reconfiguration of production. The data will become the core of the organization. The Fourth Industrial Revolution leads to the following fundamental consequences [1]:

- consumer expectations are changing;
- operating models are transformed into new digital models;
- new partnerships are formed as companies realize the importance of new forms of cooperation;
- the quality of products is improved through data that increases the productivity of assets.

That's why it is important to identify an effective management model for the industrial development under such conditions. The agile management is the most effective management methodology in the conditions of the Fourth Industrial Revolution. It is a strategy whose purpose is to achieve a sustainable company's development by adapting to all kinds of unpredictable changes. Thus, this method is able to withstand the main challenges of the new revolution. The theoretical and practical aspects of agile management are studied by many scientists. Among them the following scholars are highlighted: J. Goodpasture (2010), K. Hass (2007), J. Highsmith (2000, 2004), C. Larman (2004), K. Schwaber (2003, 2007), J. Sutherland (2014), F. Benaben (2017), H. Yang (2016), R. Sindhwani (2017), A. Gunasekaran (2014, 2017). Attention is drawn to the fact that this methodology has been widely recognized abroad, both at the theoretical and practical level. Unlike developed countries, agile management has not yet received sufficient development in Russian industry. Thus, the aim of the study is to review the current state of the Russian industry under the conditions of the Fourth Industrial Revolution with a focus on the agile management methodology.

2. Materials and Methods

Agile management method [2-13] becomes particularly relevant in the context of the formation of the Fourth Industrial Revolution. This model is able to solve the most urgent modern problems, which consist of uncertainty and rapid changes in business environment. Agile is a management strategy for achieving a sustainable development through adaptation to all kinds of unpredictable changes. Initially, the agile methodology has successfully proved itself in the field of information technologies. The aforementioned software development methodology succeeded in squeezing the waterfall model, which is traditional for this industry. The essence of the agile system is based on the following fundamental principles:

- technical units in cooperation with the business department are located in an open area;
- test scripts are developed before programming stage;
- daily morning meetings with a brief discussion of the problems encountered;
- the development process consists of "sprints" (work cycles lasting from one to four weeks). The output of each cycle is the working code. The widely-known Deming cycle (PDCA) underlies the logic of the algorithm of this process.

There are five stages of agile management [14]:

1. envision (how to determine the product vision, the project objectives and constraints, the community and how the team work together);
2. speculate (developing an ability or feature based release plan to deliver on all aspects of the vision);
3. explore (planning and delivery project running tested stories in a short iteration, constantly seeking to reduce the risk and uncertainty);
4. adapt (reviewing the obtained results, the situation and the teams' performance and adapting as necessary);
5. close (concluding the project and passing on the key learning's').

Summarizing, agile management can be defined as short cycles of product development that deliver product incremental updates rapidly based on the changing needs of the customer. This methodology is the opposite to the waterfall management, which values extensive planning and preproduction.

The functioning principle of the agile methodology in IT-sphere is transformed into its application in the most effective management strategy within the framework of the Fourth Industrial Revolution. The main regularities of the mechanism for introducing and using the model under consideration, with an emphasis on the scale of activities and taking into account modern means of developing communications are preserved. Next, it is necessary to highlight a number of characteristic features of the agile management:

- rapid reconfiguration of labor and material resources;
- avoiding multitasking;
- multidisciplinary team of the main company;

- dual operating system, consisting of the synthesis of a classical hierarchical organizational structure and a dynamic network structure;
- an extensive partner network of companies (with duplicate and complementary competences) and suppliers;
- wide application of cloud technologies for integration of interacting companies among themselves and with the external environment;
- big data;
- high-tech industry is the most suitable sphere of application;
- high level of customization;
- minimization of losses from possible, unexpected negative changes.

So, the fundamental advantages of agile management include:

- quality improvement;
- just-in-time delivery;
- flexibility of internal processes;
- cost reduction.

Thus, agile management methodology meets the requirements of modern conditions created by the new industrial revolution. It is the best method of organizing the most advanced production process.

3. Results and Discussion

The current state of the Russian industrial sector in the conditions of the Fourth Industrial Revolution is characterized as the beginning of the possible great way. The digital economy is a fundamental basis of the Fourth Industrial Revolution. As a result, the introduction of modern digital technologies in various fields of activity, primarily in industry, is a priority for the development of the national economy. Many efforts and actions need to be applied to the development of the digital economy in Russia [15-19]. For the first time the task of forming a digital economy was identified by the President of the Russian Federation V.V. Putin in his annual address to the Federal Assembly in 2016: "... launch a large-scale system program for the development of the economy of a new technological generation, the so-called digital economy". Approved by the decree of the Government of the Russian Federation of July 28, 2017, No. 1632-r, the Digital Economy of the Russian Federation program, guided by the "Strategy for the Development of the Information Society in the Russian Federation for 2017-2030", defines the digital economy as an economic activity, a key factor in production which is data in digital form, and contributes to the formation of an information space, taking into account the needs of citizens and society in obtaining qualitative and reliable information, development and the information infrastructure of the Russian Federation, the creation and application of Russian information and telecommunication technologies, as well as the formation of a new technological basis for the social and economic sphere.

According to the study of the international consulting company McKinsey in 2017, the share of the digital economy in Russia's GDP is 3.9%, which is two to three times lower than that of the leading countries, for example, the USA, China, Brazil. Digital household expenditures make up 2.6% of Russia's GDP - this is the most significant contribution to the development of new technologies, but it is still lower than the average for the leading countries (3.6%). The share of public spending and private investment in the GDP structure is also lower than in the countries under consideration, and the volume of export of digital technologies is four times less than imports. If the volume of Russian investments in information and communication technologies, including digital household spending and investment companies and the state, increases compared to the average level of countries, the share of the digital economy in Russia will grow to 5.9% of GDP, which will allow Russia to come closer to the leading countries.

The growth rates, which were fixed in 2011-2015, are encouraging. Russia's GDP for this period grew by 7%, and the volume of the digital economy increased by 59%, to 1.2 trillion rubles. (at prices of 2015). Thus, the digital economy accounted for 24% of the total GDP growth. It is noteworthy that the volume of the digital economy is growing 8.5 times faster than other sectors of the Russian economy. There is considerable potential for further growth. As a result, it is important for Russia to accelerate the pace of digitalization and by 2025 achieve an ambitious, but very realistic goal - to triple the size of the digital economy. According to the assessment of the consulting company, an additional increase from the introduction of new digital technologies will be on average up to 2025 from 0.4 to 0.9% of GDP per year. Comparison of this growth with the forecasted growth rates of the Russian economy allows us to estimate the contribution of digitalization at the level of 19 to 34% (4.1 - 8.9 trillion rubles) of real GDP growth by 2025.

The introduction of digital technologies in the industry requires the mastery of modern management models in production. Without this, the most ambitious goals, based on positive trends, will not be achieved. In the conditions of the Fourth Industrial Revolution, many countries can compete for world leadership, including Russia, which has good potential. The potential benefits of agile management in industry are undeniable and are proved in practice (in accordance with the study «Digital Russia: new reality» conducted by McKinsey & Company in 2017). There are:

- reducing the time to market for products by 20 to 50%;
- reduction of product maintenance costs by 10-40%;
- increase of productivity by 3-5%;
- reduction of equipment downtime by 30-50%;
- improving the accuracy of forecasts by up to 85%;
- reduction of quality assurance costs by 10-20%;
- reduce costs of storing inventory at 20-50%;
- increase performance of technical functions at 45-55% because of the automation of labor.

All of the above aspects will lead to a general improvement in the quality and competitiveness of Russian products on both domestic and global markets. A significant effect of the agile approach can be obtained in the manufacturing industry due to its high labor intensity and technological backwardness of Russia from the advanced countries. Here, efficiency gains are possible in all areas of the value chain, from accelerating the development and launch of new products to the market, synchronization of the chain of production and supply of components, to a significant increase in the efficiency of planning, production, quality control and level of service for the final product. Thanks to the modernization of Russian engineering in the era of the Fourth Industrial Revolution, labor productivity can be substantially increased in this sector, reducing the gap from the most industrialized countries. Agile management is effective and the most suitable method for the Russian industrial sector under the conditions of the Fourth Industrial Revolution.

4. Conclusions

Concluding the results of the study, the following features should be noted:

1. the Forth Industrial Revolution possesses a tremendous potential for transforming the Russian industry, which was traditionally considered quite conservative in the use of digital technologies;
2. the current state of the Russian industry in the conditions of the Fourth Industrial Revolution is promising, according to forecasts the contribution of digitalization at the level of 19 to 34% of real GDP growth by 2025 can be achieved;

3. implementing of agile management in Russian industrial sector is the most suitable and effective methodology according to the requirements dictated by the next industrial revolution;

4. the change of paradigms in the organization of production provokes the redistribution of the roles of world economy leaders and Russia has a chance to become this leader.

In the future, it may be of interest to study the development of the digital economy in Russia in combination with the use of the agile management. The most striking examples of successful industrial enterprises that apply this methodology can be described. Also tracing the dynamics of the state of the Russian industrial sector in the new conditions dictated by the Fourth Industrial Revolution may be a continuation of this work.

References

[1] Schwab K 2017 *The fourth industrial revolution* (Moscow: Eksmo) p 208
[2] Highsmith J 2000 *Adaptive software development: a collaborative approach to managing com-plex systems* (New York: Dorset House) p 349
[3] Cohn M and Schwaber K 2003 *Agile Times* **1(1)** pp 10-12
[4] Schwaber K 2007 *The enterprise and Scrum* (Redmond: Microsoft Press) p 147
[5] Goodpasture J 2010 *Project management the agile way: Making it work in the enterprise* (Fort Lauderdale: Ross Publishing) p 337
[6] Hass K 2007 *The Blending of Traditional and Agile Project Management* (World Today)
[7] Larman C 2004 *Agile and iterative development: A manager's guide* (Boston: Addison-Wesley) p 329
[8] Sutherland J 2014 *SCRUM. The art of doing twice the work in half the time* (New York: Crown Business) p 242
[9] Gunasekaran A 2017 *International Journal of Production Research* pp 1-13
[10] Dubey R and Gunasekaran A 2014 *The International Journal of Advanced Manufacturing Technology* **76**(9-12) pp 2147–2157
[11] Sindhwani R and Malhotra V 2017 *Benchmarking: An International Journal* **24**(2) pp 467-487
[12] Yang H, Baradat C, Krut S and Pierrot F 2016 *The International Journal of Advanced Manufacturing Technology* **85**(1-4) pp 25-35
[13] Benaben F and Vernadat F 2017 *Enterprise Information Systems* **11**(4) pp 470-473
[14] Highsmith J 2004 *Agile Project Management: Creating Innovative Products* (Boston: Addison Wesley) p 269
[15] Ilin I, Levina A, Abran A and Iliashenko O 2017 *ACM International Conference Proceeding Series*
[16] Kuladzhi T, Babkin A and Murtazaev S A 2018 *Advances in Intelligent Systems and Computing* **692**
[17] Balashova E S and Gromova E A 2017 *Journal of Advanced Research in Law and Economics* **8**(3) pp 749-753
[18] Rudskaya I and Rodionov D 2017 *Academy of Strategic Management Journal* **16**(2)
[19] Glukhov V V, Ilin I V and Anisiforov A B 2015 *ACM International Conference Proceeding Series*

Modeling of hot deformation behavior of high phosphorus steel using Johnson-Cook model

Vartika Gupta[1], S K Rajput[1,*], Yashwant Mehta[2], Shashee Kant Soni[1], Tarun Soota[1]

[1]Department of Mechanical Engineering, Bundelkhand Institute of Engineering and Technology, Jhansi, India, 284128.
[2]Department of Metallurgical and Materials Engineering, Indian Institute of Technology, Roorkee, India, 247667.
*Corresponding author. Tel.: +91-510-2320321: Fax: +91-510-2320349
E-mail address: rajput_skumar@rediffmail.com

Abstract: To describe the hot deformation behavior of high phosphorus steel, isothermal hot compression test was performed using Gleeble 3800 thermo-mechanical simulator in the temperatures ranging from 750-1050°C at constant strain rates of 0.001-10 s^{-1}. Based on the experimental results, Johnson-Cook model was developed to predict the values of flow stress. Johnson-Cook model shows a fairly good agreement with experimental and predicted values. Average absolute relative error (AARE), and relative error (RE) of JC model are calculated as 24.356%, and -19.97% to 33.179%, respectively.

Keywords: Hot deformation behavior, Johnson-Cook Model, Flow stress

1. Introduction

High phosphorus steel has widely used in structural applications due to its corrosion resistance properties. Phosphorus is added intentionally in steel in the form of ferro-phosphorus (23% to 26% P), to increase strength, machinability and corrosion resistance. Phosphorus also having some negative effects on the steel as it creates brittleness by segregating at the grain boundaries. However, this can be overcome by adding small amount of carbon because it displaces phosphorus from the grain boundaries due to site competition effect. Additional heat treatment in the dual phase region also remove phosphorus away from grain boundaries, results improved ductility and toughness [1]. Hot deformation process is a very vital step during industrial processing of any metal or an alloy. To depict the metal flow pattern and metallurgical transformations during metal forming process, understanding of flow behaviors of metal during hot deformation is very important [2-6]. Due to these complications in the path of hot deformation behavior of materials, leads to the assumptions that computer aided model can describe the material response under varying load conditions more easily and accurately. For this purpose, various constitutive equations were used to simulate and model the flow behavior of metals and alloys [7, 8]. Many researchers have tried to develop constitutive model based on the experimental results to describe the deformation behavior of metals and alloys [9]. Johnson Cook (J-C) model is the most widely used constitutive model to predict the deformation behavior of materials at varying processing parameters such as temperature, strain and strain rate [10]. Compared to other models, the J-C model can predict the deformation behaviors of materials at elevated temperatures [11, 12] at less material parameters and time [13]. Although there are some studies present on the hot deformation behavior of high P steel in the literature, but the prediction of flow behavior using Johnson-Cook model is lacking. Hence, there is need to predict the flow behavior using Johnson-Cook model in comprehensive manner. In the present study, isothermal hot compression tests were performed upto total true strain of 0.7 using Gleeble 3800 thermo-mechanical simulator in the temperatures ranging from 750°C-1050°C at constant strain rates from 0.001 to 10 s^{-1}. Johnson-Cook model is used to model and predict the flow behavior of high phosphorus steel.

2. Experimental Methodology

2.1 Material and Methods

Initially iron scrap was melted in the furnace followed by addition of ferrous-phosphorus, graphite and ferro-silicon. Solid plate of dimensions 200mm x 400mm x 40mm was obtained using sand mould casting followed by cutting into smaller pieces of dimensions 30mm x 40mm x 30mm with the help of power hacksaw [14]. Table 1 shows the chemical composition of high phosphorus steel.

Table 1. Chemical composition (in wt %) of high phosphorus steel.

Elements	Si	P	C	Mn	Cr	Ni	Cu	Al	W	Fe
Wt %	0.26	0.13	0.05	0.2	0.13	0.07	0.02	0.003	0.02	99.12

2.2 Hot Compression

IOP Conf. Series: Materials Science and Engineering **404** (2018) 012015 doi:10.1088/1757-899X/404/1/012015

Hot compression test were conducted on cylindrical specimen of length 15mm and 10mm diameter with the help of Gleeble 3800 thermo-mechanical simulator under vacuum. To control the temperature during the experiment, K-type thermocouple was spot welded at the center of the specimen. To minimize friction and temperature gradient, ISO-T anvil, nickel based lubricant and graphite foil were used. Specimens were heated upto the austenitization temperature of 1050°C with heating rate of 5°C for 10s, followed by cooling with the cooling rate of 1°Cs^{-1} to the deformation temperatures ranging from 750°C-1050°C. Specimens were deformed upto true strain of 0.7 using constant strain rates ranging from 0.001- 10 s^{-1}. In-situ water quenched the specimens to preserve and avoid and metadynamic phenomena after deformation.

3. Results and discussions

3.1 Flow behavior

Sample stress-strain curves at deformation temperature of 750°C at constant strain rates ranging from 0.001 s^{-1} to 10 s^{-1} is depicted in fig.1. It has been observed that flow stress increases with increasing strain rate for a particular temperature. Stress-strain curve at low strain rates (0.001 and 0.01 s^{-1}) shows peak stress after strain hardening, followed by softening due to dynamic recrystallization (DRX). At higher strain rates (0.1, 0.5, 1 and 10 s^{-1}), the flow stress increased in whole range of testing strain. Peak stress shifted towards higher strain with increasing strain rates.

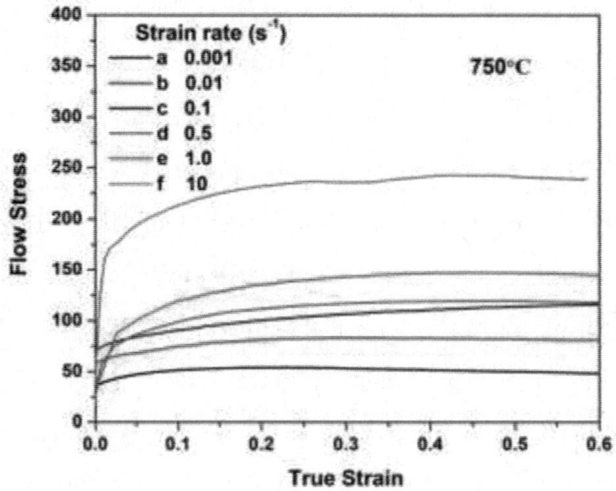

Fig.1 True stress-strain curve of high phosphorous steel at 750°C.

3.2 Johnson-Cook Model

Johnson-Cook (J-C) model was used to predict the flow behavior of high phosphorous steel during hot deformation considering the influences of temperature, strain and strain rates, and is represented by equation (1) [15, 16].

$$\sigma = (\sigma_o + B\varepsilon^n)(1 + C\ln \dot{\varepsilon}^{*})(1 - T^{*m}) \quad (1)$$

where σ is flow stress, σ_o is yield stress, ε is equivalent plastic strain. Dimensionless strain rate $\dot{\varepsilon}^* = \dot{\varepsilon}/\dot{\varepsilon}_0$, where $\dot{\varepsilon}$ and $\dot{\varepsilon}_0$, are strain rate and reference strain rate, respectively. *B*, *n*, *C* and *m* are the material constants.

$T^* = (T-T_r)/(T_\square-T_r)$, where *T* is the absolute temperature, $T_\square$ ($T_\square$=1525°C for present material) and T_r, are the melting temperature and reference temperature, respectively. To calculate the value of material constants, reference temperature is taken as 750° ($T \geq T_r$), and reference strain rate as 10 s^{-1} in the present study.

Value of yield stress σ_o is found to be 81.04137 MPa at reference temperature and strain rate (Fig.1). To determine the value of material constants *B*, *n*, *C*, and *m* at reference temperature and strain rate, equation (1) can be rewritten as,

$$\sigma = \left(\sigma_o + B\varepsilon^n\right) \tag{2}$$

On taking natural logarithm of both sides, Equation (2) can be represented as equation (3).

$$\ln(\sigma - \sigma_o) = \ln(B) + n\ln(\varepsilon) \tag{3}$$

Substituting the values of yield stress at corresponding strain, equation (3) gives the relationship between $ln\ (\sigma\text{-}\sigma_o)$ and $ln\ \varepsilon$ (Fig. 2), and the value of n and B are obtained as 0.10068 and 172.1747, respectively.

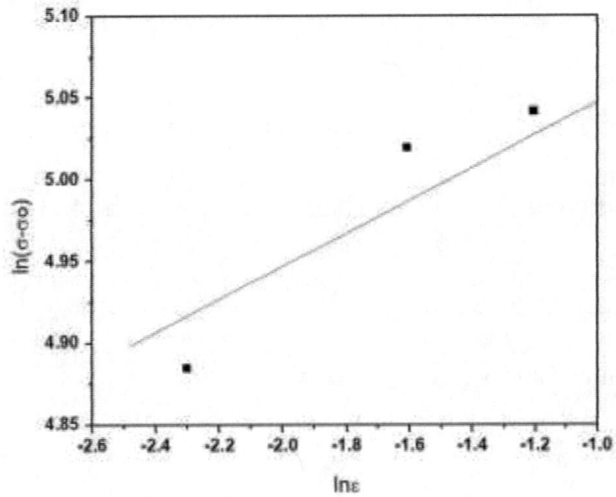

Fig.2 Relationship between $ln\ (\sigma\text{-}\sigma_o)$ and $ln\ \varepsilon$.

Substituting the values of n and B in equation (1) at reference temperature, it can be expressed as equation (4). Figure 3 shows the graph between $\sigma/(\sigma_o+B\varepsilon^n)$ and $ln\varepsilon^*$ at different strains, and obtained the average value of C as 0.07442.

$$\sigma/(\sigma_o + B\varepsilon^n) = 1 + C\ln\varepsilon^* \tag{4}$$

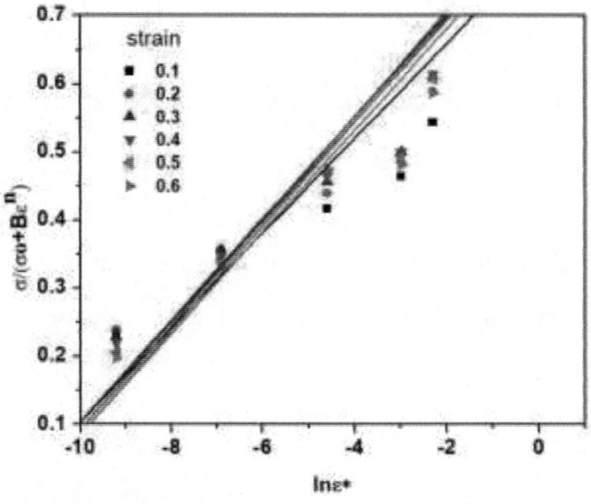

Fig.3 Relation between $\sigma/(\sigma_o+B\varepsilon^n)$ and $\ln \varepsilon^*$.

At reference strain rate equation (1) can also be rewritten as equation (5).

$$\sigma/(\sigma_o + B\varepsilon^n) = 1 - T^{*m} \tag{5}$$

Taking natural algorithm on both sides of the equation (5),

$$\ln[1-(\sigma/\sigma_o + B\varepsilon^n))] = m \ln T^* \tag{6}$$

According to equation (6), the graph between $\ln[1-(\sigma/(\sigma_o+B\varepsilon^n))]$ and $\ln T^*$ plotted (Fig. 4), and the value of m can be calculated as 0.18247.

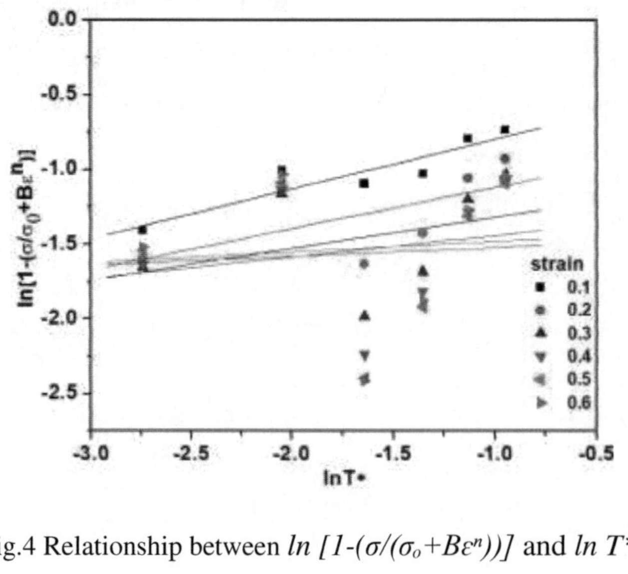

Fig.4 Relationship between $\ln[1-(\sigma/(\sigma_o+B\varepsilon^n))]$ and $\ln T^*$.

After determining all value of material constants, J-C model can be summarized as equation (7).

$$\sigma = (81.0417 + 172.1747\,\varepsilon^{0.10068})(1 + 0.07442\ \ln \dot{\varepsilon}^{*})(1 - T^{*0.18224}) \tag{7}$$

The values of flow stress are predicted at different temperatures, strain and strain rate using equation (7), and the comparison between experimental and predicted value of flow stress were plotted, as depicted in fig. 5. It can be seen from the graph, that the values of flow stress at reference strain rate are closer to that of experimental values. Further, the values of flow stress tend to move farther from the experimental values with increasing temperatures and strain rates. Thus, it can be said from fig. 5 that Johnson-Cook model gives a fairly good agreement between the experimental and predicted flow stress.

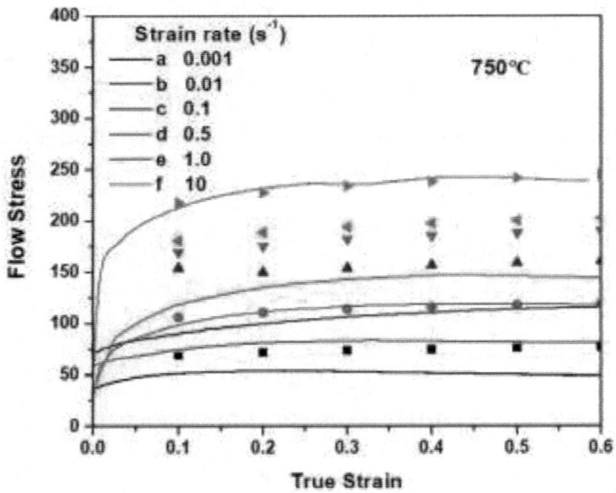

Fig.5 Comparison between experimental and predicted value of flow stress using JC model at deformation temperature 750°C.

The accuracy and predictability of the Johnson-Cook model can be evaluated by relative error (RE) and average absolute relative error (AARE), and can be expressed by equation (8) and (9), respectively.

$$AARE(\%) = \frac{1}{N}\sum_{1=1}^{N}\left|\frac{\sigma_E - \sigma_P}{\sigma_P}\right| \times 100 \tag{8}$$

$$RE = \left(\frac{\sigma_E - \sigma_P}{\sigma_E} \right) \times 100 \tag{9}$$

where σ_p and σ_E, are the predicted and experimental flow stress values, respectively. N is the total number of data which is employed for investigation purpose. Relative error and average absolute relative error are used to analyze the accuracy and predictability of the flow stress values in phenomenological model. Average absolute error is used to calculate the optimized value, whereas correlation coefficient provides the closeness information of experimental and predicted values [17]. Average absolute error is calculated step by step and is therefore considered as an unbiased statics for measuring the predictability of the model. The average absolute error, and relative error, of Johnson Cook model are 24.356%, and -19.97% to 33.179%, respectively. These values show fairly good predictability.

4. Conclusions

In the present study, isothermal hot compression tests were conducted in the temperature ranging from 750°C-1050°C at constant strain rates of 0.001-10 s^{-1}. Experimental data obtained from the isothermal test were used to predict the flow stress values using Johnson-Cook model. Following conclusions are drawn from the present study.

1. It has been observed from the flow curves that the values of flow stress increases with increasing strain rate for a particular temperature. The true stress-strain curve shows initially strain hardening followed by thermal softening due to DRX at low strain rates of 0.001-0.01 s^{-1} and deformation temperature of 750°C.
2. Johnson-Cook model shows a fairly good agreement with experimental and predicted values. Average absolute relative error (AARE), and relative error (RE), of JC model are calculated as 24.356%, and -19.97% to 33.179%, respectively.

References

[1] Sahoo G, Balasubramaniam R 2007 Mechanical behaviour of novel phosphoric irons for concrete reinforcement applications *Scr. Mater.* **56** 117-120.
[2] Lin YC, Chen MS and Zhang J 2008 Constitutive modelling for elevated temperature flow behaviour of 42CrMo steel *Comput. Mater. Sci.* **42** 470-477.

[3] Rajput SK, Chaudhari GP and Nath SK 2014 Physical simulation of hot deformation of low carbon Ti-Nb micro-alloyed steel and microstructural studies *J. Mater. Eng. Perform.* **23** 2930-2942.

[4] Rajput SK, Dikovits M, Chaudhari GP, Poletti C, Warchomicka F, Pancholi V and Nath SK 2013 Physical simulation of hot deformation and microstructural evolution of AISI 1016 steel using processing maps *Mater. Sci. Eng.* A **587** 291-300.

[5] Rajput SK, Chaudhari GP and Nath SK 2016 Characterization of hot deformation of low carbon steel using processing maps, constitutive equations and Zener-Hollomon parameter *J. Mater. Proc. Tech.* **237** 113-125.

[6] Singh K, Rajput SK and Mehta Y 2016 Modelling of the hot deformation behaviour of high phosphorus steel using artificial neural networks *Mater. Dis.* **6** 1-8.

[7] Lin YC, Xia Y, Chen XM and Chen MS 2010 Constitutive descriptions for hot compressed 2124 T851 aluminium alloy over a wide range of temperature and strain rate *Comput. Mater. Sci.* **50** 227-233.

[8] Li W, Li H, Wang Z and Zheng Z 2011 Constitutive equations for high temperature flow stress prediction of Ale14Cue7Ce alloy *Mater. Sci. Eng.* **528** 4098-4103.

[9] Ashtiani HR and Shashvari P 2016 A comparative study on the phenomenological and artificial neural network models to predict hot deformation behaviour of AlCuMgPb alloy *J. Alloys Comp.* **687** 263-273.

[10] Johnson GR and Cook WH 1985 A constitutive model and data for metals subjected to large strains, high strain rates and high temperatures *Eng. Fract. Mech.* **21** 1-31.

[11] Salari S, Naderi M and Bleck W 2015 Constitutive modelling during simultaneous forming and quenching of a boron bearing steel at high temperatures *J. Mater. Eng. Preform.* **24** 808-815.

[12] Akbari Z, Mirzadeh H and Mcabrera J 2015 A simple constitutive model for predicting flow stress of medium carbon microalloyed steel during hot deformation *Mater. Des.* **77** 126-131.

[13] He A, Xie G, Zhang H and Wang X 2014 A modified Zerilli-Armstrong constitutive model to predict hot deformation behaviour of 20CrMo alloy steel *Mater. Des.* **56** 122-127.

[14] Mehta Y, Rajput SK, Dabhade VV and Chaudhari GP 2016 Physical simulation of hot deformation and microstructural evolution of Fe-0.05C-0.13P steel *J. Mater. Eng. Perform.* **25** 1376-1383.

[15] Børvik T, Hopperstad OS and Berstad T 2003 On the influence of stress triaxiality and strain rate on the behaviour of a structural steel *Eur. J. Mech. A-Solids* **22** 15-32.

[16] Clausen AH, Børvik T, Hopperstad OS and Benallal A 2004 Flow and fracture characteristics of aluminum alloy AA5083-H116 as function of strain rate, temperature and triaxiality *Mater. Sci. Eng. A* **364** 260-272.

[17] Gang C, Wei C and Anhze G 2015 Strain-Compensated Arrhenius-Type Constitutive Model for Flow Behavior of Al-12Zn-2.4Mg-1.2Cu Alloy *Rare Metal Mater. Eng.* **44(9)** 2120-2125.

Thermal Characterization of inorganic salts as phase change materials for thermal energy storage systems

Dibakar Rakshit[1] and Ashutosh Pratap Singh[2]
Indian Institute of Technology Delhi, New Delhi-110016, India
E-mail: [1]Dibakar.Rakshit@ces.iitd.ac.in,[2]jes162135@ces.iitd.ac.in

Abstract. There is a lag between the demand and supply of energy in today's global scenario. Apart from that, the world is facing serious environmental challenges like global warming due to increased burning of fossil fuels. To mitigate this, people are focusing towards the renewable based energy technology. But harnessing energy from the renewable sources has its own challenges. They are intermittent sources of energy hence, some energy storage medium is necessary for the continuous supply of energy. The present study aims to evaluate the various characteristics of the phase change materials used as energy storage mediums. Various thermo physical properties like melting temperature, enthalpy values and specific heats capacities are calculated using Differential Scanning Calorimetry technique for high temperature PCM materials. In this study, Sodium nitrate and Potassium nitrate are taken as the PCMs because of their high melting temperature above 300 °C and they are widely used in concentrated solar powerplants as a heat transfer fluids.

Keywords: Phase change materials(PCMs), thermophysical properties, $NaNO_3$,KNO_3

1. Introduction

The present study includes characterization study of phase change materials for thermal energy storage systems. For accessing the performance of PCMs used in different applications like solar distillation, passive storage, cooling systems etc. the knowledge of properties of PCMs is very essential. For the control of the performance of the PCM based material which is used in Thermal energy storage systems, inherent properties and characters of the material should be known to us. In this study, evaluation of the thermo physical properties of some selected PCMs (Sodium Nitrate and Potassium Nitrate) is done through an experimental set up using Differential Scanning Calorimetry (DSC) technique.

2. Experimental Setup

The experimental set up for determination of thermo physical properties consists of a Differential Scanning calorimeter. The DSC Q-2000 model by TA instruments is used for the thermal characterization of PCM materials.The principle of DSC is based upon the difference in thermal response when a temperature program is applied for the PCM sample and the reference material [1]. The DSC works on two modes, the first is step mode and the other one is dynamic mode. Different modes are recommended in various papers for different PCMs like for salt hydrates step mode is there and for paraffins both the modes can be operated [1].In the present work, dynamic mode of operation is used. In the setup, Indium is used as the reference material.

2.1 Software used for the analysis of the result

There are a number of software packages supported by DSC Q-2000 model by TA instruments like TRIOS, DSC Run, Advantage /Ultimate Analysis. For thermal analysis Ultimate Analysis is good as it can be used for multiple calibrations on integration with the autosampler [2]. In the context of present study, Ultimate analysis 2000 is used in the analysis of the DSC data.

2.2 Weighing and preparation of samples

Proper weighing and preparation of samples of known masses of Potassium Nitrate (KNO_3) and sodium nitrate ($NaNO_3$) is done using the electronic weighing balance and DSC crimping press. At the time of preparation of the samples handling of the PCM salt samples should be done with utmost care as they are hygroscopic in nature and any moisture in the samples will cause deviations in the experimental results. For the removal of the inherent moisture present in the samples, they were dried in the oven at about 90-100 °C.

2.3 DSC Calibration

Firstly, DSC is calibrated for measuring the thermo physical properties. Calibration is necessary for good sensitivity. There are three types of measurements:

- DSC Baseline measurement using both the empty pans.
- Using reference material i.e indium in this case.
- The actual PCM material measurements.

The two modes of operations are dynamic mode and step mode: In dynamic mode constant heating and cooling rate while in step mode small ramps are there. After calibration, the sample is prepared and kept inside the DSC pan for measurements.

3. Results

3.1DSC measurements for potassium nitrate samples

Firstly a sample of potassium nitrate having mass of 28.3 mg is prepared and tested at different heat rates which are 10 °C/ min, 15 °C/ min and 20 °C/ min respectively. The DSC is operated in dynamic mode at the given heat rates and the data is recorded for each of the heat rates.

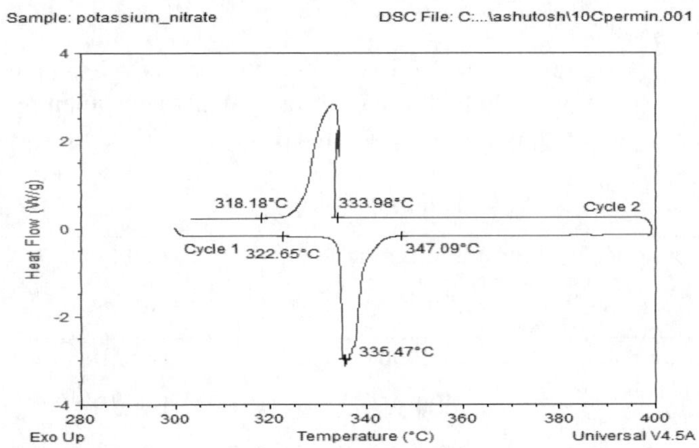

Figure 1. DSC heat flow graph for KNO_3 at heat rate of 10 °C/ min for sample mass 28.3 mg

Similarly, DSC heat flow graph is plotted for heat rates of rate of 15°C/ min and 20°C/ min for sample mass 28.3 mg. The nature of the DSC graphs for heat rates of 15°C/ min and 20°C/ min are similar to that of 10°C/ min, hence they are not shown as it would be of repetitive in nature. But all the thermophysical property values are given in the tabular form for appreciating the deviation in values. After plotting the DSC heat flow graph for sample mass of 28.3 mg , the mass of the potassium nitrate sample is changed for knowing the effect of decrease of sample mass on the DSC characteristic curves. The mass of the sample taken is 13.3 mg. In the heat flow curve in the melting region figure 1, the point of maximum slope on the leading side and then extrapolation of heat flow curve gives melting onset temperature [3]. In this way the melting temperature and onset temperature is measured. The integration of the curve within the melting range gives Latent heat of fusion of the sample (J/gm).

IOP Conf. Series: Materials Science and Engineering **404** (2018) 012016 doi:10.1088/1757-899X/404/1/012016

3.1.1Effect of temperature rate and sample mass on thermophysical properties of KNO_3 in DSC measurements

Table 1. Potassium Nitrate (KNO_3) Sample mass: 28.3 mg

S.No	Temperature Rate	Onset Temperature (in °C)	Melting temperature (in °C)	Latent heat (in J/g)
1	10 °C/min	334.65	335.58	98.71
2	15 °C/min	334.46	336.01	97.69
3	20 °C/min	334.45	336.75	96.76

Table 2. Potassium Nitrate (KNO_3) Sample mass: 13.3 mg

S.No	Temperature Rate	Onset Temperature (in °C)	Melting temperature (in °C)	Latent heat (in J/g)
1	10 °C/min	334.62	334.86	94.83
2	15 °C/min	334.47	336.18	94.38
3	20 °C/min	334.38	335.88	93.72

For the sample mass 28.3 mg the melting temperature increases on increasing the heat rate while in the second case sample having mass 13.3 mg the melting temperature is maximum for 15 °C/min heat rate. The onset melting temperature is approximately the same for both the sample masses while the melting temperature differs for both. The latent heat decreases slightly with the decrease in sample mass of KNO_3 taken for experimental testing. It decreases slightly with the increase in heat rate.

3.2 DSC measurements for sodium nitrate samples

The melting onset temperature does not differ much (306.55 ± 0.25 °C) for both the $NaNO_3$sample masses while the melting temperature shows variation in both the cases. In first case the melting temperature is maximum at 15 °C/min while for sample mass 11.6 mg it is maximum for 20 °C/min.

Table 3. Sodium nitrate ($NaNO_3$) Sample mass: 24 mg

S.No	Temperature Rate	Onset Temperature (in °C)	Melting temperature (in °C)	Latent heat (in J/g)
1	10 °C/min	306.42	309.38	173.50
2	15 °C/min	306.38	309.95	177.30
3	20 °C/min	306.51	308.62	175.50

Table 4. Sodium nitrate ($NaNO_3$) Sample mass: 11.6 mg

S.No	Temperature Rate	Onset Temperature (in °C)	Melting temperature (in °C)	Latent heat (in J/g)
1	10 °C/min	306.55	307.76	172.06
2	15 °C/min	306.43	310.50	176.50
3	20 °C/min	306.48	310.81	175.00

This deviation can be due to the impurities present in the samples. The latent heat decreases slightly with the decrease in sample mass taken for experimental testing. But the latent heat is maximum at the heat rate of 15 °C/min for both the sample masses.

The heat flow graphs for Sodium nitrate samples have almost similar nature as that of potassium nitrate in figure 1. But there are some deviations in the solidification curves which is explained in the section 3.2.2. Solidification curve is the portion of the curve when the PCM starts solidifying from the melted stage and reverses its direction in the heat flow graph.

3.2.1 Heat flow graph for sodium nitrate sample for different heat rates

The peaks of the DSC melting curves shift due to the different heat rates for the $NaNO_3$ sample as shown in figure 2. On increasing the heat rates the peak shifts towards the right. It is at minimum temperature for a heat rate of 10 °C/min and increases on increasing the heat rates. This is due to the kinetics of the melting process. Larger the heat rate broader the melting curve will be and the peak is shifted towards the right. Slower heating rate gives less noise in the signal and the resolution will be better than higher heating rates.

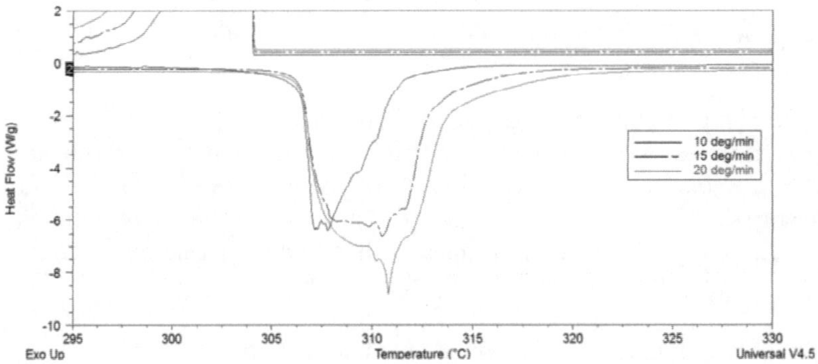

Figure 2. Combined DSC Heat flow graph for Sodium Nitrate ($NaNO_3$) for sample mass 11.6 mg

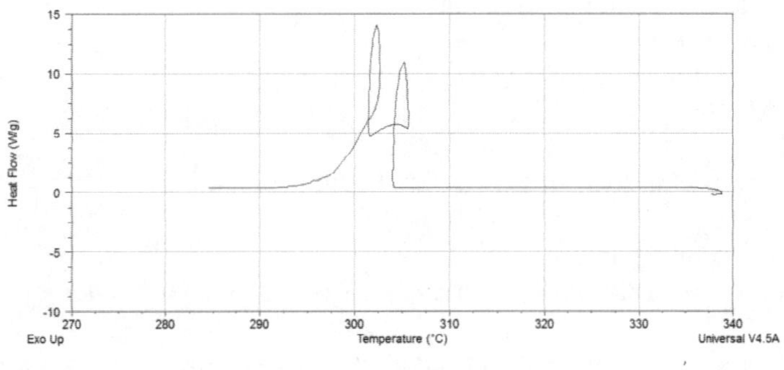

Figure 3. Solidification curve for Sodium Nitrate ($NaNO_3$) having sample mass 11.6 mg

3.2.2 Solidification Curve for Sodium Nitrate sample

The solidification curve for Sodium Nitrate ($NaNO_3$) is shown in figure 3. There is deviation from the ideal solidification graphs in this case. Normally the solidification graph consists of 1 peak but here There are 2 peaks in the solidification curve. This is due to a phenomenon known as polymorphism. This is because of the occurrence of different crystalline form of sodium nitrate in that range of temperature. The reason can be hygroscopic nature of the salt and humidity acts as catalyst for the polymorphism behaviour.

3.3Variation in Specific heat capacity of Sodium Nitrate above the melting temperature

Using the DSC measurement data, the specific heat capacity of sodium nitrate at different temperature points is found out. The temperature points at which the specific heat (C_p) values are taken are above the melting temperatures. The specific heat values show an interesting behaviour in the vicinity of the melting temperature. The function of specific heat with respect to temperature is an increasing one but the nature of increase is different for different heat rates. The variation is analysed for three heat rates which are 10 ˚C/min, 15 ˚C/min and 20˚C/min respectively, as shown in figure 4.On increasing the heat rates the specific heat of the sodium nitrate in the melted form decreases.

3.4 Empirical relationship between Specific heat (C_P) and Temperature (t) above melting temperature
Using the curve fitting technique, the set of experimental data points is fitted to a polynomial curve in accordance with the method of least square error.

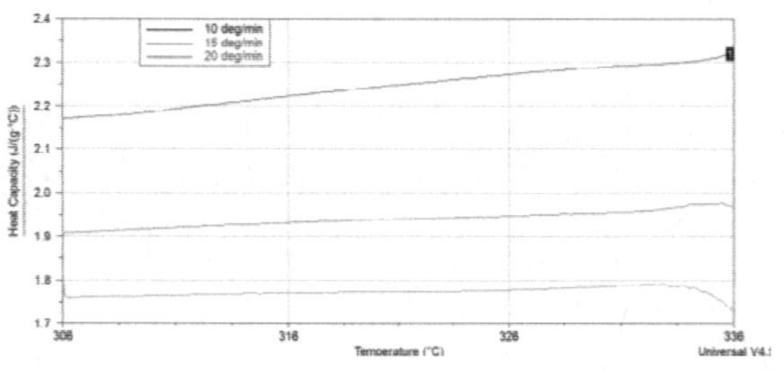

Figure 4. Variation of specific heat above melting temperature for $NaNO_3$ sample

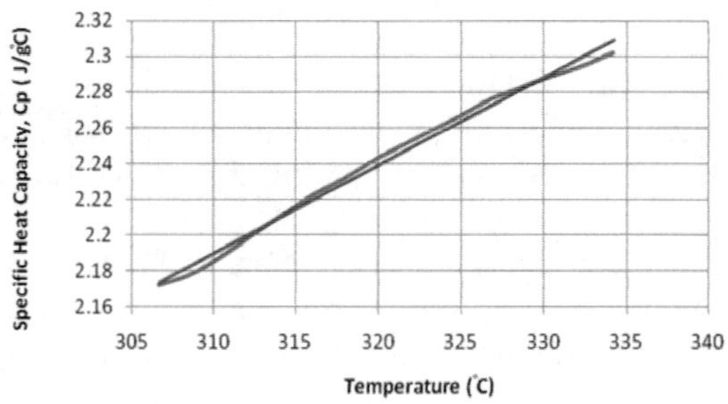

Figure 5. Fitting of curve for $NaNO_3$ at 10 °C/ min

For minimising the errors, and accurately predicting the degree of polynomial equation which is to be fitted to the data points, regression coefficients (R^2) are checked accordingly. Curves having R^2values near to one are termed as 'Good fit' and in our scope of calculations the R^2 values are in the range of 0.98-0.99. The mathematical relationship developed for specific heat (C_P) and temperature is of utmost importance when we have to predict the sensible heat storage capacity of the thermal energy storage reservoir and in its designing above the melting range of sodium nitrate.

In Figure 5, the variation of specific heat with temperature is plotted at a heat rate of 10 ˚C/ min.
The curve is fitted with a linear polynomial curve having equation

$$C_p = 0.004\, t + 0.665 \qquad .....(1)$$

Similarly, the variation of specific heat with temperature is plotted at the heat rates 15 °C/ min and 20 °C/ min. The curve is fitted with a 3rd order polynomial for 15 °C/ min heat rate having equation

$$C_p = 4\times 10^{-6}\, t^3 - 0.003t^2 + 1.254t - 132.2 \qquad ....(2)$$

The curve is fitted with a 4th order polynomial for 20°C/ min heat rate having equation

$$C_p = -3\times 10^{-7}t^4 + 0.000\, t^3 - 0.173\, t^2 + 37.10\, t - 2967 \qquad ....(3)$$

Where, C_p = Specific heat (in J/gm°C) and t= temperature in °C
$R^2 = 0.992$ for (1), $R^2 = 0.994$ for (2) and $R^2 = 0.982$ for (3) where, R^2 = Regression coefficient

4. Conclusion

The results are well in conformance with the theoretical data available within the error range of 5%.The sensitivity of DSC is high for smaller heat rates and 10 ˚C/ min heat rate gives the most accurate result among all. The present study provides valuable database regarding characteristics and properties of PCMs over varying experimental conditions. Also, there are no defined national or international standards for the characterization of PCMs. Every researcher applies his own methodology and approach for problem solving and the analysis of results.

References

[1] C. Barreneche, A. Solé, L. Miró, I. Martorell, A. I. Fernández, and L. F. Cabeza, "Study on differential scanning calorimetry analysis with two operation modes and organic and inorganic phase change material (PCM)," Thermochim. Acta, vol. 553, pp. 23–26, 2013.
[2] http://www.tainstruments.com/pdf/brochure/2012%20DSC%20Brochure%20r1.pdf, Brochure of DSC, Q-2000 by TA instruments.
[3] S. Mine, "Thermal characterization of phase change materials for thermal energy storage," 2016.

Fuzzy logic optimization of process parameters affecting surface temperature in MAF

Rajneesh Kumar Singh [1], D.K. Singh [2] and Swati Gangwar [3]
[1,2,3]Department of mechanical engineering, Madan Mohan Malaviya University of Technology, Gorakhpur, 273010, India
Email - rajneeshsingh_rsme@mmmut.ac.in

Abstract. Magnetic abrasive finishing process was introduced to boost surface quality of super-alloys, composites and ceramics. This process erode material from the surface of workpiece in micro-nano range. Surface texture achieved from this process was very excellent. Thermal stability of finishing surface of workpiece is vital aspect during the finishing of MAF process because it affects the surface texture, surface integrity, finishing accuracy and material removal. In present study, experimentation had been performed to measure the surface temperature of finishing surface of mild steel using MAF process. Taguchi L_9 orthogonal array had been employed to design the experiment. Current, working gap, abrasive weight and rotational speed were important process parameters that had been selected for analysis. Fuzzy logic had been used to develop Artificial intelligent based predictive model to predict surface temperature during finishing operation in MAF process. Surface temperature model built on Fuzzy logic had been validated.

Keywords. Magnetic Abrasive Finishing, Surface Temperature, L_9 orthogonal array, Fuzzy logic.

1.Introduction

Magnetic abrasive finishing was introduced to boost surface quality of advance and harder materials [1]. Finishing was achieved in this process by moving magnetic abrasive particle which were homogenous mixture of large ferromagnetic particles with small abrasive particles (Silicon carbide) mixed in specified ratio with help of strong magnetic field. These magnetic particles were positioned into working gap which lies between top surface of workpiece and magnetic poles [2]. Due to strong magnetic field present in working gap, magnetic particles aligned with magnetic forces forming flexible magnetic abrasive brush. Finishing of surface was achieved by relative motion between workpiece and flexible magnetic abrasive brush [3]. Flexible magnetic abrasive brush acted as multi-point cutters responsible for material removal from top surface of workpiece. Material removal was done in form of microchips which were created by tangential cutting force of magnetic brush [4]. This process causes substantial amount of heat on the surface of workpiece when finishing was performed. This happens due to deformation in material, fracture and friction when top surface of workpiece interacts with magnetic abrasive brush.

In past, two theoretical temperature analyses were done using Silicon nitride ceramic (Si_3N_4) as workpiece and Chromium oxide (Cr_2O_3) as abrasive particles. First was done by Komanduri et al. in which they derived analytical thermal solution by using standard moving heat theory of Jaeger. This model substantially calculated flash temperature produced during machining and flash time at interaction point [5]. Other attempt was made by Kumar et al. and they developed Finite element model to study the thermal behaviour of surface temperature. Conclusions of these studies were that temperature elevated due to escalation in magnetic particle velocity & magnetic potential [6]. So far, experimental work on temperature analysis was done by Mulik et al. on the workpiece magnetic brush interface during Ultrasonic MAF process (UAMAF) and they took ferrous alloy as workpiece for conducting experiment. Key finding of their study was that, temperature directly depends on voltage, weight of abrasive particles and pulse on time [7].

IOP Conf. Series: Materials Science and Engineering **404** (2018) 012017 doi:10.1088/1757-899X/404/1/012017

Literature review enlighten that so far, few experimental study had been performed for temperature analysis using MAF process. Present study focuses on surface temperature analysis of MAF process using Mild steel and compare it with artificial intelligent based fuzzy logic model developed for the surface temperature prediction [8]. In this research work, experiment had been performed on MAF set-up to measure surface temperature between mild steel workpiece and magnetic abrasive brush. Taguchi L_9 orthogonal array had been used to design the organized experiments. The main process parameters that considered were working gap, abrasive weight, current and rotational speed. Artificial intelligent based modelling had been done using Fuzzy logic for the surface temperature prediction. Finally, validation of predictive model had been done by comparing it with experimental work which was performed to study surface temperature effect on the surface of top surface of workpiece.

2. Experimentation of surface temperature of mild steel using MAF process

In this section detailed reporting had been done about the experimental set-up which had been fabricated in-house, about selection of workpiece material and important process factors on which surface temperature depends, how temperature measurement had been done and finally, thorough experimental procedure had been discussed.

2.1 Experimental setup of MAF process

The MAF set up was comprised of flat faced electromagnets having 1000 turn of copper wire wounded around iron core. This two-poled electromagnet was mounted on pillar drilling machine. Pillar drilling machine was used to provide rotational motion to electromagnet with speed range between 100-300 rpm. Direct current (3A – 6A) was provided by DC power supply which provided static direct current to electromagnet. Static direct current generated 0 - 1.02 T of magnetic flux density in working gap of 1.00 -2.00 mm. A digital gaussmeter (model EMF-PORTABLE, range 0-2.0 T) was used to measure magnetic flux density in the working gap and variation in the strength of magnetic flux density was done by varying the input current. Formation of magnetic abrasive brush in working gap took place when current was supplied to electromagnet. Mechanical mixture of iron powder (300 mesh number) and silicon carbide (400 mesh number) were used to create magnetic abrasive brush and magnetic abrasive particle used for magnetic brush was Unbounded in nature. The process parameters were based on preceding works conducted in this field and range of input parameters listed in table (1) were selected on basis of setup constrained.

Fig.1 k-type thermocouple

Table 1- Process input parameters

Sr no	Process parameter	Range
1	Working gap (mm)	1.0-2.0
2	Abrasive weight (gm)	20-30
3	Current (A)	3.0-6.0
4	Rotational speed of magnet (rpm)	100-300

2.2 Experimental procedure for surface temperature

The mild steel workpiece was fastened on tool vice mounted on the table of pillar drill machine. The required working gap between bottom electromagnet pole and top surface of mild steel workpiece was measured by using slip gauges. Unbounded magnetic abrasive particles were used to form the magnetic abrasive brush in working gap. Weight of total magnetic abrasive particles was 100 gm and silicon carbide were mixed in required proportion i.e. 20 gm to 30 gm in iron powder. For measuring temperature for present thermal experiment K-type thermocouple made of chromel and alumel had been used as shown in figure (1) [9]. For measuring surface temperature of mild steel, through holes have been drilled into the workpiece. In these holes K-types thermocouples have been inserted and actual position of holes were obtained by measuring variation in magnetic flux density in the working gap using magnetic flux meter at different position [10]. Literary, no feed had been provided to the workpiece and its value has been kept to zero. While performing pilot experiment, it was seen that raise in temperature had become constant after 20 odd minutes. Hence the experiments were conducted for 20 minutes only. The maximum surface temperature in mild steel was found to be 41° C while the ambient temperature was 23 ^{0}C as shown in table (2).

Table 2 – L_9 Orthogonal Array experimental result

Trail no	Working gap (mm)	Abrasive weight (gm)	Current (A)	Speed (rpm)	Maximum Temperature (^{o}C)
1	1.0	20	3.0	100	30
2	1.0	25	4.5	200	35
3	1.0	30	6.0	300	41
4	1.5	20	4.5	300	32
5	1.5	25	6.0	100	37
6	1.5	30	3.0	200	34
7	2.0	20	6.0	200	33
8	2.0	25	3.0	300	30
9	2.0	30	4.5	100	31

2.3 Regression Analysis using Taguchi L_9 Orthogonal Array for surface temperature

Minitab software was used for Regression analysis using Taguchi L_9 Orthogonal Array for obtaining the significant process parameters and the rank of process parameters. Rank of temperature analysis of different input parameters has been shown in table (3). The current had largest contribution in the temperature of workpiece surface. Followed by working gap and abrasive weight. Speed of the magnetic abrasive brush had least impact on surface temperature.

Table 3 - Response Table for Means

Level	Gap	Weight	Current	Speed
1	35.33	31.67	31.33	32.67
2	34.33	34.00	32.67	34.00
3	31.33	35.33	37.00	34.33
Delta	4.00	3.67	5.67	1.67
Rank	2	3	1	4

ANOVA was done to obtain the percentage contribution of the process parameters to the maximum temperature of mild steel in MAF process. Table (4) maximum contribution was of current which had 48.16% of total contribution. Then working gap with 24%, abrasive weight with 20.16% and least contribution was made done by speed with 4.16%. Finally, table (5) shows the value of coefficient which had been obtained by the regression analysis. These coefficients were used to form the regression equation of the surface temperature of mild steel.

IOP Conf. Series: Materials Science and Engineering **404** (2018) 012017 doi:10.1088/1757-899X/404/1/012017

Table 4 - Analysis of Variance

Source	DF	Adj SS	Adj MS	F-Value	P-Value
Regression	4	96.500	24.125	12.87	0.015
Weight	1	20.167	20.167	10.76	0.031
Current	1	48.167	48.167	25.69	0.007
Speed	1	4.167	4.167	2.22	0.210
Gap	1	24.000	24.000	12.80	0.023
Error	4	7.500	1.875		
Total	8	104.000			

Table 5- Coefficients

Term	Coef	SE Coef	T-Value	P-Value	VIF
Constant	20.33	3.86	5.27	0.006	
Weight	0.367	0.112	3.28	0.031	1.00
Current	1.889	0.373	5.07	0.007	1.00
Speed	0.00833	0.00559	1.49	0.210	1.00
Gap	-4.00	1.12	-3.58	0.023	1.00

Regression Equation:

$$\text{Temperature} = 20.33 + 0.367\ \text{Weight} + 1.889\ \text{Current} + 0.00833\ \text{Speed} - 4.00\ \text{Gap}$$

3.Fuzzy logic modelling of surface temperature of mild steel.

Fuzzy logic was systematic methodology introduced by Lot fi Zadeh in year 1965 to solve problem which had incomplete, ambiguous, noisy values and it provided an easy means to attain a conclusion [11]. Fuzzy logic's tactics was synonymous to a human making decisions, but in a faster mean. This System consisted of five vital parts, that were - Fuzzifier, Membership functions, Rules, Inference System, De-fuzzifier [12].

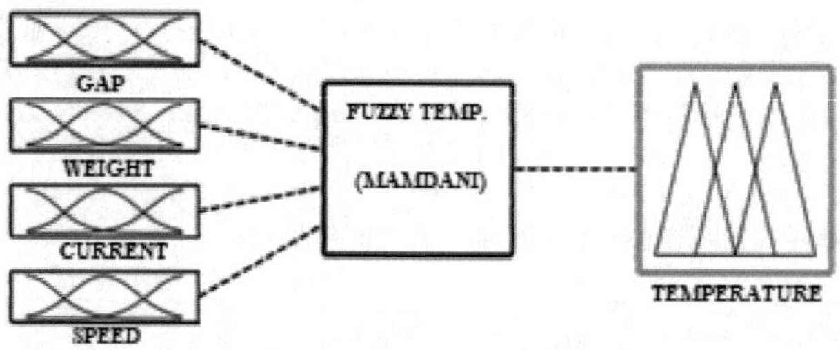

Fig.2: Fuzzy logic model

The modelling of surface temperature of mild steel workpiece in MAF process had been done by using fuzzy logic interface system of MATLAB. For present modelling, Mamdani fuzzifier had been used as shown in figure (2). Fuzzy model consisted of four process parameters (working gap, abrasive weight, current and rotational speed) as input to Mamdani fuzzifier and one response parameter i.e. Surface Temperature as out to Mamdani fuzzifier.

3.1 Membership functions option

The triangular membership functions were used for defining the range of input process and output response parameters. Range for process parameters were taken as follows, Working gap between 1.00-2.00 mm, Abrasive weight between 20-30 gm, Current between 3.0-6.0 A and Rotational speed between 100-300 rpm. Three membership functions were created for every input process parameters (working gap, abrasive weight, current and rotational speed). These three-membership function were assigned as Low, Medium and High.

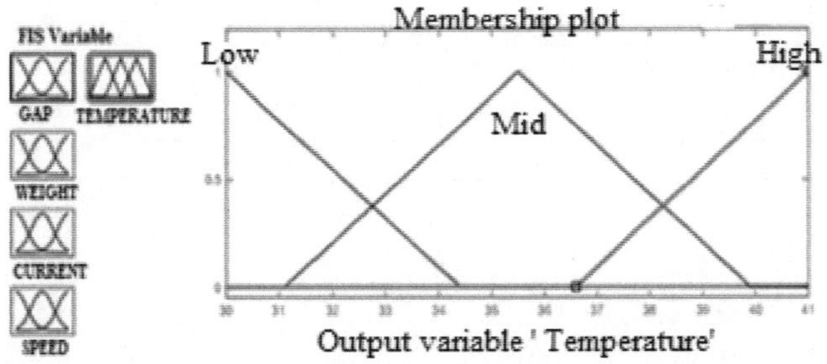

Fig. 3 Membership Functions for response parameter Surface Temperature

Membership functions for output response surface temperature of the workpiece temperature had been shown in figure (3). The range was taken between 30-41 ^{0}C for surface temperature. Again, three-membership function were allocated as Low, Medium and High.

3.2 Defining Fuzzy Rules sets

To model surface temperature in fuzzy system, rules had to be properly defined in the Mamdani fuzzifier. For obtaining optimized solution, rules had been defined by using Taguchi L_9 orthogonal arrays of experimental results. Total nine rules were defined as per fuzzy rules and these rules were fed to Mamdani fuzzy inference system in fuzzy logic toolbox, MATLAB 2014Ra. The set of rules along with membership function was seen in rule viewer of fuzzy model. Rule viewer revealed that after the devising of rules, the optimum value of surface temperature at any set between low and high bounds of the process parameter could be easily predicted.

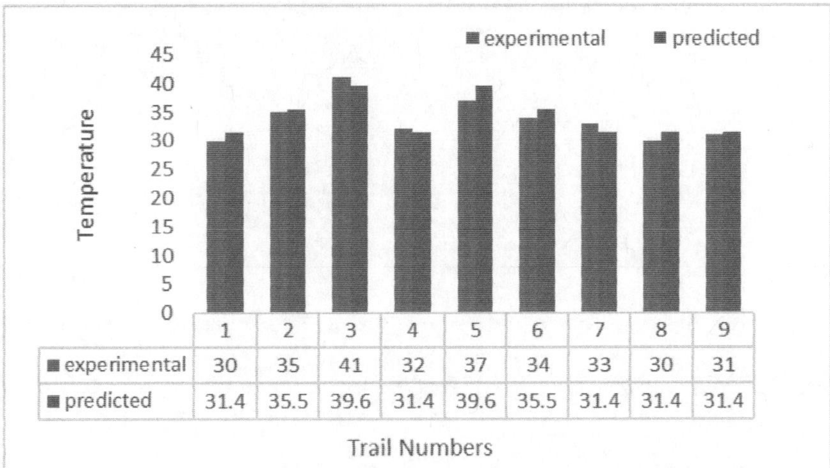

	1	2	3	4	5	6	7	8	9
experimental	30	35	41	32	37	34	33	30	31
predicted	31.4	35.5	39.6	31.4	39.6	35.5	31.4	31.4	31.4

Fig 4– predictive values of fuzzy logic to Taguchi experimental values

Figure (4) is bar chart comparison between different process parameters of experimental values to different process parameters values that had been predicted by Fuzzy logic model for same sets of process parameters. The closest predict was done by fuzzy model was at trail number 8 with error around 1.29% and farthest prediction was at trail number 5 with error -7.02%.

3.3 Validation of fuzzy model

Validation of the predictive fuzzy model had been done by conducting confirmatory test and the process parameters were taken different from Taguchi L_9 experimental test. From table (6) shows that surface temperature was 34 ^{0}C in confirmatory test. And fuzzy logic model predicted 35.5 ^{0}C and error between confirmatory test and predictive model was 4.5%. Therefore, artificial intelligent based fuzzy logic was found to be good accordance with experiment test.

Table 6- Validation of fuzzy surface temperature model

Working gap	Abrasive weight	Current	Speed	Experimental temperature	Predicted temperature
1.0	25	3	200	34	35.5

4. Results and discussion

Experimental investigation of surface temperature of mild steel at ambient temperature was 23 ^{0}C had been performed using MAF process. it revealed that finishing of mild steel generated low surface temperature because thermal conductivity of mild steel good. The maximum temperature was found to 41^{0}C when process parameters were at working gap 1mm, abrasive weight 30gm, Current 6 A and speed 300 rpm.

Taguchi experimental analysis exhibited the effect of working gap on surface temperature is shown in figure (5). With increase in working gap surface temperature decreases because magnetic flux in the working gap becomes weaker. But surface temperature increases with increase of SiC abrasive weight, Current and Rotational speed of electromagnet.

The most significant process parameters that affected the surface temperature were current and abrasive weight. When current increases, Magnetic abrasive brush becomes stiffer with increase of magnetic flux leading to increase in normal magnetic force on magnetic abrasive particles. Further, Rotational speed and working gap had less significant.

Figure (4) gives the assessment between the performed experimental results and predicted results using fuzzy logic model. Closest experimental value to predicted fuzzy logic model was found to be 1.29% for surface temperature. Thus, the predictive model gave more than 90% accuracy. Hence, there is excellent agreement between fuzzy logic model and experimental results in all cases.

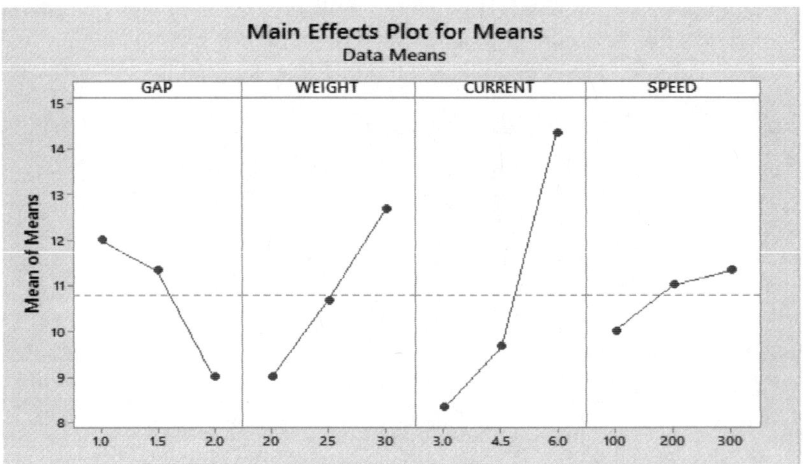

Fig 5. Taguchi mean plot of process parameters

5. Conclusions

1. The maximum surface temperature of mild steel in MAF process in present case is found to be low and the maximum temperature gained in experiment was of 41^{0}C. Reason for low surface temperature can be understand that mild steel has fair thermal conductivity which swiftly dissipate heat from the finishing surface of the workpiece.
2. Regression analysis is done to find the most significant process parameters that affect the surface temperature are Current and Working gap of magnetic brush. Abrasive weight and Rotational speed are found to be less influential factors.

3. ANOVA tells the percentage contributions of process parameters on the surface temperature. Current has 48.16% of total contribution then follow by working gap with 24%, abrasive weight with 20.16% and speed with 4.16%.
4. Artificial intelligent based fuzzy logic model had been developed for prediction of surface temperature. Fuzzy logic model was simpler to evaluate and successfully predicted the surface temperature.
5. Predictive Fuzzy logic model of surface temperature has been validated with experimental results and there has been good accordance with experimental results. The overall error was in range of -7% to 5%.

References

[1] Fox. M, Agrawal. K., Shinmura.T and Komandari R. (1994) *Magnetic Abrasive Finishing of Rollers* Annals *of the CIRP* 43/1, 181-184.
[2] Singh D.K., Jain V.K., Raghurama V. and Komanduri R. (2004) *Analysis of Surface Texture Generated by a Flexible Magnetic Abrasive Brush Wear* 259 1254–1261.
[3] Singh D. K., Jain V. K. and Raghuram V. (2004) *Parametric Study of Magnetic Abrasive Finishing Process Journal of Materials Processing Technology* 149(1-3) 22-29.
[4] Singh D K, Jain V K Raghuram V and Komanduri R (2005) *Analysis of surface roughness and surface texture generated by pulsating flexible magnetic abrasive brush World Tribology congress III WTC* 2005-63134.
[5] Hou. Z.B. and Komanduri R. (1998) magnetic *Field assisted finishing of ceramics – part III: thermal model ASME Journal of Tribolgy* 120 660-667.
[6] Kumar G. and Yadav V (2008) *Temperature distribution in the workpiece due to plane magnetic abrasive finishing using FEM International Journal of Advance Manufacturing Technology* 41 1051-1058.
[7] Mulik R S, Srivastava V and Pandey P M. (2012) *Experimental Investigations and Modelling of Temperature in the Work-Brush Interface during Ultrasonic Assisted Magnetic Abrasive Finishing Process Materials and Manufacturing Processes* 27:1 1-9.
[8] Oh J.H. and Lee S.H. (2010) *Prediction of surface roughness in magnetic abrasive finishing using acoustic emission and force sensor data fusion Proceedings of the Institution of Mechanical Engineers Part B: Journal of Engineering Manufacture Proc. IMechE* 225 853- 865.
[9] Davies M.A., Ueda. T., M'Saoubi, R., Mullan B. and Cooke A.L. (2007) *On the measurement of temperature in material removal processes Annals of CIRP* 56(2) 581–604.
[10] Shen B., Xiao. G., Guo. C., Malkin. S. and Shih A.J. (2008) *Thermocouple fixation method for grinding temperature measurement. J. Manuf. Science Engineering ASME* 130 051014.
[11] L.A. Zadeh (1965) *Fuzzy sets Information and control 8* 338–353.
[12] Kanish T.C., Kuppan P., Narayanan S., Denis and Ashok S. (2014) *A Fuzzy Logic based Model to predict the improvement in surface roughness in Magnetic Field Assisted Abrasive Finishing 12th Global Congress on Manufacturing and Management Procedia Engineering* 97 1948 – 1956.

IOP Conf. Series: Materials Science and Engineering **404** (2018) 012018 doi:10.1088/1757-899X/404/1/012018

Thermal buckling analysis of laminated hybrid beam with variable piezoelectric layer thicknesses

N Rahman[1] and M N Alam[2]
[1,2] Department of Mechanical Engineering, Zakir Husain College of Engineering and Technology, A. M. U. Aligarh, Uttar Pradesh, India.

E-mail: [1]najeebalig@rediffmail.com,[2]naushad7863@rediffmail.com

Abstract. Thermal buckling analysis of a laminated hybrid beam with surface mounted piezoelectric layers is presented in this work. A one dimensional finite element (1D-FE) model based on efficient layerwise (zigzag) theory is used for the analysis. Two noded and three noded 1D elements are utilized for interpolating electromechanical and thermal variables respectively. Piezoelectric layers are bonded to top and bottom surfaces of the elastic substrate of hybrid beam.The beams are subjected to uniform temperature with closed circuit condition at top and bottom surfaces. The thicknesses of piezoelectric layers are varied and its effect on critical buckling temperatures is studied. Results are presented for composite and sandwich beam configurations under simply supported boundary conditions. The 1D-FE results are compared with the 2D-FE results obtained using commercial FE package ABAQUS.

1. Introduction

Temperature variations cause significant shape distortions and thermal stresses in laminated composite and sandwich structures. Electro thermal load induce pre-buckling transverse normal strain. This may have significant effect on the buckling temperature depending on the material properties. Designing and analysis of such practical smart structures inevitably require the use of finite element (FE) method. Khdeir [1] studied thermal buckling of cross-ply laminated composite beams. He developed the state space concept in conjunction with Jordan canonical form. Metin Aydogdu [2] presented the thermal buckling analysis of cross-ply laminated composite beams subjected to different boundary conditions on the basis of a unified three degree of freedom shear deformable beam theory. Kapuria and Alam [3] developed an efficient electromechanically coupled geometrically nonlinear zigzag theory for buckling analysis of hybrid piezoelectric beams, under electro-thermo-mechanical loads. Zhong et al. [4] investigated the thermal buckling and postbuckling behaviour of composite plates with embedded SMA wire actuators using a finite element analysis. Chen and Chen [5] investigated thermal buckling of laminated cylindrical shells and determined the critical temperature under clamped and simply supported end conditions.

A one dimensional finite element (1D-FE) model of hybrid piezoelectric beam is presented for buckling analysis under thermal load. The coupled efficient layerwise (zigzag) theory [3] is used for making the model. The thicknesses of piezoelectric layers bonded to the top and bottom surfaces of the beam are varied and its effects on critical buckling temperatures is studied. Hybrid composite and

---

[1] To whom any correspondence should be addressed.

sandwich beam configurations under simply supported boundary conditions are considered for the analysis. 1D-FE results are compared with the 2D-FE results obtained using commercial FE package ABAQUS.

2. One dimensional Finite element model of hybrid piezoelectric beam

Consider a hybrid beam subjected to thermal load, transverse load and actuation potentials applied to some piezoelectric layers. The load distribution is independent of the width coordinate. The potential, thermal and displacement field variables ϕ, θ, u_x and u_z are assumed [3] in terms of the three primary variables of the reference plane, u_{x_0}, u_{z_0}, ψ_0 the potential variables, ϕ^l and the thermal variable, θ^j as

$$\phi(x,z,t) = \xi^l_\phi(z)\phi^l(x,t) \tag{1}$$

$$\theta(x,z,t) = \xi^j_\theta(z)\theta^j(x,t) \tag{2}$$

$$\begin{aligned} u_x(x,z,t) &= u_{x_0}(x,t) - z\frac{\partial u_{z_0}}{\partial x}(x,t) + S^k(z)\psi_0(x,t) \\ &+S^{kl}(z)\frac{\partial \phi^l}{\partial x}(x,t) + \bar{S}^{kj}(z)\frac{\partial \theta^j}{\partial x}(x,t) \end{aligned} \tag{3}$$

$$u_z(x,z,t) = u_{z_0}(x,t) - \bar{\xi}^l_\phi(z)\phi^l(x,t) + \bar{\xi}^j_\theta(z)\theta^j(x,t) \tag{4}$$

where $\xi^l_\phi(z)$ and $\xi^j_\theta(z)$ are linear interpolation functions for through the thickness variation of potential field ϕ and thermal variable θ; $\bar{\xi}^l_\phi(z)$ and $\bar{\xi}^j_\theta(z)$ are piecewise linear and quadratic functions respectively for electric and thermal fields; $S^k(z), S^{kl}(z)$ and $\bar{S}^{kj}(z)$ are the cubic functions of z. The axial displacement, u_x and transverse shear stress, σ_{zx} are continuous at the layer interfaces and the displacement field ensures shear traction free condition at the top and bottom surfaces of beam

The variational equation, using the extended Hamilton's principle [6] for the beam, is obtained as:

$$\begin{aligned} &\int_0^a \left[\left\{\delta\tilde{\varepsilon}_1^T \quad \delta\tilde{\varepsilon}_5^T \quad \frac{\partial\delta\phi^l}{\partial x} \quad \delta\phi^l\right\}\bar{F} - \delta\tilde{u}_2^T g_{u\phi} - \delta\tilde{u}_2^T\{\hat{q}_2 \quad -\hat{q}_4^l\}^T + \left(\frac{\partial\delta u_{z_0}}{\partial x}N_x\frac{\partial u_{z_0}}{\partial x}\right)\right]dx - \\ &\left[N_x^*\delta u_{x_0}^* + V_x^*\delta u_{z_0}^* - M_x^*\frac{\partial\delta u_{z_0}^*}{\partial x} + P_x^*\delta\psi_0^* + \left(H^{l^*} - V_\phi^{l^*}\right)\delta\phi^{l^*} + R_x^{l^*}\frac{\partial\delta\phi^{l^*}}{\partial x}\right]\Bigg|_0^a \end{aligned} \tag{5}$$

where $\bar{F}$ is the generalized stress vector and $g_{u\phi}$ is the electromechanical load vector of the beam; $\hat{q}_2$ and $\hat{q}_4^l$ are the damping loads; $\tilde{u}_2$ is the generalised displacement vector; $\tilde{\varepsilon}_1$ and $\tilde{\varepsilon}_5$ are generalised beam mechanical strains; N_x, M_x, P_x, R_x^l and V_x, V_ϕ^l are the beam stress resultants of σ_x and σ_{zx} respectively; H^l and G^l (discussed later) are the beam electric displacement resultants. The superscript * means values at the ends.

A two noded beam element [7] is considered for electromechanical variables and three noded element for thermal variable [8]. The primary variables $u_{x_0}, u_{z_0}, \psi_0, \phi^l$ and θ^j are interpolated in an element of length l as

$$\begin{Bmatrix} u_{x_0} \\ u_{z_0} \\ \psi_0 \\ \phi^l \end{Bmatrix} = \begin{bmatrix} \tilde{N} & 0 & 0 & 0 \\ 0 & \hat{N} & 0 & 0 \\ 0 & 0 & \tilde{N} & 0 \\ 0 & 0 & 0 & \tilde{N} \end{bmatrix} \begin{Bmatrix} u_{x_0}^e \\ u_{z_0}^e \\ \psi_0^e \\ \phi^{l^e} \end{Bmatrix}, \qquad \theta^j = \bar{N}\theta^{j^e} \tag{6}$$

where $\tilde{N}$ is a vector of linear Lagrange interpolation functions, $\hat{N}$ is a vector of cubic Hermite interpolation functions and $\bar{N}$ is a vector of quadratic Lagrange Interpolation functions; $u_{x_0}^e, u_{z_0}^e, \psi_0^e, \phi^{l^e}$ and θ^{j^e} are vectors of nodal values.The element generalized displacement vector, d^e is defined as

$$d^{e^T} = \left\{u_{x_0}^{e^T} \quad u_{z_0}^{e^T} \quad \psi_0^{e^T} \quad \phi^{l^{e^T}}\right\} \tag{7}$$

The contribution T^e of one element of length l to the integral from 0 to l in Eq. (5) for the case of static inplane electro-thermo-mechanical load is obtained as

$$T^e = \int_0^l \left\{\delta\tilde{\varepsilon}_1^T F_1 + \delta\tilde{\varepsilon}_5^T F_5 + \left\{\frac{\partial\delta\phi^l}{\partial x} \quad \delta\phi^l\right\}\{H^l \quad G^l\}^T + \delta\tilde{u}_2^T {q_4^l}^T\right\} dx \tag{8}$$

where F_1 and F_5 are the vectors of beam stress resultant of σ_x and σ_{zx} respectively and q_4^l is the electrical load.

For buckling analysis, the transverse loads applied on the bottom and top surfaces of the beam and hence total transverse mechanical load are considered as zero. With this consideration the Eq. (8) may be simplified as:

$$T^e = \int_0^l \left[\delta\bar{\varepsilon}^T\left(\bar{D}\bar{\varepsilon} + g_{\theta_2}^j \frac{\partial^2\theta^j}{\partial x^2} + g_{\theta_1}^j \frac{\partial\theta^j}{\partial x} + g_{\theta_0}^j \theta^j\right) - \delta\tilde{u}_2^T(\bar{g}_{u\phi})\right] dx \tag{9}$$

where only electric load is considered to obtain the vector $\bar{g}_{u\phi}$ as

$$\bar{g}_{u\phi} = \{0 \quad -\bar{q}_4^l\}^T \tag{10}$$

with

$$\bar{q}_4^l = b\left[D_{z_N}\delta_{ln_\phi} - D_{z_0}\delta_{l1} + q_{li}\delta_{ll_i}\right] \tag{11}$$

$\bar{D}$ is the generalized stiffness matrix of the beam, $g_{\theta_2}^j, g_{\theta_1}^j, g_{\theta_0}^j$ are the generalized stress-temperature coefficient matrices.

The generalised strain, $\bar{\varepsilon}$ and the displacement vector, $\tilde{u}_2$ are interpolated as

$$\bar{\varepsilon} = \bar{B}d^e, \quad \tilde{u}_2 = \bar{B}_{m_2}d^e \tag{12}$$

Here $\bar{B}$ and $\bar{B}_{m_2}$ are strain displacement and displacement interpolation matrices respectively.

Substituting the expressions of θ^j ,$\bar{\varepsilon}$ and $\tilde{u}_2$ from Eqs. (6) and (12) respectively, we get

$$T^e = \int_0^l \left[\delta {d^e}^T \bar{B}^T\left(\bar{D}\bar{B}d^e + g_{\theta_2}^j \frac{\partial^2\bar{N}}{\partial x^2}\theta^{je} + g_{\theta_1}^j \frac{\partial\bar{N}}{\partial x}\theta^{je} + g_{\theta_0}^j \bar{N}\theta^{je}\right) - \delta {d^e}^T \bar{B}_{m_2}^T(\bar{g}_{u\phi})\right] dx = \int_0^l \delta {d^e}^T\left[\bar{B}^T\bar{D}\bar{B}d^e + \bar{B}^T\left(g_{\theta_2}^j \frac{\partial^2\bar{N}}{\partial x^2} + g_{\theta_1}^j \frac{\partial\bar{N}}{\partial x} + g_{\theta_0}^j \bar{N}\right)\theta^{je} - \bar{B}_{m_2}^T\bar{g}_{u\phi}\right] dx = \delta {d^e}^T[K^e d^e + \bar{P}_\theta^e - \bar{P}^e] \tag{13}$$

where K^e, $\bar{P}^e$ and $\bar{P}_\theta^e$ are given by

$$K^e = \int_0^l \bar{B}^T\bar{D}\bar{B}dx, \; \bar{P}^e = \int_0^l \bar{B}_{m_2}^T\bar{g}_{u\phi}dx, \; \bar{P}_\theta^e = \left(\int_0^l \bar{B}^T\left(g_{\theta_2}^j \frac{\partial^2\bar{N}}{\partial x^2} + g_{\theta_1}^j \frac{\partial\bar{N}}{\partial x} + g_{\theta_0}^j \bar{N}\right) dx\right)\theta^{je} \tag{14}$$

Thus the general equation after integration is obtained as:

$$K^e d^e = \bar{P}^e - \bar{P}_\theta^e \tag{15}$$

where

$$\bar{P}^e = \left\{0 \quad 0 \quad a_{18}\bar{q}_\phi^{l^e}\right\}^T \tag{16}$$

with

$$\bar{q}_\phi^{l^e} = \{\bar{q}_{\phi_1}^{l^e} \quad \bar{q}_{\phi_2}^{l^e}\}^T, \qquad a_{18} = b\int_0^l \tilde{N}^T\tilde{N}dx \tag{17}$$

3. Results and discussion

3.1. Validation

The present 1D-FE formulation of zigzag theory is first validated for buckling response under thermal loading with simply supported hybrid piezoelectric beam of Ref. [3]. Beam configurations (b) and (c) and pre-buckling load case 1 of Ref. [3] is considered for validation. The results are compared (Table-1) for span to thickness ratio $S = l/h = 20$.

Critical temperature, $\bar{\theta}_{cr}$ for fundamental buckling mode, $n = 1$						
S	Beam b			Beam c		
	1D FE	Ref [3]	% error	1D FE	Ref [3]	% error
20	17.415	16.9378	2.817	51.550	49.5434	4.050

Table 1: Validation of critical temperature, $\bar{\theta}_{cr}$ with simply supported beams of Ref. [3]

The %age of difference is 4.05 which shows the correctness of the present (1D-FE) formulation.

3.2. Numerical Example

The hybrid beams (d) and (e) [9] with composite and sandwich substrate respectively are considered for the analysis. Piezoelectric layers (PZT-5A) of thicknesses h_p are bonded at the top and bottom surfaces of the substrate in both the beams. h_s is the thickness of elastic substrate and h is the total beam thickness. +z is considered to be the poling direction for piezoelectric layers. The surfaces of the substrate where PZT-5A layers are bonded are grounded. The composite substrate (d) has four numbers of $0.25h_s$ thick graphite epoxy layers with layup$[0^o/90^o/90^o/0^o]$. The sandwich substrate (e) has three layers, an intermediate soft core and top-bottom face sheets of thicknesses $0.08h_s/0.84h_s/0.08h_s$. h_p is varied from $0.1h$ to $0.3h$ keeping h as constant .

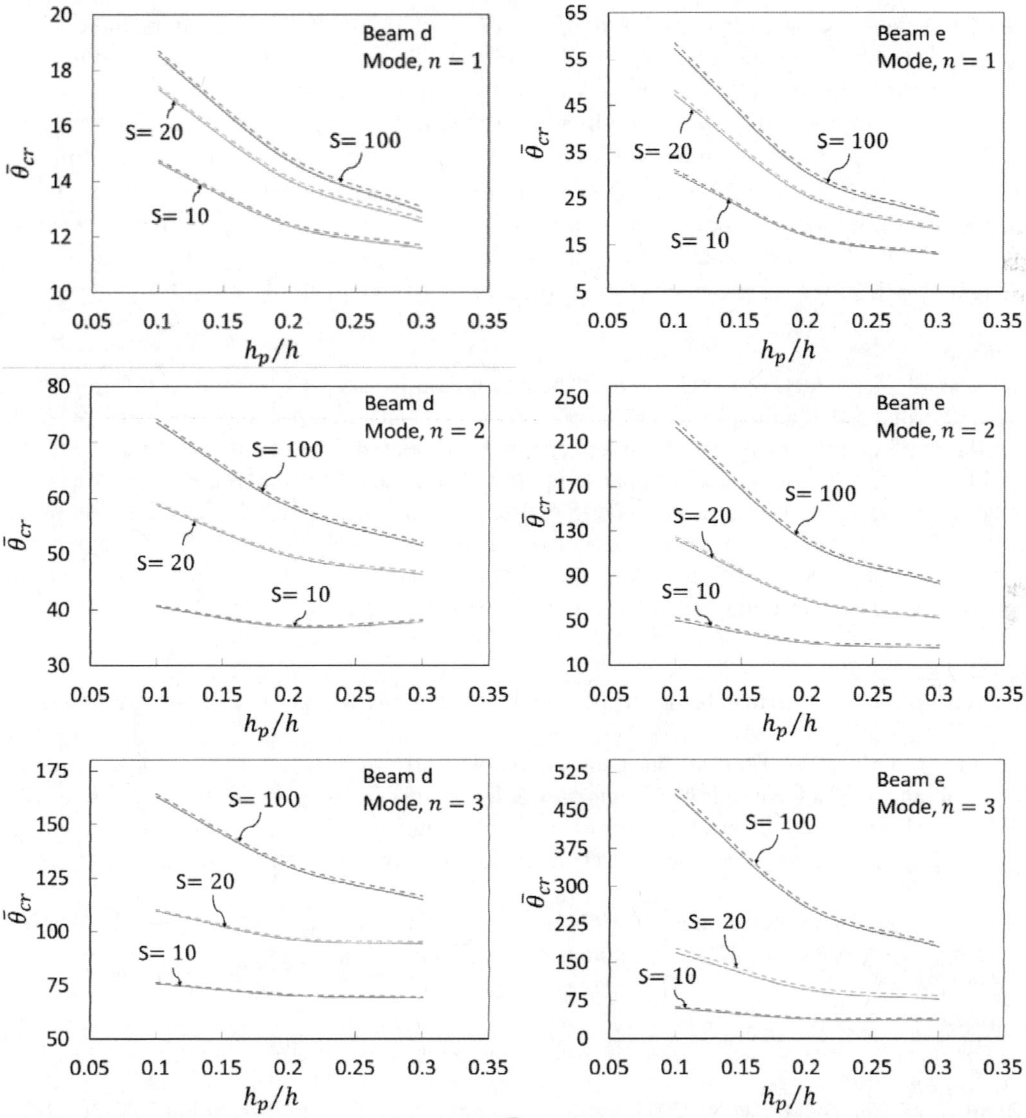

Figure 1: Variation of critical temperature $\bar{\theta}_{cr}$ with thickness ratio of piezoelectric layers

The following pre-buckling load case is considered

1. Uniform temperature rise of the beam with zero potential at top and bottom surfaces and immovable beam ends. The critical temperature for buckling is expressed as θ_{cr} which is non-dimensionalised as $\bar{\theta}_{cr} = \alpha_0 \theta_{cr} S^2$, where $S = l/h$ and α_0 is taken as $22.5 \times 10^{-6} K^{-1}$.

The critical buckling temperatures are obtained for hybrid beams under simply supported boundary conditions. The converged 1D-FE results are obtained using forty elements of equal size. The 1D-FE results are compared with 2D-FE (ABAQUS) results where the beam is first analyzed in heat transfer step which is then followed by buckling step. The temperature distribution as obtained in heat transfer step is used as a predefined field in the next step of linear perturbation buckling procedure to obtain the buckling response under thermal load. The beam has been discretized into more than 600 elements for span to thickness ratio $S = 10$. This number increases substantially for larger span to thickness ratios of hybrid beam.

The variation of critical buckling temperature for first three buckling modes ($n = 1,2,3$) with thickness ratio of piezoelectric layer, h_p/h for both the hybrid beams is shown in Fig.1.The results are presented for different span to thickness ratios, S. The critical buckling temperature increases as the beams are made thinner for the same span length. It may be observed that for the same value of S there is substantial increase in values of critical buckling temperature for higher buckling modes.

The critical temperature, $\bar{\theta}_{cr}$ decreases with increase in thickness ratio of piezoelectric layers for beam d as well as beam e. $\bar{\theta}_{cr}$ is maximum for $h_p = 0.1h$ and exponentially reduces as the thickness of piezoelectric layer increases and become $0.3h$. Further the critical temperature corresponding to the fundamental buckling mode for beam e is higher in comparison to beam d for all the thickness ratios (h_p/h) considered. The 1D-FE and 2D-FE plots are in good agreement. The critical temperatures are obtained ignoring the effect of pre-buckling deformation.

4. Conclusions

The thickness of piezoelectric layers in smart beams is significant as it results in different buckling response under thermal loading. A 1D-FE model to assess the effect of thicknesses of piezoelectric layers on thermal buckling response of a hybrid beam is presented. The results are obtained for simply supported boundary conditions. The critical temperature decreases with increase in thickness ratio of piezoelectric layers for both hybrid composite and sandwich beams. For all the thickness ratios, the critical temperature corresponding to the fundamental buckling mode is higher for sandwich beam in comparison to the composite beam. These observations may be used as a basis for experimental investigation and design for composite and sandwich hybrid structures.

References

[1] Khdeir AA 2001 Thermal buckling of cross-ply laminated composite beams *Acta Mech.* **149** 201–13

[2] Metin Aydogdu 2007 Thermal buckling analysis of cross-ply laminated composite beams with general boundary conditions *Composites Science and Technology* **67** 1096–1104

[3] Kapuria S and Alam M N 2005 Nonlinear zigzag theory for buckling of hybrid piezoelectric rectangular beams under electrothermomechanical loads *Journal of Engineering Mechanics* **131** 367–376

[4] Zhong ZW, Roger RC, Chun M 1994 Pates III CS *Proc. of the ADPA/AIAA/ASME/SPIE Conference on Active Materials and Adaptive Structures* pp 115-32

[5] Chen L W and Chen L Y 1987 Thermal buckling of laminated cylindrical plates *Composite Strut.* **8** 189-205

[6] Tiersten H.F 1969 Linear piezoelectric plate vibrations (New York: Plenum Publishing Corporation)

[7] Rahman N and Alam M N 2018 Finite element modelling to assess the effect of surface mounted piezoelectric patch size on vibration response of a hybrid beam *IOP Conf. Ser.: Mater. Sci. Eng.* **310** 012053 doi:10.1088/1757-899X/310/1/012053

[8] Rahman N and Alam M N 2015 Structural control of piezoelectric laminated beams under thermal load *Journal of Thermal Stresses* **38** 69-95

[9] Rahman N and Alam M N 2018 Effect of piezoelectric layer thickness on static response of a hybrid piezoelectric beam *Adv. Sci. Eng. Med.* **10** 455–459

IOP Conf. Series: Materials Science and Engineering **404** (2018) 012019 doi:10.1088/1757-899X/404/1/012019

Towards Effective Management of Cold Chain: A DEMATEL Approach

Mr. Shahbaz Khan[1*], Mr Mohd Imran Khan[1], Dr Abid Haleem[2]
[1]Research Scholar, Department of Mechanical Engineering Faculty of Engineering and Technology, Jamia Millia Islamia, New Delhi-110025, India
[2]Professor of Mechanical Engineering, Faculty of Engineering and Technology, Jamia Millia Islamia, New Delhi-110025, India

shahbaz.me12@gmail.com

Abstract: Information and infrastructure management is critical component in management of cold chain. Management of cold chain thrives to preserve the integrity of cold chain from the point of production and processing, to storage at the consuming household or restaurant. This paper identifies the barriers for effective management of cold chain and evaluate the inter-relationship among them. The findings suggest that unavailability of proper infrastructure is highly impacting whereas the lack of regulatory framework and commitment of top management are relatively influencing the effective management of cold chain. Government needs to upgrade the redundant infrastructure & develop robust network and impose regulatory framework to mitigate the uncertainties associated with cold chain operations. Mitigation of barriers may have implications on food security and conservation of environment resources. This work may be used policy makers of government or industries to understand the interrelationship of barriers and their influence on cold chain management.
Keywords: Cold Chain; Cold Chain Management; Barriers; Temperature Controlled Logistics, Perishable Products

1. Introduction

Perishable products such as food, flowers, pharmaceuticals, and chemicals, needs temperature-controlled logistics, to mitigate the chemical reactions in them. To ensure the safety and quality of perishable products industries are increasingly relying on the cold chain. The cold chain refers to "the transportation of temperature sensitive products along a supply chain through thermal and refrigerated packaging methods and the logistical planning to protect the integrity of these shipments" [1]. A typical cold chain infrastructure generally consists of 'pre-cooling facilities', 'cold storages', 'refrigerated carriers', 'packaging', 'warehouse', 'traceability system' and retailer & consumers, under the aegis of information management systems [2]. An efficient management of the cold chain is the key to prevent unnecessary losses and maintaining the bottom line. Due to increasing focus of developing economies towards export of agricultural products, temperature-controlled logistics is getting more vital. To keep food free from bacterial, microbial, and fungal contamination for extended period of time, for commercial as well as health reasons, the role of cold chain is becoming imperative [3]. However, developing countries, are facing diverse problems such as deficient infrastructural facilities, including lack of regulatory framework, deficient infrastructural facilities, out-of-date technologies, and shortage of competent work force in developing a robust cold chain logistics. Effective management of cold chain needs a thorough understanding of chemical and biological processes linked with perishability in addition to physical means (such as dedicated warehousing, loading/unloading facilities, and refrigeration units non-destructive examination, vacuum pre-cooling technology, automatic identification, global positioning system (GPS) transportation vehicle automatic temperature control system, and electronic data interchange) to insure appropriate temperature conditions along the supply chain. Therefore, there is a need to identify the barriers that are critical in developing and managing comprehensive cold chain.

1.1 Research Objectives

IOP Conf. Series: Materials Science and Engineering **404** (2018) 012019 doi:10.1088/1757-899X/404/1/012019

The following are the primary aims of the research undertaken and communicated through this paper:

I. To identify the barriers for effective management of cold chain.

II. To evaluate the identified barriers in order to understand their interrelationships in effective management of cold chain

It is important for the policy makers and executive involved in business of cold chain to understand characteristics and interrelationship of barriers that influence the cold chain management.

2. Major Barriers for effective management of cold chain

The understanding of the barriers for effective management of cold chain may help management, as well as policymakers to develop an effective strategy. Barriers for effective management of cold chain are identified through systematic literature review and later verified with expert's opinion are shown in Table 1.

Table 1: Barrier for effective management of cold chain

S.No	Barriers	References
B1	Lack of top management support	[4], [5]
B2	Higher number of intermediaries	[6]
B3	Improper collaboration	[7,15]
B4	High cost for installation and operation	[2,8]
B5	Lack of proper infrastructure	[9,13]
B6	Lack of government regulation/ standard	[10,11]
B7	Lack of capacity building	[12,14]
B8	Lack of effective traceability	[11]

3. Research Methodology

The aim of this work is to evaluate the barriers of for effective management of cold chain through causal relationship. The major barrier is identified through literature review and experts input. The causal relationship among these barriers are is evaluated through a DEMATEL approach.

3.1 DEMATEL Method

The DEMATEL technique is adopted as a solution methodology in this article. DEMATEL is a comprehensive method to provide the causal relationship among the complex factors through graphs. Some other system-based tool such as ISM and TISM are used for the develop the relationship among factor/ barriers [14,5,11]. However, these system-based tool (ISM/TISM) is having some limitation [11,16]. These methods are more subjective in nature and inefficient to quantify the relationships. Therefore, we adopt the DEMATEL approach for the analysing the complex causal relationship among the barrier for effective management of cold chain. The DEMATEL method is summarised step by step as follows:

Step 1: The scale may be designated four levels (i.e. 0,1,2,3,4) to construct direct influence matrix (See Table 2). Each expert was asked to evaluate the direct influence between each two-factor combination through a score 0,1,2,3,4. The notation of x_{ij} represents that the factor i influence the factor j. The diagonal element (i.e. $i=j$) of the direct inflectional matrix is zero. A non-negative n×n matrix is obtained for each expert as $X^k = [x_{ij}^k]$ where k is the no of experts $1 \leq k \leq H$. Thereafter obtain X^1, X^2, X^3.... X^H from H experts.

Table 2: Scale and their interpretation

Scale	0	1	2	3	4
Interpretation	No influence	Very Low influence	Medium influence	High influence	Very high influence

Step 2: Establish an overall direct-relation matrix from H respondents, the average matrix $A = [aij]$ can be obtained from the equation:

$$a_{ij} = \frac{\sum_{k=1}^{H} x_{ij}^k}{H} \quad (1)$$

Step 3: Obtain normalized initial direct-relation matrix, *D* using Equation (2) & (3).

$$D = A.S \quad (2) \qquad \text{Where } S = \frac{1}{max_{1 \le i \le n} \sum_{j=1}^{n} a_{ij}} \quad (3)$$

Each element in matrix *D* falls between zero and one.

Step 4: Calculate Total relation matrix "T" using Equation (4)

$$T = D.(I-D)^{-1} \qquad \text{Where "I" is the identity matrix.} \quad (4)$$

Step 5: Determine the causal parameters through Equations. (4) and (5):

$$r_i = \sum_{i=1}^{n} t_{ij} \ \forall_j \quad (5) \qquad c_i = \sum_{j=1}^{n} t_{ij} \ \forall_i \quad (6)$$

Where R_i indicate the sum of rows and C_i represent the sum of columns.

Step 6: Develop a causal and effect diagram using the dataset consisting of prominence (Pi) and net effect(Ei) using expressions (6) and (7):

$$P_i = R_i + C_j \mid i=j \quad (7) \qquad E_i = R_i - C_j \mid i=j \quad (8)$$

The difference ($R_i - C_i$) shows the net effect that factor *i* contributes to the system. Moreover, if (*Ri* – *Ci*) is positive, factor *i* is a net cause, while factor *i* is a net receiver or result if (*Ri* – *Ci*) is negative.

4. Results

The barriers for effective management of cold chain are identified through literature review. These barriers are discussed with a five members expert's group to obtaining the deeper insight of these barriers and finalised eight barriers (See Table 1). After finalising the barriers, the experts were asked to evaluate the direct influence among the barriers for effective management of cold chain on the scale of 0-4 (See Table 2). The evaluation is taken in form overall direct relation matrix is obtained through expert input and shown in table 3 using equation 1.

Table 3: Overall Direct-Relationship Matrix (Z)

Barriers	B1	B2	B3	B4	B5	B6	B7	B8
B1	0	1.8	2	1	2	0	2	3
B2	0	0	3.8	1	0	0.8	3	2
B3	1	2	0	2	0	0	2	3
B4	2	1	2	0	2	0	2.2	3
B5	2	2	1	3	0	1	2	3
B6	2	0	1	1	3	0	1	0
B7	0	2	2.2	1.8	2	0	0	2
B8	1	0	0	1	2	1	0	0

Further, the Normalized direct-relation matrix (N) is developed using Equation (2) and (3). Thereafter, this matrix is transformed into Total-Relation Matrix (T) using Equation (4).

In matrix *T*, the summation of rows and column are represented by R and C respectively using equation (5) and (6) (Please see table 4). Ri represents the total effect of barrier i to the other barrier, and Ci represents net influence received by j barrier from other barriers. After the determination of R and C for every row and column, the prominence (Pi) and net effect (Ei) is calculated using equation (7) and (8). The net cause/effect of each barrier are decided through "Ei (i.e. R-C)". If Ei is positive, then the barrier is considered to produce the net cause and if negative then the barrier is the net effect. These cause and effect are shown in Table 4

Table 4: Cause and Effect of barriers

Barriers	R	C	R+C	R-C	Cause/Effect
B1	2.860626	1.95317	4.813795	0.907456	Cause
B2	2.481888	2.277517	4.759405	0.204372	Cause
B3	2.362093	2.853708	5.2158	-0.49161	Effect
B4	2.979116	2.768657	5.747773	0.210459	Cause
B5	3.395697	2.631742	6.027439	0.763955	Cause
B6	2.268626	0.808977	3.077603	1.45965	Cause
B7	2.511215	2.914986	5.426201	-0.40377	Effect
B8	1.421412	4.071917	5.493328	-2.65051	Effect

The R+C and R-C are plotted in figure 1, which shows the causal relationship among the barriers for effective management of cold chain. These results are discussed with the experts for further insights.

5. Discussion

The cold chain management is very challenging in the developing countries due to the several barriers. The result shows that the most important barrier is the "lack of proper infrastructure". Based on the "R+C" values, the importance order of the barriers is B5 > B4> B8 >B7> B3 > B2 > B1 > B6. These identified barriers are categorised into "influential group" and "influenced group" based on the sign of the "R-C" values. Five barriers belong to the "influential group" and three from the "influenced group".

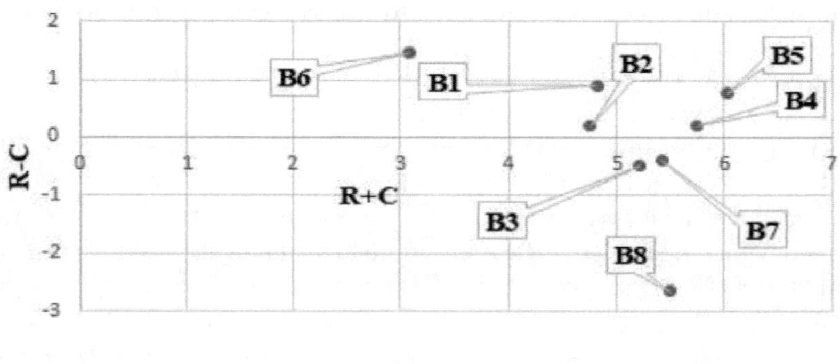

Figure 1: The cause and effect diagram of barriers for effective management of cold chain

5.1 Influential barriers

The most influential barrier is the "lack of government regulation/ standard" among all the identified barriers. Due to the lack of standard and government regulation, the cold chain partners are not able to effectively manage the cold chain practices. The Government would need to support the industries by proving some relaxation in taxes and infrastructure subsidies in their initiatives. The second most influencing barrier is the "Lack of top management support". The top management takes initiatives through collaboration with other cold chain partners in capacity building by providing training to the workforce. However, the top management is facing challenges from the other factors such as lack of proper infrastructure (poor highways, lack of cold storage, transportation vehicle automatic temperature control) and a higher number of intermediaries. Next influential barrier is the "lack of proper infrastructure". Due to the poor infrastructure, the cold chain integrity is challenged several times. In the developing countries, the poor highway condition causes the unnecessary delay in the logistics and lack of cold storage causes the wastages of a large amount of vegetables and fruits. The fourth important barrier is the "High cost of installation and operations". The installation cost of the cold chain infrastructure is very high. This factor affects the industry people to maintain the cold chain integrity. The producer (such as farmers) does not afford the higher operating cost of the cold chain. The government would need to provide cold chain facilities to the farmers. The next influential barrier is the "higher number of intermediaries". In the developing countries, a large number of middlemen, which increases the unnecessary hurdles to maintain the cold chain integrity. In India, some initiative is taken by the government to reduce the unnecessary delay through "National Agriculture Market (NAM)", "Kisan Mandi" which connect the framers to the market.

5.2 Influenced barriers

The barriers which are influenced by other barriers belong to the influenced barrier. The increasing order of the influenced barrier is the B7>B3>B8. The highest influenced barrier is the "lack of effective traceability". Effective traceability is influenced by government regulation, top management support and collaboration among cold chain partners. The second most influenced barrier is "improper collaboration". To effectively manage the cold chain collaboration is an important element. Last influenced barrier is the "lack of capacity building". Due to the uneducated farmers and semi-skilled workers, the cold chain faces many challenges.

6. Implications
This research work will be beneficial for industry professionals in developing strategies for mitigating these barriers for effective cold chain management in the developing countries. This work assists the decision makers to have a holistic insight into the inter-relationships among the various barriers instead of looking separately. These barriers are further analyses for the root cause analysis. This study also helps the government to develop the policies to prevent the wastage of perishable products.

7. Conclusion
The barriers to the effective management of cold are identified through the literature review and validated through the expert opinion. After finalizing the barriers, the interrelationship among the barrier is evaluated using the DEMATEL approach. These barriers are categorised into two groups namely "influenced" and "influential" group. the result is discussed with the expert and useful insight is provided in the discussion section. Finally, the implication of this study is provided.

References
1. Rodrigue, J.P., Comtois, C. and Slack, B. (2013), "The Geography of Transport Systems." Routledge, New York, NY.
2. Joshi R., Banwet D. K., & Shankar R. (2009). "Indian cold chain: Modeling the inhibitors." British Food Journal 111 (11)
3. Smith, J.N. (2005), "Specialized logistics for a longer perishable supply chain", World Trade, Vol. 18 No. 11, pp. 46-47.
4. Swaminathan, J.M. (2007), "Managing supply chain operations in India: pitfalls and opportunities". International Series in Operations Research and Management Science, Vol. 98.
5. Haleem, A., & Khan, M. I. (2017). "Towards successful adoption of Halal Logistics and its' implications for the stakeholders." British Food Journal, 119 (7).
6. Shukla, M., Jharkharia, S., (2013). "Agri-fresh produce supply chain management: a state-of-the-art literature review." Int. J. Oper. Prod. Manage. 33 (2), 114–158.
7. Kottila, M.R. and Ronni, P. (2008), "Collaboration and trust in two organic food chains", British Food Journal, Vol. 110 Nos 4/5, pp. 376-94.
8. Shukla, S., Shankar, R., Singh, S., (2014). "Food safety regulatory model in India." Food Control 37, 401–413
9. Mahmood, K., Garcia, O. and Saha, A.K. (2005). "A closer look on the situation of milk production in Asia", IFCN Dairy Report 2005, International Farm Comparison Network, Global Farm, Braunschweig, pp. 124-5,
10. Berger, H. (2007), "Dealing with dynamics engineering solutions to the dairy dilemma", Trends in Food Science & Technology, Vol. 18 No. 1, pp. 123-34.
11. Khan, S., Haleem, A., Khan, M.I., Abidi., M. H., & Al-Ahmari, A., (2018). "Implementing traceability systems in specific SCM through CSFs." Sustainability 10 (1)
12. Khan, A.U. (2005), "The domestic food market: is India ready for food processing?", Conference on SPS towards Global Competitiveness in the Food Processing Sector,Monday, 5 September, Pune, India, available at: www.idfresearch.org/pdf/dommarket.pdf
13. Saurikhia, A., Ahmed, S., Haleem, A., Gangopadhyay, S., Khan, M. I., (2017). "Evaluating technology management factors for fly-ash utilization in the road sector using an ISM approach." International Journal of Management Science and Engineering Management.
14. Khan, M. I., Khan, U., & Haleem, A. (2014). "Corporate Social Responsibility: Modelling of Critical Factors using an Integrated ISM and Fuzzy-Micmac approach." International Journal of Global Business and Competitiveness, 9(1).
15. Dubey, R., Gunasekaran, A., Ali, S.S., (2015). "Exploring the relationship between leadership, operational practices, institutional pressures and environmental performance: a framework for green supply chain." Int. J. Prod. Econ. 160, 120–132.
16. Asjad, M., & Khan, S. (2016). "Analysis of maintenance cost for an asset using the genetic algorithm." International Journal of System Assurance Engineering and Management, 8(2).

IOP Conf. Series: Materials Science and Engineering **404** (2018) 012020 doi:10.1088/1757-899X/404/1/012020

Detection of propanol gas using titanium dioxide based thick film

Ankit Kumar Vishwakarma[1, a], Lallan Yadav[1, b]

[1]Thin film laboratory, Department of Physics,
Deen Dayal Upadhyaya Gorakhpur University, Gorakhpur U.P.-273009 India

Email- [a]kv.ankit92@gmail.com, [b]nisaly06@rediffmail.com

Abstract: Cadmium Sulphide (CdS) doped Titanium dioxide (TiO_2) thick film sensor was produced on alumina substrate and their propanol sensing behaviour were investigated. Undoped TiO_2 and 2wt% CdS-TiO_2 thick film are fabricated on alumina substrate. Sensing properties such as, sensitivity and selectivity measurement are carried out in the range 0- 5000 ppm at room temperature. The results showed that, 2wt% CdS-TiO_2 sample gives higher response (60%) for propanol against L.P.G. and ethanol. The response of propanol is 3.75 times more than L.P.G. and 1.94 time of Ethanol. 2wt% CdS-TiO_2 sample are more (56.08 %) selective for propanol over L.P.G. and Ethanol. The result obtained indicated that CdS-TiO_2 composite thick film may used for propanol sensing.

Keywords: TiO_2, propanol, sensitivity and Thick film

1. Introduction

Sensor is used to detect physical condition of chemical compound and generally fabricated in the form of electrical devices that senses organic gases, inorganic gases and toxic gases. These gas sensors have attractive the user due to its atmospheric condition such as reliable, fast, cheap and low maintenance devices used as a gas sensing properties [1]. Selectivity, sensitivity and stability are appropriate properties of sensor. TiO_2 is one of most prefer n-type semiconductor with unique properties for development of conductomatric gas sensor due to their nontoxic nature, chemical stability, commercial availability at low cast [2]. Joy Tan et-al, studied the carbon monoxide gas sensor based on TiO_2 nano crystalline with a langastic substrate and he observed that addition of the Au on the surface of TiO_2 increase the sensitivity of the sensor toward CO [3]. Navid Alaei Sheini reported Ag-doped TiO_2 gas sensor and he found sensitivity can increase significantly by diffusing silver to the titanium dioxide in presence of ethanol conductance of the device decrease [4]. A. M. Ruiz et-al suggested that, the effect of various metal addition in gas sensing performance of the TiO_2 nano crystalline obtain from hydrothermal treatment and they showed that TiO_2 based sensor can be operate at high temperature for good chemical stability of TiO_2 [5]. M.H. Seo et-al presented that, the detection of organic gases using TiO_2 nano tube based gas sensor and they observed that TiO_2 based gas sensor prepare by hydrothermal treatment exhibited high sensitivity to toluene rather than CO and H_2 [6]. Yadava et-al have also investigated the sensing behaviour and mechanism of titanium dioxide-based MOS hydrogen sensor and

IOP Conf. Series: Materials Science and Engineering **404** (2018) 012020 doi:10.1088/1757-899X/404/1/012020

showed that the grided Pd-TiO_2-Si sensor is most appropriate detector for H_2 gas at room temperature [7].
In the present work, we investigated the response of 1wt% CdS doped TiO_2 thick film. The undoped and doped thick film samples are prepared and its suitability for detection of ethanol, LPG and propanol is studied. The measurement results and sensing mechanism for test gases are presented. The results indicate that CdS (2wt %) doped TiO_2 sensor is a suitable detection for propanol.

2. Experimental

Undoped TiO_2 and CdS-TiO_2 thick film are fabricated in laboratory. Initially, TiO_2 powder grinded with CH_3COOH acid for 45 min and few drop of labolin are added to make TiO_2 paste. This paste is rolled on the alumina substrate (finger electrode in rear side and heater in back side). The sensitivity of the fabricated sensors was measured for test gases using experimental set up shown in fig.1. It consist a locally made chamber along with DC Power Supply DL-3203 and multimeter (Aplab 107-N).

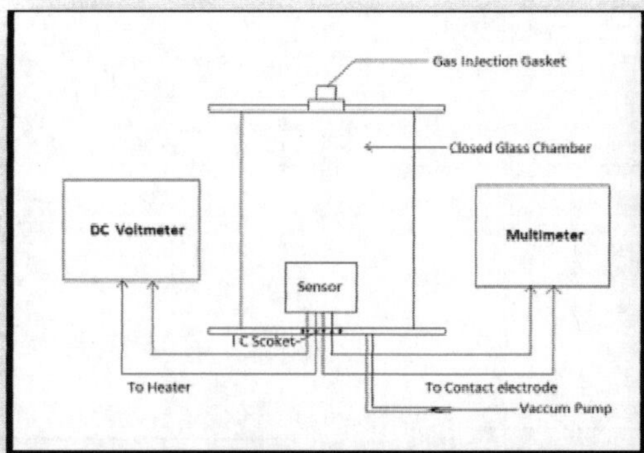

Figure1. Sensing measurement

3. Result and discussion

3.1 Gas sensing properties

The variation in the resistance of the fabricated undoped and 2 wt% CdS doped TiO_2 thick film sensors were recorded varying the concentration (0-5000 ppm) of gases, LPG, propanol and ethanol. The response of the sensor calculated using formula [8-9].

$$S = [\frac{(Ra-Rg)}{Ra}]\ x100 \qquad (1)$$

where R_a is resistance in air and R_g is resistance of present in gas. Fig.2 (A) shows the variation of response with propanol gas concentration. It is evident from this figure that response increases as concentration increases. The change in response for undoped sensor is 26% and for CdS doped it is 60%. Also, the response for ethanol is 23% and 31.1% for

IOP Conf. Series: Materials Science and Engineering **404** (2018) 012020 doi:10.1088/1757-899X/404/1/012020

undoped and CdS-TiO_2 sample are presented in fig.2 (B). fig.2 (C) shows the LPG has response 10% and 16% for undoped TiO_2 and 2wt% CdS-TiO_2. The comparative response is plotted in fig.2 (D); it is found that the response of Propanol is more sensitive and it is 3.75 times of LPG and1.94 times of ethanol for 2wt% CdS-TiO_2 samples. The response of propanol of 2wt% CdS-TiO_2 is 2.30 time of undoped TiO_2 samples. The response of LPG, ethanol and propanol (5000 ppm) for undoped and 2wt% CdS doped thick film sensor is listed in table1. The selectivity of the fabricated sample was estimated using formula.

$$(S_{ELi}) = [\frac{Si}{\sum Si}] \times 100 \qquad (2)$$

where S_i the response of gas for which relative it's and $\sum Si$ is the sum of the response of all tested gas. The selectivity of fabricated sample is shown in fig.3 (histogram). It is found that 56.08%, 28.97% and 14.95 for propanol, acetone and LPG respectively.

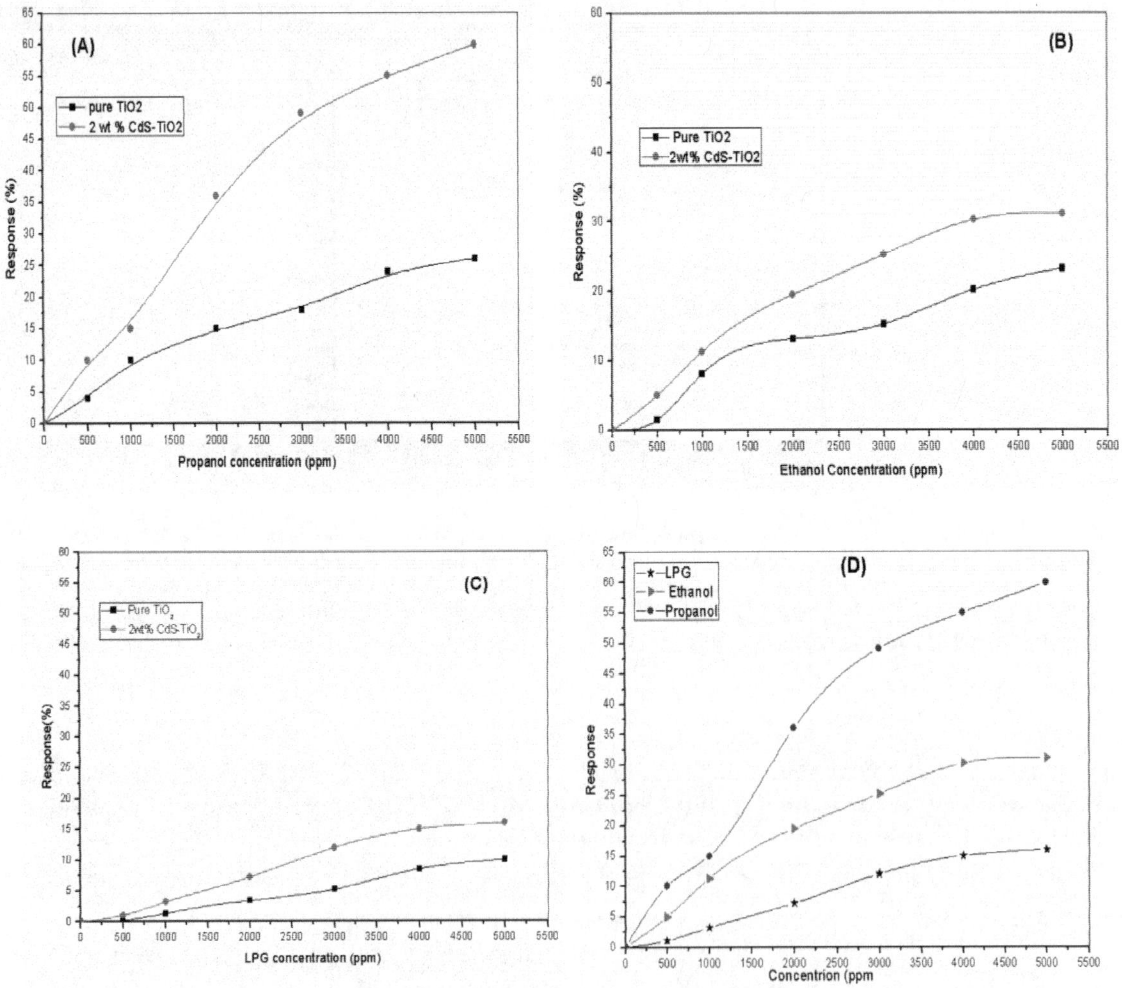

Figure2. Response curve of (A) Propanol, (B) Ethanol, (C) LPG and (D) Comparative response of LPG, Propanol and Ethanol for CdS-TiO_2 thick film sensor

3.2 Sensing mechanism

At the room temperature propanol gas sensing mechanism can be descried through dissociation and absorption of propanol molecule on useable vacant site throughout the expose surface of TiO_2. When sensor is exposed in air, pre-adsorption oxygen on the surface would trap electron due to strong electron negativity of oxygen, there by producing adsorbed oxygen ions O_2^-. Which reaction as follow:

$$O_{2\,(gas)} \rightarrow O_{2\,(ads)}$$

$$O_{2\,(ads)} + e^- \rightarrow O_2^-{}_{(ads)}$$

$$O_2^-{}_{(ads)} + e^- \rightarrow 2\,O^-{}_{(ads)}$$

The reducing gas exposed on the oxides gas surface can be explained by reaction [10].

$$R + O^-{}_{(ads)} \rightarrow RO + e^-$$

Where R is reducing gas, O^- is the oxygen adsorption, e^- are electron. When reducing gas such as propanol ethanol get in touch with TiO_2, they will be adsorbed on the surface and reaction is presented as:

$$C_3H_7OH + O^-{}_{(ads)} \rightarrow C_3H_7O^- + H_2O + e^-$$

The reducing gases interact with $O^-{}_{(ads)}$ onto surface of TiO_2 and result to increase in electron concentration and hence resistance decreases. The higher concentration of the reducing gas, and the lower the resistance, i.e., the sensitivity increases with increasing the concentrating of gases [11].

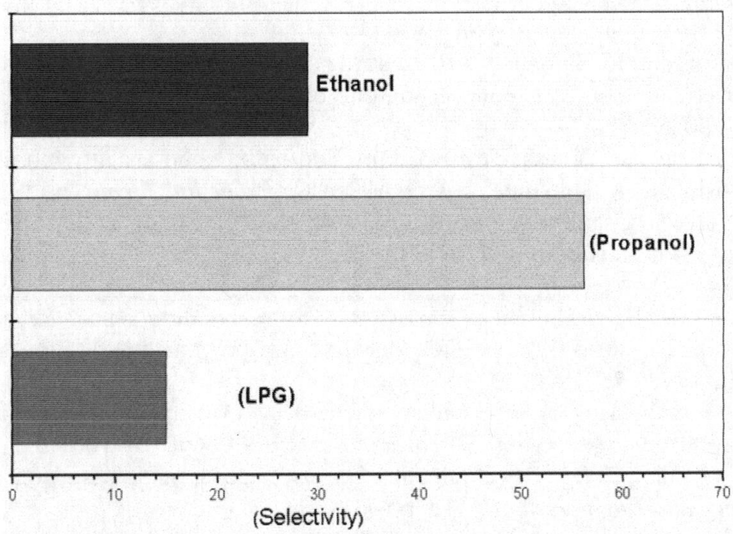

Figure3. Selectivity of CdS-TiO_2 thick film sensor for ethanol, propanol and LPG

IOP Conf. Series: Materials Science and Engineering **404** (2018) 012020 doi:10.1088/1757-899X/404/1/012020

Table.1

CdS wt% in TiO_2	Ethanol	LPG	Propanol
0	23.2	10	26
2	31.1	16	60

4. Conclusion

The effect of CdS into TiO_2 thick film gas sensor for propanol, ethanol and LPG sensing was experimentally investigated. It is observed that response of the fabricated sensors can increase significantly with CdS doping. The response of propanol is 3.75 times of LPG and 1.94 times of ethanol of 2wt % CdS-TiO_2 sample for concentration 5000 ppm. Thus, the results obtained indicated that CdS doped TiO_2 thick film sensor (2 wt% CdS-TiO_2) exhibited maximum response and it is more selective to propanol over LPG and ethanol.

References

[1] Barsan N, Koziej D, Weimar U, 2007, Metal oxide-based gas sensor research: how to, *Sensor and actuators B*, **12**, pp-18-135

[2] Brauli S, Archanjo, Pablo F. Siles, Camilla K. B. Q. M. Olviveira, 2013, Characterization of metal oxide based gas nano sensor and micro sensor fabrication via local anodic oxidation using AFM, *Advanced in material science and engineering*, 2013 898565

[3] Joy Tan, Wojtek Wlodarski and Kourrose kalantar-Zadeh, *2006,* Carbon monoxide gas sensor based on Titanium dioxide nanocrysatlline with a Langasite substrate, *IEEE Sensor 2006,* EXCO, Daeegu, Koria

[4] Navid Alaei Sheini, Mahsa Rohani, 2016, Ag doped titanium dioxide gas sensor, *IOP Conf. Series: Materials science and Engineering*, 108, 012033

[5] A. M. Ruiz, A. Corneta, K. Shinanoe, J.R. Moraute, 2005, Effect of various metal addition on the gas sensing performance of TiO_2 nano-crystalline obtain hydrothermal treatment, *Sens. Actuator, B chem.* 108, 34-40

[6] Min-Hyun Seo, Masayoshi Yuasa, Testuya, Jany Soo Huh, Noboru Yamazeo, 2009, Detection of organice gases using TiO_2 nano tubes based gas sensor, *Procedia Chemistry*, **1**, 192-195

[7] Lallan Yadav, Naveen Chandra Gupta, R. Dwivedi, Ravi S. Singh, 2007, Sensing behavior and mechanism of titanium dioxide-based MOS hydrogen sensor, *Microelectronics Journal*, **38**, 1226–1232

[8] Mohammed Mabrook, Peter Hawkins, 2001, A rapidly-responding sensor for benzene, methanol, and ethanol vapour based on films of TiO_2 disperse in a polymer operating at room temperature, *Sensor and actuators B*, **75**, 197-202

[9] Y. Yamada, Y seno, Y. Masuoko, T. Nakamura, 2000, NO_2 sensing charater of Nb doped TiO_2 films and their electronics properties, *Sensor and actuators B*, **66**, 164-166

[10] K. Arshak, I. G.aidan, 2005, Development of a novel gas sensor based on oxide thick films, *Materials Science and Engineering B*, **118,** pp-71-113

[11] Liu pengyi, Chen jungang, Sun Wangdian, 2004, Sheet resistance and gas sensing properties of tin oxide thin film by plasma enhanced chemical vapour deposition, *Plasma Science & technology*, vol. **6** No. 2

Controlling localised properties in fused deposition modelling parts: A feasibility study

Abhishek Kumar Srivastava, Prashant K. Jain
ME Discipline, PDPM Indian Institute of Information Technology, Design and Manufacturing Jabalpur, Jabalpur 482005, MP, India

E-mail: srivastavaabhishek1234@gmail.com

Abstract. Recent advancement in additive manufacturing (AM) enables us to make functional prototypes with the highest accuracy. Among various AM techniques, fused deposition modeling (FDM) is most popular due to its ability to make a complex shape in minimum time and least cost. However, the strength of the parts fabricated with FDM is always of primary concern so present approach provides a way to control localized mechanical property in FDM parts. Model of a 3D object is bifurcated into various subparts by incorporating stress data. All subparts are fabricated by using appropriate process parameters suited to their stress value. As the process parameters have a prominent effect on the strength of the part, hence the part fabricated contains better strength compared to conventional FDM approach.

1. Introduction

Additive manufacturing (AM) provides a platform to make parts with the best quality at a reasonable price within the shortest time, which is essential to survive in the present competitive manufacturing world. In order to reduce the product development cycle time, now a day's standpoint of industries has changed from conventional manufacturing technique to additive manufacturing. Initially, additive manufacturing was used to make prototypes of the parts to test its functionality, but now a day's it is used to manufacture actual machines components, as well as the medical equipment,'s so that the fabricated parts must possess enough strength to sustain during its functionality. Additive Manufacturing is a manufacturing technique in which any complex parts can be fabricated from 3D computer aided designs (CAD) model in layer-by-layer fashion with a suitable material (polymer, metal, etc.) [1]. Various processes in AM have been evolved that uses a variety of material, ranging from plastics to metals [2,3]. Most popular AM techniques are stereolithography (SLA), selective laser sintering (SLS), fused deposition manufacturing (FDM), selective laser melting (SLM). Schematic diagram of FDM is shown in figure 1. FDM uses a thermoplastic material that passes through heated liquefier, where the state of filament changes from solid to semi-molten, afterward it is deposited through a nozzle having relative motion with the base where it solidifies under a controlled environment. Most commonly used filaments materials in FDM are acrylonitrile butadiene styrene (ABS) and polylactic acid (PLA). FDM is most popular AM process which gives free control of process parameters. FDM allows controlling internal structure of 3D printed object. While the main advantage of FDM process is that it can make parts with locally controlled properties like strength, density, and porosity.

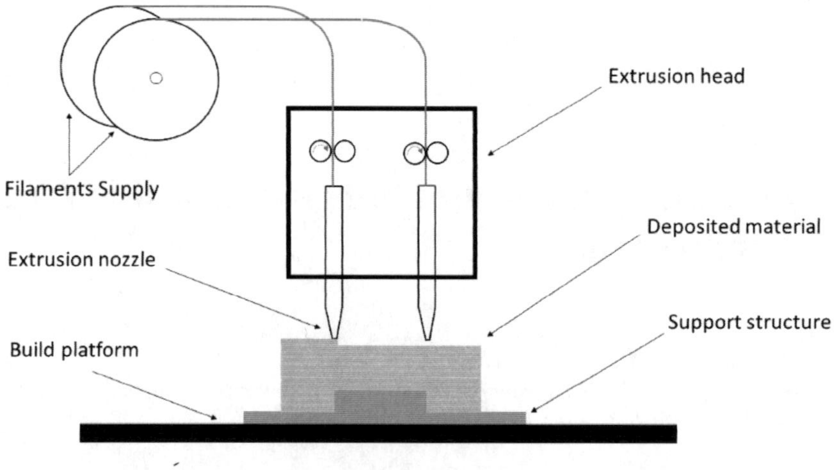

Figure 1. Schematic diagram of Fused Deposition Modelling

2. Literature review

Despite having several advantages, FDM technology has some limitation too, one of which is the poor strength of the part fabricated. Mechanical properties of the parts fabricated through FDM process are affected by design and processing conditions. Many attempts have been taken in order to improve the strength of the part such as proper material selection or process parameter optimization [4]. Rodriguez et al. [5] researched the effect of infill angle on the fracture behavior of the fabricated parts. It was found that if the infill angle is changed, fracture behavior also changes from ductile to brittle due to improper bonding between the filaments. Ahn et al. [6] have studied the effect of different process parameters like deposition temperature, air gap, bead width and raster angle on the tensile and compressive strengths of directionally fabricated specimens. It was found that parameters such as raster orientation and air gap affect tensile strength by a substantial amount. Zhong et al. [7] studied that tensile strength of ABS filament can be increased by a considerable amount if glass fiber is reinforced into ABS matrix. It was found that at the expense of handleability and flexibility strength of the abs filament can be increased by using glass fiber. Roger et al. [8] have studied about heterogeneous internal filling or multi-materials in FDM process in order to obtain targeted mechanical properties. Also, optimization of infill density distribution can be done in order to improve stiffness in stress concertation zones in the fabricated parts. Ning et al. [9] presented carbon fiber in a matrix of thermoplastic CFRP composites have a significant effect on the mechanical property of the FDM printed parts, but the limitation with this process is that FDM technology is not compatible with all kind of materials. Sood et al. [10] investigated the effect of process parameters such as layer thickness, raster angle and air gap on the tensile behavior of the fabricated specimen. If number of layers is more, it will result in high-temperature gradient towards the bottom of part thus increase the diffusion between adjacent raster and strength will improve. Due to the rapid growth of layered manufacturing now a day's design and manufacturing of heterogeneous parts are very popular. Present work gives us a way to control localized property in part and same methodology has been used to fabricate part by using FDM.

3. Methodology

Proposed methodology provides us a way for fabricating prototypes having localized mechanical property in different regions by considering stress pattern obtained from ANSYS Programming Design Logic (APDL) of ANSYS software. Procedure for multiple STL printing is shown in the figure. 2. Firstly, the solid model of an object is generated using any of the modelling software, further finite element analysis of the model is done using APDL module of ANSYS. Stress pattern obtained using

ANSYS (APDL) is shown in figure 3. A MATLAB program is developed to generate point cloud by incorporating obtained stress data from ANSYS(APDL). Point cloud for the critical region can be separated from that of the non-critical region just by giving suitable stress value. Further, a MATLAB program is developed which will generate STL file from the generated point cloud. Hence a single STL is bifurcated into two STL's by incorporating stress data. Merging of two STL's is done by using an open source software Replicator-G and printing are performed. The proposed approach can reduce the weight of the parts by assigning faster prototyping condition in the non-critical region also this approach can be used for fabrication of functionally graded material.

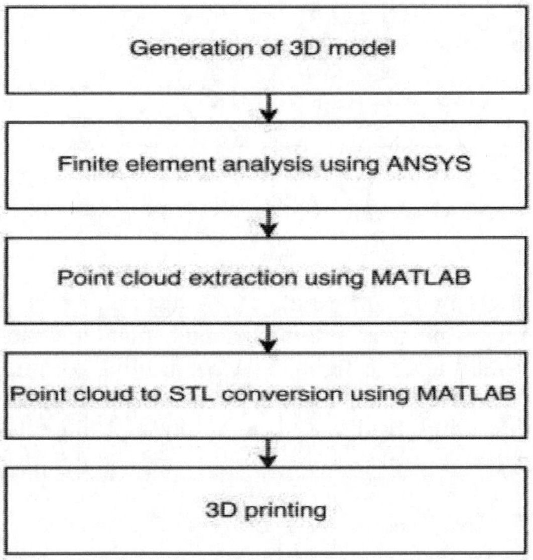

Figure 2. Proposed Methodology

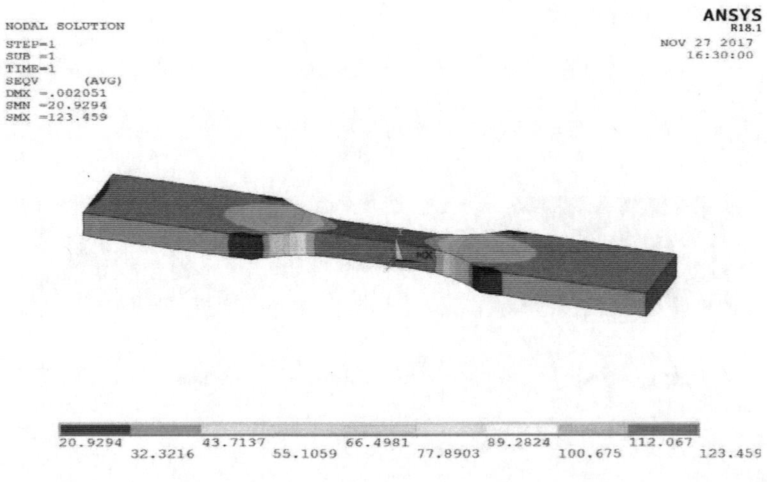

Figure 3. Stress pattern obtained from ANSYS

4. Experimental Details

Among various process parameters four parameters namely infill density, deposition temperature, travel feed and feed rate are the most significant factors impacting the strength of the printed parts. The importance of these parameters has been seen in the various literature [5–10]. As the primary aim of the work is to investigate mechanical properties of printed parts, test specimens are designed and fabricated by following ASTM D638 standards, which specifies the geometry and dimension of the specimen for tensile tests. Schematic diagram of the designed tensile specimen with along with the dimensions is shown in the figure. 4

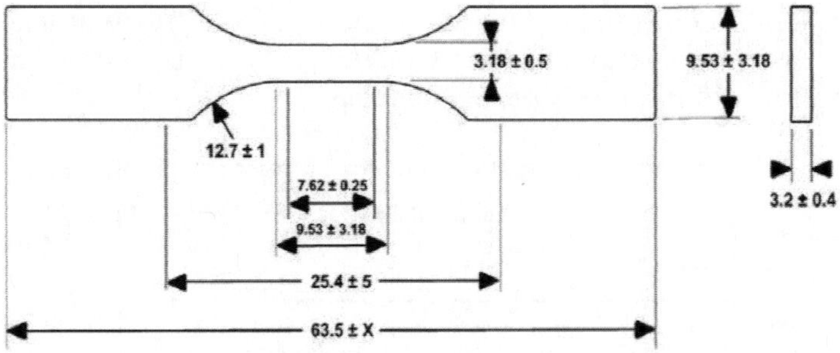

Figure 4. Schematic of ASTM 638 (Type V test specimen)

5. Tensile Testing

Following ASTM D638 standard (for ABS plastic), tensile testing of each printed specimen (shown in the figure. 4) is performed on a universal testing machine (Tinius Olsen H250K (USA)). Tensile specimen with different regions is shown in figure 5. Also, printing parameters for different regions in a specimen are given in Table 1. Outer region (i.e. non critical region) is printed with 70% infill density, 2300C temperature, 50 mm/min travel feed and 70 mm/min feed rate whereas inner region (i.e. non critical region) is printed with 100% infill density, 2500C temperature, 40 mm/min travel feed and 60 mm/min feed rate.

Table 1. Printing parameters for different regions

Process parameters	Region 1	Region2
Infill density	70%	100%
Deposition temperature	230	250
Travel feed	50	40
Feed rate	70	60

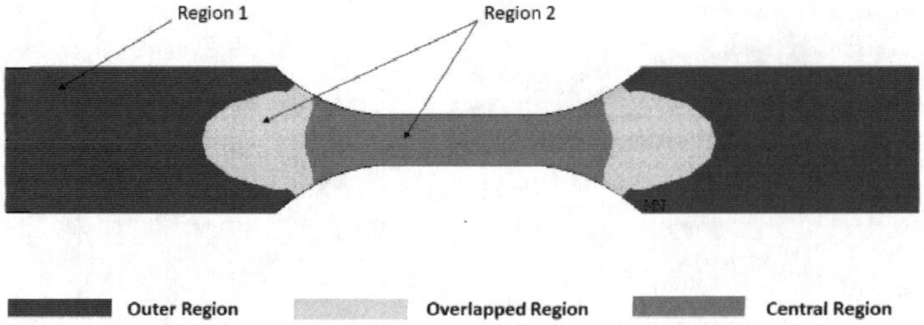

Figure 5. Tensile specimen with different regions

6. Results and discussion

Figure. 6 shows the stress-strain curve of fabricated specimens. The plot includes elastic deformation followed by failure point. Depending on the printing strategies, tensile strength, as well as failure point of the specimens, were varying. Three types of printing strategies were employed for fabrication of specimen as shown in figure 7. Specimen for the first experiment was fabricated with 70% infill density without any overlap. Likewise, other specimens were fabricated with varying infill density in the critical and non-critical region along with some overlap. Printing parameter for critical and non- critical region is given in Table 1. It can be noticed that the maximum tensile strength was achieved for a specimen with maximum overlap and as the overlap amount is decreased tensile strength is also decreased. This is due to the reason that for the specimen having maximum overlap, planes between the two regions are in perfect contact and the adhesion between the layers are stronger resulting in increased tensile strength. Interestingly, the tensile strength of less overlap parts increased by a significant amount compared to parts having no overlap (i.e. uniform infill). An interesting phenomenon that has been observed in experiments is necking at lower strains. This is because when thermoplastic polymers are subjected to deformation by stretching long chain molecule align themselves in the direction of elongation and the specimen becomes stronger and stiffer, and this process leads to early unstable necking [11]. Figure. 8 shows the failed tensile test specimens. It can be observed that all specimens were failed almost in the same fashion as the standard specimen, this shows the feasibility of the proposed methodology.

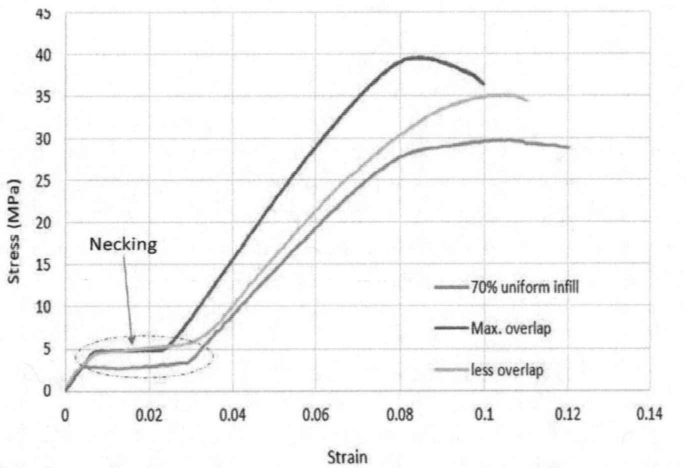

Figure 6. Representative stress-strain data of tensile specimens printed with different strategies

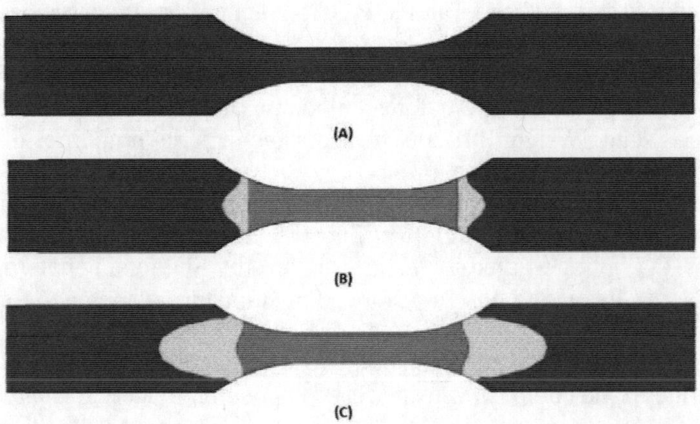

Figure 7. Specimens with different strategies (A) 70% uniform infill (B) Less overlap (C) Maximum overlap

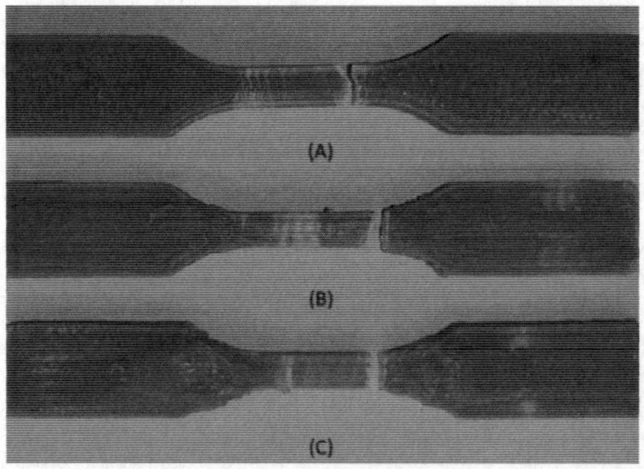

Figure 8. Failed tensile test specimens (A) 70% uniform infill (B) Less overlap (C) Maximum overlap

7. Conclusion

The main aim of the present research was to evaluate the feasibility of fabricating parts having localized property control. Specimens with localized property control have been successfully fabricated on commercial FDM machine by assigning suitable process parameters in different regions. Deposition temperature, infill density, feed rate, travel feed were selected as the most prominent factor affecting the strength of the FDM parts. Several results were drawn from the analysis. Maximum tensile strength values were reported for samples having maximum overlap and as the overlap amount was decreased tensile strength was decreased also printed specimens were failed almost in the same fashion as the standard specimen, this shows the feasibility of the process.

Reference

[1] Kim H, Park E, Kim S, Park B, Kim N and Lee S 2017 Experimental Study on Mechanical Properties of Single- and Dual-material 3D Printed Products *Procedia Manuf.* **10** 887–97

[2] Kumar N, Jain P K, Tandon P and Pandey P M 2018 Extrusion-based additive manufacturing process for producing flexible parts *J. Brazilian Soc. Mech. Sci. Eng.* **40** 143

[3] Kumar N, Jain P K, Tandon P and Pandey P M 2018 Additive manufacturing of flexible electrically conductive polymer composites via CNC-assisted fused layer modeling process *J. Brazilian Soc. Mech. Sci. Eng.* **40** 175

[4] Kumar N, Shaikh S, Jain P K and Tandon P 2015 Effect of fractal curve based toolpath on part strength in fused deposition modelling *Int. J. Rapid Manuf.* **5** 186–98

[5] Rodríguez J F, Thomas J P and Renaud J E 2003 Mechanical behavior of acrylonitrile butadiene styrene fused deposition materials modeling *Rapid Prototyp. J.* **9** 219–30

[6] Ahn S, Montero M and Wright P K 2007 Anisotropic material properties of fused deposition modeling ABS

[7] Zhong W, Li F, Zhang Z, Song L and Li Z 2001 Short fiber reinforced composites for fused deposition modeling *Mater. Sci. Eng. A301* **301** 125–30

[8] Roger F 2015 3D-printing of thermoplastic structures by FDM using heterogeneous infill and multi-materials : An integrated design-advanced manufacturing approach for factories of the future Abstract : pp 1–7

[9] Ning F, Cong W, Qiu J, Wei J and Wang S 2015 Additive manufacturing of carbon fiber reinforced thermoplastic composites using fused deposition modeling *Compos. Part B Eng.* **80** 369–78

[10] Kumar A, Ohdar R K and Mahapatra S S 2010 Parametric appraisal of mechanical property of fused deposition modelling processed parts *Mater. Des.* **31** 287–95

[11] Jain P K, Pandey P M and Rao P V M 2010 Tailoring material properties in layered manufacturing *Mater. Des.* **31** 3490–8

Sustainability Accounting and Reporting for Supply Chains in India-State-of-the-Art and Research Challenges

Kamlesh Tiwari[1], Dr. Mohd. Shadab Khan[2] and Dr. P K Bharti[3]

[1, 2, 3] Department of Mechanical Engineering, Integral University, Lucknow, India
Email: Tiwari.kamlesh@yahoo.com

Abstract- This article presents an analysis of the research opportunities in the rapidly growing field of sustainability accounting and reporting (SAR) for supply chains in India. Indian companies are gearing towards the next level of responsible growth. It was found that SAR is based on the principles of full cost accounting that includes the costing of sustainability, natural capital inventory accounting, and input-output analysis. The legal framework in India for SAR is quite old and is still evolving. Sustainability is a present challenge for India in the supply chains and value chains given their massive scales and impact on Indian and global environment, on economy and on people. This article presents an overview of the research opportunities in the field of SAR in India in the area of supply chain management.

Keywords: Sustainability, Accounting, Reporting, Standards, Disclosures, Global Reporting Initiative, Supply Chain Management

1. Introduction and Literature Review

The top Indian companies value sustainability and social responsibilities, are keen to grow responsibly, and are keen to harmonise sustainability accounting and reporting with their current accounting and reporting frameworks [1] [2] [3] [4]. The leading Indian companies have joined the network of Global Reporting Initiative (GRI) and are now able to include sustainability reporting as per GRI standards for disclosure and reporting into their reporting of business responsibilities (as required by SEBI) [1] [2]. GRI has released a comprehensive documentation of standards for integrating their sustainability accounting and reporting standards with the SEBI's requirements for reporting of business responsibilities [5]. However, some of the latest the research studies (both academic and professional) have explored and found on-the-ground challenges in SAR of Indian companies in the supply chains and also in the overall value chains [6] [7] [8]. Sustainability is often represented as a model consisting of three major areas: economics, environment, and empowerment (EEE- Triple bottom lines) so far as any supply chain are concerned. [9] [10] [11] [12]. Big supply chains are global and have great impact on the locations where they are present [13] [14]. Many researchers consider sustainability has to do a lot with capability development in fields which contribute these triple areas [15] [16] [17]. One of the specialised capabilities required is to measure or accounting the sustainable practices and their reporting [18] [19] [20] [21]. SAR is a framework of accounting and public reporting that encompasses economic, environmental, and social impacts of a business (popularly referred to as triple bottom-line reporting) that reflects the corporate social responsibility, and commitment of a company towards sustainability and sustainable development [22] [23] [24].

SAR is based on the principles of full cost accounting (or full cost-benefit analysis) that includes the costing of sustainability (triple bottom-line costing), natural capital inventory accounting, and input-output analysis [24] [25]. Sustainability costing is a measure of the economic, environmental, and social consumption by an organisation during a period that is a reflection of what it shall need to pay to restore the Earth to a state prior to the consumption [24] [25] [26]. The natural capital inventory comprises of the natural stocks acquired by a company that is critical (consumption of ozone layer, biodiversity, and environment cleanliness), non-renewable (or non-substitutable; like oil and gas, and minerals), and renewable (or substitutable; like, natural lands, ponds, fisheries, and timber) [24] [25] [26] [27]. Finally, the input-output analysis refers to the inputs consumed (from environment, society, and national economy) and outputs produced (useful products as well as wastes) by an organisation [24] [25] [26].

SAR has emerged as a global framework but still lacks corporate commitment, appropriate and clear regulations, and clear reflection of benefits for the businesses and manufacturing industries [28]. SAR requires clear shift in leadership commitment, governance, integration with business planning, on-the-ground measuring and reporting techniques (read: GRI standards are still superficial), organisational learning, cultural integration into organisational cultures, and development of effective data collection and information systems [6] [29]. A report by KPMG (2015) on corporate responsibility revealed inconsistencies in content, quality, and trends in global SAR. This report found on-the-ground challenges in clarity of materiality and data, clarity and methods for measuring sustainability indicators, clarity in formulation of sustainability targets, communication about the data collected, and analysis and reporting of SAR benefits to the business, industry and society [3]. Rambaud & Richard (2015) also stressed the need for clarity in accounting of environmental, social, and economic capital of a company and the ways these capitals have been invested to produce useful products and services for humanity [30].

Figure A.3 [31] in Appendix A presents the principles that an organisation should follow for SAR, and Figure A.4 [31] presents the basic layout of the SAR framework. The GRI SAR standards are outlined in Figure A.2 (Source: GRI Website). However, GRI only provides guiding principles and standards for SAR. The findings by KPMG (2015) of inconsistencies in global SAR may be related with the flexibility in decision-making allowed to businesses in the GRI reporting guidelines as shown in Figure A.5 [31]. For example, if the management of an organisation decides internally that their emissions are small and not worth reporting, they may exclude measuring carbon emissions and reporting justifying it as of the category "not necessary to report". Thus, legal and stakeholder pressures are important for consistency in SAR. In addition, large manufacturing organisations may not be ready to take accountability for emissions and non-compliances to environmental guidelines by their supply chain partners although there is a section in GRI standards on working with suppliers and outsourcing partners for sustainability compliance and SAR.

There are many challenges pertaining to SAR in India in the entire supply chains and the extended value chains. The next section presents a comprehensive analysis of challenges to SAR in India and the research opportunities in supply chains and value chains for strengthening it.

2. Challenges and Research Opportunities in Sustainability Accounting and Reporting in India

Figure A. 1 in Appendix A presents a report on top performing companies in India in the field of corporate social responsibility and sustainability [8]. The report found that the top listed companies are investing significantly in sustainable development as reflected in their SAR disclosures [8]. The report also found commitments to spending, management of spread of spending, corporate social responsibility matrix, insights into sustainable development, and sustainability governance. However, the SAR framework in India is highly skewed among listed companies and also among the industries [6] [7] [8].

The legal framework in India for SAR is quite old and is still evolving (KPMG, 2014). The SEBI's requirement for listed companies to report their level of business responsibilities, and integration with GRI's reporting standards and guidelines are good enablers for SAR in India, at least for the listed companies. However, the perceptions about SAR in companies are still more of an added responsibility for compliance reporting or for gaining some brand equity and reputation (KPMG, 2014). Further, the on-the-ground sustainability indicators and their measurement methods and approaches are not defined clearly [7]. Large companies are inward focussed in sustainability compliance and SAR and hence fail to apply the much required "stakeholder pressure" on their direct and indirect suppliers [7].

Adoption of GRI standards and guidelines by large companies for SAR in India is a major step. GRI opens up significant research opportunities for SAR in India. Academic and professional researchers may study the SAR standards and guidelines outlined by GRI and present on-the-ground approaches for following them. The

GRI SAR standard is divided into four sets of standards: universal standards (coded as 100), economic topics (coded as 200), and environmental topics (coded as 300, and social topics (coded as 400) (Figure A.2). The universal standards are further divided into foundation, general disclosures, and management approach. This part helps an organisation to create a scope and approach of SAR disclosures. This part is basic but opens up significant scope of study and methods that an organisation or industry can use for presenting and justifying its SAR scope and approaches.

The study and analysis of foundation, general disclosures, and management approach may result in profiles differentiated by size of organisation, nature of business, industry, sector, domains of impact, and extent of impact. The sustainability indicators may be designed to reflect the measures and measurement processes and skills to be followed by the organisation to justify the SAR requirements of economic, environmental, and social topics outlined in the GRI standards and guidelines.

The economic topics under GRI standards and guidelines cover the areas of economic performance, market presence, indirect impacts to local economy, procurement practices, anti-corruption practices, and anti-competitive behaviour. First of all, Indian researchers may like to study if these areas offer comprehensive coverage in the context of India. Many practices in India are traditional and may not be perceived as harmful while they are viewed differently in the global context. For example, India has a cash-based economy that has been thriving since independence. Even in 21st century, a significant part of Indian supply chains are informal and cash-based. There are little controls on environmental consumption practices, economic practices, and employment practices of Indian supply chains. Some good laws and institutions are in place but the enforcements are at low levels.

Recently, the decisions by Government of India to demonetise large denominations, introduce GST taxation, and put capping on the cash-based transactions have been widely discussed in both positive and negative contexts. A lot of research is needed on what is right for India. Should the supply chains continue to support cash-based economy or completely digitise everything and link with the unique identification of individuals, or maintain a balance? Should all labour and small vendor payments be routed through banks only? In addition, to economic impacts are there social impacts of such changes? These types of questions can only be answered after rigorous data collection and analysis.

The environmental topics under GRI standards and guidelines cover the areas of materials used by organisations, energy consumption (both non-renewable and renewable), water consumption, biodiversity consumption, emissions, effluents and wastes, environmental compliance, and sustainability assessment of suppliers. India is a large country with second largest population in the world. Hence, the extent of consumption and the permissible thresholds may not be in line with the globally acceptable standards. For example, the urban pollution levels in India are among the highest in the world. India may be facing a significant urgency in cleaning the local air than most of the western countries. Perhaps, whatever is done currently is dangerously inadequate. New research studies can highlight how close this country is to a massive catastrophe and what is the urgency facing us.

In addition to the areas related to social impacts of Indian supply chain and value chain systems, there is significant scope of research about Industrial engineering and processes for sustainable practices and SAR in Indian supply chains and value chains. Adoption of advanced technologies for measurement of hazardous gaseous emissions in factories is still evolving in India. Further, there is lack of reliable technologies and skills supporting safety and secure practices (related to industrial ergonomics) for workers working in hazardous environments and confined spaces. In spite of significant and rapid developmental projects happening in India, workers are still not provided adequate safety equipment that are common in western countries. For example, workers entering areas with possibility of toxic gaseous emissions still wear very basic gas masks or do not wear gas masks at all. Further, there are challenges of inadequate wages, inappropriate working conditions, child labour, etc. in Indian supply chains.

In addition, there needs to be research studies on technologies and designs for tracking and accounting consolidated data on emissions from all the vehicles and other logistics machineries and facilities used by a company (directly or through third party logistics providers). For example, there are research opportunities in the field of advanced wireless sensor networking, and Internet of Things for sensing environment-related variables.

IOP Conf. Series: Materials Science and Engineering **404** (2018) 012022 doi:10.1088/1757-899X/404/1/012022

In the era of application of device-based Internet enabling, RFID and Internet of things, closed vehicle area networking, dedicated bandwidth and frequency-band allocation for tracking and controlling supply chain operations, cloud computing technologies, advanced planning systems, cloud manufacturing, lean, agile, and quick response architectures and designs in supply chains, India can not only implement some of the best technologies but also develop local excellence centres for sustainability and SAR. India has a massive pool of highly competent scientists and engineers in these fields. These technologies have significant potential in making SAR effective and making triple bottom line a reality in India as revealed in some technical studies published in major technology journals of the world [32] [33] [34] [35]. As a fact, a large number of research papers related to these technologies in IEEE, ACM, Springer's journals, Emerald's journals, Science Direct, etc. are contributed by scientists and engineers in India or ones from Indian background.

3. Conclusions

Sustainability is a present challenge for India in the supply chains and value chains given their massive scales and impact on Indian and global environment, on economy of India, and on people. SAR is an emerging concept, which initially started informally in some countries and is now standardised through GRI. Many Indian companies have recently joined the GRI and are gradually improving their commitment to sustainability and SAR. However, India has significant sustainability challenge in supply chains and value chains at the technological level, ground level operations, and in traditional cultural practices. Supply chain researchers have significant opportunity in studying the Industrial engineering and technical aspects, economic aspects, labour safety and working conditions (ergonomic aspects), use of effective technologies for improving the conditions of labour, and economic aspects of sustainability compliance and SAR. Researchers may like to enter this area for conducting technical, architectural, design, as well as social studies on supply chain and value chain management. India has a massive pool of talent in relevant technology and management to build SAR and sustainability effectiveness and excellence locally.

Appendix A

Rank (2016)	Company	Rank (2015)	Company
1	Tata Chemicals Ltd.	1	Tata Steel Ltd.
2	Tata Steel Ltd.	2	Tata Power Company Ltd.
3	Tata Power Company Ltd.	3	UltraTech Cement Ltd.
4	Shree Cements Ltd.	4	Mahindra & Mahindra Ltd.
5	Tata Motors Ltd.	5	Tata Motors Ltd.
6	UltraTech Cement Ltd.	6	Tata Chemicals Ltd.
7	Mahindra & Mahindra Ltd.	7	ITC Ltd.
8	ACC Ltd.	8	Shree Cement Ltd.
9	Ambuja Cements Ltd.	9	Bharat Petroleum Corporation Ltd.
10	ITC Ltd.	10	Larsen & Toubro Ltd.

Figure A.1: SAR Ranking of top ten companies in India (Source: Majumdar, Rana, & Sanan, 2017)

IOP Conf. Series: Materials Science and Engineering **404** (2018) 012022 doi:10.1088/1757-899X/404/1/012022

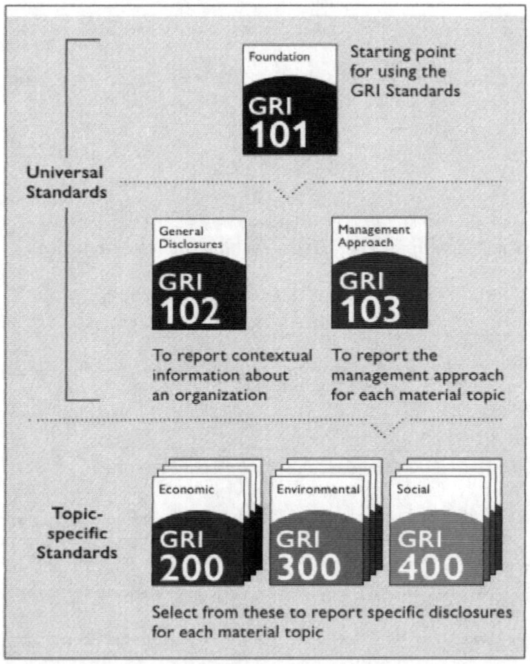

Figure A.2: Framework of GRI standards

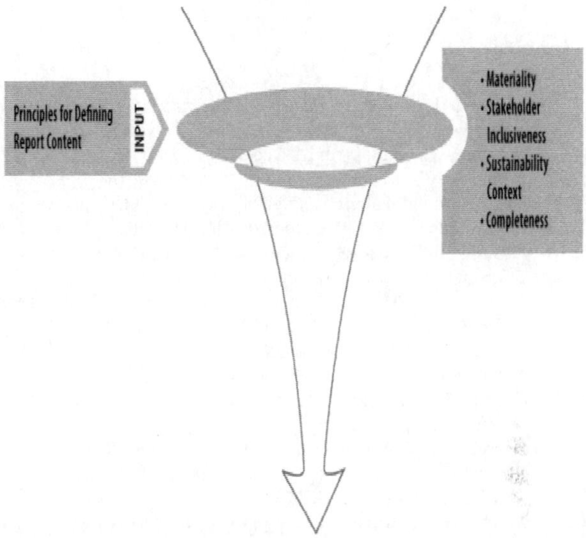

Figure A.3: SAR Principles

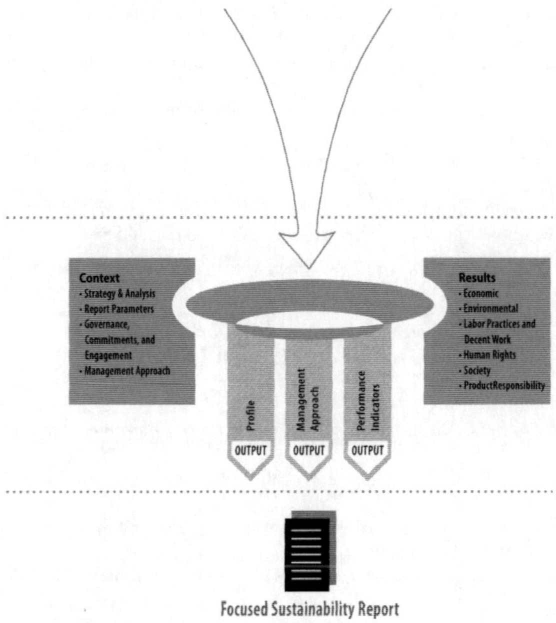

Figure A.4: SAR Framework

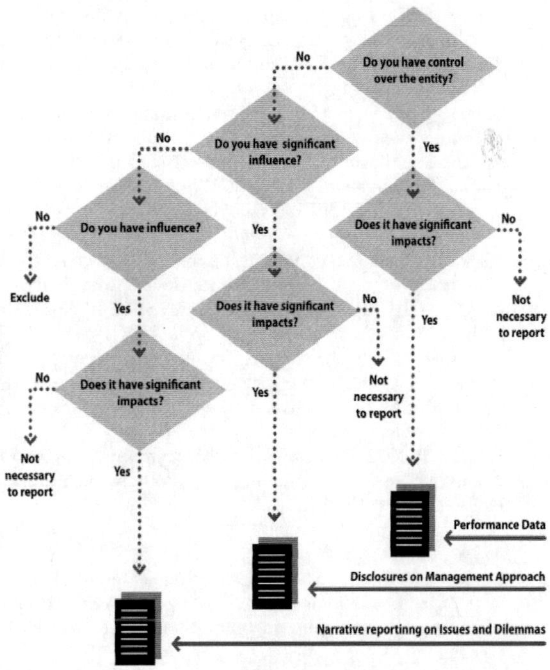

Figure A.5: Decision-making on SAR Framework

Acknowledgement

Manuscript Communication Number (MCN): IU/R&D/2018 – MCN 000318 office of research and development, Integral University, Lucknow.

References

[1] GRI 2017a *Harmonizing sustainability reporting practices in India* (GRI Press) (Online) URL: https://www.globalreporting.org/information/news-and-press-center/Pages/Harmonizing-sustainability-reporting-practices-in-India.aspx (Accessed: 21 December 2017).

[2] GRI 2017b *Leading Indian companies join GRI's reporting network* (GRI Press) [Online], URL: https://www.globalreporting.org/information/news-and-press-center/Pages/Leading-Indian-companies-join-GRI%27s-reporting-network.aspx [Accessed: 21 December 2017]

[3] KPMG 2015 *Currents of change-The KPMG Survey of Corporate Responsibility Reporting*-2015 (KPMG Survey of Corporate Responsibility Reporting) 2-48.

[4] Laskar N and Maji S G 2016 Corporate sustainability reporting practices in India: myth or reality? *Social Responsibility Journal* **12 (4)** 625-641 Emerald.

[5] GRI 2017c *Linking the GRI Standards and the SEBI BRR Framework* (GRI Standards) 3-26 (Global Reporting Initiative Netherlands).

[6] Garg P 2017 Development of sustainability reporting index (SRI) with special reference to companies in India *Decision* **44 (4)** 259-273 Springer.

[7] Kumar K S V and Devi V R 2015 *Sustainability Reporting Practices in India: Challenges and Prospects* Twelfth AIMS International. Conference on Management 2-5 January 2015 IIM, Kozhikode 1712-1717 The Association of Indian Management Scholars (AIMS).

[8] Majumdar U, Rana N and Sanan N *2017 Responsible Business Rankings: India's Top Companies for Sustainability and CSR* 2017 IIM Udaipur and Futurescape 6-58.

[9] Rauschmayer F and Lessmann O 2013 The Capability Approach and Sustainability *Journal of Human Development and Capabilitie*, **14 (1)** p 1-5 Routeledge.

[10] Christen M and Schmidt S 2010 *A Formal Framework for Theories of Sustainability* Sustainability Research Program (University of Basel Switzerland) p 1-16.

[11] United Nations Education Scientific and Cultural Organisation (UNESCO) (2006) Framework for the UN DESD International Implementation Scheme Document no. ED/DESD/2006/PI/1 [online] p 1-46 available at http://unesdoc.unesco.org/images/0014/001486/148650E.pdf [retrieved on 18 April 2016].

[12] Burger P and Christen M 2011 Towards a capability approach of sustainability *Journal of Cleaner Production* **19 (8)** p 787–795 Elsevier.

[13] Seuring S and Muller M 2008 From a literature review to a conceptual framework for sustainable supply chain management *Journal of Cleaner Production* **16** p 1699–1710 Elsevier.

[14] Seuring S and Muller M 2008a Core Issues in Sustainable Supply Chain Management – a Delphi Study *Business Strategy and the Environment* **17** 455–466 Wiley InterScience.

[15] Voget-Kleschin L 2013 Employing the Capability Approach in Conceptualizing Sustainable Development *Journal of Human Development and Capabilities* **14 (4)** p 483-502 Routeledge.

[16] Alkire S 2011 *The capability approach and human development* Oxford Poverty and Human Development Initiative (University of Oxford) [online] Retrieved from http://www.ophi.org.uk/wp-content/uploads/OPHI-HDCA-SS11-Intro-to-the-Capability-Approach-SA.pdf [retrieved on 18 April, 2016].

[17] Mahadi A 2012 *Adopting the Capabilities Approach in Developing a global Framework for measuring Sustainable Development* Published Master Thesis (Department of Earth Sciences - Uppsala University) p 1-71.

[18] Hall P 2012 USAID Environmental Compliance Reporting *Global Environment Management Support* p 1-13.

[19] Christopher M 2011 *Logistics & supply chain management* (fourth edition) Prentice Hall Financial Times (London: Pearson Education).

[20] Palevich R 2012 *The Lean Sustainable Supply Chain* (Worldwide: FT Press).

[21] Rechtschaffen C 2007 Shining the Spotlight on European Union Environmental Compliance *Pace Environmental Law Review* **24 (1)** p 161-186.

[22] Bebbington J and Larrinaga C 2014 Accounting and sustainable development: An exploration *Accounting, Organizations and Society* **39** 395–413 Elsevier.

[23] Hahn R and Kuhnen M 2013 Determinants of sustainability reporting: a review of results, trends, theory, and opportunities in an expanding field of research *Journal of Cleaner Production* **59** 5-21 Elsevier.

[24] Lamberton G 2005 Sustainability accounting—a brief history and conceptual framework *Accounting Forum* **29** 7–26 Elsevier.

[25] Bebbington J, Brown J and Frame B 2007 Accounting technologies and sustainability assessment models *Ecological Economics* **61** 224-236 Elsevier.

[26] Gray R 2013 Back to basics: What do we mean by environmental (and social) accounting and what is it for? *Critical Perspectives on Accounting* **24** 459–468 Elsevier.

[27] DEFRA 2013 *Environmental Reporting Guidelines: Including mandatory greenhouse gas emissions reporting guidance* Department for Environment Food & Rural Affairs (DEFRA) Government of UK 2-135.

[28] Stubbs W, Higgins C and Milne M 2013 Why Do Companies Not Produce Sustainability Reports? *Business Strategy and the Environment* **22** 456–470 Wiley.
[29] Smith P A C and Sharicz C 2011 The shift needed for sustainability *The Learning Organization* **18 (1)** 73-86 Emerald.
[30] Rambaud A and Richard J 2015 The Triple Depreciation Line instead of the Triple Bottom Line: Towards a genuine integrated reporting *Critical Perspectives on Accounting* **33** 92-116 Elsevier.
[31] GRI 2011 *Sustainability Reporting Guidelines* GRI Standards 3-195 (Global Reporting Initiative Netherlands).
[32] Fang S, Zhu Y, Ahati J, Pei H, Yan J and Liu Z 2014 An Integrated System for Regional Environmental Monitoring and Management Based on Internet of Things *IEEE Tranactions on Industrial Informatics* **10 (2)** 1596 1605 IEEE Xplore.
[33] Malhotra A, Melville N P and Watson R T 2013 Spurring impactful research on Information Systems for Environmental Sustainability *MIS Quarterly* **37 (4)** 1265-1274.
[34] Shrouf F, Ordieres J and Miragliotta G 2014 *Smart factories in Industry 4.0: A review of the concept and of energy management approached in production based on the Internet of Things paradigm* In 2014 IEEE International Conference on Industrial Engineering and Engineering Management (IEEM) 9-12 Dec. 2014 (Bandar Sunway- Malaysia IEEE Explore).
[35] Jones M J 2010 Accounting for the environment: Towards a theoretical perspective for environmental accounting and reporting *Accounting Forum* **34** 123–138 Elsevier.

A Comparative Study of Practices for Integration of Sustainability With Innovation for Micro, Small & Medium Scale Manufacturing Enterprises (Msmes) in India and in England

S Khurana[1] and B Mannan[2] and A Haleem[3]
[1] Research Scholar ,Jamia Millia Islamia ,Delhi
[2] Research Scholar ,Jamia Millia Islamia ,Delhi
[3] Professor ,Jamia Millia Islamia ,Delhi

E-mail: sonal.khurana@gmail.com

Abstract- The Brundtland report in 1987 has started a wide dialogue on incorporating sustainability into innovation, that is, the fusion of environmental and social aspects into products, processes, and organizational structures. While previous research has often focused on the large firms, the last decade has seen the main target of research being shifted to small industries that are now being increasingly recognized as main contributors to sustainable development. Also, in case of the Indian perspective, the Indian Government decided to declare 2010 as a Decade of Innovation. So this provides us with an opportunity to create a roadmap for Sustainable Oriented Innovation. MSME have been discussed in brief taking into account the Indian perspective. A study of literature on sustainability, sustainable development, sustainable manufacturing and innovation has been done. The need and benefit of integrating sustainability with innovation have been emphasized. Also, literature is obtained indicating how integrating sustainability with innovation acts as an enabler for MSME's to enhance their opportunities and also to increase their development potential. The manufacturing enterprises in developed (UK) and a developing country (India) which have incorporated sustainability with innovation are compared. Finally, the differences and the similarities in the magnitude of incorporation of sustainable oriented innovation are discussed.

Keywords: Sustainability, Sustainable Development, Innovation, Sustainable Manufacturing, Dynamic Capabilities, Micro Small & Medium Scale Manufacturing Enterprises (MSMEs)

1. Introduction

The year 2015 has been marked the 70th anniversary of the United Nations. It was also the year when countries assembled together to formulate the next generation of goals for our people, and our only home-planet: earth.2015 has also seen the hosting of the climate conference in Paris. This new agenda will lead the transition from millennium development goals-the MDGs to the next generation of sustainable goals, with a new time frame. There are seventeen goals covering the wider scope of sustainable development agenda, which includes more economic issues such as growth, employment, infrastructure and inequality; environmental concerns that include water, energy, terrestrial and marine ecosystems; and most important is the goal with the aim of promising more peaceful, better governed and inclusive societies. And integrating sustainability with innovation is a step in this direction only.

Various reports of media in the previous couple of years have been investigated and it has been found that enterprises from varied industries are either in the process of launching or are progressing to launch sustainable goods. For the past few years, scholarly investigations on sustainable technologies and manufacturing systems are being explored (Pujari,2006;Pujari, Peattie, & Wright, 2004) .It is not possible nowadays to think about economic development without considering preservation of the environment at the same time and giving shared advantage to the society (Khurana, Mannan & Haleem, 2014). In this way, corporations are routinely looking to exploit extensive variety of input resources which have better productivity and increased responsibility in their products and technologies that are used to manufacture the products (Kedia, Nordtvedt & Chen, 2008). In a highly competitive economic world these days, it is not necessary for the most competing organizations to utilize lower cost inputs any longer (Bosworth & Collins, 2003). Thus, the corporations that have been using performance measuring systems to find out and eliminate resource intensive operations have increased thereby moving towards efficient production models (Akoum, 2009).

Today, the latest patterns demonstrate that corporations are giving a lot of importance to sustainable product innovation; there is a considerable measure of disarray on what makes a green or sustainable good (Baumann, Boons, & Bragd, 2002). In the present study, we pursue Ottman's study who in 2006 stated that there is no user good which has no effect on the surroundings, in industrial terms, the terms 'green good' or 'sustainable good' are majorly used to portray the goods which aim in shielding or upgrading the green habitat by preserving energy and/or resources and also works hard to reduce or eliminate usage of harmful agents, pollution, and waste. This statement accentuates on the means by which varied types of sustainable goods concentrate on important environmental problems, for example, energy, materials/resources, and pollution/ harmful waste.

It is vital to take into account that not all items have a major ecological presence on every step of physical product life cycle .The ecological presence doesn't arise from each and every perspective (material, energy, and pollution) but numerous items have a noteworthy ecological effect in any of the steps. For instance, the company that manufactures furniture impacts mainly the forests (material) but the process of manufacturing washing machine has major ecological impact at the time of the usage of the goods (usage of energy, water, detergent etc). There are numerous different enterprises which have high ecological effect such as automobile and oil whose ecological presence can be substantial as they include all physical life cycle phases
(manufacturing, product usage, and disposal). Presenting radical sustainable introduction in any of the phases of product's physical life cycle or dealing with challenges related to sustainability across totally diverse areas like material choice, energy use, or pollution control may lead to major differentiation in the product and creates competition in the market for these products. Innovation in product or processes requires an improved level of corporate responsibility towards environmental issues and also an enhanced level of the companies' implementing environmental policies to initiate the ideas of sustainable product and put them into the right place and beating the difficulties and risks.

The principal motivation behind this study is to propose an integrated model that can help MSMEs develop innovation capabilities which can enable them manufacture innovative products which are environmentally and commercially sustainable. This research primarily throws light on manufacturing sector because to the following reasons: (i) increased possibility to influence environment such as air pollution, effluent run-off and improper disposal of solid wastes; (ii) novel technologies to cleaner processes as well as procedure which are not proceeding at a fast measure to portray the compelling need for environmental protection.
The main motivation behind choosing MSMEs that they are huge in numbers, survives with weak compliance to environmental norms, generally operated by owner-managers, develops or borrow technology without much concern to environment, normally there is no systematic framework in place for sustainability cross-check, paucity of sophisticated equipment and testing facility, and could be extremely polluting. Because MSMEs have tremendous potential for creating jobs for skilled and unskilled workers they enjoy tacit and explicit support of all governments across the globe. On this very count many a times the environmental agencies turn a blind eye to MSMEs' record on environmental-friendliness compliance. It is not very difficult to visualize that many MSMEs which are operating on poor environmental sustainability; they could make healthy life unsustainable in that part of the country and eventually start damaging the environment of the whole country or world. Another aspect is, if the MSMEs grow big without a concept or concern for sustainability then they will become an even bigger threat to the world environment.

2. Literature Review

2.1 Background

Recently, a survey was conducted by McKinsey and Company (2008) on companies' concern on climate change, and it was concluded that the highest rate of executives' reviewed take into account climate change essential, and about 60% consider creating and showcasing new items important. As stated in Data monitor, organizations have up till now introduced 458 goods that claim to be ''sustainable,'' ''eco friendly,'' or ''natural habitat friendly,'' and this figure will touch to 1570 new sustainable goods to be introduced in the coming year. For instance, development of soy-based seat cushion foam by Ford, elimination of the restricted ingredients by green list process as done by SC Johnson, and announcement by Lipton Tea of the enlargement of their program on sustainability, assuring to provide their tea from sustainable estates.

This encouraging attitude by the firms directs a distinct change in the environmental discussion pointing out that moving towards environmental value and social good is due to the product innovation driven by market and latest technologies instead of regulation only.

Also, the environmental issues like pollution and depletion of fossil fuel resources have become issues that are concerning the international organisations, national government and consumers. Recently, the OECD (2009) report suggested that the worldwide greenhouse gas emissions are most likely to increase by 70% by 2050.This has led to the augmented importance of sustainable development concept. This concept has been taken as a standard for the economic, environmental and social policies integration by the companies affecting their manufacturing processes and also their competitiveness (Bloom & Van Reenen, 2002). Thus innovation process incorporating development which is sustainable has received increased attention during the past years.

2.2 Micro, Small and Medium Scale Enterprises

They are a heterogeneous cluster as far as size and sector diversity is concerned (Hillary, 2004) and overall, it is not easy to clearly outline what an MSME is, because countries adopt different criteria (e.g. employment, sales, and turnover) for definition purposes. But, generally the MSMEs are outlined by the number of employees with a threshold between 100 and 500 (Ayyagari, Beck, & Demirguc-Kunt, 2007). They assume a vital part within the developing countries by adding to the national economic yield and in generating employment and are taken as drivers for expansion of the economy and advancement (Khurana, Khan & Mannan, 2013).These enterprises are there in all large production sub-divisions (food processing (tea and desiccated coconut), textile, foundry, brick and ceramic, etc.) and consist of 85% of the total number of producing enterprises in Asia. In China, over 99% of enterprises come in the category of small and medium scale enterprises. In various Asian countries, MSMEs represent 60–70% of the industrial production held domestically and contributes to 75–80% of the export revenue (Vepa, 1997). In many of the Asian countries, approximately 60–70% of the total labour force in the manufacturing sector is employed by the SMEs (Khurana, Khan & Mannan, 2012). This article focuses on to develop sustainable development through technological innovation in micro, small- and medium-scale enterprises which have got attention from the government entrepreneurs in the country.

Also, MSMEs build a large group internationally which, for instance, data for OECD economies show where MSMEs make out between 96% and 99% of the total number of enterprises (OECD, 2009). MSMEs as a group contribute to a large share of overall pollution. The EU MSMEs account for approximately 64% of pollution. They are not thought of as subparts of their larger industries (Ullah, Abbas & Akbar, 2010) and fourth 'SME peculiarities' imply that they will innovate differently for sustainability (Moore and Manring, 2009).

2.2.1 Micro, Small & Medium Scale Industries: Indian Perspective

For the last three decades, proper economic development in our country has taken place under certain severe constraints (Mannan and Khurana, 2012). The more vital of these constraints are: inadequacy of resources along with capital, equipment, technology, skills, etc., high growth rate of population, and adverse land-man ratio (Mannan, Khurana and Haleem, 2016). Besides this, the establishment of basic and heavy producer goods industries, that is an important pre-requirement for laying the base of rapid industrialisation, economic growth involves a heavy strain on restricted resources of capital and skills and provides limited employment opportunities (Husain & Sushil, 1997). It is in this context that great emphasis has been placed in our industrial

policy statements and programmes for the promotion and development of small industries which does not make high demand on scarce capital resources and also creates much larger opportunities for employment, including self-employment (Khurana, Khan &Mannan, 2012).
According to the Fourth All India statistics of MSME (Micro Small Medium Enterprises) 2006-07, the MSME sector approximately provides employment to 80.523 million people with the help of estimated 36.176 million enterprises. Of the total number of enterprises, 55.34% were in rural areas while 44.64% were in urban areas. 68.21% of the enterprises were engaged in services with remaining 31.79% engaged in manufacturing.

The MSME sector has a tremendous potential to cater to the issues of providing employment to the growing population (Mannan, Khurana & Haleem, 2012). Acknowledging the wide potential of innovative firms and agriculture based industries in MSME sector, the budget speech 2014-2015 delivered by the Hon'ble Finance Minister in the Parliament on 10-07-2014, said: "To establish Technology Centre Network to promote Innovation, Entrepreneurship and Agro-Industry, I propose to set up a fund with a corpus of INR 2000 Million." It would be taken up for incorporation of fresh ideas and to provide essential buttress for accelerating entrepreneurship." The Hon'ble Prime Minister, in his Independence Day speech in 2014 had also stressed on developing 'zero defect' products so that the probability of getting rejection from the world market is very less and 'zero effect' so that the manufacturing process does not harm the environment. Thus this provides us with a huge potential to incorporate sustainability into innovation in case of Indian MSMEs.

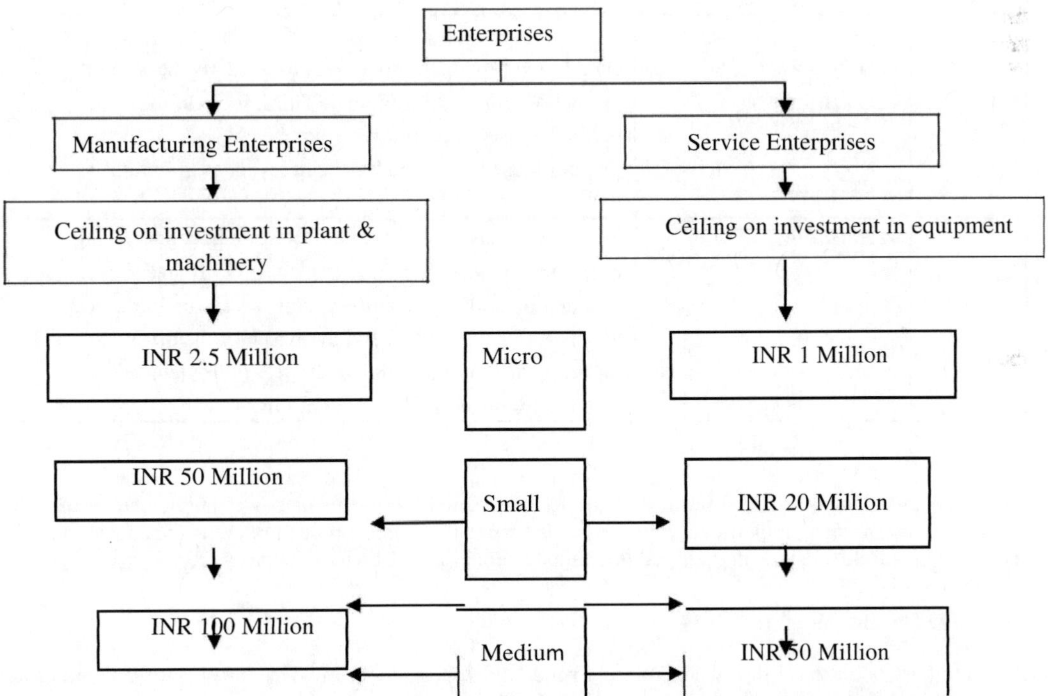

Figure 1: Structure of the Small & Medium Scale Enterprises (*Source:* Khurana, Khan &Mannan, 2012)

2.3 Sustainability
The Forestry Commission of Great Britain outlined sustainability with four goals. First, it is for social progress that acknowledges the need of everybody. Second, it is for effectiveness of the protection of the environment. Third, it ought to pursue judicious use of natural resources. At last, it is for maintaining of high and balanced levels of growth of the economy and generation of employment. The US National Research Council defined the sustainability as "the level of

human consumption and activity, which might continue into the foreseeable future, so that the systems that provides goods and services to the humans, persists indefinitely".

The above mentioned researches are mainly concerned with the definitions of sustainability and sustainable development at the conceptual and abstract perspective. And, these ideas are associated with managerial issues such as social, economic, and environmental purposes. However, the features of sustainability can be expended by adding some other factors or by specifying the in depth composition of sustainability (Mannan, Khan & Khurana, 2013). The MSA (Manufacturing Skills Australia) developed the operating definition of sustainability with additional detailed specifications in terms of environmental, social, and economic factors. It is divided into internal and external factors. As an example, the internal factors of environmental sustainability are to reduce the cost and waste. Also; Werbach outlined sustainability with four components: social, economic, environmental, and cultural. It had been outlined by adding the cultural construct, which implies, to safeguard and understand the worth of diversity through which communities manifest their identity and develop traditions across generations, to the general definition. Along with these researches and studies, sustainability can be mainly defined in social, environmental, and economic perspective (Mannan, Khan & Khurana, 2012). And it may be extended with additional general or specified factors in interdisciplinary area (Mannan, Khan&Khurana, 2013). In this paper, sustainability in manufacturing or sustainable manufacturing is reviewed based on the existing definitions in following section (Mannan, Khurana, & Haleem, 2016).

Table 1: Sustainability Definition by various researchers

S. No.	Reference	Definition of Sustainability
1	The Forestry Commission of Great Britain	First, it is for social progress that acknowledges the need of everybody. Second, it is for effectiveness of the protection of the environment. Third, it ought to pursue judicious use of natural resources. At last, it is for maintaining of high and balanced levels of growth of the economy and generation of employment.
2	Werbach, A	He outlined sustainability with four components: social, economic, environmental, and cultural. It had been outlined by adding the cultural construct, which implies, to safeguard and understand the worth of the diversity through which communities manifest their identity and develop traditions across generations.

We have referred to sustainability in total in this article, but we will allow for narrower approaches dealing with a selected sustainability dimension, particularly ecology, as firms often focus on or initiate the enhancements in either environmental or social dimensions in their path toward sustainability (Klewitz & Hansen, 2014).

2.3.1 Sustainable Development

The term got importance after 1987, when in the Brundtland report, the United Nations' World Commission on Environment and Development outlined the *sustainable development* as development that "meets the needs of the present generation without compromising the ability of future generations to meet their own needs."

Along with the greater concern to the social, environmental, and other important issues, all existing methods, technologies, and paradigms are required to be innovated in terms of the win-win approach that is sustainable for the our future(Mannan,Khan & Khurana , 2012). This idea can be applied to economy, management, policy, social operation, and manufacturing technology. The World Commission on Environment Development (WCED) report (later published as a book "Our Common Future") outlined that the sustainable development is "development that meets the needs of the present without compromising the ability of future generations to meet their own needs."

Sustainable development gives us a new direction of thinking thoroughly considering and overseeing human effect on the world – one that can create enduring advantages for all of us (Kumar, Luthra, & Haleem, 2013). Two points are important to sustainable development. In the first place, the conclusion that the growth of the economy alone is insufficient in resolving the world's issues: those related to economy and ecology and other aspects of any process are connected with each other. Taking only one of these at a time will cause errors in judgment and "unsustainable" results. Dealing solely on net revenues has conventionally prompted social and environmental harms that affect the general public in the long run (Husain, 2014).

The most important issue for the development to be sustainable is the urgency to take into account "three pillars" simultaneously: society, the economy and the environment (Lee, Kang, & Noh, 2012). Whatever the case may be, the primary logic is not changed – human beings, environment and systems related to economy are interconnected. We may not give emphasis to their interconnection for couple of years more; however history has frequently reminded us of the interconnection by some form of alarm or crisis.

Every nation's historical, economic, social and political setting is different from the other; however the fundamentals of sustainable development apply to all. Economic growth is important; however growth alone, while not seeing all the elements that add to well-being, does not decrease poverty sustainably (Husain & Sushil, 2000).

Manufacturers are planning to design greater number of products that appeal for their aesthetic qualities or their ease of usage as well as for their ecological and social sustainability (Husain, 2016). The changes in awareness of the consumer and the multiplication of increased number of sustainable items and services in the course of recent years are encouraging. Some critics and consumer advocates correctly bring up that some of this is often "fluff" or "greenwashing". Items that claim to be ecologically friendly can't actually be so once you investigate at the list of ingredients or analyse the complete product life cycle (Luthra, Kumar, Kumar, & Haleem, 2011). The fact that greater number of people and businesses perceive and even want to capitalise on this realisation demonstrates the increasing mass appeal of sustainability.

2.3.2 Sustainable manufacturing

Manufacturing industries are highly effective in terms of sustainability (Haleem, Sushil, Qadri, & Kumar, 2012). This is because various kinds of problems, which are associated with social, economical, and environmental issues, can be caused by the manufacturing process (Luthra, Kumar, Garg, & Haleem, 2015). Thus, it is very essential to implement manufacturing processes with sustainability. That is sustainable manufacturing. A lot of research has been done about definition and concept of sustainable manufacturing. National Council for Advanced Manufacturing (NACFAM) of U.S.A. outlined that the concept of sustainable manufacturing incorporates the manufacture of products which are sustainable and the manufacturing sustainably all the products. Manufacturing using renewable sources of energy, energy efficiency, green building, and other sustainable & social equity-related goods are related to sustainable product (Cooper, 1990). Also, manufacturing sustainably the majority of products involves taking into consideration the complete sustainability/full life cycle issues related to the goods manufactured.

Lowell Center for Sustainable Production (LCSP) outlined that the production which is sustainable is "*the creation of goods and services using processes and system that are: Non-polluting, Conserving of energy and natural resources, economically viable, Safe and healthful for workers, communities, and consumers, and socially and creatively rewarding for all working people*".

Park, Craggs, & Shilton (2011) researched about the technology that aims to reduce energy consumption in manufacturing sector by studying various policies, standards, and researches and their reviewing research shows various information and data which can be helpful for the development of energy-saving and sustainable technology in manufacturing sector.

LCA (Life Cycle Assessment) is established based on ISO 14040-14049 for investigating the impacts of both manufactured and consumed products on environment. And Eco-Indicator 95 and 99 are made to find out the damage impacts on human health, ecosystem, resource, etc However, these methodologies do not concentrate on the most vital issues of manufacturing company, which are on economical impact of production.

Many studies and researches on sustainable manufacturing show numerous basics and definitions of sustainable manufacturing, but they are just conceptual or constrained within the range of analysis such as damage effects on human health and ecosystem. Thus, systematic, structured, and integrated approach is required for practical application of sustainable manufacturing.

IOP Conf. Series: Materials Science and Engineering **404** (2018) 012023 doi:10.1088/1757-899X/404/1/012023

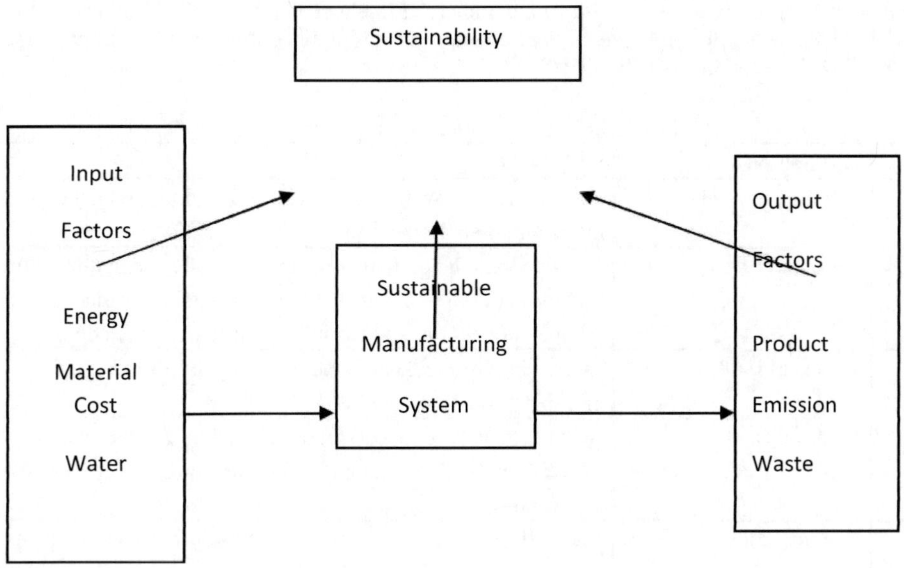

Figure 2: Sustainable Manufacturing System (*Adopted from:* Choi, Lee, & Ham, 2016)

Figure 2 depicts the composition of sustainability factors with the sustainable manufacturing system. With the existing definition of sustainability and sustainable manufacturing, various types of energy, physical materials, water, cost, package, and human resource are input factors of sustainable manufacturing system, and the quantity of manufactured product and generated emission/waste are output factors of sustainable manufacturing system.
In case of energy, it can be used such as gas, oil, electricity, etc. Material depicts the all kinds of substances to be made to product. Water is generally industrial water. The prices are fixed. Packages are various materials for packing products. Human resources include direct/ indirect labors. Emission includes various green-house gases such as carbon dioxide, methane gas, nitrous oxide, etc. Waste includes solid and liquid wastes. Thus, emission and waste should be reduced against the quantity of manufactured product.

2.4 Innovation

Schumpeter (1934) is generally regarded to be among the first to think of the procedure of innovation in organizations. According to him, innovation is considered as the formation and execution of combinations which are novel or different. These different combinations can be associated with novel goods, services, work processes, markets, delivery systems and policies.

"Innovation is the particular instrument of entrepreneurs, the method by which they exploit change as an open door for a different business or service." (Drucker, 1985)
"It is the heart and the way of thinking of entrepreneurship" (Drucker, 1998)
Innovation is economic and social achievement which is obtained by introducing novel ways or novel combining of present approaches in the input-output conversion that majorly change the relationship between utility value and price of good which is given to consumer and or user, community, society, and environment (Fontana, 2009).

Innovation is "the successful development, execution and usage of fresh or structurally improved items, procedures, services or organisational forms" (Hildrum, 2007).
Innovation is more comprehensive than creativity or Invention (Rothwell, 1992). An invention is an imagination, sketch or an idea for novel concept or a better device, good, procedure or a

system, but, an innovation is achieved with first commercial implementation involving an invention (van der Panne, van Beers, & Kleinknecht, 2003).An invention becomes innovation once it enters the economic system (Grossman & Helpman, 1994; Bowonder & Miyake, 1988).

Table 2: Definition of innovation by various Researchers

S.NO	REFERENCE	DEFINITION
1.	Schumpeter(1934)	He depicted innovation as the creation and implementation of different combinations.
2.	Drucker (1985)	Innovation is the particular instrument of entrepreneurs, the method through which they explore new ideas or innovation as an open door for a new business venture or service delivery.
3.	Drucker (1998)	Innovation is the heart and the way of thinking of entrepreneurship.
4.	Hildrum (2007)	Innovation is "the successful development, execution and usage of fresh or structurally improved items, procedures, services or organisational forms"
5.	Fontana(2009)	Innovation is economic and social achievement which is obtained by introducing novel ways or novel combining of present approaches in the input-output conversion that majorly change the relationship between utility value and price of good which is given to consumer and or user, community, society, and environment

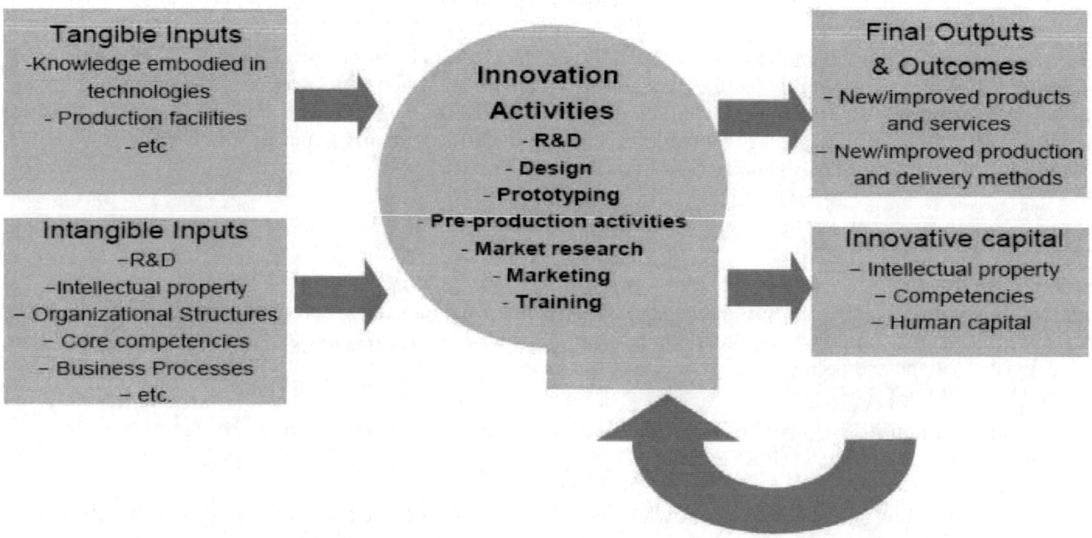

Figure 3: A typical system of innovation (Source: Rose, Shipp, Lal & Stone, 2009)

Innovation/Introduction is differentiated in the following forms:

- Good and Procedure Introduction
- Technological and Organisational Introduction
- Radical (or Revolutionary) and Incremental (or Evolutionary) Introduction
- Disruptive (or Discontinuous) and Sustaining (or Continuous) Introduction

IOP Conf. Series: Materials Science and Engineering **404** (2018) 012023 doi:10.1088/1757-899X/404/1/012023

- Open (External) and Closed (Internal) Introduction
- Component level and System level Introduction

Good and procedure introductions are considered as a very crucial innovation activity of the companies (Alum & Drucker, 1986).
Open innovation has been applied by the food industry in the recent times.
Practicing innovation is the desire on the part of policy makers to capitalize on the full potential of advanced technology that has become available in the wake of globalization of economies (Husain, 2016, Galanakis, 2006).
Depending on the stage of adoption and degree of success, the firms can be classified as (Docherty, 2006):

- If the company introduces innovation under the period of review, then the same is termed as the innovative firm. It is not necessary for the innovations to be a commercial success and many a times, innovations fail.
- If the innovation activities are going on during the period under review, then the firm is called as the innovation active firm. It is not necessary whether the exercise concluded in the execution of an introduction.
- An introductive active firm in which innovation efforts have been made but have not been successful in converting them into innovations is termed as potentially innovative firm.

2.5 Integrating Sustainability with Innovation for Indian Manufacturing Enterprises

2.5.1 Need for Integrating Sustainability with Innovation for Indian manufacturing Enterprises

According to Dangayach & Deshmukh (2003), 'reforms in the economic sector and competition globally have given an opportunity for the Indian manufacturing companies to see the key role of manufacturing'. This has given motivation to Indian enterprises to give high focus to quality and technology management.
Ganapathy (2014) argued that Indian manufacturing enterprises are working hard to enhance their sustainable performance in order to satisfy multiple stakeholders. Sustainable oriented innovation is an upcoming approach which reduces environmental impact and influences enterprises to improve the business performance.
Thirupathi & Vinod (2016) studied sustainable manufacturing practices in Indian automobile manufacturing sector and found that sustainable manufacturing practices are vital for automotive manufacturing enterprises in order to have competitive advantage in terms of cost, quality, delivery, flexibility, research and development. Thus, Indian manufacturing enterprises need a sustainable manufacturing strategy for survival in global arena.
Innovation can be considered as an important force for greater efficiency and increasing the competition in the market, and also, with the help of collaborative approaches and comprehensive growth, it helps in allaying poverty (Mannan, Khurana & Haleem, 2012; Husain
& Sushil, 2000). Besides this, innovation has always been the approach through which individuals tackled the major difficulties confronting the society (Arora & Nath, 2015; Gaynor, 2002). The probability of draining assets faced by the world has increased and demand has exceeded the supply. The worldwide food system is attempting to cater to the increasing demands and the World Bank estimates an increase of 50 per cent in global demand for food by

2030 (Bessant & Bessant, 2003). Such types of gap will be observed in the supply of water, and increased consumption of energy will lead to high pressure on the system. Energy consumed by Asia alone is approximated to become twice in the coming decade– to about 48 per cent for oil and 22 per cent for natural gas. Such major problems can only be taken care of by sustainable innovative solutions (Dangelico and Pujari, 2010). This data from the review of the literature motivated to perform this research on incorporation of sustainability with innovation for Indian manufacturing enterprises.

2.5.2 Comparative Case Study involving MSMEs in Bangalore and North East England The main aim is to investigate and compare the manufacturing enterprises in developed (UK) and a developing country (India) which have incorporated sustainability with innovation.

This research is focussed on small enterprises in the engineering industry in (1) the northeast of England comprising Durham city and Sedgefield districts of Durham county and the city of Sunderland in the UK, and (2) Bangalore urban district in Karnataka state of India. The former is known for its cluster of manufacturing enterprises, particularly that of engineering (Subrahmanya, 2001). The Bangalore urban district is an industrially better developed region in India and is known for its concentration of engineering and electronics industries (Subrahmanya, 2001). There are approximately 337 medium and large industrial enterprises and about 30,000 small-scale industrial units located in and around Bangalore (Government of Karnataka, 2000b). In Bangalore, the inclination was on Peenya Industrial Estate. Peenya Industrial Estate is one of the largest and oldest of its kind in the whole of South and Southeast Asia.

After an in depth review of the literature, case studies of two countries were compiled to draw out similarities and differences in the pattern of sustainable oriented innovation followed by the MSMEs in two countries.

The basic characteristics of small engineering enterprises of Bangalore and Northeast England are described in terms of employment size.

Table 3: Distribution of small enterprises by employment (Source: Subrahmanya, 2005)

Employment size class	Bangalore	NorthEast England
Up to 10	13	9
11-50	21	18
51-100	-	7
Total Number of enterprises	34	34

2.5.3 Sustainable Oriented Innovation in MSMEs: Analysis of patterns

2.5.3.1 Dimensions

Sustainable Oriented Innovations of micro, small and medium scale engineering enterprises may take place through the following dimensions (Subrahmanya, 2005):

- New product development
- Substitution of raw material to make it more sustainable
- Changing product design
- Training given to the employees for enhancing internal R&D
- Adoption of life cycle approach
- Energy conservation

Table 4: Sustainable Oriented Innovation in MSMEs: Bangalore (Source: Subrahmanya, 2005; Author's own elaboration)

Factors → / ↓ Dimension	Self Motivation	Customer Needs	Competition	Government Rules & Regulations
New product development	**2**	**10**	**6**	-
Use of substitute material	-	-	**10**	**9**
Changed product design	-	**19**	**11**	-
Training given to the employees	-	**8**	**9**	**12**
Adoption of life cycle approach	-	**8**	**10**	-
Energy conservation	-	-	**2**	**15**

The factors which influence the integration of sustainability with innovation are as follows:

- Self motivation
- Customer needs
- Competition
- Government rules & regulations

These factors have been selected based upon the review of literature from various international journals on sustainable manufacturing in MSMEs.

Table 5: Sustainable Oriented Innovation in MSMEs: North East England (Source: Subrahmanya, 2005; Author's own elaboration)

Factors → / ↓ Dimension	Self Motivation	Customer Needs	Competition	Government Rules & Regulations
New product development	**10**	-	**3**	**9**
Use of substitute material	**5**	-	-	**5**
Changed product design	**16**	**12**	**5**	-
Training given to the employees	**18**	-	**2**	-
Adoption of life cycle approach	**15**	**6**	-	-
Energy conservation	**20**	-	-	**6**

2.5.4. Similarities in the magnitude of sustainable oriented innovation
An increased proportion of small enterprises in North East England than in Bangalore have embarked upon sustainable oriented innovations of one kind or the other. In the same way, an increased proportion of innovating small enterprises in Northeast England (ten out of 22) than in Bangalore (two out of 18) have achieved 'radical innovation' by developing novel products and obtaining 'product patents'. But, the greater part of the small enterprises in both the regions are working in the area of 'incremental innovations', majorly characterized by changing product designs. The incremental innovators are mainly driven by customer needs and/or guidance and support. Thus, there are similarities in the patterns of sustainable oriented innovations carried out by small engineering enterprises in both Bangalore and Northeast England. But, a major proportion of these enterprises are working in the area of radical as well as incremental innovations in Northeast England than in Bangalore.

The incremental innovators, on the other hand, are mainly focussed on changing product designs. In Northeast England, incremental innovations are mainly driven by self motivation and are carried out either independently or with customer support. In Bangalore, on the other hand, incremental innovators are of two kinds: (i) those who are driven by customer needs; (ii) those who are driven because of competition in the market and innovated with customer support to meet customer needs.

2.5.5. Differences in the magnitude of sustainable oriented innovations
Though Bangalore is viewed as the only 'global hub of innovation', this designation has been given to the region more because of its advancements in the area of information and communication technologies (ICTs). Thus, it is called 'the Silicon Valley of India' (Government of Karnataka, 2000b). India is acknowledged as a 'dynamic adopter' (UNDP, 2001) of novel technologies and not a 'potential technology leader', as compared to the UK. Although countries like India are viewed as progressive adopters of new technologies, the diffusion of old inventions is slow and incomplete (UNDP, 2001).

The R&D pace of India is much lower than that of developed countries such as the UK, the US, Japan, France, Sweden and Switzerland. The average R&D intensity of these countries was 2.25 in 1981, 2.68 in 1991 and 2.80 in 1993 as against India's 0.62 in 1981, 0.85 in 1991 and 0.86 in 1993 (DST, 2000). Although countries like India have drawn much attention as potential economic powers, they have not been investing rapidly enough to improve their innovative capacity across economic sectors to levels similar to OECD countries (Council on Competitiveness, 1999).

3. Conclusions
The acknowledgement that we are living in a world with limited resources is pushing activities towards greater sustainability. The requirement to evade unsustainable procedure has been a test for various decades now, but today it is progressively being taken into account with increased direness. The requirement for carrying out increased sustainable production and consumption practices has increased. It has changed the way the solution is needed. We have to increase our pace of using sustainable practices in a relatively brief time period. And this requires changes which are radical in nature and also should take into consideration the changes in complete
product structure or arrangement. In this sense, sustainability is taking towards an increased recognition of the requirement for innovation. Enhancing innovation potential is integral to achieving increased productivity levels and is imperative for achieving and sustaining the overall competitiveness of an economy.
This paper has also tried to find out the pattern of sustainable oriented innovations followed by small enterprises in the engineering units in Bangalore as compared to those of Northeast England. The pattern is developed based on (1) the factors that influence the sustainable oriented innovations, and (2) the dimensions of sustainable oriented innovations. Based on this pattern between the two regions, the similarities and dissimilarities are analyzed. Sustainable oriented innovations of small enterprises may be carried out in different ways depending upon various factors, internal as well as external to an enterprise. But the factors and dimensions of sustainable oriented innovations, radical or incremental, in small engineering enterprises appear to be similar between the northeast of England in the UK and the Bangalore urban district in India.

IOP Conf. Series: Materials Science and Engineering **404** (2018) 012023 doi:10.1088/1757-899X/404/1/012023

References

1. Akoum, I. (2009). Business success: Does good governance matter? A theoretical framework. *Journal for Global Business Advancement*, *2*(4), 365-380.

2. Alum, R. A., & Drucker, P. F. (1986). Innovation and Entrepreneurship: Practice and Principles.

3. Arora, P. & Nath, P. (2015). Innovation in Indian Firms: Evidence from the Pilot National Innovation Survey. *ASCI Journal of Management*, *Vol. 41*(1), 75-90.

4. Ayyagari, M., Beck, T., & Demirguc-Kunt, A. (2007). Small and medium enterprises across the globe. *Small Business Economics*, *29*(4), 415-434.

5. Baumann, H., Boons, F., & Bragd, A. (2002). Mapping the green product development field:
6. engineering, policy and business perspectives. *Journal of Cleaner Production*, *10*(5), 409-425.

7. Bessant, J. R., & Bessant, J. R. (2003). *High-involvement innovation: Building and sustaining competitive advantage through continuous change*. J. Wiley.

8. Bloom, N., & Van Reenen, J. (2002). Patents, real options and firm performance. *The Economic Journal*, *112*(478), C97-C116.

9. Bosworth, B., & Collins, S. M. (2003). The empirics of growth: An update.*Brookings papers on economic activity*, *2003*(2), 113-206.

10. Bowonder, B., & Miyake, T. (1988). Measuring innovativeness of an industry: an analysis of the electronics industry in India, Japan and Korea. *Science and Public Policy*, *15*(5), 279-303.

11. Brundtland, G. H. (1987). Our common future: Report of the 1987 World Commission on Environment and Development. *United Nations, Oslo*, 1-59.
12. Choi, B., Lee, J. N., & Ham, J. (2016). Assessing the Impact of Open and Closed Knowledge Sourcing Approach on Innovation in Small and Medium Enterprises. *Procedia Computer Science*, *91*, 314-323.

13. Cooper, R. G. (1990). Stage-gate systems: a new tool for managing new products. *Business horizons*, *33*(3), 44-54.

14. Council on Competitiveness, 1999. The New Challenge to America's Prosperity: Findings from the Innovation Index, Council on Competitiveness, Washington DC.

15. Dangayach, G.S., &Deshmukh,S.G.(2003).Evidence of manufacturing strategies in Indian industry:a survey .*International Journal of Production Economics*,83(3),279-298

16. Dangelico, R. M., & Pujari, D. (2010). Mainstreaming green product innovation: Why and how companies integrate environmental sustainability.*Journal of Business Ethics*, *95*(3), 471-486.

17. Docherty, M. (2006). Primer on open innovation: Principles and practice.*PDMA Visions Magazine*, *30*(2), 13-17.

18. Drucker, P. (1998). The discipline of innovation. *Leader To Leader*, *1998*(9), 13-15. http://dx.doi.org/10.1002/ltl.40619980906

19. Drucker, P (1985). *Innovation and Entrepreneurship: practise and principles*. London: Heinemann.

20. DST (Department of Science and Technology), 2000. Science and Technology Data Book, Ministry of Science and Technology, Government of India, New Delhi, India.

21. EU European Commission (2004). *Innovation Management and the Knowledge-Driven Economy*, Directorate-General for Enterprise, ECSC-EC-EAEC, Brussels.

22. Forestry Commission Scotland, & Great Britain. Forestry Commission. (2003). *Forests and water guidelines*. Forestry Commission.

23. Fontana, A. (2009).Innovate We can! Jakarta : Grasindo

24. Galanakis, K. (2006). Innovation process. Make sense using systems thinking. *Technovation*, *26*(11), 1222-1232.

25. Ganapathy, S.P., Natarajan, J., Gunasekaran, A., &Subramanian, N. (2014).Influence of eco innovation on Indian manufacturing sector sustainable performance. *International Journal of Sustainable Development & World Ecology*,21(3),198-209

26. Gaynor, G. (2002). *Innovation by design: what it takes to keep your company on the cutting edge*. Amacom.

27. Government of Karnataka, 2000b. The City of the Future, Bangalore. Publicity material, Karnataka Udyog Mitra, Bangalore, India.

28. Grossman, G. M., & Helpman, E. (1993). *Endogenous innovation in the theory of growth* (No.

29. w4527). National Bureau of Economic Research.

30. Haleem, A., Sushil, Qadri, M. A., & Kumar, S. (2012). Analysis of critical success factors of world-class manufacturing practices: an application of interpretative structural modelling and interpretative ranking process.*Production Planning & Control*, *23*(10-11), 722-734.

31. Hildrum, J. (2007). Does the emergence of distributed innovation call for new innovation process theories.

32. Hillary, R. (2004). Environmental management systems and the smaller enterprise. *Journal of cleaner production*, *12*(6), 561-569.

33. Husain, Z. (2014). Successful technology collaborations in automobile industry–strategic implications for firms in developing countries. *International Journal of Strategic Business Alliances*, *3*(4), 221-249.

34. Husain, Z. (2016). Technology strategy framework for firms in growing economies. *Journal for Global Business Advancement*, *9*(3), 248-274.

35. Husain, Z. (1997). Sushil. Management of technology: learning issues for seven Indian

36. Companies. *Technology Management: Strategy and Applications*, *3*, 109-35.

37. Husain, Z., & Sushil. (2000). Active transfer of technology in the automobile industry: Indian experiences. *International Journal of Services Technology and Management*, *1*(2-3), 236-280.

38. Kedia, B. L., Perez-Nordtvedt, L., & Chen, J. S. (2008). Importance of international skills for international business. *Journal for Global Business Advancement*, *1*(2-3), 153-177.

39. Khurana, S., Mannan, B., & Haleem, A. (2014). Integrating innovation with sustainability: A study of practices/status for Indian manufacturing industries (SMEs). In *BOOK OF ABSTRACTS* (p. 94).

40. Khurana,S., Khan,J., Mannan,B. (2013). *A study on Technology Management: Enablers and Barriers in its implementation in the case of SMEs in India*, Sustainability and development, Bloomsbury Publishing India pvt. Ltd:754-774.

41. Khurana,S., Khan,J., Mannan,B. (2012). *Enablers and Barriers for implementing Technology Transfer Projects: A study of SMEs in India.*Emerging paradigms in marketing, Wisdom Publisher India:303-311.

42. Klewitz, J., & Hansen, E. G. (2014). Sustainability-oriented innovation of SMEs: a systematic review. *Journal of Cleaner Production*, *65*, 57-75.

43. Kumar, S., Luthra, S., & Haleem, A. (2013). Customer involvement in greening the supply chain: an interpretive structural modeling methodology.*Journal of Industrial Engineering International*, *9*(1), 1-13.

44. Lee, J. Y., Kang, H. S., & Do Noh, S. (2012). Simulation-based analysis for sustainability of manufacturing system. *International journal of precision engineering and manufacturing*, *13*(7), 1221-1230.

45. Luthra, S., Kumar, S., Garg, D., & Haleem, A. (2015). Barriers to renewable/sustainable energy technologies adoption: Indian perspective.*Renewable and Sustainable Energy Reviews*, *41*, 762-776.

46. Luthra, S., Kumar, V., Kumar, S., & Haleem, A. (2011). Barriers to implement green supply chain management in automobile industry using interpretive structural modeling technique: An Indian perspective. *Journal of Industrial Engineering and Management*, *4*(2), 231-257.

47. Mannan,B., Khan,J., Khurana,S.(2012). *Information Technology: A Green Supply Chain Enable*
48. Emerging paradigms in marketing, Wisdom Publisher India: 336-346.

49. Mannan, B., & Khurana, S. (2012). Enablers and barriers for introduction of robotics as an AMT in the Indian industries (case of SME's). *International Journal of Computer Applications*, 19-24.

50. Mannan, B., Khurana, S., & Haleem, A. (2016). Modeling of critical factors for integrating sustainability with innovation for Indian small-and medium-scale manufacturing enterprises: An ISM and MICMAC approach. *Cogent Business & Management*, *3*(1), 1140318.

51. Mannan,B. Khan,J., Khurana,S., (2013). *Enablers and barriers to KM in project based organization, Sustainability and development,* Bloomsbury Publishing India pvt. Ltd:754-774.

52. Mannan, B., Khurana, S., & Haleem, A. (2012). *Identification and analysis of critical factors of KM in project management: An interpretive structural modelling approach"*XVI Annual International conference of Society of Operation Management, Jointly hosted by IIT Delhi and IIM Lucknow.

53. Moore, S. B., & Manring, S. L. (2009). Strategy development in small and medium sized enterprises for sustainability and increased value creation.*Journal of cleaner production*, *17*(2), 276-282.

54. OECD (2009).Annual Report. Retreived from https://www.oecd.org/newsroom/43125523.pdf.

55. Park, J. B. K., Craggs, R. J., & Shilton, A. N. (2011). Wastewater treatment high rate algal ponds for biofuel production. *Bioresource technology*, *102*(1), 35-42.

56. Pujari, D. (2006). Eco-innovation and new product development: understanding the influences on market performance. *Technovation*, *26*(1), 76-85.

57. Pujari, D., Peattie, K., & Wright, G. (2004). Organizational antecedents of environmental responsiveness in industrial new product development.*Industrial Marketing Management*, *33*(5), 381-391.

58. Rose, S., Shipp, S., Lal, B., & Stone, A. (2009). Frameworks for measuring innovation: Initial approaches. *Athena Alliance, Washington*, (s 5).

59. Rothwell, R. (1992). Successful industrial innovation: critical factors for the 1990s. *R&d Management*, *22*(3), 221-240.

60. Schumpeter, J. A. (1934). *The theory of economic development: An inquiry into profits, capital, credit, interest, and the business cycle* (Vol. 55). Transaction publishers.

61. Subrahmanya, M. B. (2005). Pattern of technological innovations in small enterprises: a comparative perspective of Bangalore (India) and Northeast England (UK). *Technovation*, *25*(3), 269-280.

62. Subrahmanya, M. B. (2001). Technological innovations in small firms in the North East of England :Dimensions & Implications.International Journal of Entrepreneurship and Innovation,2(3), 141-152

63. Thirupathi, R.M., & Vinodh, S. (2016).Application of interpretive structural modeling and structural equation modeling for analysis of sustainable manufacturing factors in Indian automotive component sector. *International Journal of Production Research*,1-22

64. Ullah, F., Abbas, Q., & Akbar, S. (2010). The rationale for location preferences of technology-based small firms in the UK. *Journal for Global Business Advancement*, *3*(1), 79-93.

65. UNDP (United Nations Development Programme), 2001. Human Development Report 2001, Oxford University Press, New York.

66. Van der Panne, G., Van Beers, C., & Kleinknecht, A. (2003). Success and failure of innovation:
67. literature review. *International Journal of Innovation Management*, *7*(03), 309-338.

68. Vepa, R. K. (1997). Small Can Be Beautiful: Recommendations on Small Enterprises. *Economic and Political Weekly*, 1581-1583.

69. Werbach, A. (2009). *Strategy for sustainability*. Boston, Mass.: Harvard Business Press.

IOP Conf. Series: Materials Science and Engineering **404** (2018) 012024 doi:10.1088/1757-899X/404/1/012024

Natural fiber based adhesive butt joints as a replacement to gas welded butt joints for thin tubes: An experimental study

Anubhav Mishra[1], Suruchi Singh[2], Rajeev Kumar[3*] and Arun Mital[4]
[1,2]M.Tech Student, Department of Mechanical Engineering, Institute of Engineering & Technology, Lucknow, Uttar Pradesh, India -226021
[3*]Professor, Department of Mechanical Engineering, Institute of Engineering & Technology, Lucknow, Uttar Pradesh, India -226021
[4]Associate Professor, Department of Mechanical Engineering, Institute of Engineering & Technology, Lucknow, Uttar Pradesh, India -226021

Email: anubhavmishra.er@gmail.com, suruchimech04@gmail.com, rajeev.kumar@ietlucknow.ac.in and arun.mittal@ietlucknow.ac.in

Abstract- In this investigation, adhesive based butt joints were made between two mild steel tubular rods of 1.5 mm thickness and between mild steel and aluminium tubular rods of 1.6 mm wall thickness by winding a single sleeve of woven jute fiber. The sleeve was wetted with a suitable epoxy resin before it was wound. In few hours the winding cures to give a strong joint. The strength of the butt joint was determined for two loading conditions; axial tension and under 3-point bending. The strength of woven jute fiber butt joints was then compared with the strength of similar tubular rods joined with help of gas welding. It was observed that the strength of composite butt joints between dissimilar materials, aluminium and mild steel tubular rods, was higher in bending than butt joint between mild steel to mild steel tubular rods while the later was stronger in tension.

Keywords: Woven jute fiber, similar and dissimilar materials, epoxy resin, butt joints, adhesive joint.

1. Introduction

In product fabrication, structural and pipeline applications joining of two materials plays a very important role. There are various types of joining techniques available they can be broadly classified as (i) welding, brazing and soldering (ii) stitching (iii) adhesive bonding (iv) tying components together (v) fasteners- joining through screwing, bolting and riveting (vi) joining used by magnetic attraction. All the joining techniques have their own advantages and disadvantages. In most cases it has been often found that joints are the weakest link. Welding is the most common joining techniques in industrial applications however; the major constraint of welding is that it can only join similar materials. Further, though welding joins the steel parts effectively but it has its own limitations like subjecting components to high temperatures with associated problems such as scaling, distortion, crack in heat affected zone (HAZ) or weldment. Stitching is generally preferred for joining components like leather, fabric etc. Adhesive bonding can join dissimilar materials but it is not good in tension. Tying components has its disadvantages such as breakage of tow and outward impression of knots.

IOP Conf. Series: Materials Science and Engineering **404** (2018) 012024 doi:10.1088/1757-899X/404/1/012024

Fasteners perform very well in many applications but it is costly and decreases the strength of components.

J M Lees [1] used GFRP (Glass reinforced fiber polymer) for tubular pipe sections of small diameter bonded by adhesives and checked the performance of joint under combination of high tensile axial load and under low pressure. Jiao et al [2] studied the strength of CFRP (Carbon fiber reinforced polymer) and butt welded joint of very high strength steel tube. Two coaxial steel tubes were joined by CFRP and welding and their load capacity was observed under tension on different specimens. De Morais et al [3] carried out the investigation of strength of stainless steel pipes joined using two epoxy adhesives. Experimental test on single lap and butt joints carried out along with adherent and napkin ring tests. Result shows that tensile and shear strength of both the epoxy adhesive were similar and failure could be predicted by maximum shear strain criteria using FE Analyses. Mertiny & Ellyin [4] used a novel adhesive insertion method for joining composite pipes and studied their performance. Related damage behaviour of prototype size pipe joined by above method was also considered. Griffin et al [5] used a theoretical model to find out the strain in adhesively bonded joints of pipes. Delamination was the main mode of failure in this study. Pedro Glvez et al [6] developed a finite element model of a bus steel structures in order to obtain the forces that work on the reference node from the obtained force values, the value and type of stress at the reference node were calculated. A new carbon fiber reinforced polymer (CFRP) node was developed replacing the existing welded joint by steel – CFRP adhesive joints. This new adhesive joint showed strength values higher than 30% of the strain values along with minimizing the fatigue problems. Kumar et al [7] in their study developed T-joint between two tubular rods using glass based FRP. The GFRP was wounded diagonally, straight and circularly over the joining area to create the T-joint configuration. The strength of these joints was checked under four loading (i) tensile (ii) in-plane bending (iii) bending under transverse loading and (iv) torsion cum bending. The strength of FRP joints were compared to the strength of welded T-joints developed using similar sized rods. Further, the strength of T-joint between dissimilar tubular rods were developed and tested and compared.

In the present study, an adhesive based joint using natural fiber is used for butt joining of similar and dissimilar metals pipes. The parts were joined by using woven jute mat sleeve which was impregnated with an epoxy resin before winding. Strength of these joints were then experimentally determined and compared with the strength of two mild steel tubular rods butt joined with gas welding.

2. Materials and methods

This segment describes the material used, specimen preparation and various testing procedures used in this study.

2.1 Material for reinforcement

The use of natural fibres are increasing day by day because of their easily availability and lesser cost as compare to synthetic fibres. In the present work, woven jute fiber mat was used as reinforcing material. Jute is created from plants which are classified in the family of malvacease. It is one of the most common and widely used and easily available natural fibres after cotton fiber. It is 100% biodegradable, recyclable and thus environmentally friendly. Jute has tensile strength in the range of 393-773 MPa, Young's modulus 26.50 GPa, elongation at break 1.50-1.80 and density 1.30 g/cm^3 [8]. Jute also offers many advantageous like good insulating properties and lower thermal conductivity and it regains moisture at

IOP Conf. Series: Materials Science and Engineering **404** (2018) 012024 doi:10.1088/1757-899X/404/1/012024

moderate level. In addition, jute has ability to merge with natural and synthetic fibres and it also accepts cellulosic dye classes.

2.2 Materials for specimen

Two non-galvanized Mild Steel pipes of 24.5 mm diameter and 1.5 mm wall thickness and 120 mm in length were used for making specimens with similar materials while for joining dissimilar materials similar steel pipes along with aluminium pipes of 24.5 mm diameter and 1.6 mm wall thickness and 120 mm length were used. Figure 1 shows the schematic diagram of butt joint between two tubular rods.

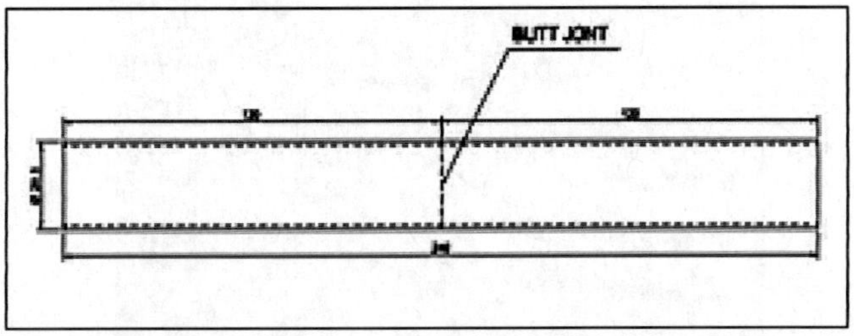

Fig 1: Schematic drawing of butt joint between pipes

2.3 Resin

Standard epoxy resin (Araldite AW 106) was used with hardener (HV 953 IN) for joining the specimens. Resin and hardener were mixed in the ratio of 5:3 and it took 3-4 hrs for curing at room temperature.

2.4 Specimen fabrication

In this study, two types of specimens were prepared. Butt joint between two similar materials using two mild steel pipes and joint between two dissimilar materials using mild steel and aluminium pipes.

To prepare the specimen with similar materials, the surfaces of each mild steel pipes of length 120 mm, diameter 24.5 mm and thickness 1.5 mm were prepared manually roughen by emery paper (grade 180) and then cleaned and degreased with acetone for proper bonding. These pipes were then clamped onto a specially fabricated fixture to ensure straightness and alignment of specimen while the joint is created. The mild steel pipes were clamped on the fixture for joining and proper alignment during the joint fabrication. Figure 2 shows the mounting fixture along with a specimen mounted over it.

Fig 2: Clamping fixture for preparing the specimen

IOP Conf. Series: Materials Science and Engineering **404** (2018) 012024 doi:10.1088/1757-899X/404/1/012024

The epoxy paste that was prepared by combining resin with hardener was applied on the surface of mild steel pipes and after applying paste, woven jute fiber sleeve was wrapped around the surface of pipe over a span length of 80 mm. Specimen was left on the fixture for 3-4 hours for curing at room temperature. After the curing, the specimens were removed from the fixture. In a similar manner, total six specimens were prepared. Figure 3 shows the specimens prepared with mild steel pipes.

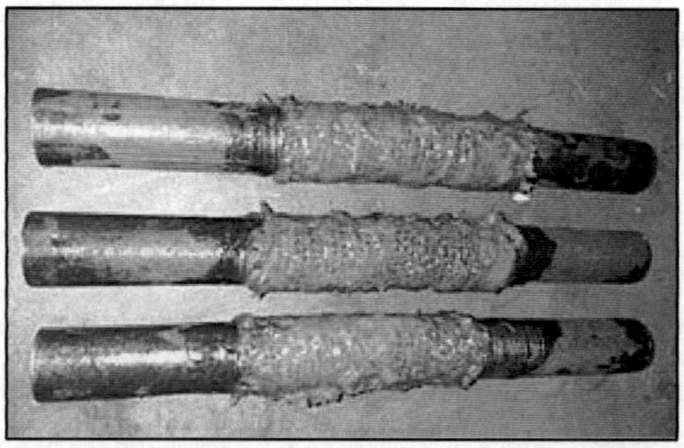

Fig 3: Specimen with similar materials

Similar procedure was followed for preparation of specimens with dissimilar materials (aluminium and mild steel). In this case also total 6 specimens were prepared. In addition, total 6 specimens were also prepared by butt joining two similar mild steel pipes with help of gas welding.

3. Experimental procedure

The specimens were tested under axial tension and under three-point bending with help of a universal testing machine to determine their breaking strengths.

3.1 Tensile testing

To test the specimen under tension, two 13mm diameter holes were drilled on both the sides of tensile specimens at a distance of 95 mm from the midpoint of the joint. The specimens were mounted between the jaws of the universal testing machine with help L shaped fixtures as shown in figure 4. Axial tension force was applied at a load rate of 2.0 kN/mm. Specimens were allowed to fail and corresponding forces were recorded.

Fig 4: Tesile specimens mounted in the universal testing machine

3.2 Three-point bending

The specimens were supported over the two rollers of the three-point bend fixture of the universal testing machine and 2.0 kN/mm lateral load was applied at the midpoint of the joint as shown in figure 5. The joints were allowed to fail under lateral loading and corresponding force values were noted.

Fig 5: Specimen under three-point bending

4. Results and Discussion

Three types of specimens were fabricated for this study; mild steel to mild steel joined in butt position using woven jute mat based composite joint, mild steel to aluminium pipes joined using same method and mild steel to mild steel pipes joined using gas welding. Each specimen was tested under axial tension and three-point bending loadings.

Figure 6 displays the breaking strengths of each type of the specimens. The average breaking force under tensile test of the gas welded joints was found to be about 12.33 kN while the failure forces for composite joints for similar materials and dissimilar materials were 2.6 kN and 1.8 kN respectively. It can be observed that the breaking strength under tensile test of the mild steel to mild steel specimen joined using gas welding was much higher as compared to the breaking strengths of similar and dissimilar materials joined using jute epoxy composite. Under tension joints made between similar materials was about 30% stronger than the joints between dissimilar materials.

IOP Conf. Series: Materials Science and Engineering **404** (2018) 012024 doi:10.1088/1757-899X/404/1/012024

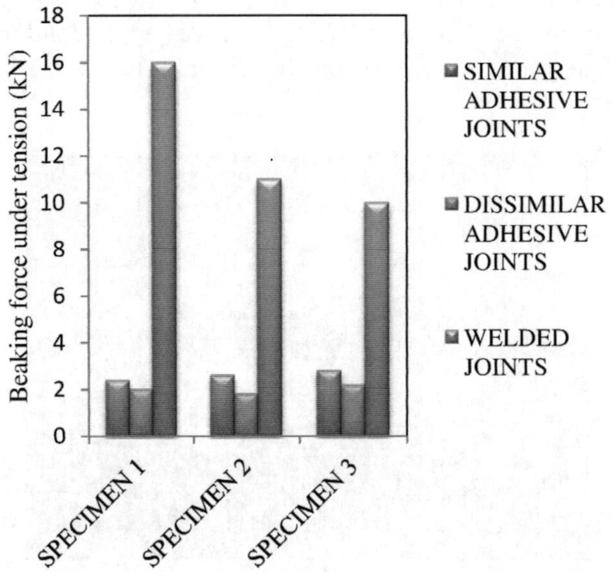

Fig 6: Breaking strengths of the specimens (Axial Tension)

In a similar way, comparison of bending strengths of the three types of the specimens can be seen from figure 7. The average breaking force under three-point bending for gas welded joints was found to be 6.53 kN while these force values for composite joint with similar material was 1.47 kN and for dissimilar materials was 1.87 kN. In this case also the breaking strength under bending test of gas welded joints was much superior to the composite joints. However, under three-point bending composite joints made for dissimilar materials has shown better strength than that of the similar materials. The strength of composite joint made between dissimilar materials was found to be about 27% higher in bending than the joint made for similar materials. While, under axial tension the strength of composite joint made for similar material has shown better strength.

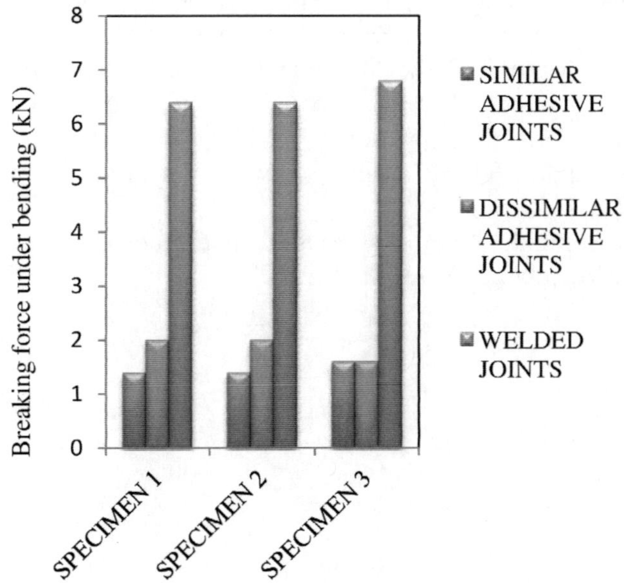

Fig 7: Breaking strength of the specimens (3-point bending)

5. Conclusion

From the above investigation it can be interpreted that the strengths of similar and dissimilar butt type adhesive joints made by natural fiber (woven jute mat) were lesser than the joints prepared with gas welding. It was expected as the breaking strengths under tensile and 3-point bending test of jute fiber which was used as reinforcement material in the present study are very less as compared to that of mild steel. Therefore, it may be concluded that such adhesive butt joint may be used where less load capacity is required. However, the results are encouraging as above conclusions are based on a preliminary study conducted. It is believed that the strength of the joints can be further increased by choosing more number of layers of reinforcing material in design or by choosing different reinforcing materials.

6. Acknowledgement

Authors would like to sincerely acknowledge Mr Virendra Mishra, Workshop Superintendent, at School of Technology, Lucknow University for providing necessary help in carrying out the experimental work in the lab.

7. References

[1] Lees JM 2006 Behaviour of GFRP adhesive pipe joints subjected to pressure and axial loadings *Compos Appl S A* **37** 1171-1179.

[2] Jiao H and Zhao XL 2004 CFRP strengthened butt-welded very high strength (VHS) circular steel tubes *Thin Wall Struct* **42** 963–978.

[3] deMorais AB, Pereira AB Teixeira JP and Cavaleiro NC 2007 Strength of epoxy adhesive-bonded stainless-steel joints *Int. J. Adhes. Adhes.* **27** 679-686.

[4] Mertiny P, and Ellyin F 2006 Joining of fiber-reinforced polymer tubes for high-pressure applications *Polym Compos* **27** 99–109.

IOP Conf. Series: Materials Science and Engineering **404** (2018) 012024 doi:10.1088/1757-899X/404/1/012024

[5] Griffin SA, Pang SS and Yang C 1991 Strength Model of Adhesive Bonded Composite Pipe Joints Under Tension *Polymer Sc & Engg* **31** 533-538

[6] Galvez P, Quesada A, Martinez MA, Abenojar J, Boada MJL and Diaz V 2017 Study of the behaviour of adhesive joints of steel with CFRP for its application in bus structures *Compos Eng* **129** 41-46.

[7] Kumar P, Singh R, Kumar R 2007 Joining similar and dissimilar materials with GFRP *Int. J. Adhes. Adhes.* **27** 68-76.

[8] Sonar T, Patil S, Deshmukh V and Acharya R 2015 Natural Fiber Reinforced Polymer Composite Material - A Review *IOSR-JMCE* 142-147.

An Analytical and Simulation Approach for Modeling Flexible Manufacturing System

Anupma Yadav[1] **and S. C. Jayswal**[2]

[1]Research Scholar, [2] Professor, Department of Mechanical Engineering, Madan Mohan Malaviya University of Technology., Gorakhpur – 273010 (U.P.)

E-mail: anupma.yadav04@gmail.com

Abstract. Varying needs of customer causes fluctuation in existing market condition. So, it's a challenge for manufacturing firms to protect their existence under every unpredicted situation. For this, there is a need of system that can handle changes smoothly. Flexible manufacturing system (FMS) is one of the solutions that enable any system to withstand the changing requirements of market. But due to high installation and capital cost of FMS, a method that can easily depicts its performance measures before its implementation is needed. In literature, there are various mathematical programming approaches for prior calculation of FMS performance. But due to their difficulty in solving complex problem, simulation technique is used for FMS analysis. Selection of performance parameters followed by system modeling using simulation technique is usually performed to see the results under varying conditions. In present work, process simulator is used for analysing FMS performance. Later, a mathematical technique is used for comparing and verifying the results obtained by simulation. Results showed it is easier to analyse the performance of complex and large FMS problems by simulation techniques in comparison to mathematical one.

Keywords. Flexible manufacturing system, modeling, process simulator, mathematical technique, performance parameter.

1. Introduction

A flexible manufacturing system comprises of certain numbers of operating workstations, that are inter-connected mechanically by a material handling system along with a storage system and electronically by a distributed computer system. Flexible automated manufacturing system is one of the most felicitous terms one can give to FMS. The word "automated" is use to differentiate FMS from those manufacturing systems which inherit the property of flexibility but are not automated, like a manned group technology machine cell. While, the word "flexible" would distinguish it from those manufacturing systems that are highly automated but are not flexible, like a conventional transfer line. Due to the capability of being flexible along with the concept of automation, it has been an area of interest for researchers from many years. Till date, there are lot of works in the literature that are concerned with design, planning, operational and implementation related issues of FMS. Various case studies, analytical methods, simulation techniques along with optimization techniques are discussed in the work of past researchers. In this paper, simulation experiments on given FMS is performed, for predicting its behaviour in real time with an advantage of no loss of money and resources. Also, mathematical modeling is performed for the same problem.

2. Literature Review

There are various works in literature that were done for analysing the behaviour of different parameters of FMS. Some of the works that are related with mathematical techniques are discussed here. Stecke (1983) calculated minimum number of machines for providing tools on the machining centers, by using an optimization formula. If number of machines exceeded the number of formed groups, then additional machines were tooled identical to few of the ones that were grouped. Major constraint of that model was its tool magazine capacity for each machine tool. Sujono and Lashkari (2007) used integer programming method for developing a model that can select machines, assign operations on them followed by selection of material handling system for transferring of parts from one machine to another. Selection was done based on material handling equipment and parts compatibility. Objective of the whole work was minimization of cost related with operating machine, material handling system, machine setups along with maximizing the compatibility of part equipment. Tiwari and Vidyarthi (2000) used Genetic

IOP Conf. Series: Materials Science and Engineering **404** (2018) 012025 doi:10.1088/1757-899X/404/1/012025

Algorithm (GA) based heuristic for solving machine loading problem in case of random FMS, with an objective of minimizing the unbalancing of system and maximizing the throughput. Constraints imposed were availability of machining time and tool slots during operation. Prakash et al. (2006) solved FMS loading problem by extending simple genetic algorithm into constraints based genetic algorithm for handling of variables that were complex in nature. It was done by introducing three new genetic operator's constraint that were based on initialization, crossover and mutation. Singholi et al. (2010) had taken case study of a firm for improving its performance. It involved mathematical models that were illustrated in literature for estimating certain performance parameters like production rate, overall utilization of the system etc. An effort was also done for presenting an improved design for the existing FMS. Erdin and Atmaca (2014) proposed that analysis of manufacturing systems should be done carefully to protect them from failure at any level. It was done by using physical, analytical or simulation methods. In that work, comparison was done between conventional, FMS and ideal system based on necessary number of servers required at each workstation by using bottleneck technique.

Certain simulation techniques were also used for FMS modeling, for studying the effect of various parameters associated with FMS performance. Caprihan and Wadhwa (2005) studied the influence of delay in information on decisions, related with selecting the best levels of routing flexibility for FMS problem. Result of simulation model indicated a tolerance limit below which if operation was carried then the objective of obtaining performance minimization was not possible. Cheng and Chan (2011) optimized part input sequence by using simulation model for production planning. Developed model simulated the FMS, which was used for producing different components. A custom-built user interface was used for inputting production and demand data from an excel spreadsheet into the developed model. Singholi (2015) evaluated routing and machine flexibility effects in different configuration of FMS by using a Taguchi method. Number of pallets and different scheduling rules were also used for finding the effects on buffer delay in the FMS considered. Simulation results indicated that number of pallets and routing flexibility are the two factors influencing average delay at buffers the most. Dosdogru et al. (2015) analysed the effect of routing flexibility by taking number of parts, number of jobs, operation number and flexibility levels as important factors. They used genetic algorithm integrated with Monte Carlo method for analysing routing flexibility. Results showed make-span was best when number of machines were at high level. Jain and Raj (2016) studied the effect of different flexibility measures on FMS performance. They used structural equation modeling, graph theory and matrix approach. Fifteen variables were identified form literature and then intensity of these variables on the quality, productivity and flexibility of FMS was studied. Arshad et al. (2016) studied effect of different scheduling rules and FMS layout on the performance of FMS. ARENA software was used for developing different models, results showed that shortest processing time scheduling rules performs best. Rybicka et al. (2016) showed how discrete event simulation software can be used in any manufacturing system for studying its performance parameter. WITNESS simulation software was used for developing model of an actual production line and studying the effect of sequence change, number of machines and number of pallets on system performance. Florescu et al. (2017) developed a methodology for studying how performance of FMS involved in processing of cylindrical parts gets affected. Reduction of waiting time of parts was the major concern. ARENA simulation software was used for analysing the existing model and validating the proposed solution. Mahmood et al. (2017) analysed FMS performance by integrated definition modeling technique. The study focused on operational study of FMS by performing simulation analysis, various performance changes were suggested in the existing model by eliminating formed bottleneck condition. Use of an efficient cutting tool was suggested for increasing the production rate of FMS. Yadav and Jayswal (2017) provided an insight on previous work in the area of FMS modeling by investigating works that are accomplished by using different modelling techniques in FMS like mathematical, artificial intelligence, hierarchical, multi criteria decision-making method, Petri Nets and simulation.

For performance evaluation of any FMS there are number of techniques which are available in literature like mathematical, artificial intelligence, hierarchical, heuristic, petri nets, MCDM method and simulation model, they have enriched the work concerned with modeling of FMS. For analysing different system performance measures of FMS and seeing how they behave with constraints in particular condition, one can consider the process of mathematical modeling and simulation as the best one. In this paper mathematical and simulation techniques are evaluated with help of an example.

3. Proposed approach

IOP Conf. Series: Materials Science and Engineering **404** (2018) 012025 doi:10.1088/1757-899X/404/1/012025

A mathematical modeling technique discussed in the work of Curry and Feldman (2009) is use for FMS modeling. Cycle time along with utilization rate of each workstation of FMS are taken as performance parameter. Simulation technique is also used later for FMS modeling. Process simulator is the simulation software which is use in the paper, due to its ability of modeling probability conditions easily. The proposed approach is explained in detail with the help of a case study one by one.

3.1. System Description

An FMS consisting of four machines on which production of two components is required (Part A and B). The first part uses four distinct machining processes while second uses only first three workstations used by part first. Process routing of both the parts is as shown in Figure 1 and 2 respectively.

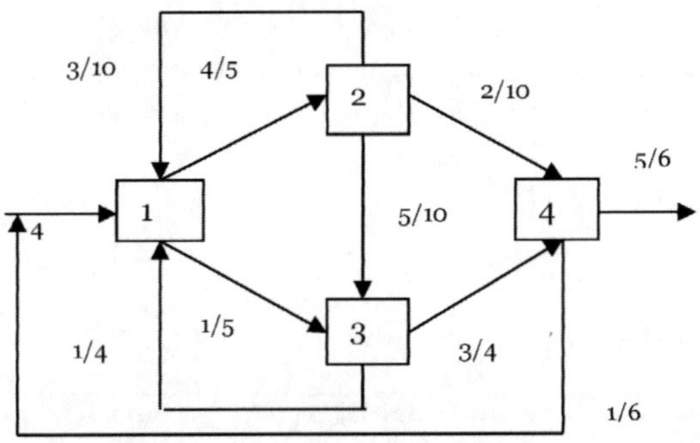

Figure 1. Process routing for part A

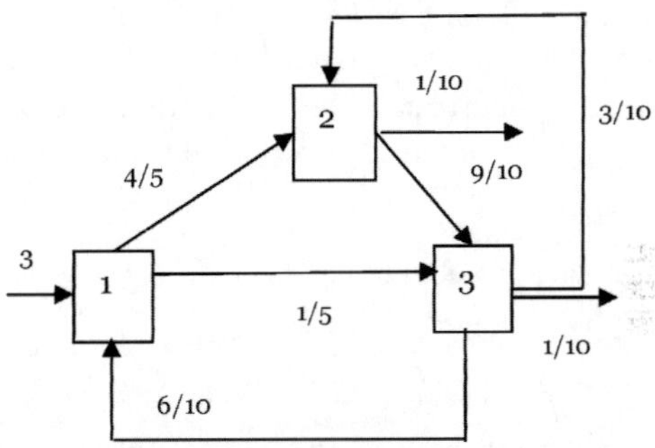

Figure 2. Process routing for part B

3.2. Mathematical modeling

From given process routing one can write system of equations that will define workstations total arrival rates.

For part A:

$$\lambda 1 = 4 + 0.3\,\lambda 2 + 0.25\,\lambda 3 + 0.166\,\lambda 4$$
$$\lambda 2 = 0 + 0.8\,\lambda 1$$
$$\lambda 3 = 0 + 0.2\lambda 1 + 0.5\lambda 2$$
$$\lambda 4 = 0 + 0.2\,\lambda 2 + 0.75\,\lambda 3$$

Now, rearrange the formed equations as,

$$\lambda 1 - 0.3\,\lambda 2 - 0.25\,\lambda 3 - 0.166\,\lambda 4 = 4$$
$$-0.8\,\lambda 1 - \lambda 2 = 0$$
$$-0.2\,\lambda 1 - 0.5\lambda 2 + \lambda 3 = 0$$
$$-0.2\,\lambda 2 - 0.75\,\lambda 3 + \lambda 4 = 0$$

One can write above euations in matrix form as:

$$\begin{bmatrix}\lambda_1\\ \lambda_2\\ \lambda_3\\ \lambda_4\end{bmatrix} = \begin{bmatrix}+1.0 & -0.3 & -0.25 & -0.16\\ -0.8 & 1.0 & 00 & 00\\ -0.2 & -0.5 & 01 & 00\\ 0.0 & -0.2 & -0.75 & 1\end{bmatrix}\begin{bmatrix}4\\0\\0\\0\end{bmatrix}$$

On solving the matrix, arrival rate for part A is obtained as:

$$(\lambda 1, \lambda 2, \lambda 3, \lambda 4) = (7.689, 6.295, 4.721, 4.800)\ /hr$$

Similarly, for part B one can write following system of equations that will lead to the given solutions:

$$\lambda 1 = 3 + 0.60\,\lambda 3$$
$$\lambda 2 = 0 + 0.8\,\lambda 1 + 0.3\,\lambda 3$$
$$\lambda 3 = 0 + 0.2\lambda 1 + 0.9\,\lambda 2$$
$$(\lambda 1, \lambda 2, \lambda 3) = (3.433, 4.044, 4.326)\ /hr$$

Now, total arrival rates for given system consisting of both parts is denoted as summation of the entire parts arrival rate at a station, as shown in equation (1).

$$\lambda_k = \sum_{i=1}^{m} \lambda_{i,k} \quad (1)$$

Where,
k is workstation.
m is total number of parts.
λi is mean arrival rate of part i.
Using equation (1) values of workstations loads are,

$(\lambda 1, \lambda 2, \lambda 3, \lambda 4) = (11.301, 10.339, 9.047, 4.800)$ /hr

Now, workstation workload is calculated by using service time value. Further, service time (dependent on the part type) is decomposed into two elements, mean and squared coefficient of variations of the processing time of part 'i' at workstation 'k' and are denoted as $E[T_s(i,k)]$ and $C_s^2(i,k)$, respectively. Workstations workload is calculated using equation (2).

$$WL_k = \sum_{i=1}^{m} \lambda_{ik}\, E[T_s(i,k)] \quad (2)$$

Processing time for both parts on different workstation are given in Table 1.

Table 1. Processing time for parts

Workstation	Mean (Part A) E [Ts(A, k)]	SCV (Part A) C_s^2(A, k)	Mean (Part B) E [Ts(B, k)]	SCV (Part B) C_s^2(B, k)
1	7.2 min or 0.12 hr	1.0	6.0 min or 0.1 hr	0.8
2	6 min or 0.1 hr	0.9	2.1 min or 0.03 hr	0.9
3	6 min or 0.1 hr	0.8	3.6 min or 0.06 hr	1.0
4	9 min or 0.15 hr	0.7	-	-

By using equation (2) and given values of processing time of both parts (Table 1), workload for all the four workstations are as follows:

WL1 = 7.689 (0.12) + 3.433 (0.10) = 1.2659
WL2 = 6.295 (0.10) + 4.044 (0.03) = 0.77104
WL3 = 4.721 (0.10) + 4.326 (0.06) = 0.73166
WL4 = 4.800 (0.15) = 0.72

For a steady state to exist, number of machines at each workstation must be strictly greater than the workload at that station. So, number of machine for workstation 1, 2, 3 and 4 is two, one, one and one respectively. Service time for an arbitrary job, independent of job type, is a random variable and is denoted by Ts (k). Mean and coefficient of variation for service time at workstation 'k' are given by equation (3) and (4) respectively.

$$E[T_s(k)] = E(S_i) = \sum_{i=1}^{m} \frac{\lambda_{i,k}}{\lambda_k} E[T_s(i,k)] = \frac{WL_k}{\lambda_k} \quad (3)$$

$$C_s^2(k) = \frac{\sum_{i=1}^{m} (\lambda_{i,k}/\lambda_k)\, E[T_s(i,k)]^2 \left(1 + C_s^2(i,k)\right)}{\left(\sum_{i=1}^{m} (\lambda_{i,k}/\lambda_k)\, E[T_s(i,k)]\right)^2} - 1 \quad (4)$$

Using equation (3) and (4), values of service time characteristics in terms of their mean and SCV values are calculated as:

E (S1) = 1.2659 / 11.301 = 0.11201 hr
E (S2) = 0.77104 / 10.339 = 4.475 hr
E (S3) = 0.73166 / 9.047 = 4.85 hr
E (S4) = 0.72 / 4.800 = 9 hr

$$C_s^2(1) = \frac{\left(\frac{7.689}{11.301}\right)(0.12)^2\,(1+1) + \left(\frac{3.433}{11.301}\right)(0.1)^2\,(1+0.8)}{(0.11201)^2} - 1 \quad = 0.997$$

$$C_s^2(2) = \frac{\left(\frac{6.295}{10.339}\right)(0.1)^2(1+0.9) + \left(\frac{4.044}{10.339}\right)(0.035)^2\,(1+0.9)}{(0.074)^2} - 1 \quad = 1.279$$

$$C_s^2(3) = \frac{\left(\frac{4.721}{9.047}\right)(0.1)^2(1+0.8) + \left(\frac{4.326}{9.047}\right)(0.06)^2\,(1+1)}{(0.0808)^2} - 1 \quad = 0.964$$

$$C_s^2(4) = \frac{\left(\frac{4.800}{4.800}\right)(0.15)^2(1+0.7)}{(0.15)^2} - 1 \quad = 0.700$$

Table 2. Calculated parameters

Workstation	1	2	3	4
Total arrival rate, λ	11.301/ hr	10.339/ hr	9.047/ hr	4.800/ hr
Workload, WL	1.2659	0.77104	0.73166	0.722
Minimum machine	2	1	1	1
Service mean, E[S]	6.72 min	4.475 min	4.85 min	9 min
Service SCV, C2[S]	0.997	1.279	0.964	0.7

Till now, parameters (Table 2) which are not subjected to any downtime or repair conditions are calculated. Now, same parameters are calculated when times are adjusted for breakdowns and repairs. Availabilities and repair time characteristics of each workstation are given in Table 3.

Table 3. Machines availability and repair time characteristics

Workstation	1	2	3	4
Availability	0.90	0.95	0.93	0.95
E[R]	1 hr	1 hr	1 hr	1 hr
C2[R]	1.75	2.00	1.50	5/3

In case of breakdowns and repairs, value of service time characteristics is calculated by using equation (5) and (6) respectively.

$$E\,[S_i] = \frac{E[T_S(i)]}{a_i} \tag{5}$$

$$C^2[S_i] = C_s^2(i) + \frac{[1 + C^2\,(R_i)]a_i\,(1-a_i)\,E[R_i]}{E[T_s(i)]} \tag{6}$$

After using equation (5) and (6), values of service time are:

E[S1] = 6.72/0.90 = 7.47 min
E[S2] = 4.47/0.95 = 4.74 min
E[S3] = 4.85/0.93 = 5.22 min
E[S4] = 9.00/0.95 = 9.48 min

$$C^2[S_1] = 0.997 + \frac{[1 + 1.75]0.90\,(1-0.90)\,60}{6.72} \quad = 3.207$$

$$C^2[S_2] = 1.279 + \frac{[1+2.00]0.95\,(1-0.95)\,60}{4.47} = 3.189$$

$$C^2[S_3] = 0.964 + \frac{[1+1.50]0.93\,(1-0.93)\,60}{4.85} = 2.977$$

$$C^2[S_4] = 0.700 + \frac{\left[1+\left(\frac{5}{3}\right)\right]0.95\,(1-0.95)\,60}{9.00} = 1.544$$

After this, utilization rate of each workstation using equation (7) is calculated.

$$U_i = \frac{E[S_i]\lambda_i}{C_i} \quad (7)$$

Where, C_i is the number of server at each workstation.

$$U1 = \left(\frac{7.47}{60}\right)\left(\frac{11.301}{2}\right) = 0.703$$

$$U2 = \left(\frac{4.74}{60}\right)\left(\frac{10.339}{1}\right) = 0.816$$

$$U3 = \left(\frac{5.22}{60}\right)\left(\frac{9.047}{1}\right) = 0.787$$

$$U4 = \left(\frac{9.48}{60}\right)\left(\frac{4.800}{1}\right) = 0.758$$

Squared coefficient of variation for the arrival stream, into each workstation is calculate by using equation (8) and is shown in Table 4.

$$C_a^2(i) = \frac{\gamma_i}{\lambda_i} C_a^2(0,i) + \sum_{k=1}^{n} \frac{\lambda_k p_{k,i}}{\lambda_i} \left[P_{k,i}(1-U_k^2)C_a^2(k) + P_{k,i}U_k^2\left(\frac{C_s^2(k)+\sqrt{C_k}-1}{\sqrt{C_k}}\right) + 1 - P_{k,i}\right] \quad (8)$$

Table 4. Squared coefficient of variation for the arrival stream

Workstation	1	2	3	4
$C_a^2(i)$	1.56	1.602	1.841	1.497

Further, calculation of queue time for single server system is done by using equation (9) and for two servers system it is calculated by using equation (10).

$$CT_q(i) = \frac{[C_a^2(i)+C_s^2(i)]}{2}\left(\frac{U_i}{1-U_i}\right)E[T_s(i)] \quad (9)$$

$$CT_q(i) = \frac{[C_a^2(i)+C_s^2(i)]}{2}\left(\frac{U_i}{1-U_i}\right)\left(\frac{U_i}{1+U_i}\right)E[T_s(i)] \quad (10)$$

After this, mean cycle time and work in process is calculated by using equation (11) and (12) respectively.

$$CT(i) = \frac{[C_a^2(i)+C_s^2(i)]}{2}\left(\frac{U_i}{1-U_i}\right)E[T_s(i)] + E[T_s(i)] = CT_q(i) + E[T_s(i)] \quad (11)$$

$$WIP(i) = CT(i)\lambda_i \quad (12)$$

Using equation (10), value of queue time for workstation one is calculated, as it has two servers at that station. For rest of the workstations (having one server), equation (9) is used for calculating queue time.

$$CT_q(1) = \left(\frac{1.561+3.207}{2}\right)\left(\frac{0.703}{1-0.703}\right)\left(\frac{0.703}{1+0.703}\right)7.47 = 17.40\ min$$

$$CT_q(2) = \left(\frac{1.602+3.189}{2}\right)\left(\frac{0.816}{1-0.816}\right)4.74 = 48.90\ min$$

$$CT_q(3) = \left(\frac{1.841 + 32.900}{2}\right)\left(\frac{0.780}{1 - 0.780}\right) 5.22 = 43.60\ min$$
$$CT_q(4) = \left(\frac{1.497 + 1.500}{2}\right)\left(\frac{0.760}{1 - 0.760}\right) 9.48 = 44.98\ min$$

Value for cycle time and work in process for each workstation is calculated by using equation (11) and (12) respectively.

$$CT\ (1) = CT_q(1) + E[T_s(1)] = 17.40 + 7.47 = 24.87\ min$$
$$CT\ (2) = CT_q(2) + E[T_s(2)] = 48.90 + 4.74 = 53.64\ min$$
$$CT\ (3) = CT_q(3) + E[T_s(3)] = 43.60 + 5.22 = 48.82\ min$$
$$CT\ (4) = CT_q(4) + E[T_s(4)] = 44.98\ + 9.48 = 54.46\ min$$

$$WIP(1) = CT(1)\,\lambda 1 = \left(\frac{24.87}{60}\right)(11.301) = 4.68$$
$$WIP(2) = CT(2)\,\lambda 2 = \left(\frac{53.64}{60}\right)(10.339) = 9.24$$
$$WIP(3) = CT(3)\,\lambda 3 = (48.82)(9.047) = 7.36$$
$$WIP(4) = CT(4)\,\lambda 4 = \left(\frac{54.46}{60}\right)(4.800) = 4.36$$

After calculating above parameters, values of factory system performance measures in terms of cycle time (CT_s), work in process (WIP_s) and throughput (th_s) is calculated. The value of factory throughput is given in terms of factory's inflow rate. From routing diagram, one can calculate total inflow rate as (4+3=7)/ hr.

$$\boldsymbol{WIP_s = \sum_{i=1}^{n} WIP_s(i)} = 4.68 + 9.24 + 7.36 + 4.36 = \mathbf{25.64}$$

And according to little's law, if a system satisfies steady state condition than the value of work in process is the product of arrival rate and cycle time, as shown in equation (13).

$$WIP\ =\ \lambda \times\ CT \qquad \mathbf{(13)}$$

By using equation (13) cycle time of both the parts is found to be 219.77 min.

$$CT_s = \frac{WIP_s}{th_s} = \frac{25.64}{7} = 3.66\ hr\ =\ 219.77\ min$$

3.3. Process Simulator

Given system is now modeled in process simulator software of Promodel corporation with file version of 9.3.0.2701, for evaluating system performance. Network for both the parts along with their obtained cycle time is shown in Figure (3). System involves various activity states that represent the machine type. Arrows were used for connecting different activities for representing the flow of process. Part A and B arrive at machine 1 according to given arrival rate and are processed further with the defined process plans. Developed model is run till production of seven parts from the system is done. One can also see the state of each entity within the system, with the help of model as shown in Figure (4).

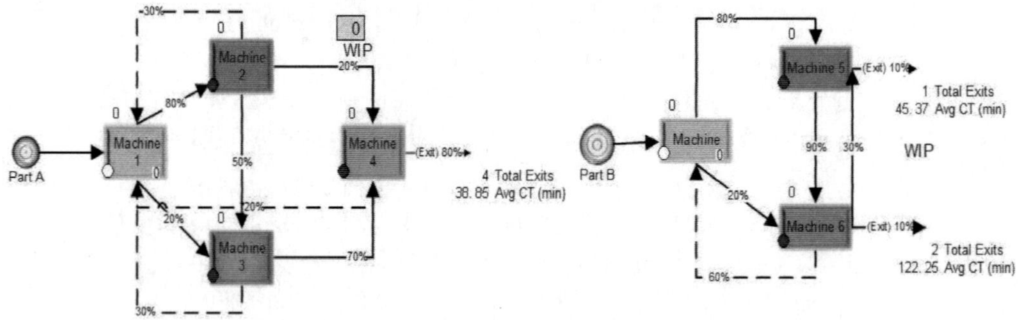

Figure 3: Simulated model of part A and B

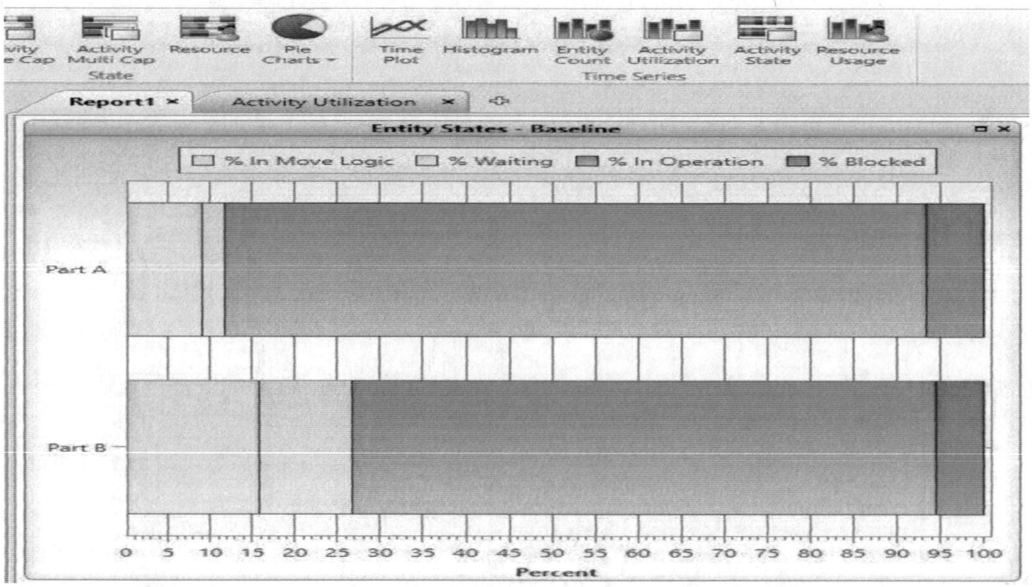

Figure 4. State of each part within the system

4. Results

On application of mathematical and simulation technique for performance evaluation of given FMS, the value of parameters obtained are as follows:

Modeling Technique	**Cycle Time**			
Mathematical	219.77 min			
Process Simulator	206.47 min			
Modeling Technique	**Utilization**			
	Workstation 1	**Workstation 2**	**Workstation 3**	**Workstation 4**
Mathematical	70.3 %	81.6%	78.7%	75.8%
Process Simulator	71.0%	76%	90%	75.0%

From obtained values, it can be seen that both techniques gave approximately same values for cycle time and workstation utilizations. The only advantage associated with simulation technique is its ease in use and less time consumption capability.

5. Conclusions

At last it can be concluded that:

1) Mathematical modeling techniques were the commonly used method for modeling purpose. Drawbacks associated with them are stated assumptions in the models, as they may not be valid in real world. Also, the process of computation becomes large with increase in the size of problem.

2) Simulation model simulate given problem in model form, so that one can study the effects of different conditions on the developed model without any loss of money and resources.

3) Results concluded that the obtained performance of given FMS by both the methods are close to each other.

4) It is also seen that mathematical modeling of complex system requires lot of time and effort, so it is better to use simulation for saving a lot of time and resources.

References

[1] Arshad M *et al* 2016 Scheduling of three layouts using four scheduling rules *Proceedings- International Conference on Industrial Engineering and Operations Management (*Kula Lumpur, Malaysia) 2539-2548.

[2] Caprihan R and Wadhwa S 2005 Scheduling of FMS with information delays: a simulation study *International Journal of Flexible Manufacturing System* **17** 39-65.

[3] Cheng H C and Chan D Y K 2011 Simulation optimization of part input sequence in a flexible manufacturing system *Proceedings of the Winter Simulation Conference*.

[4] Curry G L and Feldman R M 2009 Manufacturing systems modeling and analysis *(2rd ed.) Springer, New York*.

[5] Dosdogru A T *et al* 2015 Integration of genetic algorithm and Monte Carlo to analyze the effect of routing flexibility *International Journal of Advanced Manufacturing Technology* **81** 1379-1389.

[6] Erdin M E and Atmaca A 2014 Implementation of an overall design of a flexible manufacturing system *8th International Conference Inter disciplinarily in Engineering* (INTER-EN, Tirgu-Mures, Romania, Procedia Technology) **19** 9-10.

[7] Florescu A *et al* 2017 Operational parameters estimation for a flexible manufacturing system: A case study *MATEC Web of conference* **112** 01-06.

[8] Jain V and Raj T 2016 Modeling and analysis of FMS performance variables by ISM, SEM and GTMA approach *International Journal of Production Economics* **171** 84-96.

[9] Mahmood K *et al* 2017 Performance analysis of a flexible manufacturing system performance *The 50th CIRP Conference on Manufacturing Systems (Procedia CIRP-63)* 424-429.

[10] Prakash A K *et al* 2006 Solving machine loading problem of a FMS with constraint based genetic algorithm *European Journal of Operational Research* **175** 1043-1069.

[11] Rybicka J *et al* 2016 Testing a Flexible manufacturing system facility production capacity through discrete event simulation: Automotive Case Study *International Journal of Mechanical, Aerospace, Industrial, Mechatronics and Manufacturing Engineering* **10** 668-672.

[12] Singholi A 2015 Impact of manufacturing flexibility and pallets on buffer delay in flexible manufacturing systems *International Journal of Engineering Management and Economics* **5** 308-330.

[13] Singholi A *et al* 2010 Towards improving the performance of flexible manufacturing system: a case study *Journal of Industrial Engineering and Management* **3** 87-115.

[14] Stecke K E 1983 Formulation and solution of nonlinear integer production planning problems for FMS *Management Science* **29** 273-288.

[15] Sujono S and Lashkari R S 2007 A multi-objective model of operation allocation and material handling system selection in FMS design *International Journal of Production Economics* **105** 116-133.

[16] Tiwari M K and Vidyarthi N K 2000 Solving machine loading problems in a flexible manufacturing system using a genetic algorithm based heuristic approach *International Journal of Production and Research* **38** 3357-3384.

[17] Yadav A and Jayswal S C 2017 Modelling of flexible manufacturing system: a review *International Journal of Production Research DOI: 10.1080/00207543.2017.1387302*

IOP Conf. Series: Materials Science and Engineering **404** (2018) 012026 doi:10.1088/1757-899X/404/1/012026

Analytical Modelling and FEM Simulation of Capacitive Pressure Sensor for Intraocular Pressure Sensing

Rishabh Bhooshan Mishra, S Santosh Kumar, Ravindra Mukhiya
Smart Sensor Area, CSIR – Central Electronics Engineering Research Institute, Pilani, Rajasthan, India – 333031

Email: santoshkumar.ceeri@gmail.com

Abstract. The paper presents analytical modelling and Finite Element Method (FEM) based simulation of clamped circular capacitive pressure sensors for Intraocular Pressure (IOP) measurement. The parallel plate capacitive pressure sensor consists of a fixed backplate and a circular shaped clamped thin diaphragm made of silicon material. The plates are separated by a vacuum media. MATLAB® and COVENTORWARE® are used for analytical and FEM simulations, respectively, and a comparison of maximum deflection in the diaphragm, capacitance after pressure application and sensitivity is carried out. A pressure range of 0-60 mmHg is used for IOP measurement.

1. Introduction

Over conventional technique of Pressure measurement, MEMS pressure sensor offers several advantages such as small size, less weight, low power consumption, lower cost, and higher reliability. In these days, the research in piezoresistive and capacitive type pressure sensors are in trend. In comparison to piezoresistive sensors, the capacitive sensors offer several advantages like low power consumption, higher sensitivity to pressure, low thermal sensitivity and large dynamic range. The capacitive pressure sensors are generally used for automotive, biomedical, oceanography, aerospace application, consumer electronics etc. But miniaturization of MEMS devices for biomedical applications is very popular research area [1 - 2].

A lot of people are suffering from the irreversible vision loss eye disease i.e. glaucoma. The normal eye pressure is 1.6 to 2.8 kPa. More than normal eye pressure does not definitely cause of glaucoma disease but if the IOP is more than normal eye pressure and there are no signs of glaucoma then the problem may be ocular hypertension. The excessive pressure of fluid "aqueous humour" within eye is the major reason for the damage of the optical nerves which introduces glaucoma disease because it fills interior eye chamber. IOP increases due to several reasons like occlusion of drain ducts, closer or narrow angle between cornea and iris. The Goldmann Applanation Tonometer has been used for IOP measurement for past several decades but accuracy of this technique depends on the central corneal stiffness. For IOP measurement, the telemetry sensing technique plays important role. The measurement of IOP helps in glaucoma disease diagnosis and monitoring using telemetry principles, in which the data is monitored by implanted sensing device using wireless technique, so that further steps can be taken for medical treatment to patients [3 - 5].

2. Analytical Modeling of Capacitive Pressure sensor

The design and simulation of a circular diaphragm pressure sensor is carried out in this work for IOP measurement. Such a sensor may be fabricated using a standard fabrication process for capacitive pressure sensor.

2.1 Analytical modeling of deflection in circular shaped clamped diaphragm

The diaphragm deflection follows Hook's Law i.e. the diaphragm deflection varies linearly when pressure is applied on the diaphragm. Before defining the deflection in diaphragm, following four basic assumptions have been made for small deflection plate theory [6]:

- The diaphragm is flat and made of elastic, homogeneous and isotropic material.
- In comparison to diaphragm thickness, deflection in diaphragm is very small (deflection must be less than 1/5th of thickness in case of small deflection theory). Hence slope of deflection is very much small and the square of deflection slope is neglected.
- The normal stress to the middle plane is neglected.
- After bending, the middle surface of diaphragm remains unstrained and the strain line remains perpendicular and straight to middle surface.

The uniform pressure (P) is applied on edge clamped thin circular platewhich is made of elastic, isotropic and homogeneous material, and then plate equation can be represented by [7]:

$$\frac{1}{r}\frac{\partial}{\partial r}\left[\frac{1}{r}\frac{\partial}{\partial r}\left(r\frac{\partial w}{\partial r}\right)\right] = \frac{P}{2D} \tag{1}$$

Or

$$\frac{1}{r}\left[\frac{\partial^3 w}{\partial r^3} + \frac{1}{r}\frac{\partial^2 w}{\partial r^2} - \frac{1}{r^2}\frac{\partial w}{\partial r}\right] = \frac{P}{2D} \tag{2}$$

Here, $D = Et^3 \,/\, (12 - v^2)$ is flexural rigidity of the diaphragm, w is deflection at radius r from centre of diaphragm, t is diaphragm thickness, v is Poisson's Ratio and E is Young's modulus using which the diaphragm is made.

The following boundary conditions are used to solve the plate equation:

$$w(r = a) = 0 \tag{3}$$

$$\frac{\partial w(r = 0)}{\partial r} = 0 \tag{4}$$

$$\frac{\partial w(r = a)}{\partial r} = 0 \tag{5}$$

Here, a is diaphragm radius.

After using boundary conditions, from Eq. (3), (4) and (5), the diaphragm deflection can be given by:

$$w(r) = \frac{Pa^4}{64D}\left[1 - \left(\frac{r}{a}\right)^2\right]^2 \tag{6}$$

2.2 Analytical modeling of capacitance of circular capacitive pressure sensor

In this presented paper, we propose an equation of capacitance of circular capacitive pressure sensor after pressure application on the top of diaphragm. If A, ε and d be overlapping area between plates, permittivity of medium and separation gap between parallel plates, respectively. Then base capacitance can be given by [2]:

$$C_b = \frac{\varepsilon A}{d} \tag{7}$$

After pressure application on the diaphragm, the change in capacitance can be given by [2]:

$$C_w = \int_0^{2\pi}\int_0^{a} \frac{\varepsilon\, r dr\, d\theta}{d - w(r)} \tag{8}$$

From Eq. (6) and (8), we get:

$$C_w = \int_0^{2\pi}\int_0^{a} \frac{\varepsilon\, r dr\, d\theta}{d - \frac{Pa^4}{64D}\left[1-\left(\frac{r}{a}\right)^2\right]^2} \tag{9}$$

After solving double integral, following solution is obtained:

$$C_w = 4\pi\varepsilon\sqrt{\frac{D}{Pd}}\ln\left|\frac{a^2\sqrt{P}+8\sqrt{dD}}{a^2\sqrt{P}-8\sqrt{dD}}\right| \tag{10}$$

2.3 Sensitivity of clamped circular shaped capacitive pressure sensor

The sensitivity of sensor can be defined by ratio of change in capacitance in defined pressure range and change in pressure. The sensor's sensitivity can be defined by:

$$S = \frac{C_{max} - C_{min}}{P_{max} - P_{min}} \tag{11}$$

3. Simulation Results

3.1 Deflection in Diaphragms

Three different diaphragm thicknesses (4 μm, 5 μm and 6 μm) for 3 μm separation gap are chosen to obtain three different designs in the present work. A gap of 3 μm is chosen keeping in view the ease of fabrication. The sensors are designed for a pressure range from 0 to 8 kPa (0 to 60 mmHg) in order to measure the IOP. In each design, the maximum deflection of diaphragm at maximum pressure (8 kPa) is kept less than $1/4^{th}$ of the separation gap to ensure linear characteristics. Based on these considerations and the small deflection theory, the diaphragm radius of each sensor model is determined. The deflection results of FEM simulation are performed using MemMech module of COVENTORWARE® and compared with the analytical results obtained using MATLAB® code which is based on Eq. (6) and (10).

The simulation of diaphragm deflection using FEM based tool COVENTORWARE® for diaphragm thickness of 4 μm is shown in Fig. 1. Similarly the 5 μm and 6 μm thick diaphragms are simulated. The comparison of deflection in each 4 μm, 5 μm and 6 μm thick diaphragms using analytical equations and FEM simulations are presented in Figs. 2, 3 and 4 for specific diaphragm radiuses. The radius is obtained for separation gap of 3 μm for each diaphragm thickness. Diaphragm of particular thickness and radius has less than 0.75 μm ($1/4^{th}$ of 3 μm) deflection at 8 kPa pressure. The diaphragm deflection increases as radius of diaphragm increases. In all simulations, silicon is used as the diaphragm material and the following properties are used: E = 169.8 GPa and ν = 0.066.

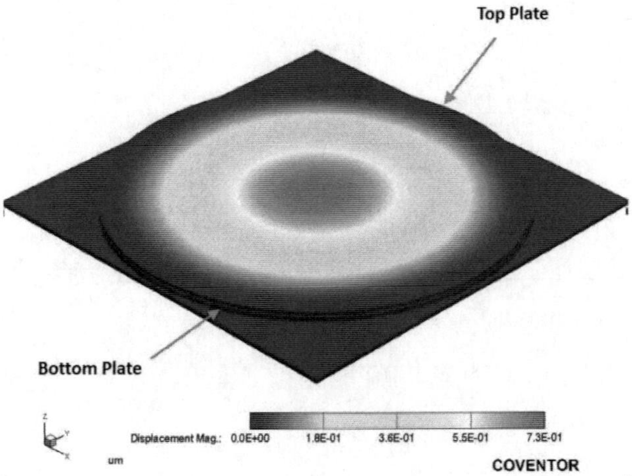

Figure 1. Maximum deflection in 4μm thick diaphragms using COVENTORWARE®.

3.2 Capacitance variation with pressure

In the present work, the capacitance variation with pressure for different diaphragm thicknesses are obtained using MemElectro module and graph between capacitance versus pressure for three sensor designs using CoSolveEM module of COVENTORWARE®. The comparison of results obtained by analytical modelling and FEM are shown in Figs. 5 to 7.

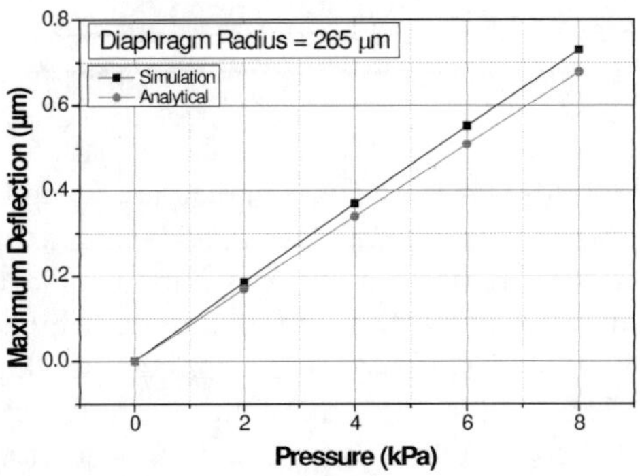

Figure 2. Maximum deflection in 4 μm thick diaphragm.

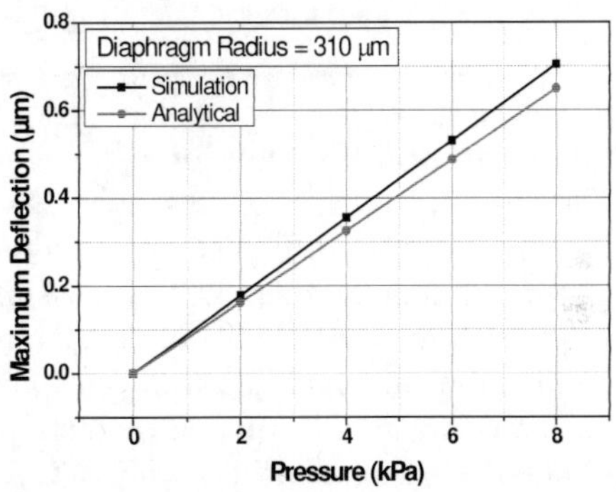

Figure 3. Maximum deflection in 5 μm thick diaphragm.

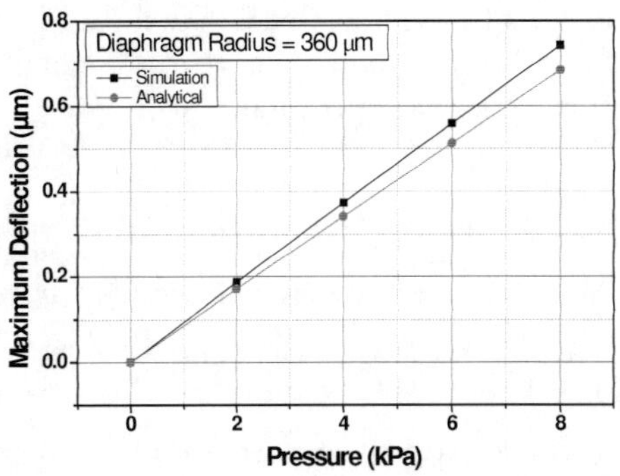

Figure 4. Maximum deflection in 6 μm thick diaphragm.

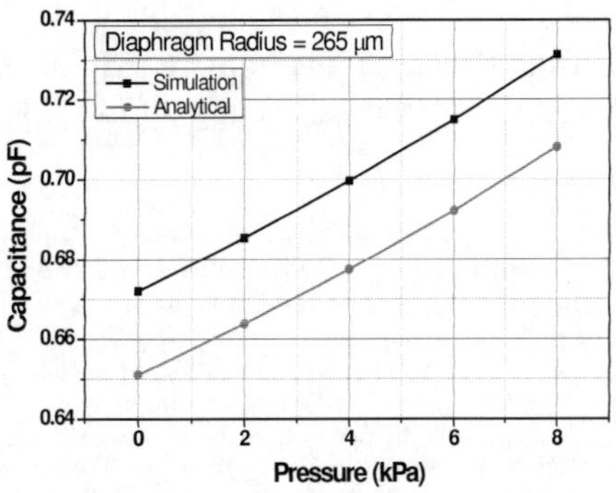

Figure 5. Capacitance vs. Pressure for 4 μm thick diaphragm.

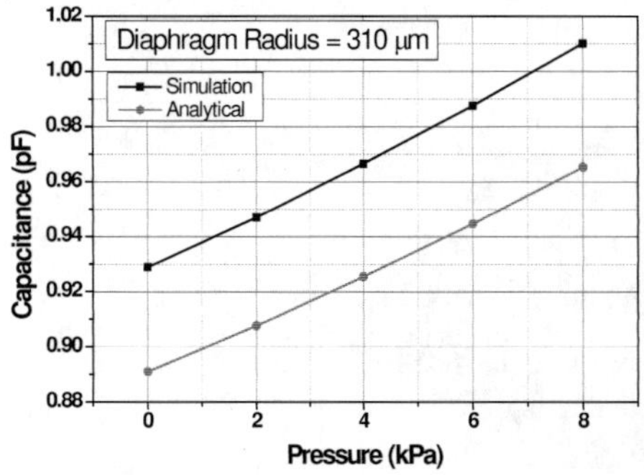

Figure 6. Capacitance vs. Pressure for 5 μm thick diaphragm.

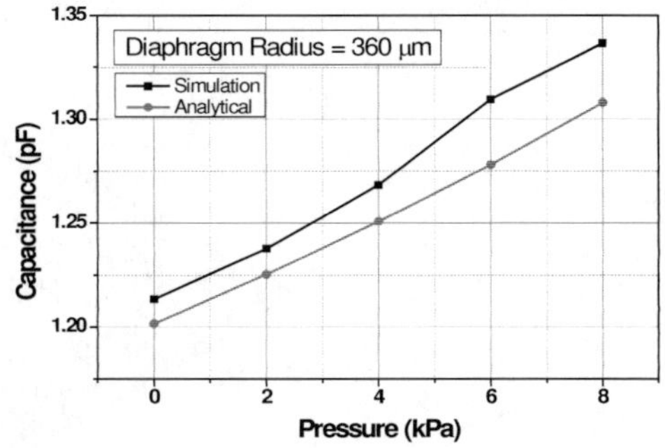

Figure 7. Capacitance vs. Pressure for 6 μm thick diaphragm.

3.3 Sensitivity

The sensitivity of all the three designs is shown in Table I.

Table 1 Sensitivity of different designs

Thickness (μm)	Radius (μm)	Sensitivity (fF/kPa) [Analytical]	Sensitivity (fF/kPa) [FEM]
4	265	7.122	7.422
5	310	9.285	10.160
6	360	10.325	15.389

4. Discussion and Conclusion

In this paper, capacitive pressure sensor is utilised for IOP sensing application using small deflection theory. The difference between the value of diaphragm deflection obtained using analytical and FEM simulation are due to certain assumptions which are practically not possible to achieve. The deviation in results in the values of capacitance obtained is due to the fringing field effect which is not considered in the analytical solution. We can conclude that the design having a diaphragm thickness of 6 μm, separation gap of 3 μm and diaphragm radius of 360 μm has the maximum sensitivity amongst all the three designs. It sensitivity is 10.325 fF/kPa and 15.389 fF/kPa using analytical equations and FEM, respectively. Therefore, this design is best suited for IOP measurement.

Acknowledgement

Authors would like to acknowledge the generous guidance of Director, CSIR – Central Electronics Engineering Research Institute (CEERI), Pilani. They would like to thanks all the scientific and technical staff of Process Technologies Group for their support and co-operation.

References

[1] Kumar S Santosh and Pant B D 2014 Design principles and considerations for 'ideal' silicon piezoresistive pressure sensor: a focused review *Microsystem Technologies* **20** pp 1213 – 1247
[2] Balavelad K, Sheeparamatti B G 2015 A Critical Review of MEMS Capacitive Pressure Sensor *Sensors and Transducers* **187** pp120-128
[3] Katuri K C, Asrani S, and Ramasubramanian M K 2008 Intraocular pressure monitoring sensors *IEEE Sensors* **8** pp 12-19
[4] Asha G K Y M 2015 Design and Simulative Analysis of a Batteryless Teflon Coated Capacitive Pressure Sensor for Glaucoma Diagnosis 19th Symposium on VLSI Design and Test no.6 pp. 1–5
[5] Xue N, Chang S P and Lee J B: A SU-8-based microfabricated implantable inductively coupled passive RF wireless intraocular pressure sensor 2012 *J. of Microelectromechanical Sy.*
[6] Ross C T F 2004 Strength of Material and Structure fourth ed. London
[7] Timoshenko S and Krigger *S W Theory of Plates and Shells* 1959 Mc. Graw-Hill

IOP Conf. Series: Materials Science and Engineering **404** (2018) 012027 doi:10.1088/1757-899X/404/1/012027

Response study of concrete gravity dam against aircraft crash

M. R. Sadique[1], M. I. Ansari[2], M.F. Athar[1*],

[1.] Department of Civil Engineering, Aligarh Muslim University, Aligarh, INDIA

[2.] Department of Civil Engineering, Jamia Millia Islamia, New Delhi INDIA

[*] faraz.athar12@gmail.com

Abstract

Concrete gravity dam is considered as acritically important mega structure and lifeline of countries with rich water resources. Its failure due to natural and manmade hazard may cause a huge loss of human life and economy. The recent terror attacks involving intentionally crash of aircrafts have raised the safety concern of these critical structures. In the present study, a finite element analysis has been carried out to analyse the behaviour of concrete gravity dams against possible aircraft crash. Reaction-time curve of fighter jet Phantom F4 has been applied on the crest of the dam from upstream face. The Concrete Damaged Plasticity model has been incorporated to predict the behaviour of concrete under high strain loading. It has been found that at a crash speed of Mach 1, fighter jet had caused severe damage to dam. Although the penetration will be small on the loading face of the dam due to massiveness of structure, however, the concrete around crest and neck will fail in tension. Further, redistribution of moments will cause failure of the structure. Maximum deformation has been found to be 8.8 mm. Stresses are quite high at the impact face, however it diminishes at a depth of 8 m from the impact face.

Keywords: Dam, Aircraft crash, Safety, Finite element

1. Introduction

Concrete gravity dams are the important infrastructures for the country because they are generally built for multiple tasks and facilities. Water resource conservation, production of hydroelectricity and flood control in the region are important among all the purposes, for those dams are generally built. The global trend is to construct either concrete gravity dam or arch concrete dam when it is needed to have multiple objectives associated with the project. Hence, one can easily understand the loss to a country associated with the destruction of such important monumental structures. Addition to this scarcity of water is one of the biggest global issues in the present world, which is the key cause for all kind of water disputes among the countries or among

IOP Conf. Series: Materials Science and Engineering **404** (2018) 012027 doi:10.1088/1757-899X/404/1/012027

states of countries all over the world [1]. Subsequently, there are possibilities that such large concrete gravity dams can be one of the key targets like nuclear contentment structures of the country for the terrorists or enemy countries and there can be air strike on them. But vulnerability study of concrete gravity dam in this regard still needs attention of the researchers. Although the safety analysis of dam has been studied by many researchers [2-7], however, the issues related to aircraft impact have never been addressed. In the present study, behaviour of concrete dams has been observed for the present warfare situation, though this kind of risk study is relatively new for concrete gravity dams. However, experience of World War II guide us for this kind of study. There are lots of researches have already been done or still going on, on other structures like nuclear contentment buildings for the risk against aeroplane or missile attack on them. But these kinds of studies are much needed for concrete gravity dams as well. The importance of the study would be clearer if we would see the post attack scenario of "Operation Chastise" during World War II, Fig 1 . It was an attack on German dams carried out on 16–17 May 1943 by Royal Air Force No. 617 Squadron, subsequently known as the "Dambusters", using a specially developed "bouncing bomb" invented and developed by Barnes Wallis. The Möhne and Eder Dams were breached, causing catastrophic flooding of the Ruhr valley and of villages in the Eder valley, while the Sorpe dam sustained only minor damage. Two hydroelectric power plants were destroyed and several more were damaged. Factories and mines were also either damaged or destroyed. An estimated 1,600 people drowned [8, 9].It means that 'failure of dam structure' and 'post failure impacts on the region' both are catastrophic. 9/11 attack on U.S.A. revealed new technique of attack for any kind of monumental or important structure. Direct hit of an aircraft can also be an option for destruction of concrete gravity dam. To understand the response of concrete gravity dam, it is quite complex to do experimental work of aircraft crash on it. Therefore, in the present study analytical technique has been chosen to simulate the problem. The studies, which have been done related to aircraft crash on nuclear contentment structure, can be considered as basis for analytical formulation.

Table 1. Configuration of Eder dam

Eder Dam	
Construction Started	1908
Completed	1914
Type	Gravity Dam
Length	393 meters
Height	44 meters
Base Width	36.2 meters
Top Width	6 meters

Fig 1. Eder dam before and after destruction.

2 Aircraft loading

Military aircrafts are fast-moving, high responsive and versatile. Among the different variety of military aircrafts, fighter jets are designed primarily for air-to-air combat and air to land target destruction. These jets are proved to be very crucial in all recent military invasions in last few decades. In the present study of safety analysis of dams against possible aircraft crash, fighter jet Phantom F4 has been considered to collide at the neck of the dam structure [10]. With a maximum velocity twice greater than that of sound, the McDonnell F-4 Phantom was one of the most versatile fighter, multipurpose strike bomber ever built. Although it is the fighter jet started in sixties of last century but still several hundreds of Phantom F4 are in service these days. A brief comparative specification of Phantom F4 with MiG-29 and F16 has been given in Table 2. It may be observed from the table that the Phantom F4 is very much competitive to fighter jets which were quite younger to it in the combat field [11].

Table 2. Comparison of the three fighter planes considered in the study

	Phantom F4	MiG-29	F-16
Wingspan(m)	11.71	13.46	9.95
Length(m)	19.20	22.70	15
Height(m)	5	6.15	4.88
Range(Km)	3000	720	4200
Weight(kg)	28,000	46,200	19,200
Max Speed(m/s)	650	830	400

In order to simplify the complexity of the finite element analysis through minimising the geometric non-linearity of the aircraft, uncoupled approach of analysis has been adopted. Hence, in the present study, the loading of the aircraft to the containment has been assigned through the well-known Riera's Approach of using reaction-time response curves of the aircraft. Reaction time response curve of Phantom F4 has been calculated in Fig. 2 for three different velocities i.e. Mach 1 (340m/s), Mach 0.735 (250 m/s) and Mach 0.632 (215 m/s). In open literature the conventional design of critical structures had been analysed by using the reaction time curve of the Phantom F4 at a speed of 215 m/s [12]. Hence, in the present study also the same curve has been used as impact load of the aircraft. However, in case of deliberate/ terror attack the aircraft may hit the target with higher speed; consequently, it will cause more disastrous effect. Hence, in the present study, analysis has also been performed for speed of 340m/s.

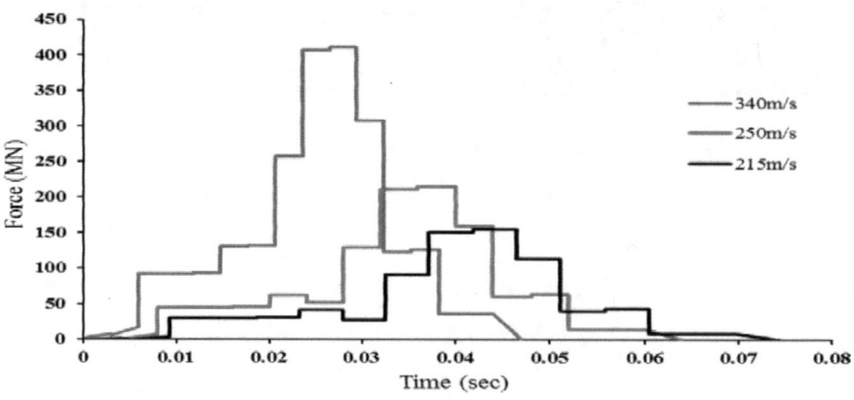

Fig. 2 Reaction time response of Phantom F4 at different velocities

In effect of an impacting projectile is also a function of the contact area. To simplify the problem, instead of providing a time dependent contact area (starting from a contact point, increasing to a circle and finally to a bird-shaped contact area) an equivalent circular area has been assumed in the present study. The force history curve of the aircraft was assigned to the dam at a given constant area equivalent to the average of total cross-sectional area of its fuselage and wings. This hypothesis has been emerged as a common and frequent opinion in available literature [13-18] has followed the same hypothesis. Accordingly, the impact zone of Phantom F4 was considered as a circular region of ϕ 6 m as discussed above. The reaction force after dividing with the contact surface area has been assigned to respective location, depicted in Fig. 4. The reaction force history was thus transformed in to pressure versus time curve varying in time domain.

3. Finite element modelling

The concrete gravity dam section considered for the modelling and simulation in this study is a 3-dimensional model of a typical monolith of concrete gravity dam with thickness about 20m. The geometrical configuration of the dam is considered similar to the geometrical configuration of tallest non-overflow monolith of Pine Flat dam [19]. Geometric parameters are presented in Fig. 3.

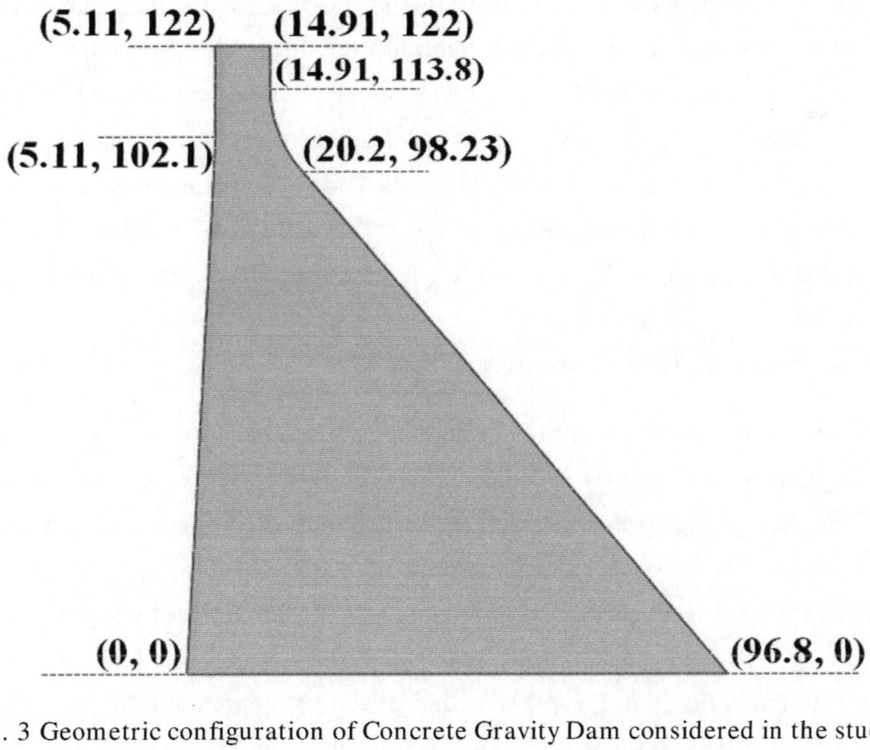

Fig. 3 Geometric configuration of Concrete Gravity Dam considered in the study

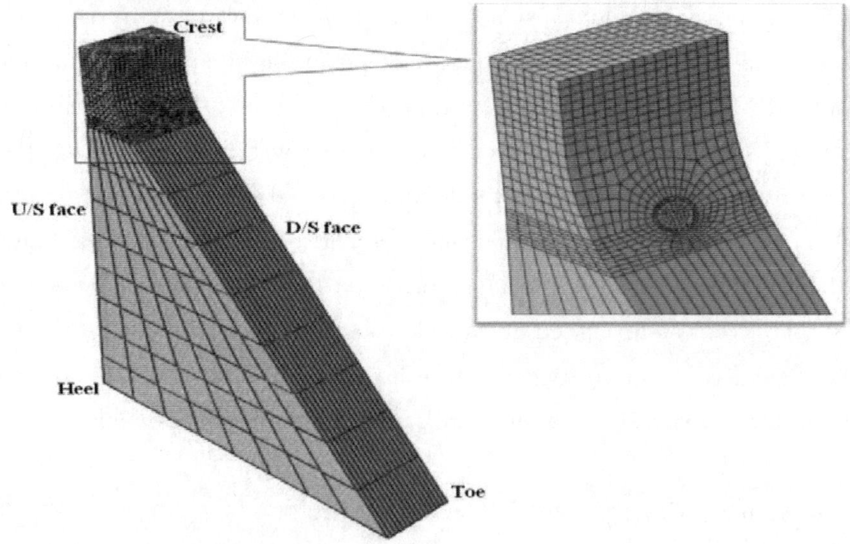

Fig. 4 Finite element discretization of Concrete Dam

In the present study, finite element tool Abaqus/Explicit [20] has been used to develop the model of the dam section. The finite element discretization of the dam section has been shown in Fig. 4. Continuum three-dimensional eight nodded reduced integration element (C3D8R) elements are adopted for discretization of elements. The selected impact zone at crest (ϕ6m) has been decided to have eight number of element along the thickness through a mesh convergence analysis. Remaining body of the dam section has been modelled with coarser elements; in order optimise the total number of element. In the present analysis, there are 6367 elements with 6599 nodes. Elements are having aspect ratio near one in the impact zone; however, away from impact zone aspect ratio may have some larger values.

4. Constitutive material model and properties

Prediction of behaviour of any structure through the finite element solver is a function of the material model opted in the analysis. Many concrete material models have been programmed and incorporated in these solvers. For concrete material models, damage and plasticity are two important features where researchers concerned. In ABAQUS, concrete damaged plasticity model (CDP) [21, 22], is widely used among researchers [23-26].In the present study also, the concrete damage plasticity model of Abaqus [20] material library was used for the simulating the behaviour of concrete in the containment structure. It is a phenomenological model which can be executed both in Abaqus standard as well as explicit analysis techniques. The material incorporates the plastic behaviour of the concrete both in compression as well as in tension. The model uses the concept of isotropic damaged elasticity in combination with isotropic tensile and compressive plasticity to represent the inelastic behaviour of concrete. It can effectively be used for plain concrete even though it is intended primarily for the analysis of reinforced concrete structures subjected to monotonic, cyclic, and/or dynamic loading. The model has the provision to use strain rate dependent behaviour of the concrete. The material parameters used to predict the behaviour of concrete are presented in Table 3.

Table.3 Material properties of concrete

Density, ρ (kg/m^3)	2400
Modulus of elasticity, E (N/m^2)	2.7386E+10
Poisson's ratio, ϑ	0.17
Dilation angle, ψ	30
Eccentricity	1.0
Initial equi-biaxial compressive yield stress to initial uniaxial compressive yield stress, f_{b0} / f_{c0}	1.16
Bulk Modulus, K	0.666
Fracture Energy, G_f (N/m)	720
Uniaxial Failure Stress (Tension), σ_{t0} (MPa)	10.8
Cracking displacement, U_{to} (m)	0.0001332
Tensile strength, σ_{st}, MPa	3.86
Compressive Strength, MPa	30

5. Results and discussion

A three-dimensional numerical modelling of a concrete dam has been performed through the finite element solver Abaqus. Fighter jet Phantom F4 has been considered to crash at a speed of mach1 near the crest of dam from upstream side. The deformations in the direction of the loading are plotted in the Fig. 5. The deformation has been found to be localised in the crest. The maximum deformation has been found to be 8 mm in the impact zone itself. However, the effect of crash was comparative low at the downstream face of the dam. Moreover, near the toe and heal of the dam there is no effect of crash due to massiveness of the structure. Tension damage in the concrete has been plotted in Fig. 6. The damage of the material under tension or compression has been assumed to occur when the corresponding damage parameter, d_t or d_c respectively, has reached a value of 0.99. The value of the damage parameter varies from 0 to 0.99 for undamaged and complete damaged material respectively. It can be visualised that the damage occurred is very widespread in and around the crest of the dam. The damage experienced by the concrete under

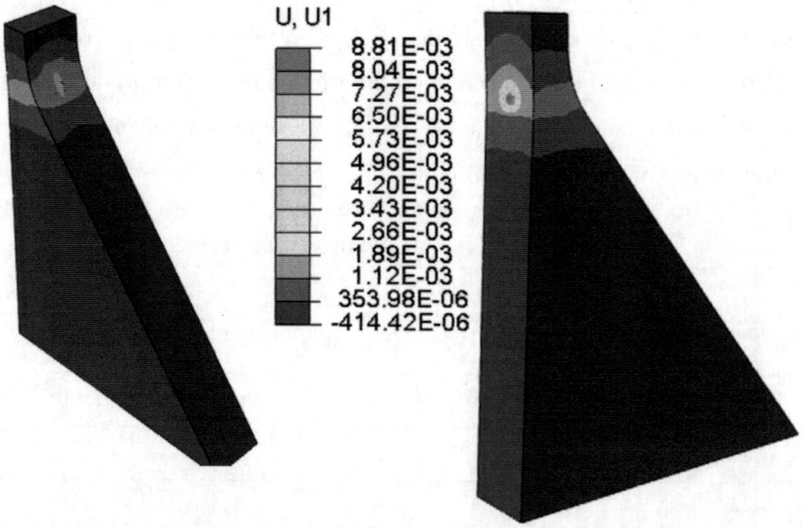

Fig. 5 Deformation along the direction of loading

tension will be highlighted more effectively with removal of fully damaged elements from the dam. In Fig. 7, different directional views of dam crest with removal of completely damaged elements have been shown. From the damage pattern reported in Fig. 7, a very good resemblance can be found with the case reported in Fig.1 i.e. Eder dam failures. Only few elements of the crest neck have been noticed undamaged in the current study. However, a redistribution of moment may lead to overturning of the above neck portion of the dam.

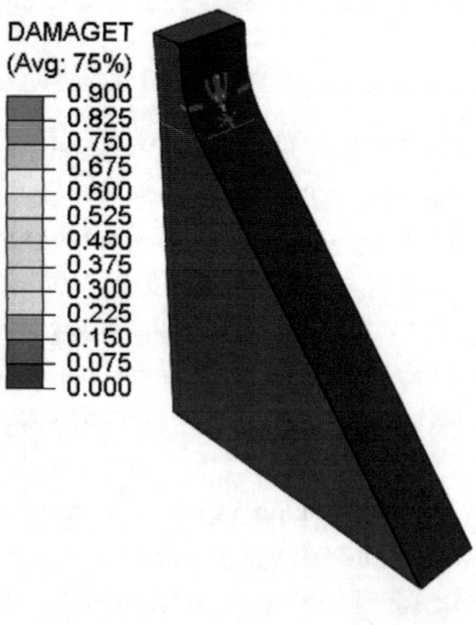

Fig. 6 Tension damage in concrete

IOP Conf. Series: Materials Science and Engineering **404** (2018) 012027 doi:10.1088/1757-899X/404/1/012027

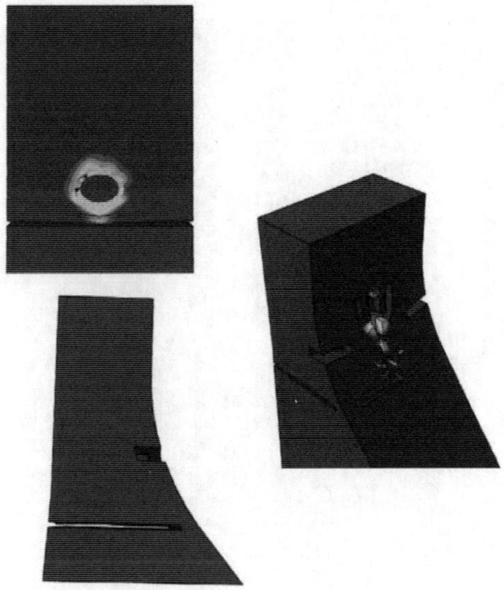

Fig. 7 Dam head after removal of completely damaged concrete element

As the behaviour of concrete is better in compression always, the compressive damage however is observed to be of lesser intensity. To review the compression and tension damage across the cross-section of the dam, damage contour plots of vertical cross-section have been shown in Fig. 8. Tension damage is more visible at the D/S side of the crest. However, in case of compression damage, complete failure of elements has not been reported.

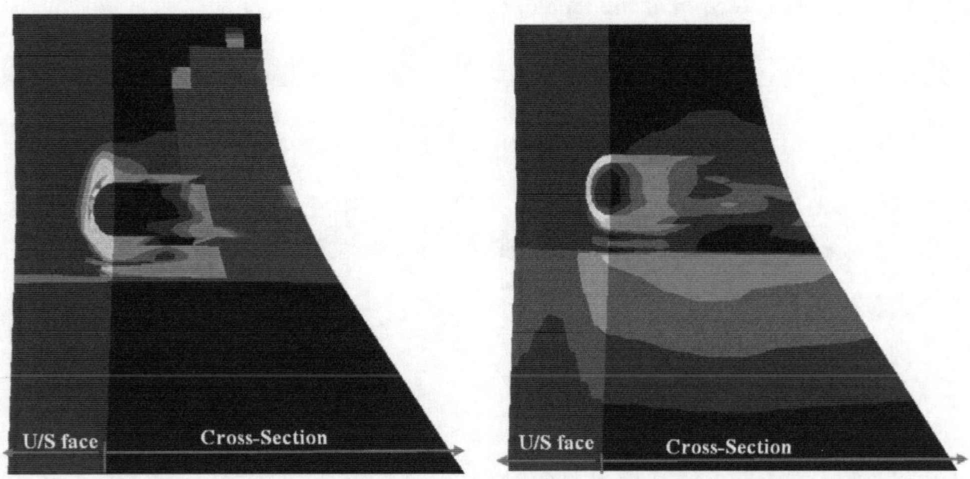

Fig. 8 Damage along the longitudinal cross-section of dam (a) Tension (b) Compression

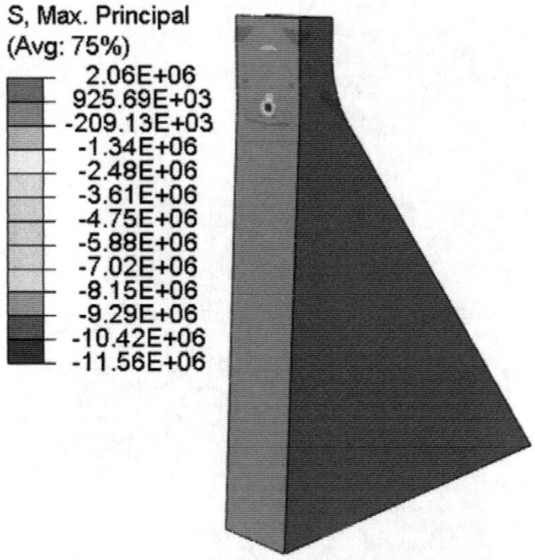

Fig. 9 Contour profile of maximum principal stress

The contour of maximum principle stresses has been plotted in Fig. 9. High concentration of compressive stress has been reported at the point of impact. A maximum compressive stress of magnitude 11.56 MPa has been observed at the centre of impact. However, patches of tensile stresses of greater than 2 MPa has been found on both the upstream and downstream face of the dam. To analyse the stress distribution across the thickness of dam, a nodal path from the centre of impact to the downstream face has been chosen as shown in Fig. 10(a). High stresses have been reported at the u/s face i.e. at the impact face, Fig. 10(b). Stresses have die down to a negligible value at a depth of 8 m inside the dam thickness. The maximum stress in the direction of loading (S_{11}) has maximum value of 38 MPa at the face of crash.

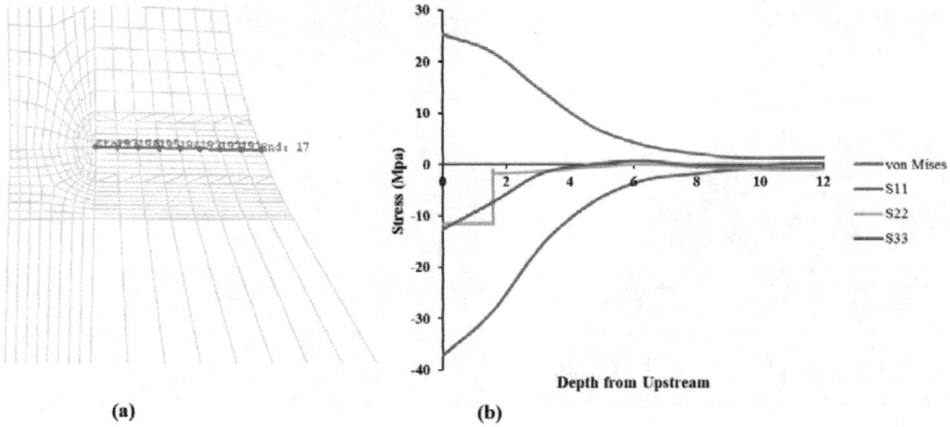

Fig. 10 (a) Path selected (b) Stress distribution along the selected path

IOP Conf. Series: Materials Science and Engineering **404** (2018) 012027 doi:10.1088/1757-899X/404/1/012027

6. Conclusions

A three-dimensional finite element simulation has been carried out using Abaqus/Explicit. A typical concrete gravity dam model similar to Pine Flat Dam has been developed. Aircraft loading has been applied with the help of Riera's Reaction time approach. A normal hit at the crest of dam from the upstream side has been considered. Deformation/ displacement has been found to be very confined in the impact zone. However, the crest of the dam has been damaged under tension. Tensile damaged was more visible on the opposite i.e. downstream face of the dam.

References

1. http://www.worldwater.org/conflict/refs.html
2. S. Zhang, G. Wang, C. Wang, B. Pang, C. Du, Numerical simulation of failure modes of concrete gravity dams subjected to underwater explosion, Structural Failure Analysis, 36 (2014) 49-64.
3. G. Wang, S. Zhang, Damage prediction of concrete gravity dams subjected to underwater explosion shock loading, Engineering Failure Analysis, 39 (2014) 72-91.
4. S. Jiang,C.Dua, Y. Hong, Failure analysis of a cracked concrete gravity dam under earthquake, Engineering Failure Analysis, 33 (2013) 265-280.
5. Md. I. Ansari, P. Agarwal, Categorization of damage index of concrete gravity dam for the health monitoring after earthquake, Journal of Earthquake Engineering,**00**(2016) 1-17.
6. N. Oudnia, Y. Bouafia, Response of concrete gravity dam by damage model under seismic excitation, Engineering Failure Analysis, 58(2) (2015) 417-428.
7. S. Zhanga, G. Wanga, X. Yu, Seismic cracking analysis of concrete gravity dams with initial cracks using the extended finite element method, Engineering Structures, 56 (2013) 528-543.
8. http://en.wikipedia.org/wiki/Operation_Chastise
9. http://www.raf.mod.uk/history/thestoryoftheraid.cfm
10. M. R.Sadique, M. A.Iqbal, P.Bhargava, Crash analysis of military aircraft on nuclear containment, Structural Engineering and Mechanics, 53(1) 2015.
11. http://www.militaryfactory.com/aircraft/fighter-aircraft.asp
12. T.Sugano, H.Tsubota, Y.Kasai, N.Koshika, S.Orui, W.A.V.Riesemann, D.C. Bickel, M.B. Parks, Full-scale aircraft impact test for evaluation of impact force,Nucl. Eng. Des. 140(1993) 373-385.
13. S.Gomathinayagam, M. V.Dharaneepathy, M. N. K. Rao, Damage-zones of containment structures under aircraft impact loads, Computer & Structures, 52(3), (1994), pp. 581-590.
14. H. Abbas, D.K. Paul, P.N.Godbole, G.C.Nayak, Reaction-time response of aircraft crash, Comp. Struct., 55(1995) 809–17.
15. H. Abbas, D.K. Paul, P.N.Godbole, G.C.Nayak, Aircraft crash upon outercontainment of nuclear power plant,Nucl. Eng. Des.,160 (1996) 13-50.

16. M. Kukreja, Damage evaluation of 500 MWe Indian pressurized heavy water reactor nuclear containment for aircraft impact,Nucl. Eng. Des., 235 (2005) 1807–1817.

17. M.A. Iqbal, S.Rai, M.R. Sadique, P.Bhargava, Numerical simulation of aircraft crash on nuclear containment structure,Nucl. Eng. Des., 243 (2012) 321-335.

18. M.R.Sadique, M.A.Iqbal, P.Bhargava, Nuclear Containment Structure Subjected to Commercial and Fighter Aircraft Crash. Nucl. Eng. Des., 260 (2013)30-46.

19. Hall, J.F., Study of the Earthquake Response of Pine Flat Dam, Earthquake Engineering & Structural Dynamics, Vol. 14, No. 2, (1986). pp. 281–295.

20. Abaqus Explicit user manuals Version 6.6.

21. J. Lee and G. Fenves, Plastic-damage model for cyclic loading of concrete structures, Journal of Engineering Mechanics, 124(8) (1998) 892–900.

22. J. Lubliner, J. Oliver, S. Oller, and E. Oate, A plastic-damage model for concrete, International Journal of Solids and Structures, 25(3) (1989) 299–326.

23. T. Jankowiak and T. Lodygowski, Identification of parameters of concrete damage plasticity constitutive model, Foundations of civil and environmental engineering, 6(1)(2005) 53–69.

24. S. V. Chaudhari, M. A. Chakrabarti, Modeling of concrete for nonlinear analysis using finite element code abaqus, International Journal of Computer Applications, 44 (2012) 14–18.

25. M. R. Mousavi and A. Khaloo, Effect of steel plate jacketing of columns in seismicbehavior of concrete beam-column connections, in 6th National Congress on Civil Engineering, 2011.

26. G. Xotta, S. Beizaee, and K. Willam, Localization analysis of coupled plasticity anddamage models for dissipative materials, Computational Modelling of Concrete Structures, 439(2014).

27. M.Hillerborg, M.Modeer, P.Peterson, Analysis of crack formation and crack growth in concrete by means of fracture mechanics and finite elements,Cem. Concr. Res., 6 (1976) 773–782.

28. B.P.Sinha, et al., Stress–strain relations for concreteundercyclicloading, J.Am.ConcreteInst.2(61) (1964) 195–210.

29. D.L.Grote,S.W.Park, M.Zhou, Dynamic behavior of concrete at high strain rates and pressures: I. experimental characterization, Int. J. of Impact Eng. 25 (2001) 869-886.

30. Y.Lu, K.Xu, Modeling of dynamic behavior of concrete material under blast loading, Int. J. of Solids Struct., 41(2004) 131-143.

IOP Conf. Series: Materials Science and Engineering **404** (2018) 012028 doi:10.1088/1757-899X/404/1/012028

Development, Characterization and Application Potential of Bio-composites: A Review

Mohammad Zahid Rayaz Khan[1] and S K Srivastava[2]

[1]Research Scholar,MED, M. M. M. University of Technology, Gorakhpur, UP 273010, INDIA
[2]Professor, MED, M. M. M. University of Technology, Gorakhpur, UP 273010, INDIA

E-mail: zahidkhan010@gmail.com

Abstract. Sustainable development of bio-composites is the solution for current ecological imbalance caused by the petroleum based synthetic materials. Synthetic fibers of conventional polymer composites are replaced by natural/biodegradable fiber resulting into bio-composites. The paper presents an overview about the development steps, manufacturing methods and characterization parameters of bio-composites. Popular manufacturing methods of bio-composites such as hand layup, injection molding, compression molding etc. have been covered briefly. It includes characterization based on mechanical, thermal and morphological parameters. The commercial and industrial application potential of bio-composites is also included.

1. Introduction

The conventional fiber reinforced polymer composites are generally made up of inorganic and organic materials reinforced into thermosetting and thermoplastic polymers. These composites had high application potential due to its better mechanical, morphological, thermal properties. However, their usage and disposal are major challenges because theyare non-biodegradable. In recent years, this problem has been overcome by using fully/partial biodegradable composite materials reinforced with the natural fiber in place of conventional fossil based fiber[1]. The researchers are being carried out in this area for the sustainable development of bio-composites which fulfill the demands of legislative authorities related to their recycling and environmental friendly products.

Natural fibers such as kenaf, sisal, jute, banana, rice hucks are used as reinforcement in the polymers for the development of bio-composites. Bio-composites are fully green, if polymer and fiber both are derived from plant or biodegradable material; however, composites are partial green if anyone out of polymer/fiber is biodegradable or obtained from the natural plant. Natural fibers are economical, low density, good specific strength properties, and biodegradability. Bio composites are suitable for delicate consumer products, packaging, disposable consumer products, and suitable for light weight applications[2][3][4][5].

In this paper, a brief review on the recent research and development carried out, characterization and application of bio-composites is presented. It includes a brief description on the various manufacturing methods based on the types of natural fibers, polymers, their geometrical properties, mechanical properties etc. A brief discussion on the characterization of bio-composites based on different destructive and non-destructive testing methods is also included. Bio-composites have shown wide potential in the development of automotive components, agricultural equipment, cabinet and cosine of different electrical and electronic equipment, and packaging etc.

2. Development of Bio-composites

The processes involved in the development of NFRP composites are: fiber and polymer preparation and amalgamation of prepared fiber and polymer by different fabrication techniques. Several methods can be used for the processing of natural fibers such as mechanical processing (for stiff and rough fibers), chemical processing (for soft, contains property of both synthetic and natural), & biological degumming[6] etc.

2.1. Basic Steps

The following four steps are generally used in the development of bio-composites such as impregnation, layup, consolidation and solidification [7].

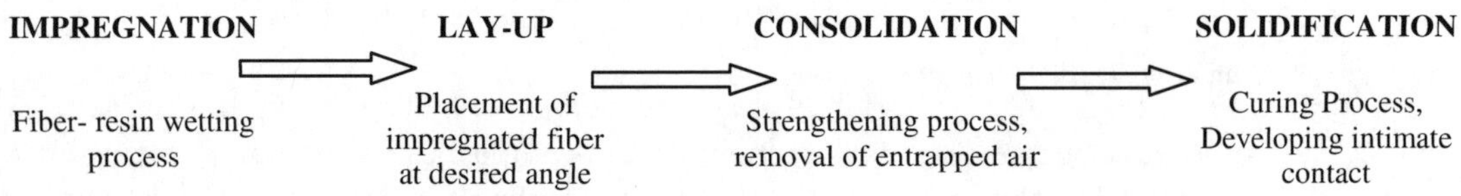

Figure 1: Steps for Development of Bio-Composites

2.2. Manufacturing Methods

There are different manufacturing methods for the successful development of NFRP composites such as injection molding for short fiber reinforced thermoplastics, open mold process (spray and hand layup), liquid composite molding process, compression molding including film staking, automated process[8][9]. The selection of particular manufacturing process depends upon the type of polymers (whether thermoplastic or thermosets) used for the development of NFRP composites [10]. Thermosets are firstly used around 1940s. Fabrication of thermosetting composite is easy because of the low tooling cost, better wettability of fiber, less heat and pressure required during curing process.In 1998, thermoplastic polymer based composite products capture around 25% markets of polymer-based composites. Thermoplastic composites are very popular in the field of aerospace and automobile industries. Few manufacturing processes are common for both types of polymers like pultrusion process, compression molding, injection molding, autoclave process, etc. Table 1 shows different manufacturing methods, processing techniques and characteristics of bio-composites.

Table 1. Manufacturing methods for bio-composites development

Methods	Operation	Characteristics
Hand Lay-Up Method[11][10]	Mixture of fiber-matrix is placed manually into open mould, mild pressure is applied by roller for the removal of entrapped air, better compatibility with thermosets.	low cost, no size constraints, suitable for academic and R&D work, mild surface finish, air pockets and long curing time are the limitation.
Filament Winding Process[10][12]	Resin impregnated continues fiber are wound on rotating mandrel, manual, computer controlled, controlled fibre orientation.	Hallow Tube like structures, continues fiber and thermoplastic resin are the constraints, high fiber volume fraction.
Pultrusion Process [13]	Resin soaked fibers are pulled out though a die of desired shape, Shape of the product depends on the die cross section.	Overall cost is low, suitable for thermosets and thermoplastics, limited for continues cross section.
Compression Moulding[10]	Fiber- resin mixture is placed into the heated closed mould; compressive load is applied through the upper half of the mould.	High production rate, good Surface finish, multiple part development, thermosets polymer, large automobile parts.

Liquid Transfer Moulding Process[14]	Resin is injected into the fiber carrying closed mould, mould cavity has vacuumed pressure or atmospheric pressure.	product having complex geometry, high fiber volume fraction, good surface finish, Resin injection, curing, and heat transfer are the major constraints.
Hybrid transfer moulding[10]	Mixture of two different resin is injected into mould containing fiber.	Low fiber volume fraction, complex shapes, thermosets polymer.
Injection Moulding[15]	Require amount of fiber resin mixture are injected into mould cavity, extruder is used.	Compatibility with both type of resin, high production rate, less curing time, different automobile parts.
Tape Winding[16][17][18]	Commingled fibers are wound on the heated mandrel for melting and strengthening of thermoplastics.	Suitable for thermoplastics polymers, single step process.
Hot Press [19]	Thermoplastic prepreg are piled up together between the heated moulds and compressed.	No need of secondary operation, popular in R&D.
Autoclave Processing[20][21]	Automatic process, prepregs are lay down in mould and entire assembly vacuumed bagged and placed for curing inside autoclave.	Higher fiber volume fraction, proper amalgamation of composite constituents, no air pockets, good interfacial adhesion, aircraft parts, both type of polymers.
Diaphragm forming [22]	Thermoplastic prepregs sheets are placed between the flexible diaphragms, diaphragms placed into the mould for curing.	only suitable for part having constant thickness and non-uniform fiber distribution.

3. Characterization Parameters

There are different tests for the mechanical characterization of bio-composites such as tensile test, flexural test; impact test, dynamic mechanical analysis etc. The scanning electron microscopy, XRD, EDX, FTIR tests are conducted for the morphology analysis of bio-composites. Table2 shows the possible number of tests, purpose, machine used (brand name), approximate cost along with capacity for the characterization of bio composites.

Table 2: Various tests for Bio -Composites Characterization

Experiments	Properties	Machine, Brands and Capacity (Cost)
Tensile Test [23]	Maximum tensile load carrying capacity, young's modulus	Universal testing machine Brands: Instron, Tinius Olsen, Zwick Capacity: 5, 15, 30, 100 KN Approx. Cost: 10-20 lakhs
Three Point Bending Test[24]	Flexural strength, maximum load at failure, flexural modulus	
Impact Test (Charpy /Izod)[25]	Impact strength	Pendulum impact test machine Brands: Instron, Tinius Olsen, Zwick, WPM Leipzig, Haida Capacity:1J-300J Approx. Cost: 3-10 lakhs
Hardness [26]	Hardness	Hardness testing machine Brands: Instron, Zwick, Buehler Capacity:10 to 150 Kg Approx. Cost: 1.5-4lakhs
Dynamic Mechanical Thermal Analysis (DMA) [27]	Fatigue strength, glass transition temperature, storage modulus	DMA analyser Brands: Instron, NETZSCH Capacity:25N to 500N Approx. Cost:10-25 lakhs
Thermogravimetric Analysis	Thermal properties, thermal	Thermogravimetric analyser

(TGA) [28][29]	degradation temperature, percentage weight loss, variation in chemical and physical properties	Brands: Perkin Elmer, Torbal Approx. Cost: 15-25 lakhs
Differential Scanning Calorimetry (DSC) 23,6 [29][30]	Observe the glass transition temperature of the polymers Tg, crystallization and melting temperatures	Thermo- Modulated calorimeter
Thermal Conductivity Test [31]	Insulation performance of materials	Manually
Thermomechanical Analysis [32]	Thermomechanical stability, coefficient of thermal expansion,	Thermomechanical analyser Brands: Thermo Scientific LTQ, Thermo Finning
Pin-on- Disc Test [33]	Friction and wear properties of material	Tribometer Approx. Cost: 5-8 lakhs
Erosion Test [34]	Erosion wear properties	Air jet erosion test rig
Fragility Test [31]	Mass loss from the surface abrasion and Impact damage	Manually
Scanning Electron Microscopy (SEM) [35][36]	Surface morphology, examine the surface of sample after deformation, physio-mechanical properties, micrographs of fractured surface	Scanning electron microscope Brands: Hitachi, JEOL, ZEISS, FEI Approx. Cost: 25-150 lakhs
X-Ray Photoelectron Spectroscopy [29][28]	Identify the crystal structure and molecule arrangement, strain, composition of material	X-Ray diffractometer Brands: PANalytical, Philips, Siemens Approx. Cost: 30-150 lakhs
FTIR [37][28]	Amount of component in the mixture,	FTIR spectrometer Brands: Nicolet, PerkinElmer, Bruker Approx. Cost: 30-100 lakhs
CT Scan Analysis [38]	Fiber distribution in the matrix	CT Scanner Approx. Cost: 15 lakhs
Elemental Dispersive X-Ray Analysis [39]	Elemental composition of composites	Nano scanning electron microscope Brands: Hitachi, ZEISS Approx. Cost: 100 Lakhs
Transmission electron microscopy Analysis (TEM) [39]	Morphology and particle size distribution in particulate composites	Transmission electron microscope Brands: CRAIC, JEOL Approx. Cost: 10 lakhs
Atomic Force Microscopy [32]	Surface morphology	Atomic force microscope Brands: Veeco, RMC Approx. Cost: 20-100 lakhs
Polarized Optical Microscopy [40]	Crystallization behaviour	Polarized optical microscope Brands: Motic, Olympus, Nikon Approx. Cost: 5-15 lakhs
Soil Burial test [28][41]	Biodegradability of Bio-Composite, weight loss	Manual Process

4. Applications

The bio-composites materials have wide domestic and industrial applications. The different types of natural fibers such as sisal, hemp, wood saw particles, bamboo, kenaf, oil palm, bagasse, animal hair etc. and polymers such as epoxy, polyester, polylactic acid(PLA), etc. are used for the sustainable development of bio-composites. Bio-composites are largely used in automobile industries, structural components, packaging, electronics industries, electrical appliances, sports, aerospace structures, etc. for the last two to three decades because it possesses high strength to weight ratio, substantially low cost, corrosion resistant, high fatigue strength, biodegradability, good surface finish, etc. These composites have the demerits such as poor hydrophilic resistance, moderate temperature sustainability and ageing effect [42]. Major industrial applications of bio-composites are summarized in table 3.

Table 3 : Applications of Bio-Composites

Area	Applications	Company
Automobile Industries[43]	Rear storage shelf/panel, door panels, spare tyre lining, noise insulation panels, boot lining, seat backs, internal engine cover, sun visor, bumper, wheel box and other parts of automobiles	Opel, Volkswagen, Audi, BMW, General Motors, Toyota, Volvo, Mercedes, Renault, Mitsubishi, Rover, Lotus
Packaging[44]	Cosmetics, laptop casings, bottles, food and beverages, mobile casings, electronic goods packaging, laptop and mobile packaging, projector and voltage stabilizer cover	FS Korea, Rexam, hp, Apple, Thomson Lamp Bulb
Sports	Safety helmets, surfing boards, rackets, polo balls, bicycle frame	-
Construction	Door panels, decking, railing, window frames	-

5. Conclusions

The waste disposal problem, limited resources and unbalance among the ecology, technology and economy associated with the conventional polymer composites led to the development of nonconventional, viz. natural and biodegradable composites (bio-composites). The paper describes different manufacturing methods for the sustainable development of bio-composites. It covers the discussion on major characterization parameters and application potential of bio-composites in different areas of engineering components. There is always a good scope for designing and exploring the new biodegradable fiber/polymer which competes with non-biodegradable fiber/polymer.

References

[1] Goda K and Cao Y 2007 Research and development of fully green composites reinforced with natural fibers *Journal of Solid Mechanics and Materials Engineering* 1 1073-1084.

[2] Mohanty A K *et al* 2000 Influence of chemical surface modification on the properties of biodegradable jute fabrics polyester amide composites *Composites Part A: Applied Science and Manufacturing* 31 143-150.

[3] Mohnaty A K *et al* 2000 Surface modification of jute and its influence on performance of biodegradable jute-fabric/Biopol composites *Composites Science and Technology* 60 71115-1124.

[4] Li X H *et al* 2004 Completely biodegradable composites of poly(propylene carbonate) and short, lignocellulose fiber Hildegardia populifolia *J of Polymer Science: Part B* 42 666-675.

[5] Shibata M *et al* 2003 Biocomposites made from short abaca fiber and biodegradable polyesters *Macromoleculer Materials and Engineering* 288 35-43.

[6] Fu J *et al* 2012 Bio-processing of bamboo fibres for textile applications: a mini review *Biocatalysis and Biotransformation* 30 141-153.

[7] Pakmer R J and Moore W E 2017 Resin impregnation process for producing a resin-fiber composite . [Online]. Available: https://google.com/patents/us5281388.

[8] Summerscales J and Grove S 2014 Manufacturing methods for Natural fiber composites *Natural fiber Composites: materials, processes and properties Woodhead Publishing Limited* 176-215.

[9] Bajpai P K *et al* 2017 Processing and characterization of biocomposites *Handbook of Ecomaterials* 1-18.

[10] Mazumdar S K 2002 Composites Manufacturing Materials, Product, and Process Engineering.

[11] Cairns D S and Shramstad J D 2017 Evaluation of hand lay-up and resin transfer molding in composite wind turbine blade manufacturing 2000. *[Online]. Available: http://montana.edu/composites/documents/jon skramstad thesis.pdf. [Accessed 23 11 2017].*

[12] Cohen D 1997 Influence of filament winding parameters on composite vessel quality and strength *Composites Part A: Applied Science and Manufacturing* 28 1035-1047.

[13] Fairuz A M *et al* 2014 Polymer composite manufacturing using a pultrusion process: a review *American Journal of Applied Sciences* 11 1798-1810.

[14] OliveiraI R D *et al* 2013 Resin transfer molding process: a numerical and experimental investigation *The International Journal of Multiphysics* 7 125-136.

[15] Kauffamn A J *et al* J 1987 Apparatus for injection molding fiber-reinforced thermoset plastic articles *[Online]. Available: http://google.com/patents/us4752201.* [Accessed 2 12 2017].

[16] Mazumdar S K and Hoa S V 1996 Determination of manufacturing conditions for hot-gas-aided

thermoplastic tape winding *Journal of Thermoplastic Composite Materials* 9 35-53.

[17] Royre D S *et al* 1989 Test methodology for the determination of optimum fusion welding conditions of polyethylene *Journal of Applied Polymer Science* 38 147-162.

[18] Mazumdar S K and Hoa S V 1995 Manufacturing of non-axisymmetric thermoplastic composite parts by tape winding technique *Materials and Manufacturing Processes* 10 47-56.

[19] Chang I and Lees J 1988 Recent development in thermoplastic composites: a review of matrix systems and processing methods *Journal of Thermoplastic Composite Materials* 1 277-296.

[20] Hubert P 201 2Autoclave processing for composites. [Online]. Available: http://sciencedirect.com/science/article/pii/b9780857090676500134. [Accessed 2 12 2017].

[21] Drakonakis V M *et al* 2013 Curing pressure influence of out-of-autoclave processing on structural composites for commercial aviation *Advances in Materials Science and Engineering* 2013 1-14.

[22] Mallon P and Bradaigh C O' 1988 Development of a pilot autoclave for polymeric diaphragm forming of continuous fibre-reinforced thermoplastics*Composites* 19 37-47.

[23] Shah P, Prajapati R and Singh P 2017 Enrichment of mechanical properties of biodegradable composites containing waste cellulose fiber and thermoplastic starch *European Journal of Advances in Engineering and Technology* 4 282-286.

[24] Graupner N,Labonate D and Mussig J 2017 Rhubarb petioles inspire biodegradable cellulose fiber-trinforced PLA composites with impact strength *Composite:PartA* 98 218-226.

[25] Senthilkumar P and Ravi S 2017 The effect of alkali treatment on the mechanical properties of sisal fiber reinforced epoxy composites *International Journal of Advanced Scientific Research & Development* 4 31-39.

[26] Xia C *et al* 2017 Natural fiber and aluminum sheet hybrid composites for high electromagnetic interference shielding performance *Composites Part B* 114 121-127.

[27] Dayo A Q *et al* 2017 Natural hemp fiber reinforced polybenzoxazine composites: Curing behavior, mechanical and thermal properties *Composites Science and Technology* 144 114-124.

[28] Prachayawarakorn J and Hwansanoet W 2012 Effect of silk protein fibers on properties of thermoplastic rice starch *Fibers and Polymers* 13 606-612.

[29] Haque P *et al* 2010 Influence of compatibilizing agent molecular structure on the mechanical properties of phosphate glass fiber-reinforced pla composites *Journal of Polymer Science: PART A Polymer Chemistry* 48 3082-3094.

[30] Berthet M A *et al* 2017 Sorting natural fibres: A way to better understand the role of fibre size polydispersity on the mechanical properties of biocomposites *Composites: Part A* 95 12-21.

[31] Ma Y *et al* 2017 Effect of fiber surface treatments on the properties of wood fiber- phenolic foam composites *BioResources* 12 4722-4736.

[32] Jang Y H *et al* 2013 Pretreatment effects of seaweed on the thermal and mechanical properties of seaweed/polypropylene biocomposites *Composite Part A* 47 83-90.

[33] Bajpai P K *et al* 2013 Tribological behavior of natural fiber reinforced PLA composites *Wear* 297 829-840.

[34] Latha P S *et al* 2015 Evaluation of mechanical and tribological properties of bamboo-glass hybrid fiber reinforced polymer composite *Journal of Industrial Textile* 0 1-16.

[35] Mathew L and Joseph R 2007 Mechanical properties of short-isora-fiber-reinforced natural rubber composites: effects of fiber length, orientation, and loading; alkali treatment; and bonding agent *Journal of AppliedPolymerScience* 103 1640-1650.

[36] Ramachandran M *et al* 2015 Analysis of bamboo fiber composite with polyester and epoxy resin *International Journal on Textile Engineering and Processes* 1 2395-3578.

[37] Li M E and Alabdulkarem A 2017 On thermal characteristics and microstrusture of a new insulation material extracted from date palm trees surface fibers *Construction and Building Materials* 138 276-284.

[38] Danso H D *et al* 2017 Mechanisms by which the inclusion of natural fibres enhance the properties of soil blocks for construction *Journal of Composite Materials* 0 1-11.

[39] N. Saba *et al* 2017 Mechanical, morphological and structural properties of cellulose nanofibers reinforced epoxy composites *International Journal of Biological Macromolecules* 97 190-200.

[40] Yao L Y *et al* 2017 Thermal properties and crystallization behaviors of polylactide/ redwood flour or bamboo fiber composites *Iran Polym J* 26 161-168 .

[41] Sukhtesaraie A and Hosseini S B 2016 Effect of Silicon carbide and pulping processes on physical and mechanical properties of pulp plastic composites (PPCs) *Journal of Asian Ceramic Societies* 4 112-119.

IOP Conf. Series: Materials Science and Engineering **404** (2018) 012028 doi:10.1088/1757-899X/404/1/012028

[42] Gallo E *et al* 2013 Tailoring the flame retardant and mechanical performances of natural fiber-reinforced biopolymer by multi-component laminate *Composites Part B-engineering* 44 112-119.

[43] Mohanty A K *et al* Natural fibers, biopolymers, and biocomposites: *An Introduction* 15-50.

[44] Fazita M R N *et al* 2016 Green composites made of bamboo fabric and poly (lactic) acid for packaging applications-A Review *Materials* 9 1-29.

[45] Khalil H P S A *et al* 2012 Bamboo fibre reinforced biocomposites: A review *Materials & Design* 42 353-368.

IOP Conf. Series: Materials Science and Engineering **404** (2018) 012029 doi:10.1088/1757-899X/404/1/012029

Static Response of Functionally Graded Plates using Higher Order Theories

Mayank Jain* **and M.Y. Yasin**
Department of Mechanical Engineering, Z. H. College of Engineering and Technology, Aligarh Muslim University, Aligarh, Uttar Pradesh, India.

*Email: jainmayank135@gmail.com

Abstract. This work presents an assessment of higher order theories for the static response of functionally graded plates. The governing equations, variationally consistent boundary conditions, force and moment resultants of higher order shear deformable theories are derived using principle of virtual work. Material properties are assumed to vary smoothly in the thickness direction with power law variation of ceramic volume fraction. The effective material properties of FG material are obtained using rule of mixture as well as Mori-Tanaka homogenization scheme. Analytical solution using a double Fourier series expansion is obtained for simply supported plates. The computer program has been developed in MATLAB and the results are validated by comparing with the results available in the literature. The values of deflection, inplane and transverse stresses are presented for different theories, span to thickness ratio and inhomogeneity parameter.

Keywords: FGM plates, Higher order theory, Rule of Mixture (ROM), Mori-Tanaka Method (MTM), Analytical Fourier series solution

1. Introduction

Functionally graded materials (FGMs) were introduced as engineering materials by Cavanagh and his co-workers [1], for developing a crack-free thermal barrier coating of gas turbine blades. FGMs are essentially composite material where the composition of constituent materials gradually varying from one surface to another. The large jumps in the inplane and transverse shear stresses due to varied material properties in the adjacent layers, which have been found as a leading cause of delamination in bonded composite structures when they are used in thermal environment, have been eliminated using FGM concept. Theoretical modelling of FGM plates require a micromechanical model for computing effective material properties, a suitable and accurate 2D plate theory for the kinematic analysis and a solution technique. Detailed literature on the modelling and analysis of FGM plates can be found in [2]. Mostly, Voigt's rule of mixtures (ROM) has been employed as an effective and easy means for obtaining the effective properties [3 - 6].

For kinematic modelling, various 2D and 3D solutions with varied accuracy are available in the literature. The accuracy has been primarily depends on the assumptions employed in deriving the theory and the degree of the functions for higher order theories considered in defining the displacement field. The most simpler theory is the classical plate theory (CPT) where the shear deformation effects are completely neglected, which has been employed by many researchers [7, 8]. This theory gives accurate results for thin plates but unable to predict accurate results for thick plates. The absence of the shear deformation effects in the CPT have been rectified by formulating first order shear deformation theory (FSDT) [4] and various higher order shear deformation theory (HOT) [9-12] by adding first and higher order shear deformation terms in the displacement field. These theories give accurate results for thick plates but their accuracy depends on the order of the functions considered in the displacement field. However, 3D exact [15- 17] and FE solutions found to yield the most accurate results.

This work aims to present short comparison of various recently developed higher order theories for the static analysis of FGM plates. The system equations have been derived using principle of virtual work. The effective properties are obtained using rule of mixture (ROM) and Mori-Tanaka homogenization schemes. Analytical solution has been obtained for simply supported plates considering a double Fourier series expansion with $(m \times n)$ terms.

2. Problem Formulation

Consider an FGM plate having dimensions (length x width x height $=axbxh$) (Fig. 1) made with two distinct isotropic materials by evenly mixing them so that the volume fractions of the constituents are smoothly changing in the thickness direction. The effective values of Young's modulus and Poisson's ratio at any thickness point in the FGM plate are determined using the rule of mixtures (ROM)

$$Y = Y_m V_m + Y_c V_c; \qquad \nu = \nu_m V_m + \nu_c V_c \tag{1}$$

where V_c is the volume fraction of ceramic phase and V_m is the volume fraction of metal phase and $V_c + V_c = 1$. In the Mori-Tanaka homogenization scheme, the effective values of shear modulus (G) and bulk modulus (K) are obtained using

$$\frac{K - K_m}{K_c - K_m} = \frac{V_c}{1 + (1 - V_c)\dfrac{K_c - K_m}{K_m + 4G_m/3}}; \qquad \frac{G - G_m}{G_c - G_m} = \frac{V_c}{1 + (1 - V_c)\dfrac{G_c - G_m}{G_m + f_1}} \tag{2}$$

where $f_1 = G_m(9K_m + 8G_m)/6(K_m + 2G_m)$. The effective values of Young's modulus (Y) and Poisson's ratio (ν) are computed from K and G as

$$\mathrm{Y} = \frac{9KG}{3K + G}; \qquad \nu = \frac{3K - 2G}{2(3K + G)} \tag{3}$$

In this study, we consider ceramic volume fraction (V_c) vary smoothly in the thickness direction according to the relation as

$$V_c = (0.5 + z/h)^p \tag{4}$$

where p is known as inhomogeneity parameter or power law index. The kinematic relations of the plate are obtained using third order theory. Let u_x, u_y *and* w be the in-plane and transverse displacements, the strains are expressed in terms of displacements as

$$\begin{aligned} &\varepsilon_x = u_{x,x}, \varepsilon_y = u_{y,y}, \gamma_{xy} = u_{y,x} + u_{x,y}, \\ &\gamma_{zx} = w_{,x} + u_{x,z}, \qquad \gamma_{yz} = w_{,y} + u_{y,z} \end{aligned} \tag{5}$$

where $()_{,x} = \partial\,()/\partial x$. Mostly, 2D plate theories neglect transverse normal stress σ_z in comparison to other stress components, the constitutive equations of linear elasticity reduce to

$$\boldsymbol{\sigma} = \boldsymbol{Q}(z)\varepsilon, \quad \tau = \widehat{\boldsymbol{Q}}(z)\gamma \tag{6}$$

where;

$$\boldsymbol{\sigma} = \begin{Bmatrix} \sigma_x \\ \sigma_y \\ \sigma_z \end{Bmatrix}, \qquad \varepsilon = \begin{Bmatrix} \varepsilon_x \\ \varepsilon_y \\ \varepsilon_z \end{Bmatrix}, \qquad \boldsymbol{Q}(z) = \begin{bmatrix} Q_{11}(z) & Q_{12}(z) & 0 \\ Q_{12}(z) & Q_{22}(z) & 0 \\ 0 & 0 & Q_{11}(z) \end{bmatrix} \tag{7}$$

$$\boldsymbol{\tau} = \begin{Bmatrix} \tau_{zx} \\ \tau_{yz} \end{Bmatrix}, \qquad \boldsymbol{\gamma} = \begin{Bmatrix} \gamma_{zx} \\ \gamma_{yz} \end{Bmatrix}, \qquad \widehat{\boldsymbol{Q}} = \begin{bmatrix} Q_{55}(z) & 0 \\ 0 & Q_{44}(z) \end{bmatrix}$$

$Q_{ij}(z)$ are the reduced elastic stiffness coefficients defined for isotropic materials as

$$\begin{gathered} Q_{11}(z) = Q_{22}(z) = Y(z)/(1-\nu(z)^2); \; Q_{12}(z) = \nu(z)Y(z)/(1-\nu(z)^2) \\ Q_{44}(z) = Q_{55}(z) = Q_{66}(z) = Y(z)/\big(1+\nu(z)\big) \end{gathered} \tag{8}$$

$Y(z)$ and $\nu(z)$ are functions of z, which are computed using the micromechanical model described in Eqs. 1 and 2 for a given profile of volume fraction V_c.

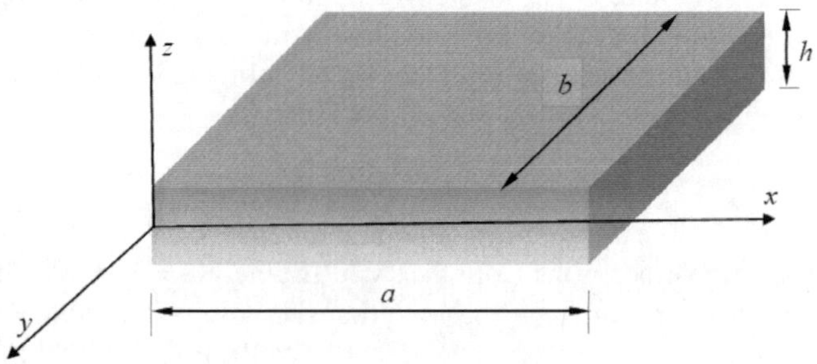

Figure 1. Geometry and coordinate systems used in FGM plate.

For the higher order theories (HOTs), transverse deflection w is taken independent of z and the in-plane displacements are assumed to follow higher order functions in z.

$$w(x,y,z,t) = w(x,y,t) \tag{9}$$

$$\boldsymbol{u} = \boldsymbol{f_0}(z)\overline{\boldsymbol{u}}_{\boldsymbol{1}} \tag{10}$$

where $\overline{\boldsymbol{u}}_{\boldsymbol{1}} = [u_{0x} \quad u_{0y} \quad -w_{0.x} \quad -w_{0.y} \quad \psi_{0x} \quad \psi_{0y}]^{\mathrm{T}}$, $f_0(z) = [I_2 \quad zI_2 \quad R(z)I_2]$, u_0 and w_0 denote the mid-surface displacements and shear strains of the plate, I_2 is 2nd order identity matrix and $R(z)$ is a higher order function which is defined as

$$R(z) = \begin{cases} z - 4z^3/3h^2 & \text{(TOT)} \\ \sin(\pi z/h) & \text{(Sinusoidal theory)} \\ (z - (h/\pi)\sin h(\pi z/h))/\cosh(\pi/2) - 1 & \text{(Hyperbolic theory)} \\ zexp\big(-2(z/h^2)\big) & \text{(Exponential theory)} \end{cases} \tag{11}$$

The strains are obtained for the HOT by substituting the displacement field of Eq. (10) in Eq (6) as

$$\boldsymbol{\varepsilon} = \boldsymbol{\varepsilon}^{\mathbf{0}} + z\mathfrak{k}^{\mathbf{0}} + R(z)\Psi_{\mathbf{0d}}, \qquad \gamma = R'(z)\Psi_{\mathbf{0}} \tag{12}$$

where $R'(z)$ is the first derivative of the higher order function $R(z)$ with respect to z and

$$\boldsymbol{\varepsilon}^{\mathbf{0}} = \begin{bmatrix} u_{0x,x} \\ u_{0y,y} \\ u_{0y,x} + u_{0x,y} \end{bmatrix}, \; \mathfrak{k}^{\mathbf{0}} = -\begin{bmatrix} w_{0,xx} \\ w_{0,yy} \\ 2w_{0,xy} \end{bmatrix}, \; \Psi_{\mathbf{0d}} = \begin{bmatrix} \psi_{0x,x} \\ \psi_{0y,y} \\ \psi_{0x,y} + \psi_{0y,x} \end{bmatrix}, \Psi_{\mathbf{0}} = \begin{bmatrix} \psi_{0x} \\ \psi_{0y} \end{bmatrix} \tag{13}$$

We use variational principle to derive the governing equations, force, moment and higher order moment resultants for 2D theories of FGM plates. The governing equations for the FGM plates are be expressed as

$$\begin{aligned} &N_{x,x} + N_{xy,x} = 0, \qquad N_{xy,x} + N_{y,y} = 0, \\ &M_{x,xx} + M_{y,yy} + 2M_{yx,yx} + p_3 = 0, \\ &P_{x,x} + P_{xy,x} - Q_x = 0, \; P_{xy,x} + P_{y,y} - Q_y = 0 \end{aligned} \tag{14}$$

where $\boldsymbol{N}, \boldsymbol{M}, \boldsymbol{Q}$ and $\boldsymbol{P}$ consist of the in-plane stress resultants, bending moment resultants, shear resultants and the higher order bending moments, respectively, which are defined by

$$\begin{bmatrix} N \\ M \\ P \end{bmatrix} = \begin{bmatrix} A & B & E \\ B^T & D & F \\ E^T & F^T & H \end{bmatrix} \begin{Bmatrix} \varepsilon^0 \\ k^0 \\ \Psi_{0d} \end{Bmatrix} \tag{15}$$
$$Q = \bar{A}\Psi_0$$

where

$$\begin{aligned} A &= \langle Q(z) \rangle; \; E = \langle R(z) Q(z) \rangle; \; \bar{A} = \langle R'(z) \hat{Q}(z) \rangle \\ B &= \langle z Q(z) \rangle; \; F = \langle z R(z) Q(z) \rangle \\ D &= \langle z^2 Q(z) \rangle; \; H = \langle R(z)^2 Q(z) \rangle \end{aligned} \tag{16}$$

3. Analytical Solution for Simply Supported Plates

The solution of the resulting equations is expanded in the following Double Fourier series, which identically satisfies the simply supported boundary conditions at the four edge:

$$\begin{bmatrix} w_0 & P_3 \\ u_{0x} & \psi_{0x} \\ u_{0y} & \psi_{0y} \end{bmatrix} = \sum_{m=1}^{\infty} \sum_{n=1}^{\infty} \begin{bmatrix} (w_0 \quad P_3)_{mn} \sin(\alpha x) \sin(\beta y) \\ (u_{0x} \quad \psi_{0x})_{mn} \cos(\alpha x) \sin(\beta y) \\ (u_{0y} \quad \psi_{0y})_{mn} \sin(\alpha x) \cos(\beta y) \end{bmatrix} \tag{17}$$

where $\alpha = m\pi/a$ and $\beta = n\pi/b$. Substituting the expansions given by $Eq.(17)$ in $Eq.(14)$ yields for $(m,n)^{th}$ Fourier component, the following equation

$$K\bar{U}^{mn} = \bar{P}^{mn} \tag{18}$$

where

$$\begin{aligned} \bar{U}^{mn} &= \left[u_{0x} \quad u_{0y} \quad w_0 \quad \psi_{0x} \quad \psi_{0y} \right]_{mn}^{T} \\ \bar{P}^{mn} &= \left[P_1 \quad P_2 \quad P_3 \quad P_4 \quad P_5 \right]_{mn}^{T} \end{aligned} \tag{19}$$

where K is symmetric stiffness matrix, $\bar{U}$ is the displacement field and $\bar{P}$ is the load vector.

4. Results and Discussions

To check the correctness and accuracy of the present formulation, we compare our results with those available in the literature. For this purpose, we consider FGM plate made of Al/ZrO_2, simply supported at all the four edges and bending due to applied uniform pressure of intensity p_0 on the top surface. The non-dimensionalized deflection and stresses are $\hat{w} = 100 Y_m h^3 \, w/12a^4(1-\nu_m{}^2)p_z{}^2$ and $\bar{\sigma}_x = \sigma_x \, h/a^2 p_0$ respectively, where Y_m is the Young's modulus of the metal phase. The material properties of ceramic and metal phases considered in this study are (a) Aluminium: Y=70 GPa, ν= 0.3 (b) Zirconia(ZrO_2): Y=200 GPa, ν = 0.3 (c) Silicon Carbide: Y=302 GPa, ν=0.17. The non-dimensional values of central deflection $\hat{w}$ and stress $\bar{\sigma}_x$ for the FG plate with different values of thickness and inhomogeneity parameter are presented and compared in Table 1 with the published results. For ensuring the convergence, the present results are reported for 100 number of Fourier terms in our analytical solution. It can be seen from the Table 1 that the results obtained using TOT, sinusoidal and exponential theories are very close to the reported values presented in the literature for both $a/h = 5$ and 10. However, the hyperbolic shear deformation theory does not yield good results for thick plates but its accuracy increases as the plate becomes thinner.

The variation of centre point deflection $\hat{w}$, stresses $\bar{\sigma}_x$ and $\bar{\tau}$ with thickness ratio a/h has been presented in Fig. 2 for $p = 1$ and 5 using TOT. It is seen that the values of $\hat{w}$ and $\bar{\sigma}_x$ decrease as a/h increases and attains a constant value after $a/h = 40$. However, the values of $\bar{\tau}_{zx}$ increase with a/h. Next, we present the variation of deflection $\hat{w}(x/a, 0.5b)$, stresses $\bar{\sigma}_x(x/a, 0.5b)$ & $\bar{\tau}(x/a, 0.5b)$ for $Al/ZrO_2$and Al/SiC plates under uniform pressure loading in Fig. 3 and 4 respectively for different values of p. The values of defections as well as stresses are changing when the value of p changes. The effect of p is large when the difference in the properties of the constituent properties is large.

Table 1. Non-dimensional deflection $\hat{w}$ and inplane stress $\bar{\sigma}_x$ of a square Al/ZrO_2 FG plate under uniform pressure loading and simply supported at the ends.

				Analytical Solution				FE solution		
p	Entity	a/h	Homog.	Hyp	Sine	Exp	TOT	Poly	Sine	TOT
1	$\hat{w}$	5	ROM	0.1963	0.2319	0.2317	0.2318	0.2330	0.2310	0.2313
			MTM	0.2158	0.2554	0.2552	0.2555			0.2548
		10	ROM	0.1952	0.2041	0.2041	0.2040			0.2035
			MTM	0.2146	0.2245	0.2244	0.2245			0.2239
	$\bar{\sigma}_x$	5	ROM	0.3484	0.3603	0.3609	0.3593			0.3595
			MTM	0.3718	0.3861	0.3868	0.3854			0.3839
		10	ROM	0.3497	0.3527	0.3528	0.3524			0.3524
			MTM	0.3733	0.3769	0.3771	0.3767			0.3760
1	$\hat{w}$	5	ROM	0.2300	0.2718	0.2716	0.2717	0.2730	0.2729	0.2714
			MTM	0.2480	0.2955	0.2953	0.2956			0.2950
		10	ROM	0.2287	0.2392	0.2391	0.2390			0.2388
			MTM	0.2465	0.2585	0.2584	0.2585			0.2580
	$\bar{\sigma}_x$	5	ROM	0.3853	0.3990	0.3997	0.3979			0.3985
			MTM	0.4097	0.4264	0.4271	0.4255			0.4244
		10	ROM	0.3868	0.3902	0.3904	0.3900			0.3903
			MTM	0.4114	0.4156	0.4158	0.4154			0.4152
2	$\hat{w}$	5	ROM	0.2616	0.3135	0.3133	0.3133	0.3147	0.3147	0.313
			MTM	0.2752	0.3335	0.3334	0.3335			0.3328
		10	ROM	0.2601	0.2730	0.2730	0.2769			0.2726
			MTM	0.2734	0.2881	0.2880	0.2880			0.2875
	$\bar{\sigma}_x$	5	ROM	0.4220	0.4383	0.4391	0.4369			0.4376
			MTM	0.4497	0.4691	0.4700	0.4681			0.4674
			MTM	0.4516	0.4565	0.4567	0.4562			0.4566

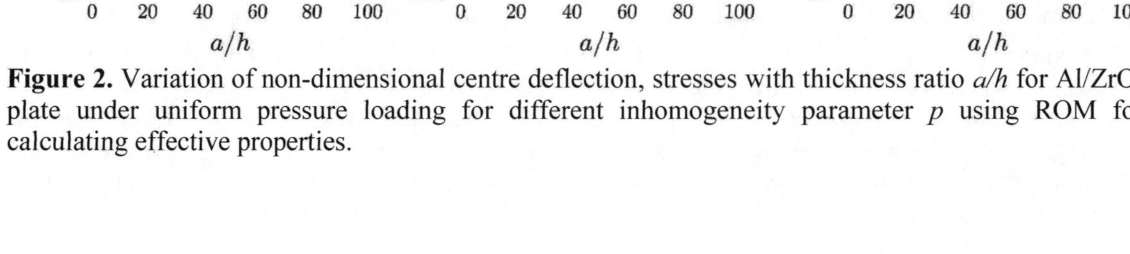

Figure 2. Variation of non-dimensional centre deflection, stresses with thickness ratio *a/h* for Al/ZrO_2 plate under uniform pressure loading for different inhomogeneity parameter *p* using ROM for calculating effective properties.

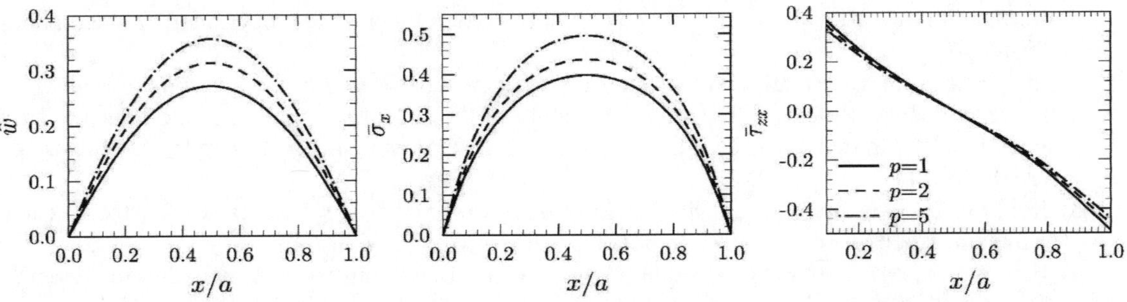

Figure 3. Variation of $\hat{w}(x/a, 0.5b)$, stresses $\bar{\sigma}_x(x/a, 0.5b)$ and $\bar{\tau}_{zx}(x/a, 0.5b)$ for Al/ZrO_2 plate under uniform pressure loading for different values of p using ROM for calculating effective properties.

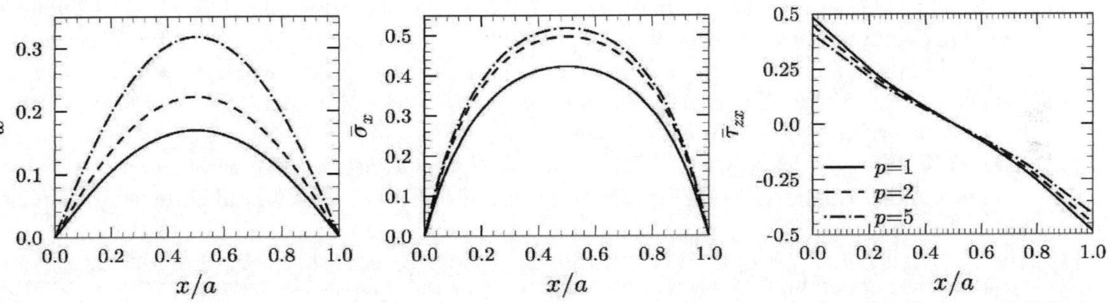

Figure 4 Variation of $\hat{w}(x/a, 0.5b)$, stresses $\bar{\sigma}_x(x/a, 0.5b)$ and$\bar{\tau}_{zx}(x/a, 0.5b)$ for Al/SiC plate under uniform pressure loading for different values of p using ROM for calculating effective properties.

5. Conclusion

Higher order theories presented above yield good accuracy. The results obtained using third order theory, exponential theory and sinusoidal theory are close to each other for very thick and thick plates. These values are very well compared with TOT FE and other FE solutions available in the literature. The hyperbolic shear deformation theory yields similar values of deflection and stresses for both very thick ($a/h = 5$) and thick ($a/h = 10$) FGM plate which should be different for different thickness. Therefore, hyperbolic shear deformation theory does not give accurate results for very thick plates.

Acknowledgments

This work is a part of the project under UGC start-up grant. The authors are thankful for the UGC for its financial support.

References

[1] Cavanagh, J.R., Cross, K.R., Newman, R.L., Spicer, W.C., 1972 *In: Proc. Thirteenth U.S. Structural Dynamics and Materials Conference* The graded thermal barrier-a new approach for turbine engine cooling **1**.

[2] Birman, V., Byrd, L.W., 2007 *Applied Mechanics Reviews* Modeling and analysis of functionally graded materials and structures **60**, 195-216.

[3] Abrate, S., 2006 *Composite Science and Technology* Free vibration, buckling, and static deflections of functionally graded plates **66**, 2383-2394.

[4] Croce, L.D., Venini, P., 2004 *Computer Methods in Applied Mechanics and Engineering* Finite elements for functionally graded Reissner- Mindlin plates **193**, 705-725.

[5] Zenkour, A.M., 2006 *Applied Mathematical Modelling* Generalized shear deformation theory for bending analysis of functionally graded plates **30**, 67-84

[6] Thai, H.T.,Choi, H. H., 2013 *Finite Elements in Analysis and Design* Finite element formulation of various four unknown shear deformation theories for functionally graded plates **75**, 50-61.
[7] Shi-Rong Li, XuanWang and Romesh C. Batra, 2015 *Journal of Applied Mechanics* Correspondence Relations Between Deflection, Buckling Load, and Frequencies of Thin Functionally Graded Material Plates and Those of Corresponding Homogeneous Plates **82,** article no. 111006.
[8] Reid, R.G., Paskaramoorthy, R., 2011 *Composite Structures* An extension to classical lamination theory for use with functionally graded plates **93**, 639-648.
[9] Ferreira, A.J.M., Batra, R.C., Roque, C.M.C., Qian, L.F., Martins, P.A.L.S., 2005 *Composite Structures* Static analysis of functionally graded plates using third-order shear deformation theory and a meshless method **69**, 449-457.
[10] Ferreira, A.J.M., Batra, R.C., Roque, C.M.C., Qian, L.F., Jorge, R.M.N., 2006 *Composite Structures* Natural frequencies of functionally graded plates by a meshless method **75**, 593-600.
[11] Gilhooley, D.F., Batra, R.C., Xiao, J.R., McCarthy, M.A., Gillespie, J.W., 2007 *Composite Structures* Analysis of thick functionally graded plates by using higher-order shear and normal deformable plate theory and MLPG method with radial basis functions **80**, 539-552
[12] Mantari, J.L., Soares, C.G., 2013 *Composites Part B* A novel higher-order shear deformation theory with stretching effect for functionally graded plates **45**, 268-281.
[13] Jha, D.K., Kant, T., Singh, R.K., 2012 *Nuclear Engineering and Design* Higher order shear and normal deformation theory for natural frequency of functionally graded rectangular plates **250**, 8-13
[14] Jha, D.K., Kant, T., Singh, R.K., 2013b *Composite Structures* Free vibration response of functionally graded thick plates with shear and normal deformations effects **96**, 799-823.
[15] Kashtalyan, M., 2004 *European Journal of Mechanics/ A Solids* Three-dimensional elasticity solution for bending of functionally graded rectangular plates **23**, 853-864.
[16] Yang, B., Ding, H.J., Chen, W.Q., 2012 *Applied Mathematical Modelling* Elasticity solutions for functionally graded rectangular plates with two opposite edges simply supported **36**, 488-503.
[17] Vel, S.S., Batra, R.C., 2004 J. *Sound Vibration* Three-dimensional exact solution for the vibration of functionally graded rectangular plates **272**, 703-730.
[18] Kapuria, S., Patni, M., Yasin, M. Y., 2015 *European Journal of Mechanics A/Solids* A quadrilateral shallow shell element based on the third-order theory for functionally graded plates and shells and the inaccuracy of rule of mixtures. **49**, 268-282

Analysis of Laminated and FGM Beams using Various Theories

Mirza Shariq Beg, M. Yaqoob Yasin, Hasan M. Khalid
Department of Mechanical Engineering, Z. H. College of Engineering and Technology, Aligarh Muslim University, Aligarh, Uttar Pradesh, India

E-mail: mirzaamu187@gmail.com

Abstract- In this work, we present a comparison of some recently developed higher order theories for static and free vibration responses of laminated and functionally graded material (FGM) beams. The equations of motion are derived using Hamilton's principle. Analytical Fourier series solution taking first n terms is derived for beams having simply supported end conditions. A computer code in MATLAB has been developed and the results obtained are compared with 2D elasticity/finite element solution for laminated beam to assess the accuracy of various theories. A parametric study has been presented for deflection, stresses and natural frequencies of FG beams.

Keywords: Functionally Graded Beams, Higher Order Theories, Voigt's ROM, Analytical Solution.

1. Introduction

Laminated/functionally graded material (FGM) beams have been extensively used as structural member in many engineering applications. The popularity of these structures increases day by day due to their tailorable material properties which has intrigued researchers in the last few decades. In FGM beams, the material properties are smoothly varying along a particular direction (preferably transverse direction) which results in continuous normal stresses and makes them utile for the components operating in severe thermal environment where the laminated composite structures fail due to delamination. The concept of FGM was given by group of material scientists in Japan around mid-eighties. Sankar [1] presented static analysis of FG beam using Euler-Bernoulli beam theory. This theory is suitable for thin beams but the discrepancies evolve as the beam gets thicker. Li [2] presented static and dynamic analysis of FG beams considering rotary inertia and shear deformation effects. Sina et.al [3] developed a refined first order shear deformation theory to analyze free vibration of FG beams. These theories, although consider the shear deformation effects, yet they yield unnatural shear stress distribution across the thickness. Higher order theories (HOTs) considering third or higher order polynomial/function of thickness coordinate have been developed for FG beams. These theories yield accurate results and the distributions of shear stresses are comparable to 2D elasticity solution. Thai and Thuc [4] employed various higher order shear deformation theories for bending and free vibration analysis of FG beams. Simsek [5] presented natural frequencies of FG beam using classical, first order and higher order shear deformation theories. Vo et. al [6] proposed a quasi-3D polynomial shear and normal deformation theory taking into account the effects of transverse shear and normal deformation in the displacement field. Kadoli et.al [7] carried out finite element analysis based on third order shear deformation theory of FG beams to study their static response.

In this work, we present a comparison of various HOTs for the static and free vibration responses of laminated and FGM beams. The effective properties are calculated using Voight's rule of mixtures (ROM). The Fourier series solution has been obtained considering first 85 odd terms. The results obtained are validated with existing literature and commercially available FE package ABAQUS. Deflection, stresses and natural frequencies are presented for laminated and FGM beams of different *a/h* values.

2. Problem formulation

Consider an FGM beam of length L width b and thickness h having metallic top surface and ceramic bottom surface. The concentration of metal and ceramic is smoothly varying with the variation of ceramic volume fraction with power law

$$V_c = \left(0.5 + \frac{z}{h}\right)^p \tag{1}$$

The geometry of the beam and the coordinate axes has been shown in figure 1. The bottom surface of the beam lies at $z = -h/2$ and top surface lies at $z = h/2$. The reference plane lies at $z = 0$. The beam is loaded with transverse pressure P_z at top surfaces without any variation along y axis. For the beam having plane stress condition $(\sigma_y = \tau_{yz} = \tau_{xy} = 0)$ and considering $\sigma_z \simeq 0$, the material constitutive behaviour for FGM beam is expressed as

$$\sigma_x = Q_{11}(z)\varepsilon_x \qquad \tau_{zx} = Q_{55}(z)\gamma_{zx} \tag{2}$$

Where σ_x , τ_{zx} axial and transverse shear stresses and ε_x, γ_{zx} are axial and transverse shear strains, $Q_{11}(z) = Y(z)$ and $Q_{55} = Y(z)/2(1 + \nu(z))$ are reduced stiffness coefficients. The effective values of Young's modulus and Poisson's ratio are obtained using ROM

$$E(z) = V_c Y_c + V_m Y_m \qquad \nu(z) = V_c \nu_c + V_m \nu_m \tag{3}$$

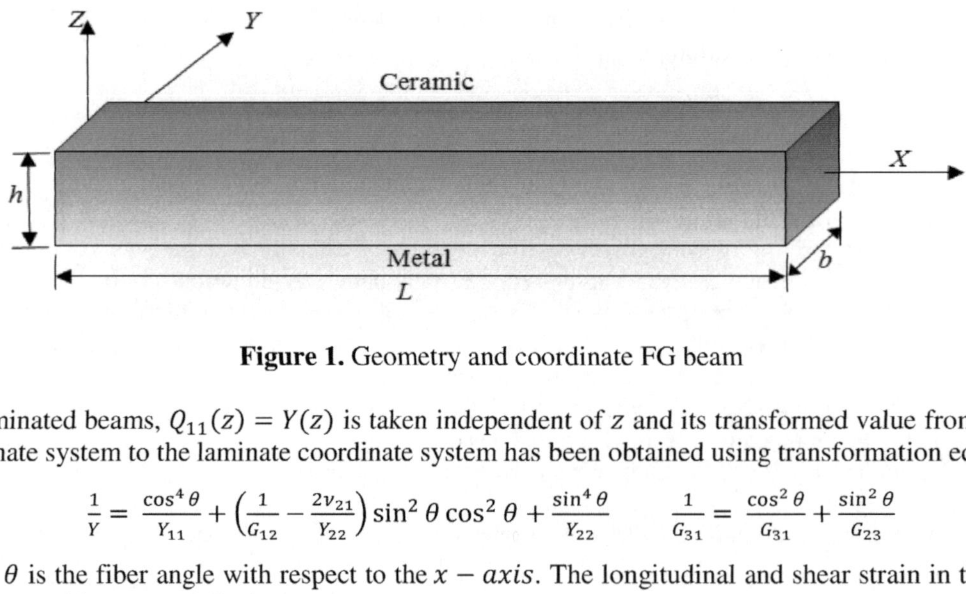

Figure 1. Geometry and coordinate FG beam

For laminated beams, $Q_{11}(z) = Y(z)$ is taken independent of z and its transformed value from lamina coordinate system to the laminate coordinate system has been obtained using transformation equation

$$\frac{1}{Y} = \frac{\cos^4\theta}{Y_{11}} + \left(\frac{1}{G_{12}} - \frac{2\nu_{21}}{Y_{22}}\right)\sin^2\theta\cos^2\theta + \frac{\sin^4\theta}{Y_{22}} \qquad \frac{1}{G_{31}} = \frac{\cos^2\theta}{G_{31}} + \frac{\sin^2\theta}{G_{23}} \tag{4}$$

Where θ is the fiber angle with respect to the $x - axis$. The longitudinal and shear strain in the beam are expressed in terms of displacements as

$$\varepsilon_x = u_{,x} \qquad \gamma_{zx} = w_{,x} + u_{,z} \tag{5}$$

In present analysis, the kinematic relations of beams are obtained using higher order theories (HOTs). The displacement field for the HOTs [8] can be expressed as

$$u = f(z)\bar{u} \qquad w = w_0 \tag{6}$$

Where $\bar{u} = \left[u_0 \;\; w_{0,x} \;\; \psi_0\right]^{\mathrm{T}}$ is the displacement vector and $f(z)$ is a higher order function of z in matrix form which takes the following form

$$f(z) = \begin{cases} [1 \quad -z \quad z - 4z^3/3h^2] & \text{(TOT)} \\ [1 \quad -z \quad \sin(\pi z/h)] & \text{(Sinusoidal theory)} \\ [1 \quad -z \quad (z - (h/\pi)\sinh(\pi z/h))]/(\cosh(\pi/2) - 1) & \text{(Hyperbolic theory)} \\ [1 \quad -z \quad z\exp(-2z^2/h^2)] & \text{(Exponential theory)} \end{cases} \tag{7}$$

Using equations Eq. (5) and Eq. (6), the strain ε_x and γ_{zx} are expressed in functional form as

$$\varepsilon_x = f_1(z)\,\bar{\varepsilon}_1 \qquad \gamma_{xz} = f_2(z)\,\bar{\varepsilon}_2 \tag{8}$$

Where $\bar{\varepsilon}_1 = [u_{0,x} \;\; w_{0,xx} \;\; \psi_{0,x}]$, $\bar{\varepsilon}_2 = \psi_0$ and $f_1(z) = f(z)$ and

$$f_2(z) = \begin{cases} 1 - 4z^2/h^2 & \text{(TOT)} \\ (\pi/h)\cos(\pi z/h) & \text{(Sinusoidal theory)} \\ \cosh(\pi z/h) - 1/(\cosh(\pi/2) - 1) & \text{(Hyperbolic theory)} \\ 1 - 4(z^2/h^2)\exp(-2\,z^2/h^2) & \text{(Exponential theory)} \end{cases} \tag{9}$$

The equations of motion based on different HOTs of Eq. (7) are derived using Hamilton's principle. Using$\langle \dots \rangle = \sum_{\mathrm{K}=1}^{\mathrm{L}} \int_{z_{\mathrm{k}-1}^{+}}^{z_{\mathrm{k}}^{-}} (\dots) b dz$, the Hamilton's principle takes the following form

$$\int_x \left[\langle(\rho^k \ddot{u}\delta u + \rho^k \ddot{w}\delta w + \sigma_x \delta\varepsilon_x + \tau_{zx}\delta\gamma_{zx})\rangle - bP_z \delta w(x, h/2, t)\right]\mathrm{d}x - \langle \sigma_x \delta u + \tau_{zx}\delta w\rangle|_x = 0 \tag{10}$$

$\forall\ \delta u_0$, δw_0 and $\delta\psi_0$, where$(\ddot{\ }) = (\ )_{,tt}$. Substituting the displacements and strains and integrating Eq. (10) by parts yields the following equations of motion

$$\begin{gathered} I_{11}\ddot{u_0} + I_{12}\ddot{w_{0,x}} + I_{13}\ddot{\psi_0} - N_{x,x} = 0 \\ I_{21}\ddot{u_{0,x}} - I_{22}\ddot{w_{0,xx}} + I_{23}\ddot{\psi_{0,x}} + \hat{I}\ddot{w_0} - M_{x,xx} = P_2 \\ I_{31}\ddot{u_0} + I_{32}\ddot{w_{0,x}} + I_{33}\ddot{\psi_0} - P_{x,x} + Q_x = 0 \end{gathered} \tag{11}$$

Where I_{ij} and $\hat{I}$ are the inertia parameters $N_x, M_x, P_x$and Q_x are the force, moment, higher order moment and shear resultants respectively and P_2 is the load vector. These resultants are expressed in terms of field variables as

$$\begin{aligned} N_x &= A_{11}u_{0,x} + A_{12}w_{0,xx} + A_{13}\psi_{0,x}\,, \quad & M_x &= A_{12}u_{0,x} + A_{22}w_{0,xx} + A_{23}\psi_{0,x} \\ P_x &= A_{31}u_{0,x} + A_{32}w_{0,xx} + A_{33}\psi_{0,x}\,, \quad & Q_x &= \bar{A}\psi_0 \end{aligned} \tag{12}$$

Where the beam stiffnesses A_{ij} and the beam inertia coefficients I_{ij} are defined as

$$\begin{bmatrix} A_{11} & A_{12} & A_{13} \\ A_{21} & A_{22} & A_{23} \\ A_{31} & A_{32} & A_{33} \end{bmatrix} = \langle f^{\mathrm{T}}(z) f(z) Q_{11}\rangle \qquad \begin{bmatrix} I_{11} & I_{12} & I_{13} \\ I_{21} & I_{22} & I_{23} \\ I_{31} & I_{32} & I_{33} \end{bmatrix} = \langle \rho^k f^{\mathrm{T}}(z) f(z)\rangle \tag{13}$$

Substituting the Eq. (12) in Eq. (11), yield the equations of motion

$$\bar{L}\ddot{\bar{U}} + L\bar{U} = \bar{P} \tag{14}$$

Where $\bar{U} = [u_0 \quad w_0 \quad \psi_0]^T, \bar{P} = [0 \quad P_2 \quad 0]^{\mathrm{T}}$. $\bar{L}$ and L are symmetric matrices containing differential operators.

We derive exact solution for the coupled differential equations of motion having simply supported boundary condition. Considering n terms of Fourier series, the primary and secondary variables are expressed as function of sine and cosine as

$$\begin{bmatrix} u_0 \\ w_0 \\ \psi_0 \end{bmatrix} = \sum_{n=1}^{\infty} \begin{bmatrix} (u_o)_n \cos\bar{n}x \\ (w_o)_n \sin\bar{n}x \\ (\psi_o)_n \cos\bar{n}x \end{bmatrix} \quad \begin{bmatrix} N_x \\ M_x \\ P_x \\ Q_x \end{bmatrix} = \sum_{n=1}^{\infty} \begin{bmatrix} (N_x)_n \sin\bar{n}x \\ (M_x)_n \sin\bar{n}x \\ (P_x)_n \sin\bar{n}x \\ (Q_x)_n \cos\bar{n}x \end{bmatrix} \quad P_2 = (P_2)_n \sin\bar{n}x \tag{15}$$

Substituting the solution in Eq. (14) yield the equation of motion,

$$M\ddot{\bar{U}}^n + K\bar{U}^n = \bar{P}^n \tag{16}$$

Where $\bar{U}^n = [u_0 \quad w_0 \quad \psi_0]_n^{\mathrm{T}}$ is the displacement vector for the nth Fourier term, $\bar{P}^n = [0 \quad -P_2 \quad 0]^{\mathrm{T}}$ is the corresponding load vector. M and K are the mass and stiffness matrices. For performing the free vibration analysis, $\bar{P}^n$ is set to zero to transform Eq. (16) to a generalized eigenvalue problem.

3. Results and Discussions

For assessing the different HOTs presented above for their accuracy, four layers symmetric $0^\circ/90^\circ/90^\circ/0^\circ$ and antisymmetric $[0^\circ/90^\circ/0^\circ/90^\circ]$ laminated beam have been considered. Each lamina of the laminated beam has equal thickness. The values of engineering constants are:
Young's modulii $[Y_1 \quad Y_2 \quad Y_3] = [181 \quad 10.3 \quad 10.3]$ GPa, shear modulii $[G_{23} \quad G_{31} \quad G_{12}] = [2.87 \quad 7.17 \quad 7.17]$GPa and Poisson's ratios $[\nu_{23} \quad \nu_{31} \quad \nu_{12}] = [0.33 \quad 0.25 \quad 0.25]$. The uniformly distributed load P_z is applied on the top surface of the beam for obtaining the static response by different HOTs. Non-dimensionalised formulae used are: transverse displacement ($\bar{w}_0 = 100w_0Y_2/hS^4 P_z$), longitudinal shear stress $\bar{\sigma}_x = \sigma_x/S^2P_z$ and transverse shear stress $\bar{\tau}_{zx} = \tau_{zx}/SP_z$.

Table 1. Comparison of results for symmetric and antisymmetric laminated composite beams.

		$0^\circ/90^\circ/90^\circ/0^\circ$				$0^\circ/90^\circ/0^\circ/90^\circ$			
S	Method	$\bar{W}_0(0.5l)$	$\bar{\sigma}_x^t(0.5l)$	$\bar{\sigma}_x^b(0.5l)$	$\bar{\tau}_{zx}^c(0)$	$\bar{W}_0(0.5l)$	$\bar{\sigma}_x^t(0.5l)$	$\bar{\sigma}_x^b(0.5l)$	$\bar{\tau}_{zx}^c(0)$
5	Exact [9]	-2.6748	-1.0602	1.0711	-0.5688	-3.7943	-1.4862	0.1582	-0.7667
	TOT	-2.5530	-1.0097	1.0097	-0.5168	-3.2514	-1.4345	0.1270	-0.7236
	Sine	-2.5815	-1.0221	1.0221	-0.5427	-3.2559	-1.4430	0.1280	-0.7449
	Hyperbolic	-1.1633	-0.7875	0.7875	-0.0000	-2.1297	-1.2153	0.1099	-0.0000
	Exponential	-2.6045	-1.0341	1.0341	-0.5693	-3.2568	-1.4508	0.1289	-0.7663
10	Exact [9]	-1.4343	-0.9031	0.9059	-0.6093	-2.4461	-1.3442	0.1268	-0.8242
	TOT	-1.3975	-0.8901	0.8901	-0.5374	-2.2976	-1.3307	0.1188	-0.7444
	Sine	-1.4054	-0.8932	0.8932	-0.5669	-2.2991	-1.3328	0.1191	-0.7685
	Hyperbolic	-1.0473	-0.8346	0.8346	-0.0000	-2.0162	-1.2759	0.1146	-0.0000
	Exponential	-1.4120	-0.8962	0.8962	-0.5972	-2.2996	-1.3348	0.1193	-0.7928
20	Exact [9]	-1.1152	-0.8636	0.8641	-0.6295	-2.0958	-1.3081	0.1188	-0.8529
	TOT	-1.1056	-0.8602	0.8602	-0.5477	-2.0579	-1.3048	0.1168	-0.7547
	Sine	-1.1076	-0.8610	0.8610	-0.5789	-2.0583	-1.3053	0.1169	-0.7802
	Hyperbolic	-1.0178	-0.8463	0.8463	-0.0000	-1.9875	-1.2911	0.1157	-0.0000
	Exponential	-1.1093	-0.8617	0.8617	-0.6112	-2.0584	-1.3058	0.1169	-0.8060

Deflection and stresses of laminated beam with different HOTs are enumerated in Table 1. Results display almost similar accuracy with exponential shear deformation theory yielding closer results to 2D elasticity solution [9]. To further elaborate the layer-wise mechanics, through-the-thickness variations of longitudinal and transverse shear stresses are presented in Figure.2, for symmetric and antisymmetric laminates along with the 2D FE results obtained using ABAQUS. Unlike other theories, the hyperbolic shear deformation theory yields zero shear stress at mid-plane and non-zero values at top and bottom of the beams.

Figure.3 exhibits the variation of effective properties namely volume fraction and Young's modulus with varying power-law exponent ($p = 0.25, 0.5, 1, 2, 4$) for the FG beam having aluminium at the bottom [$Y = 70\ GPa, \nu = 0.3$] and alumina at the top [$Y = 380\ GPa, \nu = 0.3$] [10]. ROM is used as homogenization scheme to evaluate the effective properties.

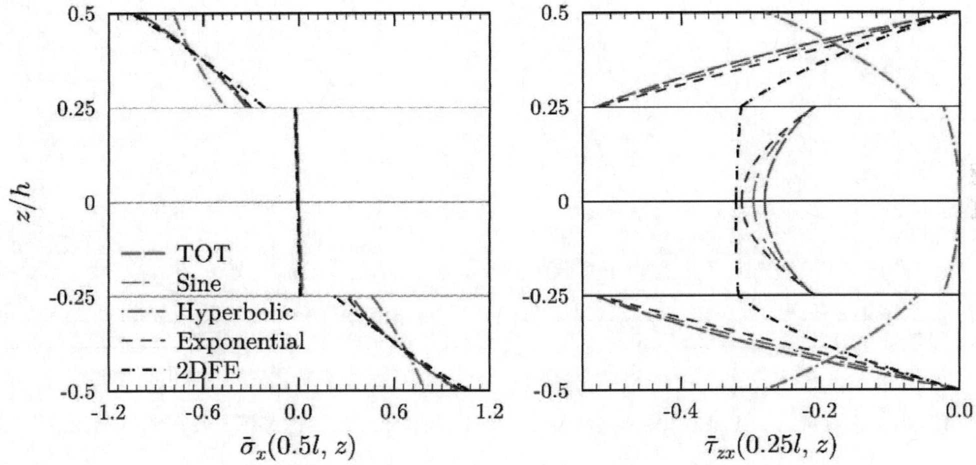

(a) Symmetric ($0^o/90^o/90^o/0^o$)

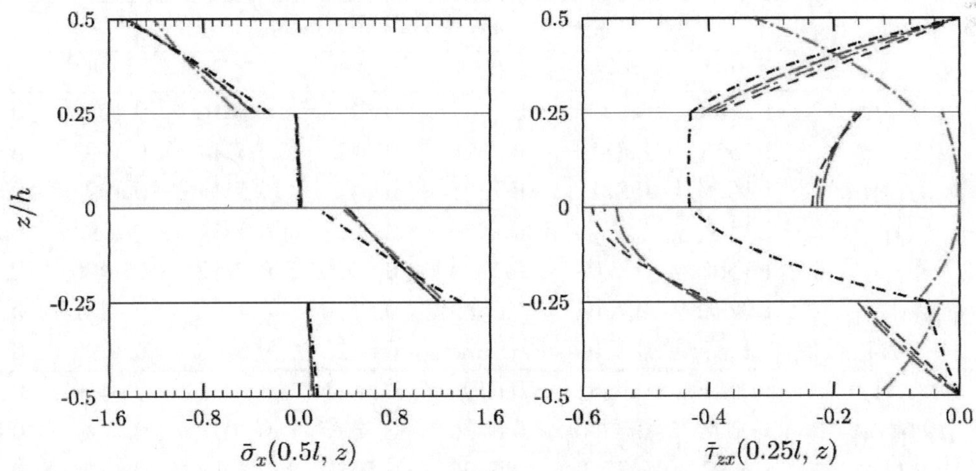

(b) Antisymmetric ($0^o/90^o/0^o/90^o$)

Figure 2. Variation of stresses in laminated beams across the thickness (S=5).

Now we analyze FG beam having aluminum at the top and zirconia at the bottom [10] using different theories for static response. The beam is simply supported and under a uniform pressure P_z at the top surface. Results for non dimensionalised transverse deflection ($w_0 = 348 w_0 Y_{Al}/60hS^4P_z$) at

IOP Conf. Series: Materials Science and Engineering **404** (2018) 012030 doi:10.1088/1757-899X/404/1/012030

the mid span of the beam using different theories are presented in Table 2. Different span to thickness ratios ($S = a/h = 4, 16$) are considered for the analysis. It can be observed that the $\bar{w}_0$ decreases with increase in the power-law exponent and the trend are validated with the existing literature [10]. Also, for higher span to thickness ratio i.e. S = 16, results are in close proximity with the results available in the existing literature [10]. Through the thickness variation for the different stresses developed under uniform loading in FG beam are presented in Figure.4. As predicted earlier for laminated composites, here also, hyperbolic shear deformation theories display distant behavior for the variation of transverse shear stress. However minimal difference is found in longitudinal stresses for the theories presented.

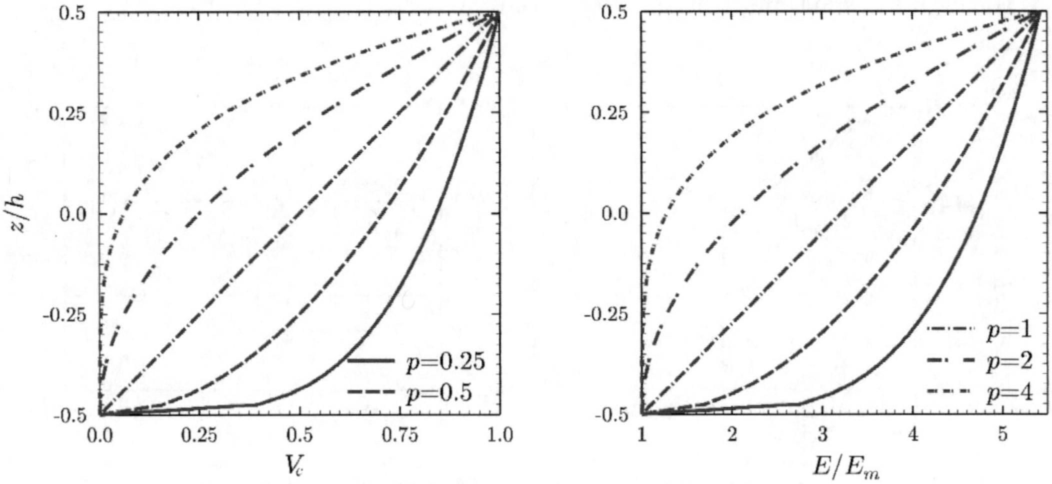

Figure 3. Variation of effective properties across thickness of FG Al/Al_2O_3beam.

Table 2. Non-dimensional central deflection of FG beams for various values of power law exponent.

S	Theory	Power-law exponent "p"						
		0	0.2	0.5	1	2	5	∞
4	FSDT[10]	1.1300	0.8486	0.7147	0.6293	0.5616	0.4918	0.3955
	HSBT[4]	1.1558	0.8710	0.7326	0.6427	0.5714	0.4998	0.4045
	HSDT[10]	1.1558	0.8710	0.7325	0.6427	0.5714	0.4998	0.4045
	TOT	1.1558	0.8710	0.7326	0.6427	0.5714	0.4998	0.4045
	Sine	1.1555	0.8708	0.7324	0.6426	0.5712	0.4996	0.4044
	Hyperbolic	1.0231	0.7664	0.6483	0.5739	0.5142	0.4501	0.3581
	Exponential	1.1547	0.8703	0.7320	0.6422	0.5708	0.4993	0.4042
16	FSDT[10]	1.0081	0.7555	0.6394	0.5661	0.5072	0.4439	0.3528
	HSDT[4]	1.0098	0.7570	0.6406	0.5670	0.5078	0.4444	0.3534
	HSDT[10]	1.0097	0.7569	0.6406	0.5670	0.5078	0.4444	0.3534
	TOT	1.0098	0.7569	0.6406	0.5670	0.5078	0.4444	0.3534
	Sine	1.0097	0.7569	0.6406	0.5670	0.5078	0.4444	0.3534
	Hyperbolic	1.0014	0.7504	0.6353	0.5627	0.5042	0.4413	0.3505
	Exponential	1.0097	0.7569	0.6406	0.5670	0.5078	0.4444	0.3534

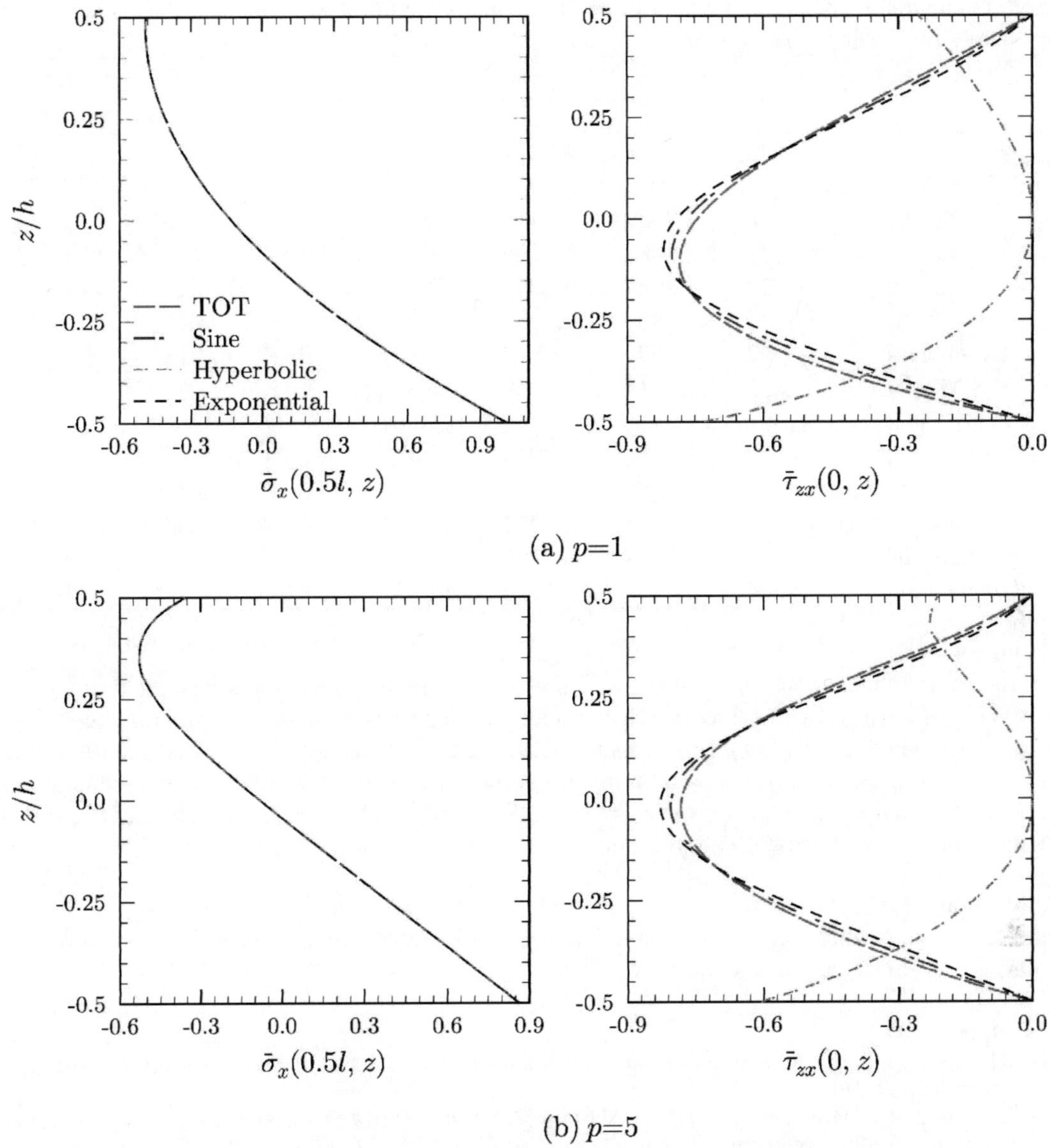

Figure 4. Variation of stresses in FG beams through thickness for (S=4).

Next, free vibration analysis is carried out for FG beam having aluminium at the bottom and alumina at the top [10]. Fundamental frequencies are evaluated for different theories against varying power-law exponent for two different span to thickness ratios ($S = 5, 20$). Results for non dimensionalised fundamental frequencies $\left(\bar{\omega} = \omega h S^2 \sqrt{\rho_{Al}/Y_{Al}}\right)$ of FG beam for simply-supported end conditions from different theories are presented in Table 3. It is very much evident from Table 3 that the results obtained from the present formulation for the fundamental frequencies are coherent with the results in the literature [10]. In addition, results from the theories presented here are displaying the same trend for the variation of fundamental frequencies with power-law exponent as displayed in Table 3.

Table 3. Comparison of the non-dimensional fundamental natural frequencies of FG beams with various values of power-law exponent.

S	Theory	Power-law exponent "*p*"						
		0	0.2	1	2	5	10	∞
5	FSDT[10]	5.1526	4.8033	3.9711	3.6050	3.4025	3.2963	2.6773
	HSBT[4]	5.1528	4.8081	3.9904	3.6264	3.4012	3.2816	2.6773
	HSDT[10]	5.1528	4.8059	3.9716	3.5979	3.3743	3.2653	2.6773
	TOT	5.1527	4.8081	3.9904	3.6264	3.4012	3.2816	2.6773
	Sine	5.1531	4.8084	3.9907	3.6263	3.3998	3.2811	2.6775
	Hyperbolic	5.3573	4.9866	4.1238	3.7577	3.5731	3.4687	2.7836
	Exponential	5.1542	4.8093	3.9914	3.6267	3.3990	3.2813	2.6781
20	FSDT[10]	5.4603	5.0812	4.2039	3.8349	3.6490	3.5405	2.8371
	HSDT[4]	5.4603	5.0815	4.2051	3.8361	3.6485	3.5390	2.8371
	HSDT[10]	5.4603	5.0814	4.2039	3.8343	3.6466	3.5379	2.8371
	TOT	5.4603	5.0815	4.2051	3.8361	3.6485	3.5390	2.8371
	Sine	5.4603	5.0815	4.2051	3.8361	3.6484	3.5389	2.8371
	Hyperbolic	5.4751	5.0944	4.2147	3.8457	3.6613	3.5530	2.8448
	Exponential	5.4604	5.0816	4.2051	3.8361	3.6483	3.5390	2.8372

4. Conclusions

This study dealt with static and free vibration analysis of simply supported laminated and functionally graded (FG) beam using various theories. It is found that exponential shear deformation theory yields closer results to 2D elasticity solution. Unlike other theories, the hyperbolic shear deformation theory yields zero shear stress at mid-plane and non-zero values at top and bottom of the laminated and FG beam. Also, for higher span to thickness ratio i.e. $S = 16$, results are in close proximity with the results available in the existing literature.

Acknowledgments

The authors are thankful to the Science and Engineering Research Board (SERB) for supporting this work through a project grant.

5. References

[1] Sankar B V. 2001 *Composites Science and Technology* An elasticity solution for functionally graded beams **61** 689-696.
[2] Li X F. 2008 *Journal of Sound and Vibration* A unified approach for analyzing static and dynamic behaviors of functionally graded Timoshenko and Euler-Bernoulli beams **318** 1210-1229.
[3] Sina S A, Navazi H M and Haddadpour H. 2009 *Materials and Design* An analytical method for free vibration analysis of functionally graded beams **30** 741-747.
[4] Thai H T and Vo T P. 2012 *International Journal of Mechanical Sciences* Bending and free vibration of functionally graded beams using various higher-order shear deformation beam theories **62** 57-66.
[5] Simsek M. 2010 *Nuclear Engineering and Design* Fundamental frequency analysis of functionally graded beams by using different higher-order beam theories **240** 697-705.
[6] Vo T P, Thai H, Nguyen T, Inam F and Lee J. 2014 *Composites: Part B* Static behaviour of functionally graded sandwich beams using a quasi-3D theory **68** 59-74.
[7] Kadoli R, Akhtar K and Ganesan N. 2008 *Applied Mathematical Modelling* Static analysis of functionally graded beams using higher order shear deformation theory **32** 2509-2525.
[8] Sayyad A S and Ghugal Y M. 2017 *Composite Structure* Bending, buckling and free vibration of laminated composite and sandwich beams: A critical review of literature **171** 486-504.
[9] Kapuria S, Dumir P C and Jain N K. 2004 *Composite Structure* Assessment of Zigzag theory for static loading, buckling, free and forced response of composite and sandwich beams **64** 317-327.
[10] Vo T P, Thai H T, Nguyen T K and Inam F. 2014 *Meccanica* Static and vibration analysis of functionally graded beams using refined shear deformation theory **49** 155-68.

IOP Conf. Series: Materials Science and Engineering **404** (2018) 012031 doi:10.1088/1757-899X/404/1/012031

Improvement of Inter-face Layer Coating by Plasma Treatment of Carbon Fiber for Carbon Fiber Reinforced Silicon Carbide Composite Applications

Sonam H. Suthar[1,2], N. Chandwani[1] and C. Jariwala[1*]

[1]Institute for Plasma Research, Bhat, Gandhinagar-382428, India
[2]Sh. M. M. Patel Institute of Science and Research,
Kadi Sarva Vishvidhyalaya, Sector-15, Gandhinagar-382015, India

Email: [*]chetan@ipr.res.in

Abstract. Carbon fiber (CF) reinforced SiC (CF/SiC) composites are sought for high temperature application in areas of automobiles, aerospace and nuclear reactors due to their outstanding properties. However, CF/SiC composites have low durability owing to oxidation of CF upon air-exposure during high temperature applications. The inter-phase layer between fiber and SiC matrix is used in order to protect fiber from oxidation. Here, we have aimed to study the effect of air RF (13.56 MHz) plasma treatment on CF tow for improvement of SiO_2 inter-face layer coating by dip coating technique. The air RF (13.56 MHz) plasma treatments on CF tows and graphite proxy substrates (for 20, 40 and 60 min) have been analyzed by Scanning Electron Microscopy (SEM) and contact angle measurement, respectively confirmed the enhancement of wettability. Such plasma treated CF tows have been further used for deposition of SiO_2 coating and uniform deposition has been achieved for 20 min plasma treated CF as evident from SEM surface analysis. Whereas, Energy Dispersive X-ray Analysis (EDXA) and Fourier Transform Infra-Red spectroscopy (FTIR) measurements have been used to get elemental and bonding information of coated silicon substrates confirmed the high quality of SiO_2 coating. While tensile strength measurement verified the enhancement of tensile strength for plasma treated SiO_2 coated CF tows as compare to untreated CF tow, and optimum strength (1449 MPa) has been obtained for 20 min plasma treated SiO_2 coated CF tow with uniform coating properties. Hence, 20 min plasma treatment on CF tow is confirmed the beneficial effect on SiO_2 inter-face layer coating for CF/SiC Composites Applications.

1. Introduction

Carbon fibers (CFs) are widely used as reinforcements in composite materials due to their excellent properties such as low coefficient of thermal expansion, high modulus and strength, low density, good electrical conductivity and relative high flexibility [1-2]. In particular, CF/SiC (in which SiC act as a matrix) is well known for its high strength and stiffness as well as its low density, which makes it an attractive candidate for the aerospace, automobile industries, gas turbine engine and nuclear reactors applications [3-4]. However, continuous use of CF reinforced composite in oxidising environment is imposed temperature limit to lower than 500° C for high temperature applications. Such drawback could be overcome by protecting CF by an oxidative protection coating when they are exposed to oxidative and chemically aggressive environments. There are numerous research work reported for the

IOP Conf. Series: Materials Science and Engineering **404** (2018) 012031 doi:10.1088/1757-899X/404/1/012031

oxidative protective coating layer on CF such as carbides, nitrides and oxides by various coating techniques [5-6]. Among them, SiO_2 based inter-face layer is one of the potential candidate which improves the oxidation resistance property of the CF tow and thus act as protective layer coating. SiO_2 has also high melting point and an exceptional oxygen barrier property.

At same time, CF surface is also chemically inert in nature and has lower wettability, which leads to the most potential problem of inadequate adhesion of coating and hence weaker bonding between the fiber and coating. Poor adhesion and lower bonding strength of coating with fiber are resulted in insufficient oxidation protection of CF. Therefore, the surface treatment of CF is necessary, which increase the surface active sites as well as surface roughness for improvement of bonding strength between the fiber and coating. Various methods have been used for surface modification of CF, such as acid treatment, plasma treatment, rare-earth treatment and gamma irradiation [7]. Among those, plasma treatment is widely used to modify fiber surface chemically and structurally without much influencing their bulk mechanical properties.

In the present work, we have studied the effect of plasma treatment on CF tow for improvement of SiO_2 inter-face layer coating carried-out by dip coating technique. We have attempted surface modification of CF by air plasma using RF (13.56 MHz) source for 20, 40 and 60 min prior to SiO_2 deposition. The SiO_2 coating on untreated and plasma treated CF tows are achieved by a dip coating technique with fixed 3 Nos. dipping cycles of coating to evaluate the surface modification effects. Surface morphology of plasma treated and SiO_2 coated CF tows are investigated by SEM analysis. The contact measurement of untreated and plasma treated proxy graphite samples are carried out using optical contact angle goniometer, while chemical composition and boding information of SiO_2 coating on silicon substrate are obtained by EDXA and FTIR measurement, respectively. The tensile strength testing of the plasma treated coated and uncoated CF tows are also investigated in order to see the improvement of quality SiO_2 coating as inter-face layer for CF/SiC composites.

2. Experimental

2.1. Materials

The CF tow used in this study is Poly-Acrylo-Nitrile (PAN) based "TAIRYFIL TC – 33 (6K)" CF tow purchased from Formosa Plastics Corporation (Taiwan). Tetraethyl Orthosilicate (TEOS, 98%) [Si $(OC_2H_5)_4$] is used as SiO_2 coating precursor obtained from Acros Organics (USA). Propan-2-ol (99.99%) for solvent and Hydrochloric Acid (HCl) for catalyst purpose are purchased from Rankem Chemicals (India).

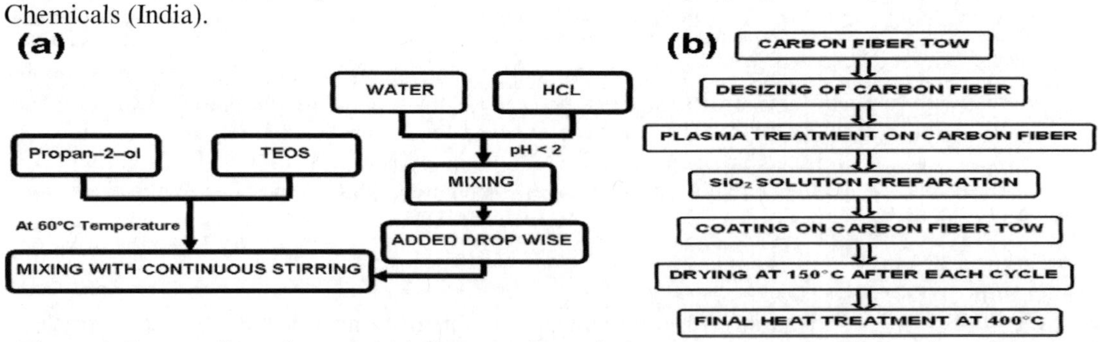

Figure 1. Process flow charts for (a) SiO_2 coating solution preparation for dip coating of CF tow and (b) Dip coating processing steps for SiO_2 coating on CF tow.

2.2. Coating Solution Preparation

The SiO_2 solution is prepared by the hydrolysis-condensation (*sol-gel* process) of SiO_2 containing inorganic-organic compounds in which TEOS is used as the SiO_2 precursor. In a beaker, known quantity of TEOS is added with propan-2-ol and the solution is continuously stirred for uniform mixing. In another beaker, HCl is added in known volume of water such that pH of the solution is kept

IOP Conf. Series: Materials Science and Engineering **404** (2018) 012031 doi:10.1088/1757-899X/404/1/012031

below 2 pH. This prepared pH solution is added drop wise into the mixture of TEOS and propan-2-ol while stirring. The final prepared solution is kept stirred for 3 hours at room temperature with fixed molar ratio of TEOS: Propan-2-ol: Water as 1: 4: 8. The whole process of preparation of SiO_2 coating solution is shown schematically by the flow chart in Fig. 1 (a).

2.3. Plasma Surface Treatment of CF

Prior to plasma treatment, all CF tows are de-sized using acetone for removal of any surface contamination as well as sizing layer. Than after, all de-sized CF tows are plasma treated in air RF (13.56 MHz) plasma formed between capacitively coupled disc electrodes (size = 22 cm dia) in a high vacuum system. All de-sized CF tows are treated for 20, 40 and 60 min time duration at pressure 5 x 10^{-1} mbar and 25 watts coupled power. Looking to difficulty in contact angle measurement and hence wettability properties, high density graphite discs (Size = 10 mm dia) are used as proxies of CF tows to get relative trends of plasma treatment effect.

2.4. SiO_2 Coating on CF

Dip Coating unit (Model No. Ho-TH 02B) with infrared heater is used for SiO_2 coating, which is supplied by the Holmarc Opto-Mechtronic Pvt. Ltd (India). SiO_2 coating on all CF tows (de-sized, 20, 40 and 60 min plasma treated) are applied using dip coating technique with fixed process parameter such as 3 Nos. dipping cycles, 8 mm/sec withdrawal speed and 5 min dipping time. After each coating cycle, CF tow is dried at 150° C by infrared heater attached to the dip coating unit, followed by final heat treatments at 400° C for 60 min. The entire process steps of the SiO_2 coating on the fiber tow are represented by the flow chart in Fig. 1(b).

2.5. SiO_2 Coating Chracterization

The surface morphology and chemical composition for plasma treated and SiO_2 coated CF tow are analysed using SEM and EDXA attachment (440i, LEO, UK), respectively. The wettability of graphite proxy substrates before and after plasma treatments are estimated by means of contact angle measurements, which are carried out using optical contact angle goniometer (OCA15, Dataphysics, Germany) as per the sessile drop method. The detail bonding information for SiO_2 coatings on silicon substrates are obtained by FTIR (Nicolet 6700, Thermo Electron Corp., USA) measurement. The Instron Universal Testing System (Model 5982) is used to measure tensile strength of all CF tows.

3. Results and Discussion

3.1. Effect of Plasma Surface Treatment on CF

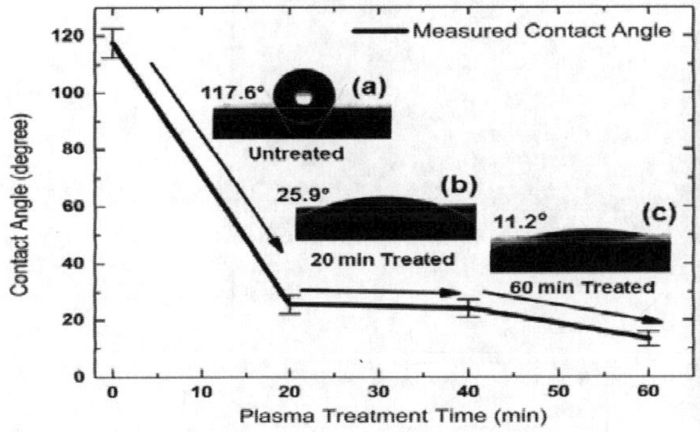

Figure 2. Contact angle measurements of plasma treated graphite samples. Inset images show for (a) untreated (b) 20 min and (c) 60 min plasma treated graphite samples.

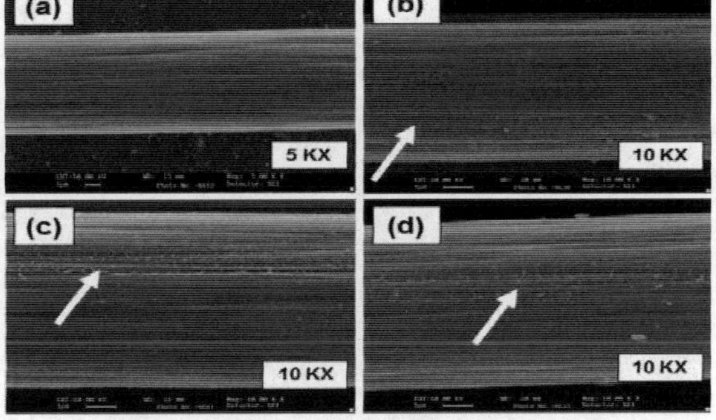

Figure 3. SEM images of CF for (a) untreated, (b) 20 min, (c) 40 min and (d) 60 min plasma treated (Arrow marks show pitting effect on CF surfaces).

Plasma is defined as an electrically conducting medium generally consisting of electrons, charged ions, and neutral atoms or molecules or both. Plasma treatment may not only enhance surface roughness (due to ion-bombardment) but also alter surface functional groups of treated CF. The effect of Air RF (13.56 MHz) plasma treatment on wettability by measuring contact angle for varying treatment time 20, 40 and 60 min on graphite proxy substrates are showed in Fig 2. The measured contact angle is drastically reduced from 117.6° (untreated) to 25.9° (20 min treated) and further reduced to 11.2° (60 min treated) confirmed the improvement of wettability (Fig 2 (a), (b) and (c)) for graphite proxy substrates and hence similar effect is assumed for CF tows. The detail SEM surface morphology analysis is revealed the clear micro-pitting effect for plasma treated surface of CF with increasing in nature for higher and higher treatment time duration (Fig. 3 (b), (c) and (d)) as compare to untreated CF (Fig.3 (a)). The surface roughness increment increases the surface area which is major contributor to improvement of wettability and hence enhancement of mechanical interlocking between fiber and coating. Further, it is also noted that with increase in treatment time the enhancement of pitting structure are observed for 40 & 60 min plasma treated CF (Fig. 3 (c) & (d)).

3.2. SiO_2 coating Effect on CF

All CF tows (untreated and plasma treated) are coated by dip coating technique with fixed 3 Nos. of dipping cycles, 8 mm/s withdrawal speed, and 5 min dipping time in order to check plasma treatments effect on coating quality. SEM images are analysed in order to study the surface morphology of the SiO_2 coating on these CF tows (Fig 4 (a), (b) and (c)). SEM surface analysis is revealed that the uniformity and continuous coating is observed for 20 min (Fig. 4 (b)) in comparison to untreated CF tow (Fig 4 (a)). It is assumed that active species in the plasma aggressively attack the edges of tissue like structure of CF and creates micro-pitting region, essentially responsible for improvement of wettability and hence uniformity of coating. Whereas, deterioration of coatings for 40 min (SEM image not shown) and 60 min (Fig. 4 (c)) plasma treated CF tows are observed due to excessive coating because of intense plasma treatment effect as evident from SEM analysis (Fig. 3 (c) & (d)). The thickness values of coating on plasma treated CF are estimated in range of 2.74 to 3.02 μm by cross-section SEM analysis. Here, these values are obtained from plasma treated silicon substrates, which are coated with identical deposition parameters and such typical SEM cross-section measurement image is shown in Fig. 4 (d). At same time, chemical composition analysis by EDXA is clearly identified major *Silicon* and *Oxygen* elements peaks with well-defined stoichiometry in EDXA spectra (Fig. 4 (e)) and also atomic concentration analysis confirmed the SiO_2 coating. Whereas, FTIR measurement on coated silicon substrate is showed the presence of characteristic Si-O rocking bond peak at 453 cm^{-1} apart from Si-O-Si asymmetric (1090 cm^{-1}) and Si-O bending (812 cm^{-1}) peaks confirmed the SiO_2 coating by dip coating process Fig. 5 (a).

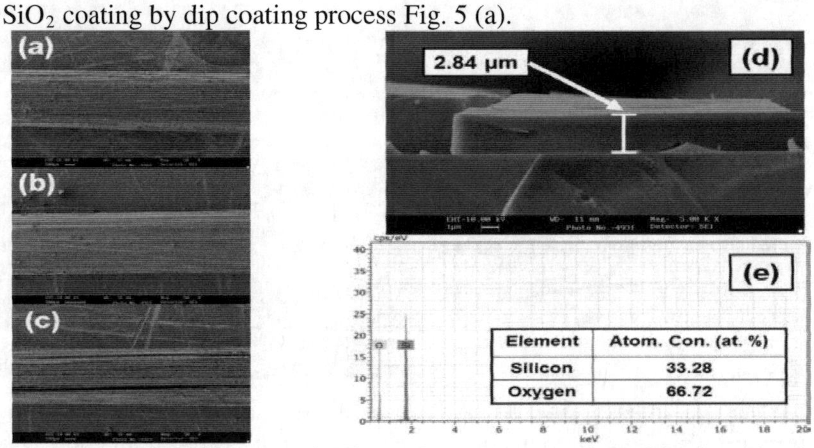

Figure 4. SEM micrograph of SiO_2 coated CF tow for (a) untreated (b) 20 min treated and (c) 60 min treated. Whereas, (d) typical SEM cross-section image of SiO_2 coated silicon substrate and (e) EDXA spectra with inset table for obtained atomic concentration values.

The mechanical property of SiO_2 coated and un-coated CF tows are measured using Instron Universal Testing System. A total of five CF tows of each set of SiO_2 coated and un-coated CF tows are measured as per ASTM standard for better accuracy (Fig 5 (b)). The tensile strength of uncoated CF tow is 1036 MPa, while tensile strength of plasma treated for 20 min, 40 min and 60 min with SiO_2 coated CF tows are 1449, 1582 and 1294 MPa, respectively. It is observed that there is continuous increase in tensile strength with increase in plasma treatment time up to 40 min, which is due to improvement of SiO_2 coating. However, for 60 min, the tensile strength decreases due to the deterioration of coating as evident from the CF tow SEM image, as shown in Fig. 4 (c). So, the optimum tensile strength with uniform SiO_2 coating has been obtained for 20 min plasma treated SiO_2 coated CF tow with fixed 3 Nos. dipping cycles, withdrawal speed 8 mm/sec and dipping time 5 min.

4. Conclusion

Plasma treatment of CF tow for improvement of SiO_2 inter-face layer coating has been successfully attempted for composite application. The effect of plasma treatment on CF has been studied by SEM and contact angle measured on graphite substrates have confirmed the enhancement of wettability. Plasma treated CF tows have been further effectively used for deposition of SiO_2 coating and uniform deposition has been achieved for 20 min plasma treated CF tow as evident from SEM surface analysis. EDXA and FTIR measurements have been used to get elemental and bonding information of coated silicon substrates confirmed the high quality of SiO_2 coating. While tensile strength measurement verified the enhancement of tensile strength for plasma treated CF as compare to untreated CF tow, and optimum strength (1449 MPa) has been obtained for 20 min treated SiO_2 coated CF tow with uniform coating properties.

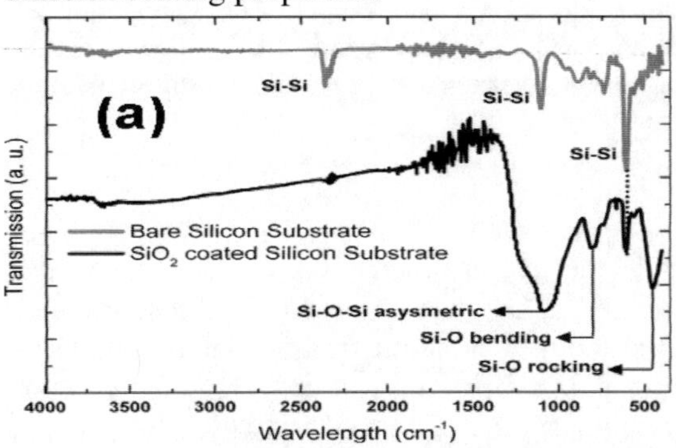

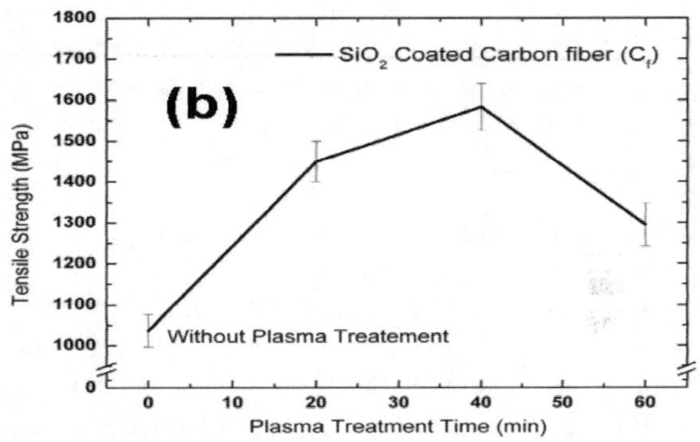

Figure 5. (a) FTIR spectra taken for SiO_2 coated silicon substrate along with bare silicon substrate. (b) Tensile measurement of untreated and plasma treated SiO_2 coated CF tows.

Acknowledgments

One of the authors (SHS) would like to express sincere thanks to Institute for Plasma Research (IPR), Gandhinagar for this work. The authors (SHS & CJ) also thank to Mr. Narendrasingh Chauhan (IPR, Gandhingar) and Mr. Hardikkumar Tailor (IPR, Gandhinagar) for SEM analysis and carbon fiber tow tensile strength measurements, respectively.

References

[1] Chawla K K 2012 *Composite Materials* (New York: Springer-Verlag).
[2] Morgan P 2005 Carbon Fibers and Their Composites (Florida: Taylor-Francis).
[3] Padmavathi N, Ghosal P, Prasad N, Subramanyam J, Ray K K 2012 *Sadhana* **37** 493.
[4] Srivastava V K 2012 *American Journal of Materials Science* **2** 1.
[5] Li S, Zhang Y M, Zhou Y F 2012 *Surf. Coat. Technol.* **206** 4720.
[6] Xu B, Hong C, Qu Q, Zhang X, Han J, Li N, Gui K, Zhao G, Liu C 2015 *Surf. Coat. Technol.* **270** 109.
[7] Tiwari S, Bijwe J 2014 *Procedia Technol.* **14** 505.

IOP Conf. Series: Materials Science and Engineering **404** (2018) 012032 doi:10.1088/1757-899X/404/1/012032

Study of mixing in circular unbalanced split & recombine micromixer

J Husain[1] and M A Ansari[1]

[1] Department of Mechanical Engineering, Z.H.C.E.T, Aligarh Muslim University, Aligarh, India.

E-mail: mubashshir@gmail.com

Abstract- In the present work, numerical simulation has been performed to analyze the mixing of fluids inside a passive circular unbalanced split & recombine micromixer. The continuity, Navier-Stokes & mass diffusion equations were solved and numerical simulation were performed for aspect ratios of 3.00. The width ratio (w1/w2) was varied for seven different values between 1.90 & 2.20 and numerical simulations were performed for five different Reynolds numbers (10, 20, 40, 60 & 80) at each value of width ratio. The mixing phenomenon was explained with the help of stream lines plots, contour plots etc. The generation of secondary flows across the transverse plane and their contribution in achieving better mixing has been explained. The mixing performance of the micromixer is quantified with the help of mixing index.

1. Introduction

Micromixer is a microfluidic device used to mix fluids in the micro-dimensions. In micromixer, the Reynolds number is low therefore viscous forces dominates rather than the inertial forces. Another important factor for consideration in the micromixers is ubiquity of laminar flow. High surface to volume ratio, smaller size, less energy consumptions, cost effectiveness are some features of micromixers. Based on the source of energy, the micromixer can be classified in two major groups: active & passive micromixer. In active micromixer, the external source of energy is used in the mixing of fluid streams while in passive micromixer, geometrical features of the system are utilized for the same process [1-2].

Ansari et al designed a passive micromixer which is fundamentally based on the principle of unbalanced splits and cross-collisions of fluid streams. In the proposed design the main fluid channel was split into two sub-channels of unequal widths that further recombine after a certain distance and again split and recombine. The proposed design gave best results for width ratio of 2.0 for Reynolds number higher than 20 and poor performance for width ratio of 1.0 [3]. Afzal et al proposed a passive micromixer with convergent divergent walls based on the principle of unbalanced split and recombine. [4]. Changing the inlet and outlet configurations of mixing chambers also improved the performance of the passive micromixer. Alam et al designed a micromixers with four mixing chambers and four different arrangements, for inlet and outlet of fluid streams [5]. Shakawat et al increased the number of splits in split and recombination based rhombic micromixers. The rhombic angle 90° gave the best mixing performance except at the low Reynolds number [6]. Alam et al placed circular obstruction in the path of fluid flow in planer curve micromixer. [7]. In simple straight channel passive micromixer, the mixing was dominated by molecular diffusion only, irrespective of Reynolds number. But in case of curved channel, Dean vortices were generated due to the combined effect of inertial and centrifugal forces. At low Reynolds number, the mixing was dominated by molecular diffusion only but as the

Reynolds number was increased, the inertia effect started to take effect as well due to continuous change in the curvature, the centrifugal effect also came into picture creating secondary Dean vortices in the micromixers, which helps in the improvement of performance of passive micromixer at higher Reynolds numbers [8].

2. Physical model of the micromixer

The schematic diagram of the circular passive unbalanced split & recombine micromixer is shown in the figure 1. The micromixer is consist of T-joint followed by six circular unbalanced and split and recombine (SAR) units. The width of the main channel (W) is kept constant at 300 μm. The diameter of the base circle of each SAR unit is 900 μm. In each SAR unit the main channel is split into two sub channel such that the sum of the width of both sub-channels equal to width of the main channel i.e. $w_1 + w_2 = 300$ μm. The total length (L) of the micromixer is equal to 10350 μm. The pitch (p_i) of the micromixer is equal to 1200 μm. The height (h) of the micromixer is kept equal to 100 μm. The aspect ratio (AR) of the micromixer is defined as the ratio of the width of the main channel (W) to the height (h) of the micromixer.

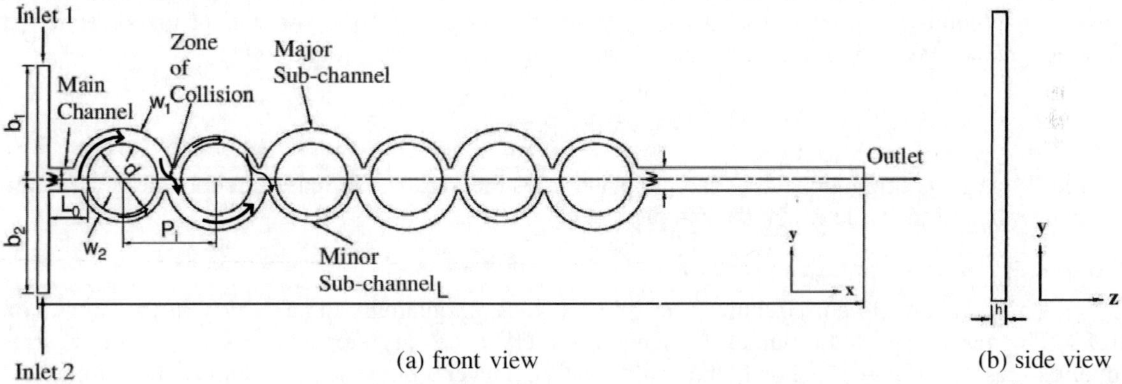

Figure 1. Schematic diagram of the micromixer (a) front view (b) side view

For aspect ratio of 2.50, the earlier results showed that mixing efficiency of circular unbalanced split & recombine micromixer was maximum at width ratio (w_1/w_2) of 2.0 for Reynolds number (Re) lying between 10 & 80 [3]. In this present work, further study was done in investigating the optimum width ratio $(w_1/w_2)_{optimum}$ for the aspect ratio (AR) of 3.00.

3. Numerical methodology

The general transport equations in numerical solver poses the following general form:

$$\frac{\partial(\rho\phi)}{\partial t} + \nabla(\rho\phi V) = \nabla(\Gamma\ grad\ \phi) + S_\phi$$

Here, ρ is the density, V is the velocity, S_ϕ is the source term, μ is the dynamic viscosity, p is pressure field, c is the mass fraction & D is the diffusion coefficient. ANSYS-CFX is cell-vertex based finite volume, coupled implicit pressure based commercial CFD code. CFX is based on finite volume method, the partial differential equations of flow as represented above is integrated over each of the control volume in the computational domain. The pressure velocity decoupling in the CFX is handled by using Rhie-Chow approach [9]. For numerical simulation, hexahedral grids had been generated in the entire numerical domain using ANSYS ICEM 18. For the same amount of cells, hexahedral cells have high accuracy in 3-D space.

Numerical simulations were performed by studying the mixing between water and ethanol. The dynamic viscosity of water & ethanol was taken as 1.002×10^{-3} Pa-sec 1.187×10^{-3} Pa-sec and density as 998.20 kg/m^3 & 789.7 kg/m^3 respectively [10]. At 20°C, the diffusivity of water and ethanol was taken as 1.2×10^{-9} m^2/s [3]. The flow inside the micromixer is considered as steady, laminar and incompressible. The isothermal condition was maintained in the micromixer. [11]

IOP Conf. Series: Materials Science and Engineering **404** (2018) 012032 doi:10.1088/1757-899X/404/1/012032

In the numerical solver, different boundary conditions were specified at both inlets, outlet and wall of the micromixer. There were two inlets and one outlet in the considered design of micromixer. At both inlets, normal inlet velocity and ethanol mass fraction was specified as boundary conditions. At outlet of micromixer, average static pressure was specified as boundary condition. No slip condition was applied at the wall of the micromixer. The numerical solution was converged when the RMS value of the residue at each node points attains a value of 10^{-6}.

The effectiveness of the micromixer was quantified by calculation of the mixing index. A plane perpendicular to direction of flow inside the micromixer and near the outlet of the micromixer was selected and mass fraction of ethanol has been taken from ample numbers of sample data points. Then, the variance of mass fraction from these data points was calculated using the following equation

$$\sigma = \sqrt{\frac{\Sigma(c_i - \bar{c}_m)}{N-1}}$$

Here, σ is the standard deviation of sample data points, c_i is the mass fraction of ethanol at the *i*th sample point and c_m is the mean mass fraction of ethanol along the sample plane and *N* is the number of the sample points taken from the micromixer. After calculation of the variance of mass fraction at the sample plane, the mixing was calculated by applying the following equation [3].

$$M = 1 - \sqrt{\frac{\sigma^2}{\sigma_{max}^2}}$$

Here, *M* is the mixing index at the sample plane, σ^2 is the variance of mass fraction and σ_{max}^2 is the maximum value of the variance on the sample plane.

4. Validation & grid independency test

Adopting an efficient and correct numerical strategy does not guaranty the accuracy of the simulation results. The validation of the numerical scheme is very necessary step in order to obtain accurate numerical results. Figure 2 shows the validation of current numerical solutions. The numerical simulations were performed for Re 80 and width ratio (w1/w2) of 1.0, 1.5, 2.0 and 2.5 at aspect ratio of 2.50 and the results were compared with the earlier published experimental results of Ansari et al [3].

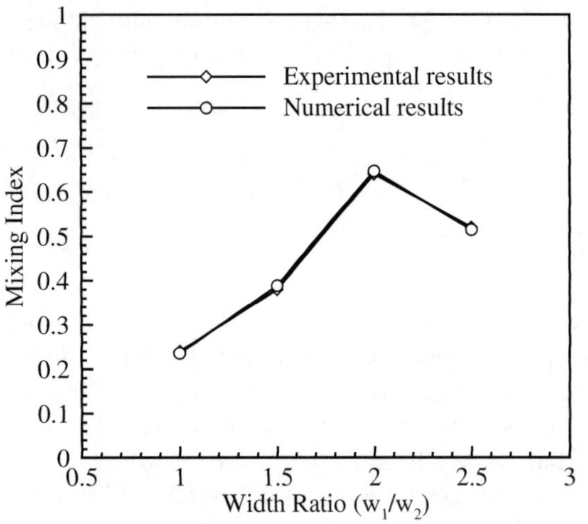

Figure 2. Validation of numerical result from experimental result [3] at Re 80 for aspect ratio 2.50

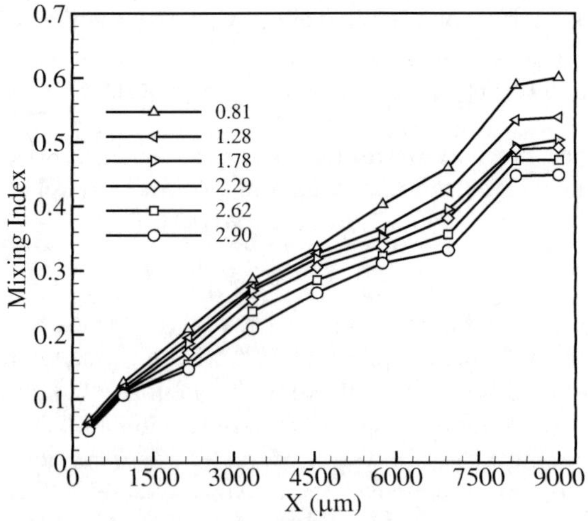

Figure 3. Grid independency test at width ratio 2.00 & Re 80 for aspect ratio 3.00

The refinement of the grid is very important factor in the generation of optimum grid. Therefore, a detailed grid independency test was performed at width ratio of 2.00 & Re 80 for six different grid sizes. The result of grid independency test is shown in the figure 3. On the basis of results obtained from the grid independence test, grid size of 2.29 million was selected for numerical simulations in the current design of micromixer.

5. Results & discussions

The mixing of fluid streams is better understand with the help of the ethanol mass fraction contour plots. The figure 4 gives a comparison of contour plots of balanced collision (w_1/w_2=1.00) & unbalanced collision (w_1/w_2=1.95). In case of balanced collision of fluid streams, the different fluid streams came into contact with each other at the junction of two mixing units only and almost flow separately in the sub-channels.

In case of unbalanced collision of fluid streams, the ethanol stream entering from inlet 1 was flowed through major sub-channel and water stream entering from inlet 2 was flowed through both major and minor sub-channels. In case of unbalanced collision, the major sub channel has large area compare to minor sub-channel, therefore, the water streams were split into two sub streams, one entered in to the major sub-channel and other entered into the minor sub-channel of the first mixing unit. While in the other hand the ethanol stream entering from inlet 1 flowed through only major sub-channel. In the first mixing unit, the ethanol streams had higher mass flow rate in the major sub channel therefore it pushed the water steam towards the inner wall of the unit resulting in the collision at the junction of second mixing unit. Due to this, ethanol stream entered in the both sub channel of the second mixing unit. This process was repeated during the entire mixing length of the miciromixer and leads to better mixing compared to the balanced collision.

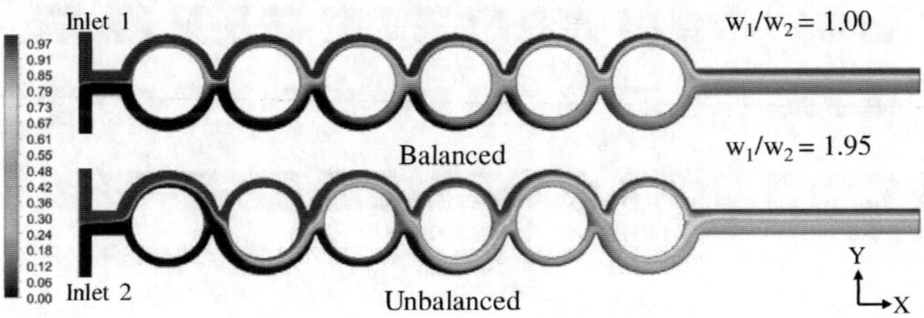

Figure 4. Comparison of contour plots of ethanol mass fraction in xy-plane at Re 80 & aspect ratio of 3.00 for balanced (w_1/w_2=1.00) and unbalanced (w_1/w_2=1.95)

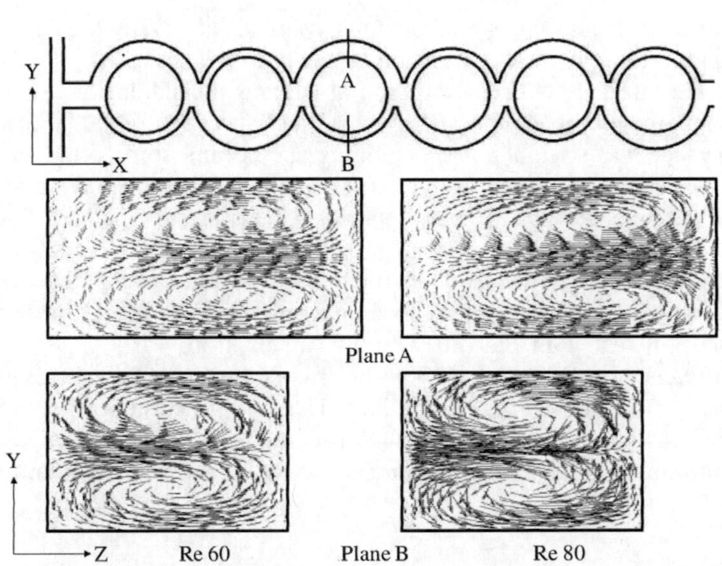

Figure 5. Velocity vector at plane A & B at Re 60 & 80 for aspect ratio (AR) of 3.00 & width ratio (w_1/w_2) of 1.95

At low Reynolds number (Re=10), the mixing is achieved due to the high residence time of fluid flow inside the micromixer as velocity was less at low Re. As the Reynolds number increases further

(Re=20), the inertia forces began to dominate the viscous forces. Therefore, when the fluid streams flowed across the circular sub-channels, the combined effect of centrifugal forces and inertia forces generated the secondary flows in the transverse plane. The counter rotating pair of dean vortices started to form in the both sub-channels and can clearly observed in the figure 5 for Re 60 & Re 80. The dean vortices caused in the increase of the interfacial area between the mixing streams along the length of the circular channel symmetrically. Therefore, the mixing interface became more and more distorted along the middle axis of transverse plane. As the Reynolds number increased further, the secondary flows played an important contributing factor in achieving good mixing efficiency.

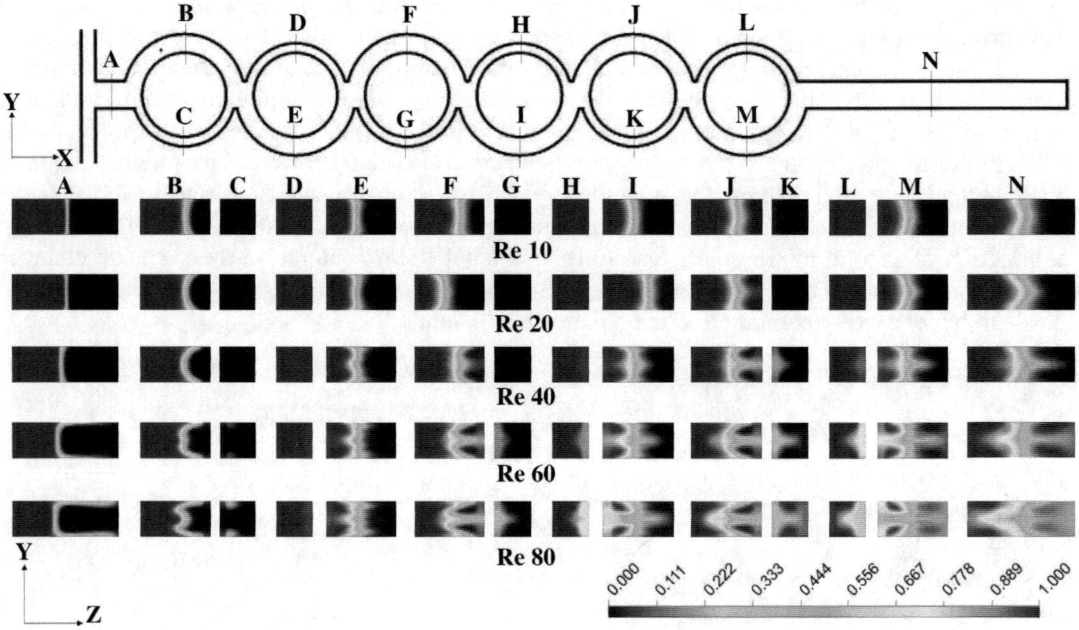

Figure 6. Ethanol mass fraction profile at various cross section along yz- plane at different axial positions along the x-axis at width ratio (w_1/w_2) of 1.95 and aspect ratio (AR) of 3.00

From the figure 6 we can see that the interface between the mixing streams played an important role in the understanding of mixing phenomenon inside the micromixer. For aspect ratio 3.00, at low Reynolds number 10, 20 & 40, the interface was straight at the beginning and became to deform from plane B in the major sub-channel and it became developed from plane E. While in case of Re 60 & 80 it started to deform from starting and became developed in the plane B of major sub-channel. For minor sub-channel, the mixing interface did not form and no mixing took place in the entire length for Re 10 & 20. While in case of Re 40 the interface was formed from the plane K and in case of Re 60 & 80, it formed from plane C and became developed from plane G for minor sub-channel. Therefore, the effect of major sub-channel was also propagated in the minor sub-channels from Re 40. As the Reynolds number was increased, the influence of transverse flow in the circular channel also increased resulting in the development of secondary flows inside the micromixer. The mass fraction profile was symmetric about the middle plane due to the formation of counter rotating pair of dean vortices. Therefore, for aspect ratio 3.00, the micromixer gave a poor performance at Re 10 and its performance was improved from Re 20.

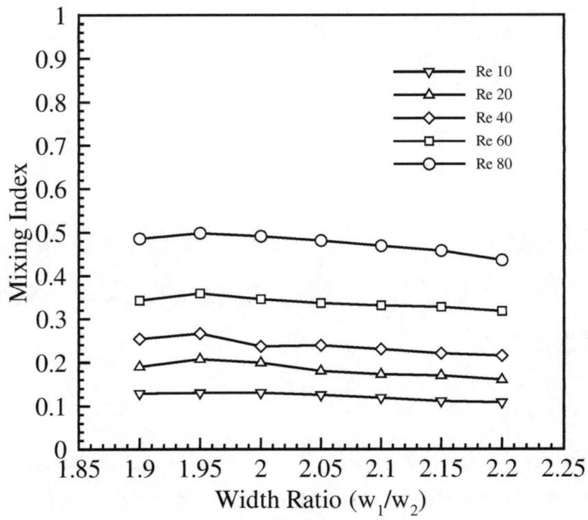

Figure 7. Mixing index for different Reynolds numbers (Re) at various width ratios (w_1/w_2) for aspect ratio of 3.00

The overall performance of micromixer was evaluated with the help of mixing index. Form the figure 7, it is quite evident that the performance of the micromixer was improved with the increase in the Reynolds number. The results of the mixing index indicate that the mixing performance of the circular unbalanced split & recombine passive micromixer become maximum at a width ratio $(w_1/w_2)_{optimum}$ of 1.95.

6. Conclusion

The present work was aimed to investigate the mixing inside the passive circular unbalanced split & recombine micromixer. The numerical simulations were performed to calculate the optimum width ratios for five different values of Reynolds numbers (10, 20, 40, 60 & 80). The mixing performance of the unbalanced split & recombine micromixer is better compared to the balanced collision micromixer. At low Reynolds numbers (Re=10), the fluid mixing is due to the high residence time of the fluids inside the micormixer. In such case, mixing is mainly due to the diffusion of the fluid particles, which is quiet slow process. When the Reynolds number was increased further the secondary flows were established in the form of counter rotating pair of Dean vortices resulting in the better mixing performance. For aspect ratio (AR) of 2.50, the mixing performance is reported to be highest at width ratio (w_1/w_2) of 2.00 but in our study we found for aspect ratio (AR) of 3.00, the optimum width ratio $(w_1/w_2)_{optimum}$ comes out to be 1.95.

References

[1] Lee C Y, Chang C L, Wang Y N and Fu L M 2011 *Int. J. Mol. Sci.* **12** 3263-87.
[2] Mansur E A, Mingxing Y E, Yundong W and Youyuan D A I 2008 *Chinese Journal of Chemical Engineering* **16(4)** 503-16.
[3] Ansari M A, Kim K Y, Anwar K and Kim S M 2010 *J. Micromech. Microeng.* **20** 055007.
[4] Afzal A and Kim K Y 2012 *Chemical Engineering Journal* **203** 182–92.
[5] Alam A and Kim K Y 2013 *Sensors and Actuators* **B 176** 639–52.
[6] Hossain S and Kim K Y 2014 *Micromachines* **5(4)** 913−28.
[7] Alam A, Afzal A and Kim K Y 2014 *Chemical Engineering Research and Design* **92** 423-34.
[8] Sudarsan A P and Ugaz V M 2006 *Lab Chip* **6** 74–82.
[9] Versteeg H and Malalasekra W 2013 *An introduction to computational fluid dynamics* vol 2 (Harlow: Longman Scientific & Technical) pp 85-154.
[10] Gonza´lez B, Calvar N, Go´mez E and Domı´nguez A 2007 *J. Chem. Thermodynamics* **39** 1578–88.
[11] Klavs F Jensen 2001 *Chemical Engineering Science* **56(2)** 293-303.

IOP Conf. Series: Materials Science and Engineering **404** (2018) 012033 doi:10.1088/1757-899X/404/1/012033

Processing and Evaluation of Microstructure & Physical Properties of Al7075 Reinforced with Cermets by Stir Casting Route

Gopal Krishna U B[1], Ajay Kumar N L[2], Vishal V.K[3], Vrushank V Salimath[3], Shashidhar S Kappadi[3], Sachin S Hadimani[3], Auradi V[4], Vasudeva B[4]

[1]Research Scholar, R&D Centre, Dept. Of ME, SIT, Tumakuru, Karnataka, IN

[2]PG Scholar, Dept. Of ME, SIT, Tumakuru, Karnataka, IN

[3]UG Scholar, Dept. Of ME, SIT, Tumakuru, Karnataka, IN

[4]Associate Professor, Dept. Of ME, SIT, Tumakuru, Karnataka, IN

[4]Associate Professor, Dept. Of ME, SIT, Tumakuru, Karnataka, IN

E-mail: gopalkrishnaub@gmail.com

Abstract: The combination of two or more materials with dissimilar characteristics leads to the formation of Metal matrix composites (MMC). Among MMCs large interest is being given to improve aluminium based MMCs and to develop high temperature ceramics. The present paper discusses the processing and physical properties of Al7075 reinforced with (WC-Co) cermet particulates prepared by stir casting route. An average particle size of 10µm WC-Co particulates produced by planetary ball milling process is used as reinforcement in Al7075 matrix. Microstructure evaluation of the prepared composite is carried out using SEM/EDX. Physical properties of the matrix and prepared composite are evaluated.

Keywords: Cermet, composite, ceramics, ball milling

1. Introduction

The conventional metals, alloys, ceramics and polymeric materials which are used in general engineering applications are exhibiting a less impressions in their properties in concern with the modern technology required materials which are able to exhibit an unusual contribution of properties. To achieve the modern materials to give a combinations in the betterment of properties development of the composite materials come into the existence. Replacement for the conventional materials can be made by Metal matrix composites implementation in various applications like machine tools, connecting rod, aerofoil structures and in the field of defense. For engineering applications MMC plays a major role and aluminium is an excellent material in the non ferrous family because of its excellent properties like low density, high strength to weight ratio, high thermal and electrical conductivity and fluidity [1]. In the view of all types of MMCs aluminium matrix composites plays a major role in automobile and aerospace industries because of their good electrical and thermal conductivity, high strength and stiffness, low coefficient of thermal expansion and good wear resistance. Utilization of ceramic particles in the aluminium matrix composites are the recent field of research. The attractive combination of ceramic materials in particulate form into a metal matrix composite will result in the increment of mechanical properties which is highly difficult to extract from the monolithic materials [2]. The processing methods used in the present technology requires the knowledge about the various processing techniques and selection of proper processing methodologies for the specific materials based on the behavior of particulate MMCs is a view of their wider commercial application along with the mass production. Due to the easy availability of particulates and economic techniques adopted for producing the particulate reinforced MMCs there is an increase in the interest for the same [3]. Cermets like WC-Co (Tungsten carbide and Cobalt) have been widely used and studied because of their good corrosion resistance and unique combination of high hardness and toughness, which results from the mixture of brittle (hard) and ductile(soft) phases. Out of many carbide materials Tungsten carbide is a well known ceramic material which is abundantly used because of its high hardness and wear resistance. Being a metallic phase the addition of cobalt which improves the toughness of the composite so that brittle fracture appearance can be reduced [4]. Researches taking place in the present day shows a less work has been conducted on the combination of both ceramic and metallic phase elements in aluminium matrix material for the preparation of the composites which may exhibit low wear resistance, corrosion resistance and high temperature applications.

2. Experimental details

In the current work, Al7075 alloy is used as a matrix. An atomic absorption spectroscopic analyzed composition of Al7075 is tabulated in Table 1. Cermet mixtures of Tungsten Carbide (WC) and Cobalt (Co) particulate with a mean size of 10 μm which are prepared using ball milling are used as reinforcing materials.

Table 1. Composition of Al7075 matrix

Alloys	Si	Fe	Cu	Mn	Mg	Cr	Zn	Ti	Al
%	0.4	0.5	1.6	0.3	2.5	0.15	5.5	0.2	Balance

2.1 Cermet preparation

Planetary Ball milling process is chosen for the preparation of the cermet because of efficient and simple method to prepare the cermet with sub-micron or micro structured powder form, which can be used as a particulate reinforcement in the preparation of the composites [5]. The rotational effect generated in the ball mill equipment produces a strong mechanical energy which results in formation of extremely fine powder [6-7]. A fairly distribution of hard phases in ductile metallic matrices is due to counteract breaking and cold welding process during milling, breaking of particle's is mainly caused by shear load transferred by grinding balls colliding under a low angle, while cold welding is dominate for perpendicular impact of grinding balls and the crystallite size decreases with proceeding milling time [8-9].

A batch of WC and Co powders is added to the container and the process of milling is carried out for 6 hr with a speed of 300rpm of rotational speed. Further, the addition of iron particles can happen during the milling process which leads to the contamination of the cermet powder. To prevent the iron contamination Alumina medias are used instead of steel medias. The homogeneous distribution of hard phases in ductile metallic matrices takes place due to counteract breaking and cold welding process. The WC-Co cermet after ball milling has an average particle size of 10 μm analyzed via sieve analysis and SEM images of the same are presented in Figure 1-2. The prepared cermets were used as reinforcement for the preparation of composite.

IOP Conf. Series: Materials Science and Engineering **404** (2018) 012033 doi:10.1088/1757-899X/404/1/012033

(1)

(2)

Figure 1-2. SEM microphotographs of WC- Co cermets powder produced by planetary ball milling at different magnifications.

2.2 Preparation of Al7075 & WC-Co Cermet Composites

Composites containing Al7075 matrix with 9wt% of WC-Co Cermet reinforcement were prepared using stir casting technique at temperature of 750°C. A batch of 400 grams of Al7075 matrix is melted in a graphite clay crucible using resistance furnace. When liquid matrix reaches 750°C the presence of adsorbed gaseous content from the molten aluminium is been removed with the addition of solid hexachloroethane (C_2Cl_6) which acts as a degassing agent. To generate vortex in the liquid matrix, stirring is carried out using zirconia coated steel rod with a spindle speed of 300 rpm for 10 min. Preheated WC-Co cermet in an amount equivalent to 9 wt% is then introduced into the liquid matrix. After introduction of the reinforcement, again stirring is carried out for a period of 60 sec to ensure the clear and fair distribution of the reinforcement. Into a preheated permanent die (125 mm×Φ15 mm) the liquid metal of composite is poured at 750°C.

3. Results and discussions

3.1 Microstructure characterization using SEM/EDAX

The prepared composites are subjected for the preparation of the test samples in the size of 5×10×5mm were taken from the center portion of the work piece. Optical microscopy and SEM equipped with EDX is employed for determining the presence and distribution of WC-Co cermet particles in Al matrix. Optical microstructure shows the formation of inter dendrite precipitates in a matrix of aluminium solid solution and also intermetallic particles are seen in the matrix. A fairly distribution of reinforcement can be observed in the SEM micrographs shown in Figure 2 (a-d). EDX spectra taken at the centre of the particle shows the presence of Al, WC and Co materials shown in Figure 3-8.

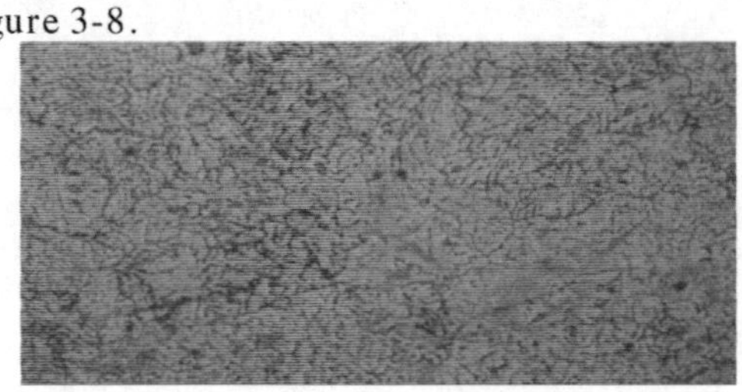

(3)

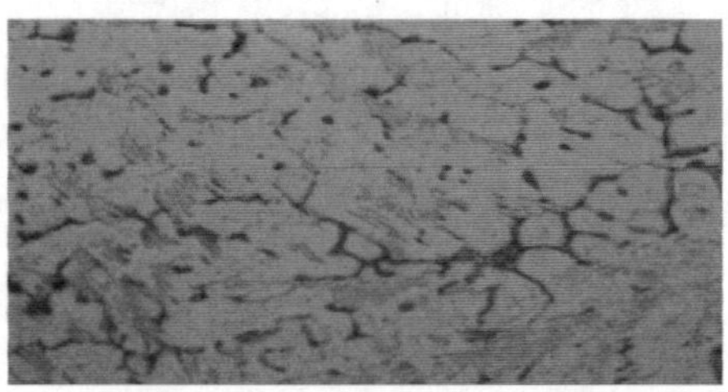

(4)

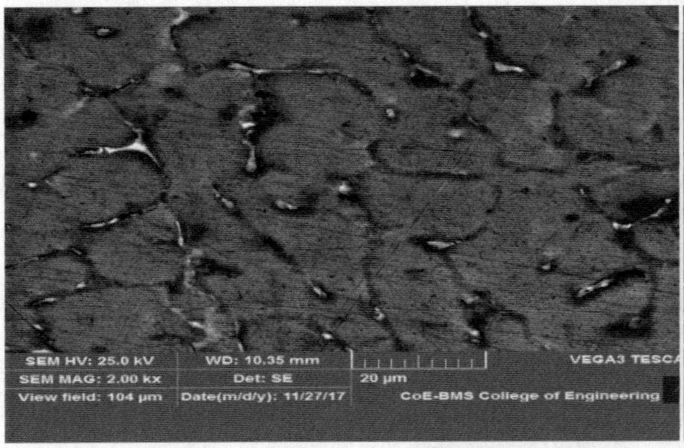

(5)

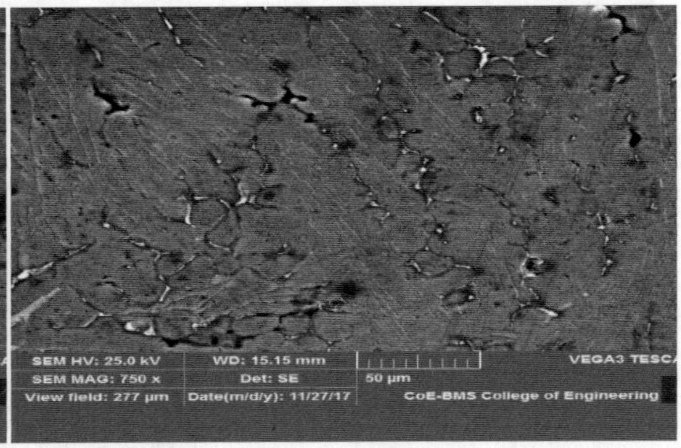

(6)

(7)

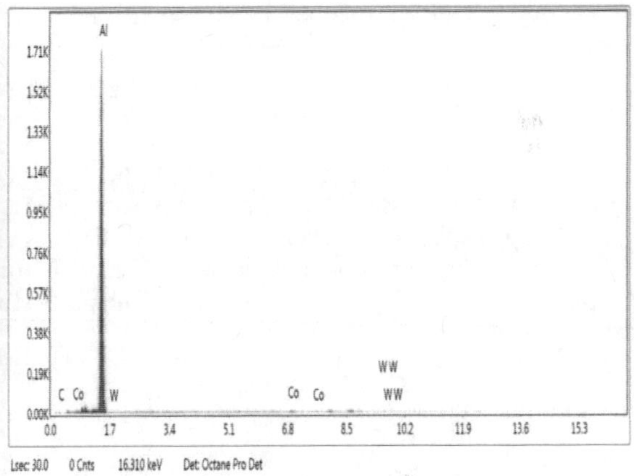

(8)

Figure 3-8. (3-4) Optical micrographs of Al7075+9% WC-Co composite at 100X and 500X magnifications. (5-6) SEM micrographs of Al7075+9% WC-Co composite at 2.00KX and 700X magnifications. (7-8) EDX spectrum of Al7075+9% WC-Co composite.

3.2 Density and Hardness measurements

Table 2. Ascast Matrix and Composite specimens Physical Properties

	Ascast matrix Specimen	Composite Specimen
Theoretical Density (g/cc)	2.92	3.66
Experimental Density (g/cc)	2.87	3.51
Porosity (%)	1.71	4.09
Vickers micro hardness (VHN)	60	156

Archimedes principle is used to determine the experimental density of the specimen while rule of mixture is used to evaluate theoretical density. From Table 2 it is clear that measured density of the composite is came in between the base Al7075 matrix and the reinforcement densities. The porosity level of the composite is higher when compared with the base matrix Al7075. The increase in porosity of the composite is due to the entrapment of the gas during the action of stirring, pouring distance and shrinkage during solidification

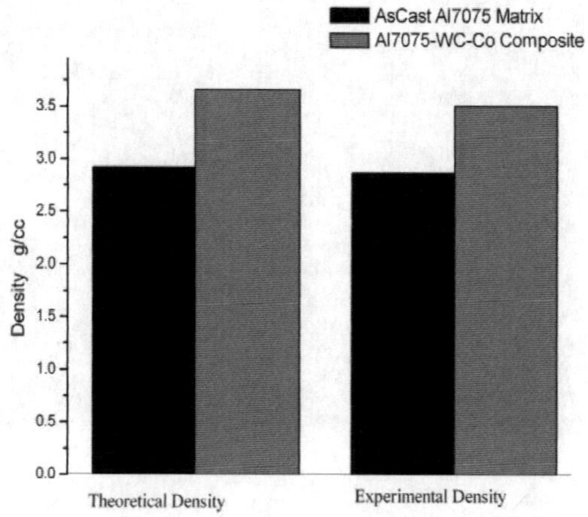

Figure 9. Variation in densities of ascast matrix and composite prepared at 750⁰C.

The prepared composites as well as matrix were subjected to hardness measurements using Zwick micro hardness tester. Measurements were done at a load of 20N on a polished surface with a settle time of 10sec and an average value of 20 readings was reported. Table 2 shows the comparison between Vickers micro hardness values of matrix and composite. Figure 10 shows the drastic improvement in the hardness of the composite due to the fair distribution of the hard ceramics in the matrix.

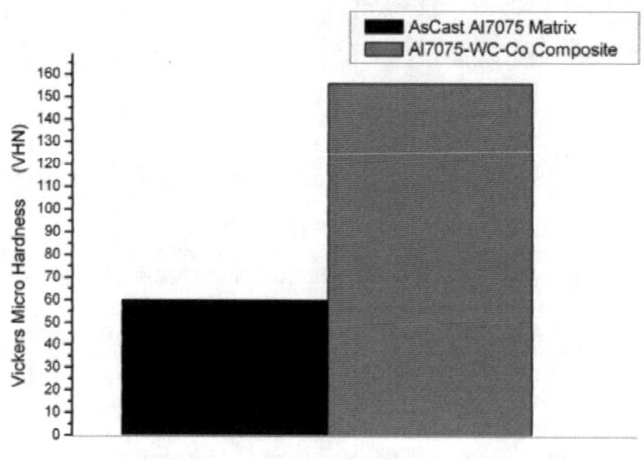

Figure 10. Vickers micro hardness of matrix before and composite prepared at 750⁰C

4. Conclusions

- Cermet reinforced aluminium matrix composite was successfully synthesized by conventional stir casting technique.
- Microstructural evaluation using OM, SEM/EDX studies have indicate fair distribution of the cermet reinforcement particles in the matrix system.
- Major improvement in hardness found in the prepared composite.

IOP Conf. Series: Materials Science and Engineering **404** (2018) 012033 doi:10.1088/1757-899X/404/1/012033

References

[1] V yuvaraja *et al* 2015, Indian journal of Engineering & Material Science, vol. **22**, pp. 51-61
[2] Jerry andrews fabian and b.selvim , IOSR Journal of Mechanical and Civil Engineering (IOSR-JMCE) e-ISSN: **2278-1684**, PP 24-29
[3] Hartajsingh *et al* 2011 journal of engineering research and studies, **P 72-78**
[4] N. Al-aqeeli *et al* 2014 Journal of Nano materials, article **id640750**, 16 pages
[5] Zhang, m.zhu and c.y.wang, 2008, International Journal of refractory metals and hard materials, vol **26**, 329-333
[6] A.wank and b.wielage, 2003, conference on modern wear and corrosion resistant coatings obtained by thermal spraying, Warsaw P November 20-21
[7] Balaji, V *et al*, 2015, Materials Today: Proceedings, Vol **2**(4–5)
[8] Sarada, B. N, Murthy, P. L. S and Ugrasen, 2015, Materials Today: Proceedings, Vol 2(4–5), 2878–2885.
[9] Suresh, S and Moorthi 2013, Procedia Engineering, Vol **64**, 1183–1190.

IOP Conf. Series: Materials Science and Engineering **404** (2018) 012034 doi:10.1088/1757-899X/404/1/012034

Temperature and thermal stress distribution in underwater friction stir welding of aluminium plates

Anil Kumar Maharia, Sonika Sahu and Mohd. Zahid Ansari

Mechanical Engineering Discipline, Indian Institute of Information Technology, Design and Manufacturing, Airport Road, Jabalpur, MP 482 005, India

E-mail: zahid@iiitdmj.ac.in

Abstract. This paper studies the effect of ambient operating conditions on temperature and thermal stresses generated in friction stir welding (FSW) of aluminium alloy plates. AA1050 plate was used with 120×60×3 mm dimensions. Both plates were joined with butt weld in air and underwater ambient conditions. Finite element analysis software COMSOL Multiphysics was used to obtain the transient temperature and thermal stresses induced inside the plates during welding under the two conditions. The effect of frictional heat generated on different parts of weld tool was determined. The results show that weld tool pin has the highest temperature in both the ambient conditions and that the thermal stress and heat affected zone was reduced in case of water.

1. Introduction

Metal joining process is a very important process in any manufacturing industry. The reliability of product depends on the mechanical properties of the joint. Joining process can be carried out either by use of fasteners or by welding. Most of the welding techniques involve local melting of base metals with little or no pressure to form the weld joint. Friction stir welding (FSW) belongs to another class of metal welding called solid-state welding wherein the base metals do not reach up to their melting. Both heat and pressure are applied in this process. This process was developed by The Welding Institute (TWI) in 1991[1] to weld materials that were difficult to weld by conventional methods.This technique was initially applied to aluminium alloys but now can be implemented over a wide range of materials [2-3].

In FSW, a rotating tool with a specially designed pin and shoulder is inserted into the border onedges of sheets or plates to be joined and subsequently traversed along the joint line. The required heat is generated by the friction between the surface of tool and workpiece which are coming in contact. FSW process parameters which affect the joint qualities are tool shape, tool travel speed, spindle rotation speed and applied force [4]. The material which gets soften due to heat, is stirredwith the help of rotating pin.Due to low temperatures involved in FSW, the weld quality is superior toother processes [5]. Though heat generated is less compared to conventional process but it is enough to soften the material. This softening causes decrease in themechanical properties by increasing the dissolution of strengthening precipitate [6-7]. To decrease this phenomenon additional cooling methods are used.By using these coolingmethods, themaximum temperature is decreased,resulting in theenhancedmechanical properties. Water is a easily available fluid with high convective heat transfer coefficient which can be used as a cooling medium for friction stir welding.

Underwater friction stir welding is a variation of normal friction stir welding in which water works as an ambient medium. Tokisue et al. [8] were first to use water in a friction joining technique. In their research, they joined Al-6061 in rotary friction weld in underwater condition and showed that enough heat could be generated even in underwater conditions. They further showed that underwater weld strength could reach up to 86% of base metals while in case of air they were able to get only 82%. Zhang et al. [9] studied the effect of water cooling on dissimilar welding and concluded that stir zone showed much smoother interface and less intermixing.

The temperature distribution is very important phenomena in deciding the different zones of weld-plate. During experimental work, the temperature ismeasured with the help of thermocouple, but the temperature of stir zone or thermo-mechanical affected zone could not be measured because of limitations in placing the thermocouples. Thermal numerical simulation provides a very effective tool for predicting the temperature of these zones. Therefore, researchers are focussing more on numerical simulation using finite element analysis software to characterize the thermal effects in different zones in FSW. Schmidt et al. [10] studied and established a model for heat generation based on the different assumptions of contact condition between plate and shoulder surface. Hongjun et al. [11] studied thermo-mechanical model and presented a methodology for predicting transient thermal and mechanical response. Armansyah et al. [12] worked on temperature distribution in friction stir welding using Altair Hyper Works and compared the simulated results with experimental.

Literature reveals that most of experimental and simulation work is done in the area of friction stir welding in natural conditions and there is not much work reported on underwater simulation. Water can decrease the heat affected zone and soften of material, which can increase the weld strength. Water has a high convective heat transfer capacity which can absorb excess heat generated during welding. This process is widely used in fabrication and repair of ships, submarine, oil transporter tank etc. Thus, the objective of the present work is to study the effect of ambient operating conditions on temperature change and the stresses generated in the plates during FSW involving two thin plates of aluminium Al 1050. The considered ambient is air and water by including the effects of convective heat transfer coefficients in numerical analysis. The simulation is conducted using COMSOL Multiphysics software. Results for transient temperature change and generated thermal stresses are obtained and analyzed.

2. Theory and numerical analysis

Figure 1 shows the schematic of FSW process and the related different thermo-mechanical zones. The total heat (Q_{total}) generated during FSW constitutes of heats generated by the frictionat shoulder surface (Q_{sh}), tool pin side surface (Q_{pin}) and pin tip surface (Q_{tip}) given as: $Q_{total} = Q_{sh} + Q_{pin} + Q_{tip}$.

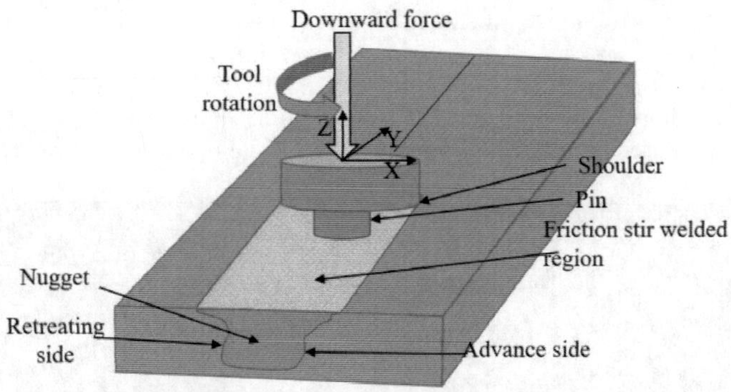

Figure 1. Schematic diagram of friction stir welding

Schmidt et al. [10] gave the following expressions to predict the heat generated by different constituents as:

$$Q_{s\square} = 2\pi\tau(R^3_{s\square} - R^3_{pin}) \quad (1)$$

$$Q_{pin} = 2\pi\omega\tau R^2_{pin} \quad (2)$$

$$Q_{tip} = \frac{2}{3}\pi\omega\tau R^3_{pi} \quad (3)$$

n

IOP Conf. Series: Materials Science and Engineering **404** (2018) 012034 doi:10.1088/1757-899X/404/1/012034

where, ω is angular rotation speed and $\tau = \mu P$ is shear stress generated at tool/shoulder surface, μ is the coefficient of friction and P is the applied contact pressure. R is the radius of different tool sections.

Numerical analysis was performed using commercial finite element analysis software COMSOL Multiphysics V5. From probe tool of software temperature variation in the tool is determined and the thermal stresses were obtained from post-processing of results. Figure 2 shows the geometric model of FSW used in the analysis. The welding was performed virtually by a tool travelling at a speed of 1.75 mm/s on two rectangular aluminium Al-1050 plates each of size 120×60×3 mm. The tool is made of mild steel with geometric properties: R_{sh}= 10 mm, R_{pin} = 3 mm and tip height H_{pin} = 2.7 mm. The tool pin is cylindrical. The welding was done at spindle rotation ω = 1200 RPM with an applied force of 25 kN. The edges of the plates were fully constrained to prevent any motion. Different boundary conditions of convective medium were applied. The convective heat transfer coefficient is taken as 2000 W/(m^{2}K)[13]for water and 15 W/(m^{2}K) [14] for air.

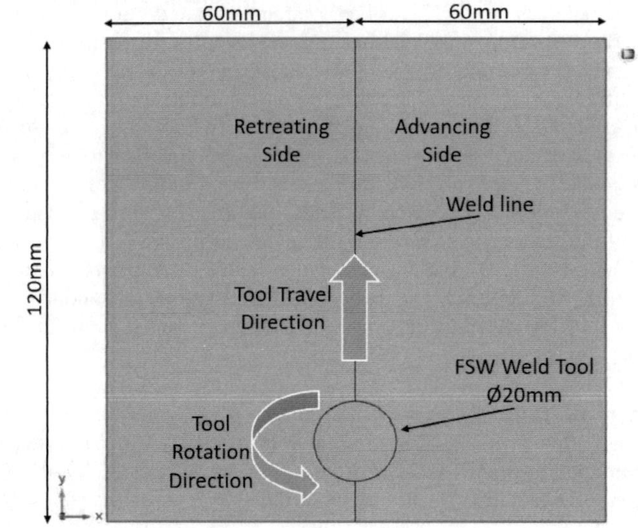

Figure 2. Geometric model of AA 1050 aluminium plates used in numerical analysis

3. Result and discussion

Figure 3 presents the temperature distribution during welding in case of water and air. The maximum temperature is more in case of air which is around 900 K, while in water it is around 700 K.

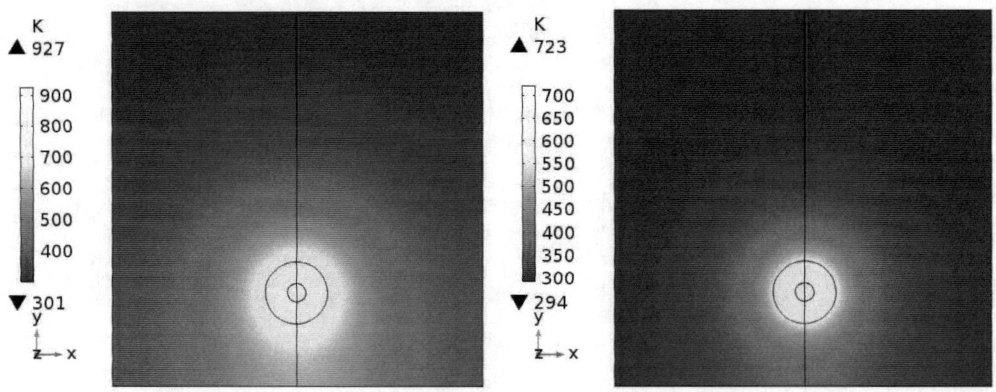

Figure 3. Temperature distribution in aluminium plates in air (left) and underwater (right) conditions

The temperature is less in case of water as excess heat is absorbed by water due to its high convective heat transfer coefficient [4,15]. Water cannot reach below the tool shoulder hence the temperature is more concentrated in this area. Due to high heat absorbing capacity of water there is decrease in peak temperature of weld which decreases the softening phenomena, resulting in enhanced mechanical properties. [6-7].

Figure 4 presents the variation in temperature of different parts of the tool. In tool, pin side will attain maximum temperature in both cases of the operating conditions. It gets heat from flux generated due to tool shoulder, heat due to its own friction and due to plastic deformation. Tool shoulder will always have minimum temperature due to direct contact with convective medium. Figure 5 shows the top view of stresses distribution in aluminium plates. Stresses distribution is mainly dependent on the thermal cycle. This cycle is highly affected by cooling medium as excess heat is absorbed by this medium. The maximum stress in concentrated around the tool pin where the plastic deformation occurs. From the figure 5, we can conclude that aluminium plates have less thermal stresses in case of water [4,14].

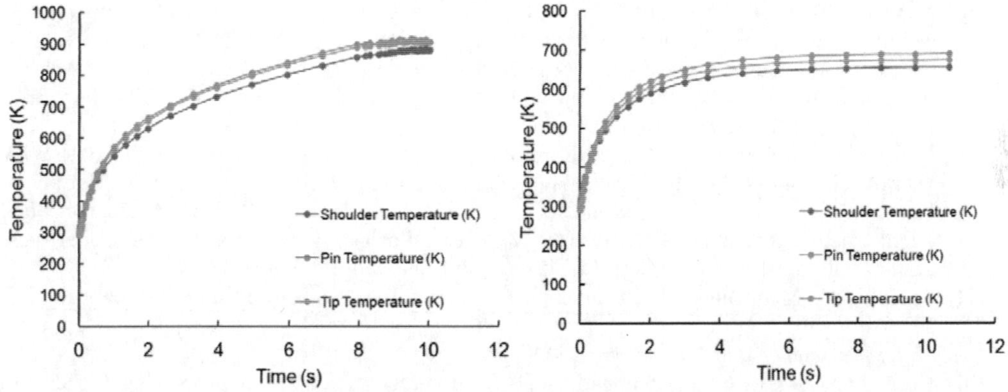

Figure 4. Temperature variation in FSW tool in air (left) and underwater (right) conditions

In case of water, the maximum stress below the tool shoulder is about 786 MN/m^2 while in air it is around 1042 MN/m^2. Initially the stresses generated are almost same in both cases. But when there is a significant rise in temperature heat absorbed by water will be more resulting in less thermal stresses compared to air. The maximum value of thermal stresses in plates was found to be about 270 MN/m^2 in case of air and around 90 MN/m^2 for water.

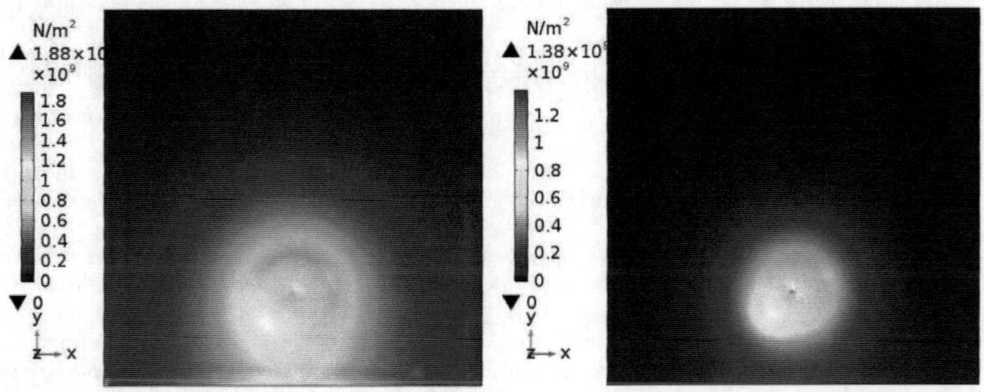

Figure 5. Thermal stresses in weld plates in air (left) and underwater (right) conditions

4. Conclusion

This study investigated the effect of performing FSW process in air and underwater ambient conditions using COMSOL Multiphysics. The frictional heat generated and its flow to different tool parts was also studied. Temperature and thermal stress distributions in plates were obtained. Results showed that compared to the maximum temperature of about 900 K obtained in air, the temperature of about 700 K was found during underwater FSW. Further, it was found that the maximum stress below the tool shoulder in underwater friction stir welding was around 25 percent less as compared to air. By using water as the convective medium excess heat was absorbed. It resulted in less heat affected zone and less thermal stresses. Thus, we can conclude that underwater FSW is better and can improve thermo-mechanical properties.

Acknowledgement

This study was supported by IIITDM, Jabalpur.

References

[1] Thomas W M, Nicholas E D,NeedhamJC, MurchMG, Templesmith P and Dawes C J1991 Patents Friction stir butt weldingGB Patent application no. 9125978.8

[2] Lienert T J, Stellwag J W L, Grimmett B and Warke R W2003 Friction stir welding studies onmild steel *Welding Journal-New York* **82**1-9

[3] NandanR, DebRoy T and Bhadeshia H 2008 Recent advances in frictionstir welding process weldment structure and properties *Progress in Materials Science* **53** 980-1023.

[4] Wahid M A, Khan Z A and Siddiquee A N 2017 Review on underwater friction stir welding A variant of friction stir welding with great potential of improving joint properties *Transaction of Nonferrous Metal Society China***28** 193-219

[5] DawoodH I, Mohammed K S, and Rajb MY 2014 Advantages of the green solid state FSW over the conventional GMAW process *Advances in Materials Science and Engineering*

[6] Khan N Z, Siddiquee A N, Khan Z A and Mukhopadhyay A K 2016 Mechanical and microstructural behavior of friction stir welded similar and dissimilarsheets of AA2219 and AA7475 aluminium alloys*Journal of Alloys and Compounds***695**2902-08

[7] Cabibbo M, Mcqueen H J, Evangelista E, SpigarelliS, Paola M D and Falchero A 2007 Microstructure and mechanical property studies of AA6056 friction stir welded plate *Materials Science and Engineering***460** 86-94

[8] Hofmann, Douglas C, and Kenneth S 2005 Submerged friction stir processing (SFSP) An improved method for creating ultra-fine-grained bulk materials *Materials Science and Engineering***402** 234-41

[9] Zhang H J, Liu H J and Yu L 2012 Effect of water cooling on the performances of friction stir welding heat-affected zone*Journal of materials engineering and performance* **21** 1182-87

[10] Schmidt H, Hattel J and Wert J 2003 An analytical model for the heat generation in friction stir welding *Modelling and Simulation in Materials Science and Engineering* **12** 143-157

[11] Hongjun Li and Di Liu 2014 Simplified Thermo-Mechanical Modeling of Friction Stir Welding with a Sequential FE Method *International Journal of Modeling and Optimization* **4** 410-16

[12] Armansyah I P, Almanar M,Shaari M S B, Jaffarullah M S, Busu N, AbidinM A F Z A and Amlie M A 2014 Temperature Distribution in Friction Stir Welding Using Finite Element Method *International Journal of Mechanical, Aerospace, Industrial and Mechatronics Engineering* **8** 1620-25

[13] Song M and Kovacevic R 2003 Thermal modeling of friction stir welding in a moving coordinate system and its validation *International Journal of Machine Tools and Manufacture* **43** 605-15

[14] Papahn H, Bahemmat P and Haghpanahi M 2016 Effect of cooling media on residual stressesinduced by a solid-state welding: underwater FSW *International Journal of AdvancedManufacturing Technology***83** 1003-12

[15] Zhao Y, Wang Q, Chen H and Yan K 2014 Microstructure and mechanical properties ofspray formed 7055 aluminum alloys by underwater friction stir welding *Materials andDesign***56** 725-30

IOP Conf. Series: Materials Science and Engineering **404** (2018) 012035 doi:10.1088/1757-899X/404/1/012035

Indian based fly ash & international based fly ash: A review paper

A Fuzail Hashmi[1,] and Moin ul Haq[2]

[1,2]Research Scholar, Department of Civil Engineering, AMU, Aligarh, India.

E-mail: fuzailhashmi121@gmail.com

Abstract. Large amount of fly ash are produces from various thermal power plant and their disposal is becoming one of the major environmental challenge. Till few years back, fly ash generated by the burning of coal fired power plants was considered as material of very low value, useful only for landfill. But its usefulness as pozzolanic additive to cement is an important discovery. Therefore, studies have been carried out to reuse high volume fly ash as cement replacement in building materials. This paper presents an overview of the previous studies carried out on the use of high volume Indian & international fly ash as a partial replacement of cement in concrete mixtures. Fresh properties, mechanical properties, abrasion resistance, porosity, water absorption, compressive strength, flexural tensile strength of mortar and concrete mixtures containing high volume fly ash as cement replacement have been reviewed.

1. Introduction

In India, most of the fly ash are of class F type. Out of which 20-25% is being utilized in cement-based materials. In order to increase its percentage utilization, an investigation was carried out for its large scale utilization. Concrete mixtures were prepared by replacing cement with 40, 50, and 60% of fly ash. The concrete construction industry is not sustainable as it consumes huge quantities of important raw materials and the principal binder in concrete is Portland cement, the production of which is a major contributor to greenhouse gas that results in global warming and other climatic change. Many concrete structures suffer from lack of durability which has an adverse effect on the resource productivity of the industry. Because the high-volume fly ash concrete system addresses all these sustainability issues, its use will enable the concrete construction industry to become more sustainable. Continuous research studies by various engineering research laboratories revealed its varied usefulness as an additive for enhancing the various qualities of concrete including its workability, strength and durability. Partial replacement of cement with fly ash in concrete save much of the energy required for production of OPC and also facilitates the economical disposal of millions of tons of fly ash. It has been found that in order to improve the other qualities of concrete like resistance of sulphate attack and thermal cracking, larger percentage of fly ash is to be used in concrete. The fly ash is less permeable to water, protecting reinforcing steel from corrosion and adding to the concrete's durability. It improves the workability of concrete because it aids in the placement of concrete into formwork and around reinforcing steel. Initially, the strength of high volume fly ash concrete is low but with the passage of time, it gains required strength. This can cause problems when slow strength gain means delays in construction. Fly ash admixtures can lengthen the time it takes for concrete to set. Sometimes this is desirable, particularly in hot weather which speeds up concrete set times, but at other times it is inconvenience and can cause delays in construction.

2. General properties of fly ash

Thermal power plants in India are primarily dependent on the combustion of bituminous coal. The low-lime fly ash similar to class F is the primary variety generated in India and significantly smaller volumes of high lime fly ash i.e. class C are also available. Indian low-lime fly ashes are characterized by a relatively higher concentration of SiO_2 and Al_2O_3 and lower contents of Fe_2O_3. The reactivity of fly ashes is dependent on their glass content and other mineral phases present. Indian fly ashes are more crystalline than those obtained in other countries, the glass content ranges from 47.0 to 60.9% [2]. The properties of the Indian fly ashes are highly crystalline, relatively coarse and widely variable, the size of fly ash particles remains in the range of 3–9 μm.

International fly ash consists of inorganic, incombustible matter present in the coal that has been fused during combustion into a glassy, amorphous structure. These fly ash particles are generally spherical in shape and range in size from 2 μm to 10 μm consist mostly of silicon dioxide (SiO_2), aluminium oxide (Al_2O_3) and iron oxide (Fe_2O_3). Fly ash like soil contains trace concentrations of the following heavy metals nickel, vanadium, cadmium, barium, chromium, copper, molybdenum, zinc and lead. The particle sizes in fly ash vary from <1 μm up to more than 100 μm with the typical particle size measuring under 20 μm.

3. Abrasion resistance

R Siddique [7] reported that that abrasion resistance of concrete having cement replacement up to 40% Indian fly ash was comparable to the normal concrete. Beyond 40% fly ash content, concretes exhibited slightly lower resistance to wear as compare to normal concrete. The abrasion resistant is primarily dependent on compressive strength of concrete. Abrasion resistant of concrete was found to increase with increase in age for all mixtures. Siddique [14,15] reported a reduction in the abrasion resistance of concrete specimens with the inclusion of 45% and 50% FA as cement replacement. This reduction increased with increasing fly ash content. Siddique *et al.* [14] reported 73%, 35% and 37.3% reduction in the 7, 28 and 56 days abrasion resistance of concrete with the inclusion of 50% fly ash as cement replacement. Kumar *et al.* [13] reported a reduction in the abrasion resistance of concrete specimens with the inclusion of 50% and 60% fly ash as cement replacement. However, abrasion resistance of high volume fly ash concrete mixture with 60% fly ash at 0.30 w/c ratio was adequate from concrete pavement consideration. Yasasvi Challapalli *et al.* [9] reported that abrasion resistance of fly ash concrete with 30 to 50 percent cement replacement was lower than the no-fly ash concrete. Abrasion resistance of concrete was strongly affected by its compressive strength, irrespective of fly ash content. Fly ash concrete up to 30% cement replacement exhibited abrasion resistance similar to the concrete without fly ash at 28 and 70 days.

The abrasion resistance of international fly ash increases as compressive strength increased. Cengiz Duran Atis [23] reported that for high strength grades (more than 40 MPa) the abrasion resistance of high volume fly ash concrete with 70% replacement of cement was found to be higher than that of counterpart normal concrete and concrete made with 50% fly ash. High volume fly ash concrete showed better abrasion resistance than NPC concrete, particularly at high compressive strengths. Langan *et al.* [10] reported a reduction in the abrasion resistance of concrete specimens with the inclusion of 50% fly ash cement replacement. Nassar [17] examined the abrasion resistance by incorporating 25% and 50% class C fly ash. The samples containing 50% fly ash showed better resistance to abrasion as compared to 25% fly ash samples.

4. Porosity and water absorption

Filho *et al.* reported 32% increment in the concrete specimen porosity with the inclusion of 50% fly ash as cement replacement. The porosity in the concrete containing fly ash is more as compared to normal concrete. This reduces the pore size, and changed the transport properties of chloride ions and reduced the pass charge and the chloride ion diffusion coefficient. P. Nath and Sarker reported that incorporation of fly ash reduced the sorptivity of concrete in early age and it decreased further at six months. Poon *et al.*[11] reported 16% and 47% increment in the porosity of paste specimens at ages of 28 and 90 days respectively by partially replacing cement with 45% fly ash when the w/c ratio was 0.24. The inclusion of HVFA in the matrix increased its porosity and water absorption. Both the porosity and water absorption increased with increasing fly ash content.

5. Compressive strength

Srila Dey [12] had reported that the compressive strength of high volume fly ash concrete decreases with the increase in Indian fly ash content. Replacement of cement by fly ash in different proportions has resulted considerable variation in the properties of fresh concrete like cohesiveness, flow ability of the mix, reduced segregation and bleeding etc. Rafat Siddique [14] found that the compressive strength of fly ash concrete mixes with 10%, 20%, 30%, 40%, and 50% fine aggregate replacement with fly ash, were higher than the control mix at all ages. However, the rate of increase of strength decreases with the increase in fly ash content. Siddique [15] reported a reduction in the compressive strength of concretes with the inclusion of 45% and 50% fly ash as cement replacement, by volume. The reduction in the compressive strength was 40%, 33%, 23% and 18% at ages 7, 28, 91 and 365 days respectively, with the inclusion of 45% fly ash, while this reduction was 42%, 37%, 30% and 23% respectively, with the inclusion of 50% fly ash. Compressive strength of HVFA concrete decreased with the increase in fly ash content at all ages, as shown in Fig. 1. However, at each replacement level of cement with fly ash, an increase in strength was observed with the increase in age due to the pozzolanic reaction of fly ash. Compressive strength of fly ash concrete containing up to 50% cement replacement could be useful in most structural applications.

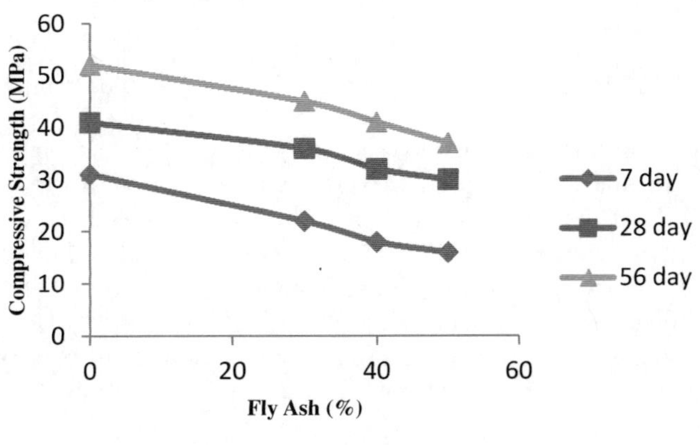

Fig. 1. Effect of FA content on the compressive strength.

Atis [23] reported that HVFA concrete attained satisfactory or higher compressive strength when compared to OPC concrete at later ages but however the rate of gain of strength in initial ages were slow. Nassar [17] observed the compressive strength at 7, 28 and 90 days of moist curing. Mix containing 25% class C fly ash exhibited higher strength than mix with 50% fly ash replacement. However, there was a continuous development of strength with age. At the age of 90 days, 50% fly ash mix surpasses the strength of mix with 25% replacement. Wang and Park [18] observed that at early stage the compressive strength of the fly ash blended concrete is less than the normal concrete. With the replacement of 15% and 25% fly ash in concrete, the strength of fly ash concrete surpasses the strength of normal mix whereas for high volume replacement of 45%-55% of fly ash, it was found that for higher water to cement ratio the compressive strength was less as that of normal concrete. But for less water-cement ratio it surpasses the strength of control mix.

IOP Conf. Series: Materials Science and Engineering **404** (2018) 012035 doi:10.1088/1757-899X/404/1/012035

6. Flexural Tensile Strength

Huang et al. [19] reported that the flexural strength of concrete decreased by incorporating 20% to 80% fly ash. The flexural strength of all mixes were developed continuously with age. However, the flexural strength of all mixes decreased as the percent replacement of fly ash increased as shown in Fig. 2. All the mixtures showed larger increase in the flexural strength between 28 days and 365 days. Jiang *et al.* reported 17% and 23% reduction in the flexural strength of concretes at ages of 56 and 120 days with the inclusion of 60% fly ash as cement replacement respectively. Sofi *et al.* [21] reported 5% reduction in the 28 days flexural strength of concrete specimens with the inclusion of 60% fly ash as cement replacement. Naik *et al.* [22] observed the flexural strength of concrete made with inclusion of 15-70% class C fly ash. Concrete containing fly ash showed continuous development of flexural strength with age. At the age of 3 days, the fly ash mixture (containing fly ash upto 30% cement replacement) showed lower flexural strength than reference mix and at 7 days age, it attained similar flexural strength as compared to reference mix.

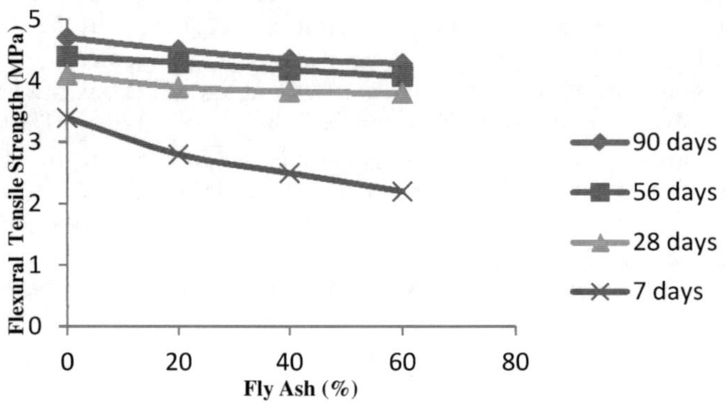

Fig. 2. Effect of fly ash content on the flexural tensile strength.

Kumar *et al.* [13] proportioned the mix by replacing class F fly ash upto 60%. Fly ash obtained from thermal power plant of Dadri in India. Mixtures were prepared at three different water cement ratio (0.3, 0.34 and 0.40). At the age of 7 days and 28 days, the fly ash mixtures showed continuous decrease in flexural strength with the increase in fly ash content for all water cement ratios. In general, the mixtures containing 40% fly ash showed the maximum flexural strength at the age of 90 days and beyond. Siddique examined the properties of concrete by replacing 40%, 45% and 50% of cement by class F fly ash. It was observed that there is a continuous development of flexural strength with age. However, flexural strength of mixtures containing fly ash was less than the control mix.

7. Conclusions

- Indian fly ash are mostly of class F which is a low lime fly ash characterised by a relatively higher concentration of SiO_2 and Al_2O_3 and lower contents of Fe_2O_3 whereas international fly ash are mostly of class C which is a high lime fly ash, generally spherical in shape and range in size from 2 μm to 10 μm consist mostly of silicon dioxide (SiO_2), aluminium oxide (Al_2O_3) and iron oxide (Fe_2O_3).
- High volume of fly ash in the concrete result in the reduction of drying shrinkage in the concrete.
- High volume fly ash concrete (upto 40% cement replacement) showed better abrasion resistant particularly at high compressive strength.
- The inclusion of high volume fly ash in the concrete increases the porosity and the percentage of water absorption, but decreases the permeability. Both the porosity and water absorption increased with increasing fly ash content.

- The compressive strength of high volume fly ash concrete is low at initial ages but with the passage of time i.e. beyond 28 days, it gains satisfactory or better compressive strength as compare to normal concrete.
- High volume fly ash concrete achieve satisfactory tensile strength as compare to normal concrete with the passage of time. However, the flexural strength of all mixes decreases as the percent replacement of fly ash increases.

References

[1] Mehta P.K., "High performance, high volume fly ash concrete for sustainable development", Proceeding of the International Workshop on sustainable development and concrete technology, Sept. 2004, pp. 3-14.

[2] Anjan K. Chatterjee, "Indian Fly Ashes: their characteristics and potential for mechan-ochemical activation for enhanced usability", ASCE-2011.

[3] Das S. K., and Yudhbir "Chemistry and mineralogy of some Indian fly ashes." Indian Concrete Journal, 2003.

[4] Fournier B., Lu D., Charland J. P. and Li, J., "Evaluation of Indian fly ashes for use in HVFA concrete", Indian Concrete Journal, 2004.

[5] Khairul Nizar Ismail, Kamarudin Hussin and Mohd Sobri Idris, "Physical, Chemical & Mineralogical Properties of Fly ash", Journal of Nuclear and Related Technology Vol. 4, Special Edition 2007.

[6] Bisallion A., Rivest M. and Malhotra V. M., "Performance of high-volume fly ash concrete in large experimental monoliths'' ACI, 1994.

[7] Siddique R., "Wear Resistance of High-Volume Fly Ash Concrete", Leonardo Journal of Sciences -2010.

[8] Rafat Siddiquea, Kushal Kapoora, El-Hadj Kadrib and Rachid Bennacerc, "Effect of polyester fibres on the compressive strength and abrasion resistance of HVFA concrete", Construction and Building materials, 2012.

[9] Yasasvi Challapalli and K.Prudhvi, "Strength And Abrasion Resistance of High Volume Fly-Ash Concrete Pavements", IOSR Journal of Mechanical and Civil Engineering, 2016.

[10] Langan, B.V., Joshi, R.C., Ward, M.A., "Strength and durability of concrete containing 50% Portland cement replacement by fly ash and other materials", Canadian Journal of Civil Engg., 1990.

[11] Poon C.S., Lam L. and Wong, Y.L., "A study on high strength concrete prepared with large volumes of low calcium fly ash", Cement Concrete Research, 2000.

[12] Srila Dey, "Study of compressive strength of high volume fly ash concrete with varying proportion of fly ash and silica fume", SSRG International Journal of Civil Engineering (SSRG-IJCE), volume 3, Issue 3, March 2016.

[13] Binod Kumar, G. K. Tike and P. K. Nanda, "Evaluation of Properties of High-Volume Fly-Ash Concrete for Pavements" Journal of Material in Civil Engineering, 2007.

[14] Siddique Rafat, "Effect of fine aggregate replacement with Class F fly ash on the mechanical properties of concrete", Cement and Concrete Research, 2003.

[15] Siddique Rafat, "Performance characteristics of high-volume Class F fly ash concrete", Cement Concrete. Research, 2004.

[16] Siddique Rafat, "Properties of concrete incorporating high volumes of class F fly ash and san fibers", Cement Concrete. Research, 2004.

[17] Nassar R.U.D, Soroushian P. and Ghebrab T. "Field investigation of high volume fly ash pavement concrete", Rresources, conservation and recycling, 2013.

[18] Wang X.Y. and Park K.B., "Analysis of compressive strength development of concrete containing high volume fly ash", Construction and Building Materials, 2007.

[19] Huang C.H, Lin S.K, Chang C.S, Chen, H.J, "Mix properties and mechanical properties of concrete containing very high-volume of Class F fly ash" Construction Building Materials, 2013.

[20] Sofi A., Swathy K. and Srija G., "Toughness study on fly ash based fibre reinforced concrete", International Journal Advance Structural Engineering, 2013.

[21] Naik T.R, Singh S.S, Hussain M.M, "Properties of high performance concrete systems incorporating large amounts of high lime fly ash", Construction and Building Materials, 1995.

[22] Cengiz Duran Atis, "High-volume fly ash concrete with high strength and low drying shrinkage", ASCE, 2002.

Improving Vibration analysis of laminated composite plate by using WU-C2 RBF based Meshfree Method

Peeyush Chandra Vishwakarma[1], Jay Damania[2] and Jeeoot Singh[3*]
[1]PG Student, Mechanical Engineering Department, MMMUT, Gorakhpur
[3*]Associate Professor, Mechanical Engineering Department, MMMUT, Gorakhpur
[2]B.Tech student, Mechanical Engineering Department, SVNIT, Surat
Corresspoonding author e-mail-jeetan_singh@gmail.com

Abstract. The paper presents free vibration analysis of laminated composite plates using WU-C2 RBF based meshfree method. Analysis of vibration response of laminated plate using three transverse shear functions is carried out using simply supported boundary conditions. The governing equations are based on HSDT and Hamilton principle. The effect of side to thickness ratio can be accurately analysed using the present meshless method. Performance of the present method is illustrated by investigating characteristic properties of laminated composite through numerical examples.

Keywords: Meshfree, RBF, laminated composite, vibration

1. Introduction

Composite materials have been playing an essential role in various research areas owing to development of material science and technology. To achieve desired properties, layers of orthotropic materials are assembled systematically in different possible orientation and thus laminated composites are manufactured. Aeronautical industry, Structural industry, Power generation plants, Automobile industry etc. including various other domains involve intensive utilization of various laminated materials. Dai et.al[1] used HSDT for analysis of static deflection and vibration of laminated composite plates . Bui et.al [2] carried out vibration analysis of a laminated plate by application of Kriging interpolation method. Fallah N. et.al[3] studied the natural frequency of laminated plates using a meshless finite volume formulation. Ferriera et.al[4] analyzed the vibration of thick and symmetric laminated plates using FSDT based on MQ radial basis function. Wang Xinwei[5] used discrete singular convolution (DSC) method for investigation of free vibration analysis of three-layer angle-ply symmetric laminated plates. Bouazza et.al[6] analyzed natural frequency of laminated plates by employing nth order HSDT. Kulikov GM et.al[7] implemented the strong formulation for the 3D vibration analysis of laminated plates.

2. Mathematical Formulation

A rectangular laminated composite plate of length 'a', breadth 'b' along x, y-axes respectively and thickness 'h' in z direction. The midplane coincides with x-y plane of the coordinate system. It is assumed that there is perfect bonding between the layers of laminated plates. Figure 1 shows the geometry of laminated plate in the rectangular coordinate system.

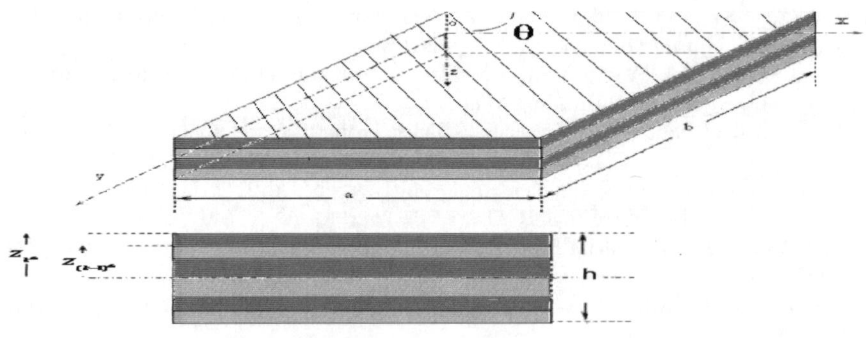

Figure 1 Geometry of laminated plate in rectangular coordinate system

Displacement field at any point on the plate, made up of uniform thickness, is expressed as

$$\begin{Bmatrix} U_x \\ U_y \\ U_z \end{Bmatrix} = \begin{Bmatrix} u_x(x,y) - z\dfrac{\partial u_z(x,y)}{\partial x} + f(z)\,\psi_x(x,y) \\ u_y(x,y) - z\dfrac{\partial u_z(x,y)}{\partial y} + f(z)\psi_y(x,y) \\ u_z(x,y) \end{Bmatrix}$$

here $f(z)$ is the transverse shear function.

Table 1. Transverse shear function $f(z)$.

S.No.	f(z)	Abbreviation	
1	$z.\left[1-\frac{4}{3h^2}z^2\right]$	HSDT 1	Levinsion [8], Reddy[9]
2	$z.m^{(-2(z/h)^2}$, m = 3	HSDT 2	Mantari et.al [10]
3	$z.e^{(-2(z/h)^2}$	HSDT 3	Karama et.al [11]

U_x,U_y and U_z are the in-plane and transverse displacements in the plate at any point (x, y, z) in x, y and z-direction respectively. And u_x, u_y and u_z are the displacements at midplane of the plate at any point (x, y) in x, y, respectively. The functions ψ_x and ψ_y are the higher rotations of the normal to the midplane due to shear deformation about x and y-axes, respectively.

Governing differential equations of the plate are obtained by collecting coefficients of δu_x, δu_y, δu_z, $\delta\psi_x$ and $\delta\psi_y$ which can be expressed as:

$$\delta u_x : \quad \frac{\partial N_{xx}}{\partial x} + \frac{\partial N_{xy}}{\partial y} = \left(I_0 \frac{\partial^2 u_x}{\partial \tau^2} - I_1 \frac{\partial^3 u_z}{\partial x \partial \tau^2} + I_3 \frac{\partial^2 \psi_x}{\partial \tau^2} \right)$$

$$\delta u_y : \quad \frac{\partial N_{xy}}{\partial x} + \frac{\partial N_{yy}}{\partial y} = \left(I_0 \frac{\partial^2 u_y}{\partial \tau^2} - I_1 \frac{\partial^3 u_z}{\partial y \partial \tau^2} + I_3 \frac{\partial^2 \psi_y}{\partial \tau^2} \right)$$

$$\delta u_z : \frac{\partial^2 M_{xx}}{\partial x^2} + \frac{\partial^2 M_{yy}}{\partial y^2} + 2\frac{\partial^2 M_{xy}}{\partial x \partial y} - N_x{}^b \frac{\partial^2 w_0}{\partial x^2} - N_y{}^b \frac{\partial^2 w_0}{\partial y^2} - 2N_{xy}{}^b \frac{\partial^2 w_0}{\partial x \partial y}$$

$$= I_0 \frac{\partial^2 w_0}{\partial t^2} + I_1 \left(\frac{\partial^3 u_x}{\partial x \partial \tau^2} + \frac{\partial^3 u_y}{\partial y \partial \tau^2} \right) - I_2 \left(\frac{\partial^4 u_z}{\partial x^2 \partial \tau^2} + \frac{\partial^4 u_z}{\partial y^2 \partial \tau^2} \right) + I_4 \left(\frac{\partial^4 \psi_x}{\partial x^2 \partial \tau^2} + \frac{\partial^4 \psi_y}{\partial y^2 \partial \tau^2} \right)$$

$$\delta\psi_x : \quad \frac{\partial M^f_{xx}}{\partial x} + \frac{\partial M^f_{xy}}{\partial y} - Q^f_x = \left(I_3 \frac{\partial^2 u_x}{\partial \tau^2} - I_4 \frac{\partial^3 u_z}{\partial x \partial \tau^2} + I_5 \frac{\partial^2 \psi_x}{\partial \tau^2} \right)$$

$$\delta\psi_y : \quad \frac{\partial M^f_{xy}}{\partial x} + \frac{\partial M^f_{yy}}{\partial y} - Q^f_y = \left(I_3 \frac{\partial^2 u_y}{\partial \tau^2} - I_4 \frac{\partial^3 u_z}{\partial y \partial \tau^2} + I_5 \frac{\partial^2 \psi_y}{\partial \tau^2} \right)$$

Force and moment resultants in the plate are expressed as:

$$N_{ij}, M_{ij}, M^f_{ij} = \int_{-h/2}^{+h/2} (\sigma_{ij}, z\sigma_{ij}, f(z)\sigma_{ij})\,dz \quad \text{And}$$

$$Q^f_x, Q^f_y = \int_{-h/2}^{+h/2} (\sigma_{xz}, \sigma_{yz}) \left(\frac{\partial f(z)}{\partial z} \right) dz$$

Laminated plate stiffness coefficients are expressed as:

$$A_{ij}, B_{ij}, D_{ij}, E_{ij}, F_{ij}, H_{ij} = \sum_{k=1}^{nn} \int_{z_k}^{z_{k+1}} \bar{Q}_{ij} \left(1, z, z^2, f(z), z\,f(z), f^2(z)\right) dz \qquad \text{i, j = 1, 2, 6}$$

$$A_{ij} = \sum_{k=1}^{nn} \int_{z_k}^{z_{k+1}} \bar{Q}_{ij} \left(\frac{\partial f(z)}{\partial z} \right)^2 dz \qquad \text{i, j = 4, 5}$$

2.1 Boundary Conditions

Consistent boundary conditions for an arbitrary edge with simply supported edge conditions are obtained and expressed as:

$$at \quad x = 0, a: \; N_{xx} = 0, \; u_y = 0, \; u_z = 0, \; M_{xx} = 0, \; \psi_y = 0$$

$$at \quad y = 0, b: \; u_x = 0, \; N_{yy} = 0, \; u_z = 0, \; \psi_x = 0, \; M_{yy} = 0$$

2.2 Solution Methodology

WU-C2 Radial Basis Function (RBF) $g = (1-cr)^5 \left(8 + 40cr + 48(\mathrm{cr})^2 + 25(\mathrm{cr})^3 + 5(\mathrm{cr})^4\right)$ is used to discretize the governing differential equations in the space domain. Where, $r = \left\|X - X_j\right\| = \sqrt{\left(x - x_j\right)^2 + \left(y - y_j\right)^2}$ is the radial distance between nodes and 'm' and 'c' are shape parameters. In this work, in order to obtain the shape parameter 'c', an expression is as followed: $c = \dfrac{1.01 N^{(1/25)N}}{\sqrt{m\left(\frac{N}{m}\right)}}$

Value of 'm' is taken as 7.8 for WU-C2 RBF after validation and convergence study. The unknown field variables u_x, u_y, u_z, ψ_x and ψ_y appearing in governing differential equations is assumed in terms of radial basis function as:

$$u_x(x,y) = \sum_{j=1}^{N} \alpha_j^{u_x} g\left(\left\|X - X_j\right\|\right); \quad u_y(x,y) = \sum_{j=1}^{N} \alpha_j^{u_y} g\left(\left\|X - X_j\right\|\right) \quad u_z(x,y) = \sum_{j=1}^{N} \alpha_j^{u_z} g\left(\left\|X - X_j\right\|\right)$$

$$\psi_x(x,y) = \sum_{j=1}^{N} \alpha_j^{\psi_x} g\left(\left\|X - X_j\right\|\right); \quad \psi_y(x,y) = \sum_{j=1}^{N} \alpha_j^{\psi_y} g\left(\left\|X - X_j\right\|\right)$$

The governing differential equations are discretized and finally expressed in compact matrix form as:

$$[[\mathrm{K}]]\{\delta\} = [\mathrm{M}]\{\delta\}$$

Here,

$\{\delta\} = [\alpha^{u_x} \; \alpha^{u_y} \; \alpha^{u_z} \; \alpha^{\psi_x} \; \alpha^{\psi_y}]^T$, $[\mathrm{K}]$ is Stiffness Matrix and $[\mathrm{M}]$ is mass matrix

The discretized governing equations for linear free vibration analysis can be written as:

$$\left[[K]_{5N\times5N} + \omega^2 [M]_{5N\times5N}\right]\{\delta\}_{5N\times1} = 0$$

Using standard eigenvalue, the frequency is calculated as:

[V, D] = eig ([K], [M]);

Frequency (ω) = $\sqrt{D}$

3. RESULT AND DISCUSSION

To demonstrate the accuracy of present formulation, an RBF based meshless code in MATLAB is developed following the analysis procedure as discussed above. Several examples have been analyzed and compared with the published results. Based on convergence study, a 15×15 node is used throughout the study.

Table 2 Convergence study for free vibrations of simply supported laminated plate. (E1/E2=40, HSDT 2)

a/h	13x13	15x15	17x17	19x19	Matsunaga[12]
5	10.45818	10.47012	10.49048	10.50049	10.6876
10	15.01458	15.04751	15.05954	15.06507	15.0721
20	17.60653	17.63393	17.64396	17.64824	17.6369
25	18.02616	18.05104	18.06009	18.06392	18.0557
50	18.64462	18.66276	18.66904	18.67157	18.6702
100	18.81421	18.82859	18.83314	18.83473	18.8352

For verification of solutions obtained for vibration from the proposed WU-C2 RBF based meshfree method, a simply supported laminated plate is considered. A four layer [0/90/90/0] laminated composite plate used, has the material parameters: E1 = 40E2, G12 = G13 = 0.6E2, G23 = 0.5E2, υ12 = 0.25, ρ = 1. The convergences of non-dimensional free vibration parameters $\varpi = \left(\frac{\omega a^2}{h}\right)\sqrt{\frac{\rho}{E2}}$ for different theories are shown in Table 2. It can be seen that convergence is achieved within 2 % at 15×15 nodes.

Table 3 Comparison study of non-dimensional frequency parameters $\varpi = \left(\frac{\omega a^2}{h}\right)\sqrt{\frac{\rho}{E2}}$ of simply supported square plate.

Method	a/h					
	5	10	20	25	50	100
HSDT 1	10.449	15.022	17.622	18.042	18.660	18.828
HSDT2	10.470	15.048	17.634	18.051	18.663	18.829
HSDT3	10.459	15.036	17.629	18.048	18.662	18.828
Matsunaga, (M1) [12]	10.688	15.072	17.637	18.056	18.670	18.835
Wu et.al,(W1) [13]	10.682	15.069	17.636	18.055	18.670	18.835
Zhen W et.al, (Z1) [14]	10.729	15.166	17.804	18.240	18.902	19.157
Cho KN et.Al, (C1) [15]	10.673	15.066	17.535	18.054	18.670	18.835

A comparison study of non-dimensional frequency parameter with various span to thickness ratios (a/h) is presented in Table 3. It can be seen that the result of the non-dimensional natural frequency with various span to thickness ratio (a/h) are compared with Matsunaga (M1)[12], Wu et.al (W1)[13] , Zhen W et.al (Z1)[14],Cho et. al(C1)[15]. The present results of different theories HSDT 1, HSDT 2, HSDT 3 have good agreement with other published results.

Table 4 A non-dimensional frequency parameter $\varpi = \left(\frac{\omega a^2}{h}\right)\sqrt{\frac{\rho}{E2}}$ of a [0/90/90/0] SSSS laminated square plate (a/h = 10).

Theory	E1/ E2			
	10	20	30	40
HSDT 1	9.813367	12.16927	13.79573	15.02168
HSDT2	9.818596	12.18196	13.81549	15.04751
HSDT3	9.816299	12.17657	13.8071	15.03642

In Table 4, the first normalized frequency obtained from the present different theories HSDT1, HSDT2 and HSDT3 is listed with respect to $a/h = 10$ and various modulus ratios $E1/E2 = 10$, 20, 30, 40.

4. CONCLUSION

Using different HSDT, free vibration response of laminated plate is determined. The present WU-C2 RBF based meshfree method is in good agreement with the published results. For different theories, in case of thin plates the effect on frequency parameter is negligible, while there is considerable effect in case of thick plates. The effect of the span to thickness ratio decreases for $a/h \geq 40$.

REFERENCES

[1] Dai KY, Liu GR, Lim KM, Chen XL 2004 A mesh-free method for static and free vibration analysis of shear deformable laminated composite plates *Journal of Sound Vibration* 269 633

[2] Bui TQ, Nguyen MN, Zhang C 2011 An efficient meshfree method for vibration analysis of laminated composite plates *Computational Mechanics* 48 175

[3] Fallah N, Delzendeh M.2017 Free vibration analysis of laminated composite plates using meshless finite volume method. *Eng Anal Bound Elem 2018* 88 132

[4] Ferreira AJM, Roque CMC, Jorge RMN 2005. Free vibration analysis of symmetric laminated composite plates by FSDT and radial basis functions *Comput Methods Appl Mech Engg* 194 4265

[5] Wang X 2017. Free vibration analysis of angle-ply symmetric laminated plates with free boundary conditions by the discrete singular convolution. *Composite Structure* 170

[6] Bouazza M, Kenouza Y, Benseddiq N, Zenkour AM 2017 A two-variable simplified nth-higher-order theory for free vibration behavior of laminated plates. *Composite Structure* 182 533

[7] Kulikov GM, Plotnikova SV, Kulikov MG 2017 Strong SaS formulation for free and forced vibrations of laminated composite plates. *Composite Structure* 180 286

[8] Mark Levinson 1980 An accurate simple theory of the statics and dynamics of elastic plates *Mechanics Research Communications* 7(6) 343

[9] Reddy JN 1984 A Simple Higher-Order Theory for Laminated Composite Plates *Journal of Applied Mechanics* 51 745

[10] Mantari JL, Oktem AS, Guedes Soares C 2011 Static and dynamic analysis of laminated composite and sandwich plates and shells by using a new higher-order shear deformation theory *Composite Structure* 94 37

[11] Karama M, Afaq KS, Mistou S 2009 A new theory for laminated composite plates *Proc Inst Mech Eng Part J Mater Des Appl* 223 53

[12] Matsunaga H 2000 Vibration and stability of cross-ply laminated composite plates according to a global higher-order plate theory *Composite Structure* 48 231

[13] Wu Z, Chen R, Chen W 2005 Refined laminated composite plate element based on global-local higher-order shear deformation theory *Composite Structure* 70 135

[14] Zhen W, Wanji C 2000 Free vibration of laminated composite and sandwich plates using global–local higher-order theory *Journal of Sound Vibration* 298 333

[15] Cho KN, Bert CW, Striz AG 1991 Free vibrations of laminated rectangular plates analyzed by higher order individual-layer theory *Journal of Sound Vibration* 145 429

[16] Wu C-P, Chen W-Y 1994 Vibration and stability of laminated plates based on a local high order plate theory *Journal of Sound Vibration* 177 503

IOP Conf. Series: Materials Science and Engineering **404** (2018) 012037 doi:10.1088/1757-899X/404/1/012037

Buckling Analysis of FGM Plates by thin plate spline RBF based Meshfree Approach

Amit Kumar[1], Rahul Kumar[2], Jay Damania[3] and Jeeoot Singh[4*]
[1]PG Student, Mechanical Engineering Department, MMMUT, Gorakhpur
[2]Reasearch Scholar, Mechanical Engineering Department, SVNIT,Surat
[3]B.Tech Student, Mechanical Engineering Department, SVNIT,Surat
[4*]Associate Professor, Mechanical Engineering Department, MMMUT, Gorakhpur
Corresponding author e-mail: jeetan_singh@yahoo.com

Abstract- Present article attempts to analyze the buckling response of FGM plates. Hamilton Principle is employed to derive the governing differential equations of theories. Discretization of governing differential equation is done using a thin plate spline RBF based meshless method. For analysis of buckling load, numerical results are obtained. Further, for purpose of validation, obtained results are compared with analytical results wherein effect of grading index and span to thickness ratio is observed.

Keywords: Meshfree, Buckling, RBF, FGM Plate

1. Introduction

FGM's have played a very intensive role in the recent development of structure and materials. They are heterogeneous materials in which the material properties vary continuously from one phase to other phase. High resistance against heat and capacity to withstand high temperature gradient has led to its intensive engineering application in the field of aircraft industry, power generation plants, computer parts, robot, nuclear power plants, automobile industry, biomedical industry etc. Owing to their unique properties many researchers have inclined their efforts on Buckling analysis of FGM materials.Yang et al.[1] carried out buckling analysis of FG plates using FSDT. Bodaghi and Saidi [2]analyzed buckling behavior of FGM plates by implementation of higher order shear deformation theory. Hosseini-Hashemi et al. [3], by implementationof Mindlin plate theory, studied buckling of in-plane loaded isotropic rectangular plates using various boundary conditions. Ghannadpour et al. [4], using finite strip method analyzed buckling behaviour of rectangular FG plates with thermal load. Thai and Choi [5],by implementation of a simple refined theory, analyzed the buckling behaviour of functionally graded plates using classical plate theory.Implementing small strain elasticity theory, Uymaz and Aydogdu [6] analyzed buckling response of rectangular FG plates usingvarious axial loadings and boundary conditions. Chu et al. [7] used RBF collocation method and Hermite collocation method, initially developed for buckling analysis of functionally graded thin plates with inplane material inhomogenity. Do and Thai [8] used radial point interpolation function to combine Kirchhoff plate theory and meshfree method and hence determined various results for isotropic and sandwich plates.

2. Mathematical Formulation

The coordinate system is considered consists of a rectangular plate with edge lengths 'a' and 'b' along x and y-axes, and thickness 'h' along z-axis respectively. Here the mid plane coincides with x-y plane. Figure 1 shows the geometry of FGM plate.

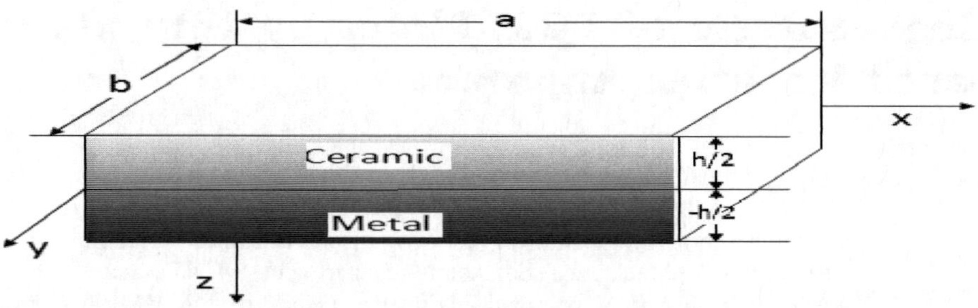

Figure 1 Geometry of rectangular FGM plate in rectangular coordinate system

The displacement field at any point in plate of uniform thickness is expressed as Reddy[9]

$$\begin{Bmatrix} U \\ V \\ W \end{Bmatrix} = \begin{Bmatrix} u_0(x,y) - z\dfrac{\partial w(x,y)}{\partial x} + z.\left[1 - \dfrac{4}{3h^2}z^2\right]\psi_x(x,y) \\ v_0(x,y) - z\dfrac{\partial w(x,y)}{\partial y} + z.\left[1 - \dfrac{4}{3h^2}z^2\right]\psi_y(x,y) \\ w_0(x,y) \end{Bmatrix} \quad (1)$$

The governing differential equations of plate using Hamilton's principle is obtained by collecting the coefficients of displacement field variables and expressed as:

$$\delta u_0 : \frac{\partial N_{xx}}{\partial x} + \frac{\partial N_{xy}}{\partial y} = 0 \quad (2)$$

$$\delta v_0 : \frac{\partial N_{xy}}{\partial x} + \frac{\partial N_{yy}}{\partial y} = 0 \quad (3)$$

$$\delta w_0 : \frac{\partial^2 M_{xx}}{\partial x^2} + \frac{\partial^2 M_{yy}}{\partial y^2} + 2\frac{\partial^2 M_{xy}}{\partial x \partial y} = N^b_{xx}\frac{\partial^2 w}{\partial x^2} + N^b_{yy}\frac{\partial^2 w}{\partial y^2} + 2N^b_{xy}\frac{\partial^2 w}{\partial x \partial y} \quad (4)$$

$$\delta \psi_x : \frac{\partial M^f_{xx}}{\partial x} + \frac{\partial M^f_{xy}}{\partial y} - Q^f_x = 0 \quad (5)$$

$$\delta \psi_y : \frac{\partial M^f_{xy}}{\partial x} + \frac{\partial M^f_{yy}}{\partial y} - Q^f_y = 0 \quad (6)$$

Where, N^b_{xx}, N^b_{yy} and N^b_{xy} are the applied in-plane compressive loading in x and y direction and shear loading respectively.

$$N_{ij}, M_{ij}, M^f_{ij} = \int_{-h/2}^{+h/2} (\sigma_{ij}, z\sigma_{ij}, f(z)\sigma_{ij})dz \quad (7)$$

$$Q^f_x, Q^f_y = \int_{-h/2}^{+h/2} \left(\sigma_{xz}, \sigma_{yz}\right)\left(\frac{\partial f(z)}{\partial z}\right)dz \quad (8)$$

The material property gradation considering power low is expressed as:

$$E(z) = [E_c - E_m]\left(\frac{2z+h}{2h}\right)^n + E_m \quad (9)$$

Here 'E' denotes the modulus of elasticity, 'h' is the thickness of the plate, E_m and E_c are the corresponding Young's modulus of elasticity of metal and ceramic, respectively, and 'z' is the thickness coordinate $(-h/2 \le z \le h/2)$, 'n' is volume fraction exponent or grading index

The stiffness coefficients are:

$$A_{ij}, B_{ij}, D_{ij}, E_{ij}, F_{ij}, H_{ij} = \int_{-h/2}^{h/2} \left\{ \left(\left([Q_{ij}^c - Q_{ij}^m] \right) \left(\frac{2z+h}{2h} \right)^n + Q_{ij}^m \right) \left(1, z, z^2, f(z), z f(z), f^2(z) \right) \right\} dz \quad \textbf{(10)}$$

for i, j = 1,2,6

$$A_{ij} = \int_{-h/2}^{h/2} \left\{ \left(\left([Q_{ij}^c - Q_{ij}^m] \right) \left(\frac{2z+h}{2h} \right)^n + Q_{ij}^m \right) \left(\frac{\partial f(z)}{\partial z} \right)^2 \right\} dz \quad \textbf{(11)}$$

for i, j = 4, 5

Here superscript c stands for ceramic and m stands for metal.

The boundary conditions for simply supported edges are:

$$x = 0, a : v = 0; \psi_y = 0; w = 0; M_{xx} = 0; N_{xx} = 0$$
$$y = 0, b : u = 0; \psi_x = 0; w = 0; M_{yy} = 0; N_{yy} = 0 \quad \textbf{(12)}$$

3. Solution methodology

Governing differential equations Eqn. and boundary conditions Eqn. (12) are discretized using polynomial radial basis function. A two dimensional rectangular domain with any arbitrary boundary having NB boundary nodes and NI domain interior nodes is shown in **Figure 2**. For present analysis thin plate spline radial basis function

$$g\left\| X - X_j, m \right\| = \log(r) r^{2m} \quad \textbf{(13)}$$

Where, $r = \left\| X - X_j \right\| = \sqrt{(x - x_j)^2 + (y - y_j)^2}$

and 'm' is shape parameter. In present analysis value of m=3 is taken after validation and convergence study. The field variables (displacements) in terms of radial basis function are expressed as:

$$u = \sum_{j=1}^{N} \alpha_j^u g\left(\left\| X - X_j \right\| \right), v = \sum_{j=1}^{N} \alpha_j^v g\left(\left\| X - X_j \right\| \right), w = \sum_{j=1}^{N} \alpha_j^w g\left(\left\| X - X_j \right\| \right), \quad \textbf{(14)}$$
$$\psi_x = \sum_{j=1}^{N} \alpha_j^{\psi_x} g\left(\left\| X - X_j \right\| \right), \psi_y = \sum_{j=1}^{N} \alpha_j^{\psi_y} g\left(\left\| X - X_j \right\| \right)$$

$\delta = \alpha_j^u, \alpha_j^v, \alpha_j^w, \alpha_j^{\psi_x}, \alpha_j^{\psi_y}$ are unknown coefficient to be determined. In the eigenvalue problems, the objective is to obtain eigenvalues (λ) and corresponding eigenvectors. The eigenvalue problem can be expressed as:

$$\begin{bmatrix} L \\ B \end{bmatrix}_{5N \times 5N} \{\delta\}_{5N \times 1} = \lambda \begin{bmatrix} A \\ 0 \end{bmatrix}_{5N \times 5N} \{\delta\}_{5N \times 1} \quad \textbf{(15)}$$

Here $\{\delta\} = [\alpha^{u_0} \ \alpha^{v_0} \ \alpha^{w_0} \ \alpha^{\psi_x} \ \alpha^{\psi_y}]^T$

Equation (15) is solved by standard eigen solvers of computational software in order to obtain eigenvalues and eigenvectors.

4.Results and discussion

A simply supported isotropic plate has been taken for convergence and validation purpose. The buckling load parameter is non-dimensionalised as $\lambda = N_x a^2 / \pi^2 (Eh^3 / 12(1 - \nu^2))$ where

$E=3\times10^6$ psi and $\nu=0.316$. The results are obtained using AHSDT. From table 1 which is depicted in Figure 2 ,it can be seen that for all the four modes, a good convergence is achieved which is within 1 percent for most of the cases beyond 13x13 nodes.

Table 1 Non dimensional buckling load parameters $\lambda = N_x a^2 / \pi^2 (Eh^3 / 12(1-\nu^2))$ of simply supported isotropic square plate under uniaxial load.

Mode	Number of Nodes						
	7x7	9x9	11x11	13x13	15x15	17x17	Timoshenko[10]
1	44.094	40.629	39.959	39.723	39.616	39.559	39.4761
2	59.491	60.879	61.294	61.448	61.515	61.548	61.9814
3	90.064	103.186	106.801	108.101	108.658	108.931	109.656
4	164.761	161.655	159.523	158.620	158.191	157.965	157.904

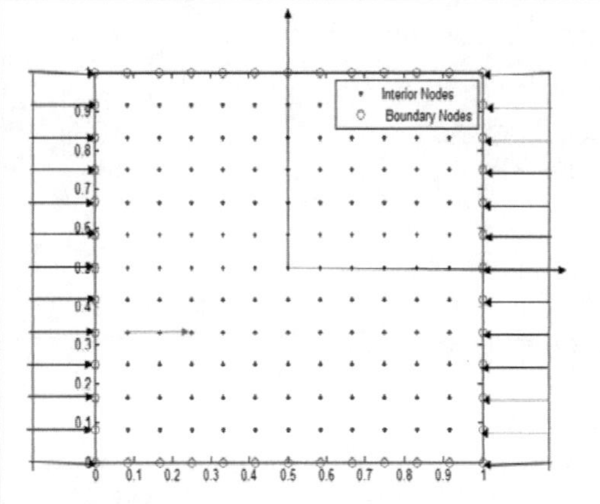

Figure-2: Uniaxial compressive load on FGM plate.

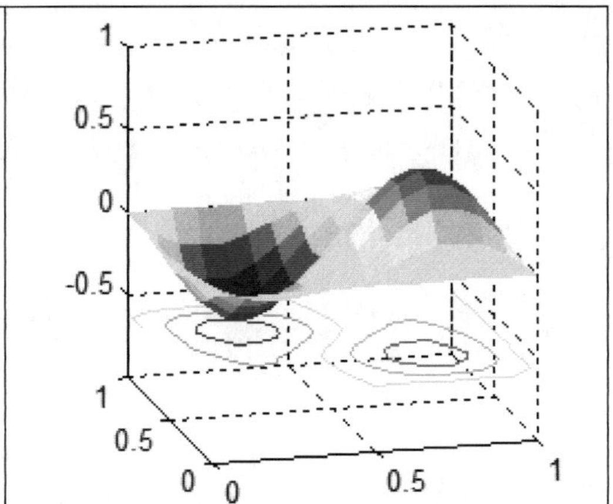

Figure-3: Deformed shape of mode-4 for FGM plate.

Figure 2 Non dimensional buckling load parameter(λ) of simply supported isotropic square plate under uniaxial load.

For further analysis functionally graded materials are taken with following properties:
FGM: Ceramic- $E_c = 200 GPa, \nu_c = 0.3$, Aluminum (Al) - $E_m = 70 GPa, \ \nu_m = 0.3$
The plate is of square shape with unit dimensions.
The buckling load parameter is non-dimensionalised as $\lambda = N_x a^2 / (E_m h^3)$

It can be seen from **Figure 4** that the critical buckling parameter increase from thick to thin FGM plate and after a/h=40, the effect of critical buckling parameter almost negligible. Buckling load parameter decreases as grading index increases and effect decreases as the value of 'n' increases.

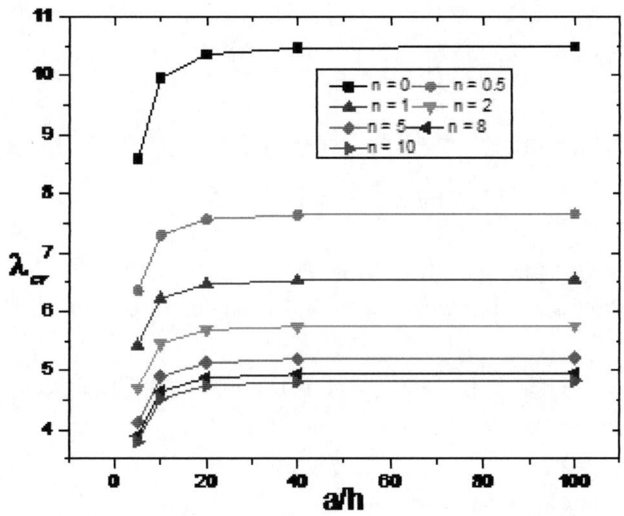

Figure 4 The critical buckling load parameter λ_{cr} of square FGM simply supported plate under uniaxial load.

5. Conclusion

Buckling loads of thick to thin FGM plates are obtained. The results herein agree well with the analytical results indicating the efficacy of present solution methodology. Effect of span to thickness ratio becomes less dominant after a/h=40.Buckling load parameter decreases as grading index increases. It is more dominant initially and effect decreases as index increases.

6. References

[1] Yang J, Liew KM, Kitipornchai S 2005. *Second-order statistics of the elastic buckling of functionally graded rectangular plates. Compos Sci Techno* **65** P 1165–75.

[2] Bodaghi M, Saidi AR 2010. *Levy-type solution for buckling analysis of thick functionally graded rectangular plates based on the higher-order shear deformation plate theory. Appl Math Model* **34** P 3659–73.

[3] Hosseini-Hashemi S, Khorshidi K, Amabili M 2008. *Exact solution for linear buckling of rectangular Mindlin plates. J Sound Vib* **315** p 318–42.

[4] Ghannadpour SAM, Ovesy HR, Nassirnia M 2012. *Buckling analysis of functionally graded plates under thermal loadings using the finite strip method. Comput Struct* **108–109** P 93–9.

[5] Thai H-T, Choi D-H 2012. *An efficient and simple refined theory for buckling analysis of functionally graded plates. Appl Math Model* **36** P 1008–22.

[6] Uymaz B, Aydogdu M 2013. *Three dimensional mechanical buckling of FG plates with general boundary conditions. Compos Struct* **96** P 174–93.

[7] Chu F, He J, Wang L, Zhong Z 2016. *Buckling analysis of functionally graded thin plate with in-plane material inhomogeneity. Eng Anal Bound Elem* **65** P 112–25.

[8] Do VNV, Thai CH 2017. *A modified Kirchhoff plate theory for analyzing thermo-mechanical static and buckling responses of functionally graded material plates. Thin-Walled Struct* **117** P 113–26.

[9] Reddy JN 1984. *A Simple Higher-Order Theory for Laminated Composite Plates. J Appl Mech* **51** P 745–52.

[10] S._Timoshenko_and_J.M._Gere__1963_*Theory_of_elastic_stability_*.**pdf n.d.**

Analysis of Abrasive Wear Characteristics of Brass 60/40 Using MATLAB

M S Khan[1], P Maurya[2]and K Moeed[2]
[1]Associate Professor, Department of Mechanical Engineering, Integral University, Lucknow, U.P, India
[2]Assistant Lecturer, Mechanical Engineering Department, NITK, Surathkal, Karnataka, India [3] Principal University Polytechnic, Integral University, Lucknow, U.P, India
E-mail: preetimaurya.maurya4@gmail.com

Abstract. Abrasive wear behavior of Brass 60/40 was investigated using pin on disc wear testing machine. The abrasive wear test of brass was conducted by taking two variables viz. orientation and normal load. The normal load varies from 5N to 35N with different orientation angles (0°, 30°, 45°, 60°, and 90°), the speed of rotating abrasive disc kept constant at 1500 rpm. The experimental data was obtained from 5N to 20N only, by using pin on disc wear testing machine, however the data beyond 20N were validated with help of MATLAB. For proper validation of result for wear (mass loss), the exploratory data analysis (EDA) and regression analysis (RA) are used. These two analyses were conducted on MATLAB by using curve fitting tool and surface fitting tool. After the completion of experimental work, the results showed reduction in wear (mass loss) under constant loading condition with increase in orientation angle while increased in wear (mass loss) under constant orientation angle with increase in normal load.

Keywords: Abrasive wear, MATLAB, EDA.

1. Introduction

Brass is used in wide range products due to its excellent electrical and thermal conductivity, good strength, good formability and resistance to corrosion. Pipe and pipe fittings are commonly manufactured from the same due to its excellent corrosive resistance property. [1]. Wear leads to loss of material and increase in clearance in moving parts. In general, wear properties play a major role in deciding the life of movable parts which are in contact with each other, and so, it must be considered in the design and development of engineering components. Improvement in wear resistance under heavy loading conditions can help in application of copper alloy (brass) for bearing material used in several industries like in automobile and power plant industries. [2]

This paper is mainly focused on several factors, which directly affects the working condition of bearing brass. The main objective of this paper to use pin-on-disc setup, which can check wear rate or loss of mass of the selected specimen at different normal loading condition (5N-20N) and orientation angle (0°, 30°, 45°, 60°, 90°) [2-4].

IOP Conf. Series: Materials Science and Engineering **404** (2018) 012038 doi:10.1088/1757-899X/404/1/012038

2. Experimental Material and procedure: -

Brass 60/40 was obtained in the form casted ingots; the dimensions of the specimens are 1cm x 1cm with a length of 4.5 cm. To wear the surface of the specimen stone abrasive disc was used. After preparation of the specimen, wear test analysis was carried with pin-on-disc apparatus by considering several normal loading conditions in steps of 5 N, 10 N, 15 N & 20 N under different orientation angle. Mass of the tested specimens was measured using weighing machine with an accuracy of 0.001g. After a travel time of 5 min. against the grinding disc, the sample was taken out carefully and debris's were removed from the valleys of the specimen to measure the exact wear (mass loss). This was continued for five times with same specimen and at same position. After that next specimen was taken for next test condition i.e. 30° angle and 5N applied load and so on. The tests were conducted for five different orientations namely 0°, 30°, 45°, 60°, 90°. Thus a total of 25 readings are taken for one particular load. This was continued for load of 10N, 15N and 20N respectively. The cumulative effect of mass loss was taken for calculating the wear mass. [5-13].

3. Results and Discussion

3.1Wear mass loss and wear morphology

For complete analysis of wear morphology, the experimental and theoretical results were compared.

3.2Experimental Results

After, analyzing the mass loss (gm) for different loading condition (5N, 10N. 15N and 20N) following results are come out from the experimental data-

- While exceeding the load beyond 5N amount of wear loss also increases.
- While considering different orientation angle from 0° to 90°, the amount of mass loss decreases.
- After analyzing the whole process, it seems that at 5N loading condition and at 90° orientation angle the possibility of mass loss is very less (experimental result – approximate 0.303mg).

3.3Theoretical results – using MATLAB

Obtained experimental results were verified using exploratory data analysis (EDA) and regression analysis (RA). For EDA, curve fitting tool implemented with the help of MATLAB. Following steps mainly used to validate the result by comparing with the experimental data and exceeding the value of load beyond 20N [19].

- Exploratory data analysis using curve fitting tool with the help of MATLAB.
- Mathematical equation formulation (Regression analysis) using curve fitting tool and calculate the theoretical value for each loading condition.
- Plot a curve for complete data using surface fitting tool.
- Testing of adequate of the model developed (Statistical significance model).
- Validate the result by exceeding the value of load beyond 20 N.

3.3.1Curve Fitting and surface fitting tool (MATLAB)

This function provides an application and functions for fitting curves and surfaces to data. Exploratory data analysis, pre-process and post process data, compare candidates' models and remove outliners were perform on MATLAB. To achieve the desire result following theories are mainly used in curve fitting tool-

- Exploratory Data Analysis (EDA)
- Regression Analysis (RA) (Mathematical equation formulation)

Firstly, EDA method has been performed for detail analysis. For this first, Frame regression table that, which model (Linear or Non-Linear Model), is best suited for experimental results by comparing the value of R-square and Adjusted R-Square and then by using that model plot a best fitted curve for required model.

3.3.1.1. EDA for 5N load. From Table 1 it is observed that wear mass decreases from 0.916 gm to 0.303 gm as applied orientation angle on the specimen increases from 0° to 90°when loading condition is 5N.

Table 1. Regression table for 5N load

Angle (degree)	Mass Loss (mg)	Equation on name						
		SSE	R-Square	DFE	AdjRSuare	RMSE	Coefficient	Type of fit
0	0.916							
30	0.908							
45	0.75							
60	0.589	$1.5361*10^{-4}$	0.9994	1	0.9977	0.0124	4	Polynomial
90	0.303							

3.3.1.2. EDA for 10N load. From Table 2 it is observed that wear mass decreases from 1.027 gm to 0.405 gm as applied orientation angle on the specimen increases from 0° to 90° when loading condition is 10N. But the only difference is that its value is more than the previous result.

Table 2. Regression table for 10N load

Angle (degree)	Mass Loss (mg)	Equation name(Poly3/Poly4)						
		SSE	R-Square	DFE	Adj R-Suare	RMSE	Coefficient	Type of fit
0	1.027							
30	0.999	$3.2171*10^{-30}$	0.9929/1	1/0	0.9717/NaN	0.0445/NaN	04/05	Polynomial
60	0.641							
90	0.405							

IOP Conf. Series: Materials Science and Engineering **404** (2018) 012038 doi:10.1088/1757-899X/404/1/012038

3.3.1.3 EDA for 15 N Loads. From Table 3, it is observed that wear mass decreases from 1.098 gm to 0.512 gm as applied orientation angle on the specimen increases from 0° to 90° when loading condition is 15N. Again the value of wear loss is more than 5N and 10N load.

Table 3. Regression table for 15N load

Angle (degree)	Mass Loss (mg)	Equation name(Poly3/Poly4)						
		SSE	R-Square	DFE	Adj RSuare	RMSE	Coefficient	Type of fit
0	1.098							
30	1.068							
45	0.968	$6.7300*10^{-30}$	0.9875/1	1/0	0.9502/NaN	0.0566/NaN	04/05	Polynomial
60	0.703							
90	0.512							

3.3.1.4. EDA for 20 N Loads. From Table 4, it is observed that wear mass decreases from 1.262 gm to 0.638 gm as applied orientation angle on the specimen increases from 0° to 90° when loading condition is 15N. Again the value of wear loss is more than 5N and 10N load.

Table 4. Regression table for 20N load

Angle (degree)	Mass Loss (mg)	Equation name(Poly3/Poly4)						
		SSE	R-Square	DFE	Adj RSuare	RMSE	Coefficient	Type of fit
0	1.262							
30	1.247							
45	1.031	$1.2942*10^{-30}$	0.9968/1	1/0	0.9872/NaN	0.0299/NaN	04/05	Polynomial
60	0.865							
90	0.638							

IOP Conf. Series: Materials Science and Engineering **404** (2018) 012038 doi:10.1088/1757-899X/404/1/012038

After successful completion of EDA method, the RA (regression analysis) method implemented. It is mainly established the relationship between two variables. After detail analysis, Non linear third degree polynomial equation whose value of R-Square and adjusted R-Square is quite good as compare to forth degree polynomial equation-

$$W = 1.016 - 0.07447 \times L + 0.02484 \times A + 0.01064 \times L^2 - 0.0004322 \times L \times A - 0.0008015 \times A^2 - 0.0003211 \times L^3 + 1.583 \times 10^{-5} \times L^2 \times A - 1.344 \times 10^{-7} \times L \times A^2 + 5.319 \times 10^{-6} \times A^3 \quad \text{Eq. (1)}$$

Here w= wear loss in gm, A= orientation angle in degree, L= Applied load in N

After formulation of mathematical equation, detail comparison between experimental and theoretical results. Best comparison between experimental and theoretical result is visible in curve 5, curve 6, curve7 and curve 8.

3.3.1.5- RA for 5N load.

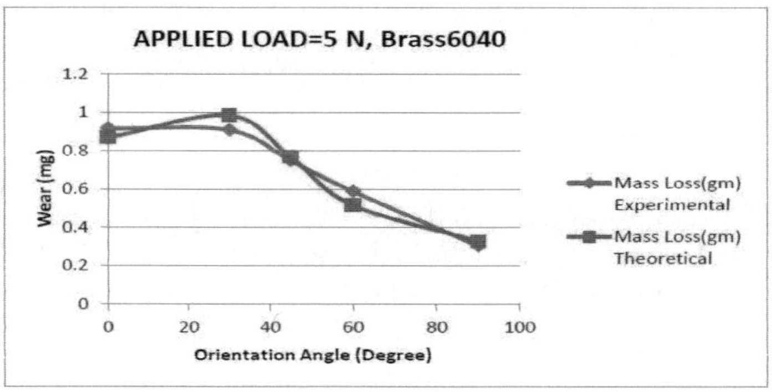

Curve 1: Variation of wear rate with different orientation angle of the work piece for 5N applied load

3.3.1.6- RA for 10N load.

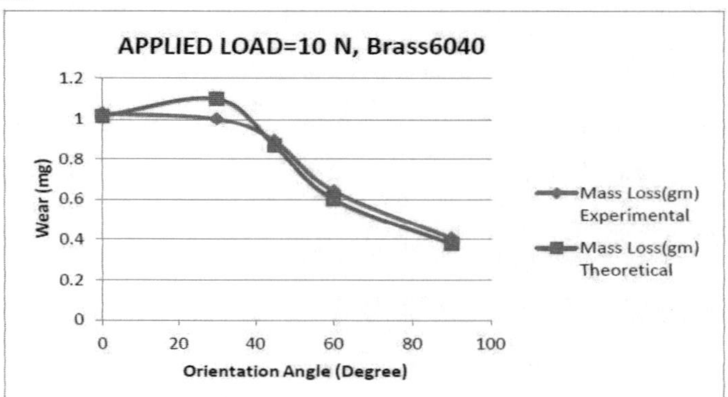

Curve 2: Variation of wear rate with different orientation angle of the work piece for 10N applied load

3.3.1.7- RA for 15N load.

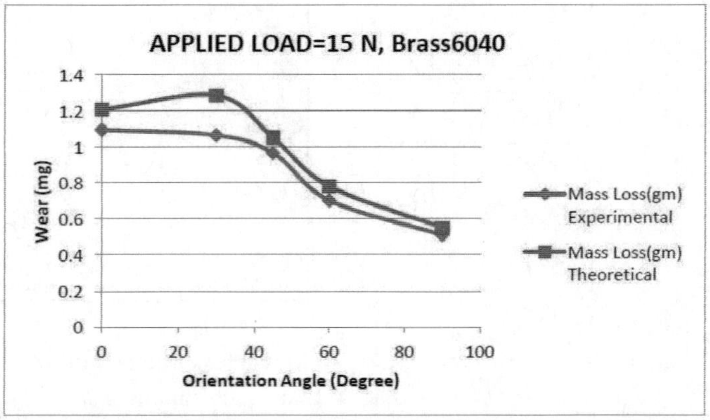

Curve 3: Variation of wear rate with different orientation angle of the work piece for 15N applied load

3.3.1.8- RA for 20N load.

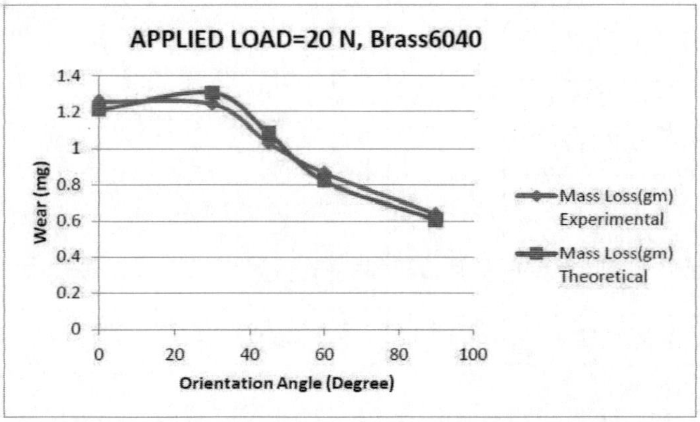

Curve 4: Variation of wear rate with different orientation angle of the work piece for 20N applied load

3.4Validate the result by exceeding the value of load beyond 20N-

For better result, exceed the value of load beyond 20N whether the required equation give the best result for this or not. Table (8, 9, 10) and curve (8, 9, 10) simply signifies that at initial stage it increases very rapidly up to 30° than decreases up to 90° angle and then increases at a very higher rate at an angle of 180°.

3.4.1- Under 25N loading condition.

Angle(degree)	Mass Loss (mg)
0	0.7871
30	0.9242
60	0.4742
90	0.2988
120	1.2598
150	4.2186
180	10.0372

Table 5. Theoretical values for 25N load

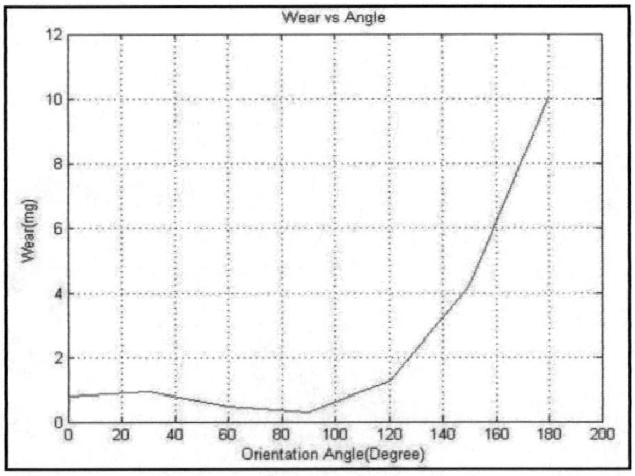

Curve 5. Variation of wear with respect to orientation angle

3.4.2- Under 30N loading condition

Angle(degree)	Mass Loss (mg)
0	-0.3118
30	-0.1095
60	-0.4956
90	-0.6082
120	0.4143
150	3.4335
180	9.3112

Table 6: Theoretical Values for 30N load

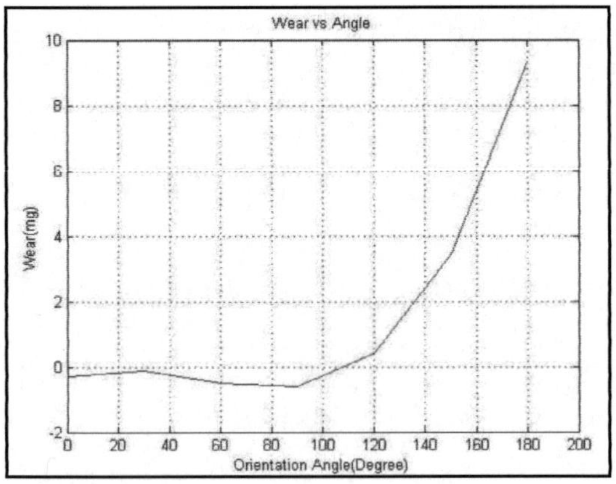

Curve 6: Variation of wear with respect to orientation angle.

3.4.3- Under 35N loading condition

Angle (degre)	Mass Loss (mg)
0	-2.3236
30	-2.0324
60	-2.3308
90	-2.3569
120	-1.2491
150	1.8541
180	7.8146

Table 7: Theoretical Values for 35N load

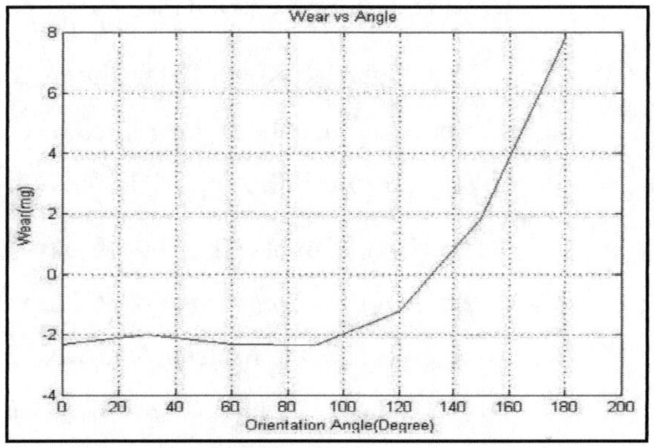

Curve 7: Variation of wear with respect to orientation angle

4. Conclusions

The abrasive wear of Brass 60/40 processed by pin-on-disc apparatus have been studied with varying loads (5N to 35N) and different orientation angles (0°, 30°, 45°, 60°, 90°), and the conclusions are summarized as follows:

- Non-Linear third order polynomial equation gives the best fit for the data which has been taken experimentally.
- The R-Square value of the empirical model and Adjusted R- Square value are very close to each other, which give the significant amount of variability to the wear loss. Predicted value of Rsquare is 0.9933 and adjusted R-square is 0.9872.
- A value closer to 1 indicates a better fit. Here the value of R-square is 0.9872 which indicates the good fit.
- Percentage deviation between measured wear loss and predicted wear loss is 7.27% which is 0.2711% in magnitude, which is less than the similar results reported in previous paper [6].
- As the fit is good, the equation (1) can be used for wide range of applied loads, hence it can be used to predict the wear of brass 60/40 beyond 20 N.
- When applied load increases, then mass loss (wear) first slowly increases up to 30° then increases very rapidly at higher rate by varying orientation angle from 0° to 90°.
- After exceeding the loading condition beyond 20N, wear (mass loss) increases at a very fast rate.
- For future references some modifications are required in the model so that after exceeding the loading condition beyond 35N load, it will give the better result.
- In case of brass the chances of wear is less under varying loading condition than other material because of its physical properties.

References:

1. Megh H.J, 2000. *Resistance to wear of aluminium bronze*, Cda publication 126.
2. MohdShahdab Khan, Zahir Hasan, Yaqoob Ali Ansari, June 3, 2014. "Statistical analysis for the Abrasive wear behavior of Al6061". *Journal of Minerals and Materials Characterization and Engineering,* **Vol. 2**, pp. 292-299. doi:10.4236/jmmce.2014.24034.
3. Yu.H.J and Bhok S.D, 1990. "Development of prototype abrasive wear for tillage tool material". *Tribology International,* **Vol. 23**, Issue 5, pp. 309-316.
4. Megh H.J, 2000. *Resistance to wear of aluminium bronze*, Cda publication 126.
5. Satpal Sharma, March 12, 2014. "Parametric Study of Abrasive wear of Co-CrC based flame spayed coating by response surface methodology". *Tribology International*, **Vol. 75**, pp. 39-50.
6. MohdShadab Khan, Zahir Hasan, Syed Mohd Farhan, Dec 30, 2013. "Effect of orientation and applied load on abrasive wear property of BRASS 60:40". *Journal of Minerals and Materials Characterization and Engineering,* **Vol. 2**, pp. 49-53.doi.org/10.4236/jmmce.2014.21008.
7. BeleteSirahbizuYigezu, M. M Mahapatra, P. K. Jha, March 7, 2013. "On modeling the abrasive wear characteristics of *in situ* Al–12%Si/TiC composites". *Material and design*, **Vol. 50**, pp. 277284.
8. AyhanOrhan, September, 2012. "Modelling of abrasive wear resistance by means of neural network of Al-SiCp composites produced by cold pressing method". *Australian Journal of Basic & Applied Sciences*, **Vol. 6**, Issue 9, pp. 264.
9. Y. Sahina, V. Kilicli, May 26, 2011. "Abrasive wear behaviour of SiCp/Al alloy composite in comparison with ausferritic ductile iron". *Wear International Elsevier*, **Vol. 271**, Issues 11-12, pp. 276274.
10. C.S. Ramesh, R. Keshavamurthy and S. Pramod, August, 2011. "Abrasive Wear behaviour of NiP coated Si_3N_4 reinforced Al6061 composites". *Journal of Materials Processing Technology*, **Vol. 211**, Issue 8, pp. 1423-1431.
11. M.D. Bermúdez, F.J. Carrión, C. Espejo, E. Martínez-López, J. Sanes, August, 2011. "Abrasive wear under multiscratching of polystyrene plus single-walled carbon nanotube nanocomposites". *Applied Surface Science*, **Vol. 257**, Issue 21, pp. 9073-9081.

IOP Conf. Series: Materials Science and Engineering **404** (2018) 012038 doi:10.1088/1757-899X/404/1/012038

12. N.S.M. El-Tayeb, T. C. Yap, P. V. Brevern, September 17, 2010. “Wear characteristics of titanium alloy Ti54 for cryogenic sliding applications”. Tribology International, **Vol. 43**, Issue 12, pp. 2345-2354.

13. Mohit Sharma, JayashreeBijwe , Peter Mitschang, September 25, 2010. “Abrasive wear studies on composites of PEEK and PES with modified surface of carbon fabric”. *Tribology International*, **Vol 44**, Issue2, pp. 81-91.

IOP Conf. Series: Materials Science and Engineering **404** (2018) 012039 doi:10.1088/1757-899X/404/1/012039

Control of semi-active suspension system using PID controller

Mohammad Saad[1], Shagil Akhtar[2], A.K. Rathore[3], Qudsiya Begume[3] and Mohd Reyaz-ur-Rahim[4]
[1]Undergraduate B.Tech, Department of Mechanical Engineering, Shri Ramswaroop Memorial Group of Professional Colleges, Lucknow
[2]B.Tech, , Department of Mechanical Engineering, Integral University, Lucknow
[3]Assistant Professor, Department of Mechanical Engineering, Shri Ramswaroop Memorial Group of Professional Colleges, Lucknow
[4]Assistant Professor, Departmentof Mechanical Engineering, Integral University, Lucknow

E-mail:saad.sidd29@gmail.com,shagilakhtar@gmail.com,akhil_lko@rediffmail.com, qudsiya.03@gmail.com, rahimreyaz@gmail.com

Abstract. This paper aims to scrutinize the performance of semi-active suspension system using a PID controller by applying Bouc-Wen model for magneto-rheological fluid. The mathematical model is being simulated in the MATLAB Simulink environment. Two crucial road excitation were given to the suspension system namely sine and random input. The results of the both have been presented and their hysteric behaviour has been studied. Results show that by applying a controller the desired effect can be reached.

1. Introduction

The classification of a vehicle suspension is done in three types i.e. passive suspension system, active suspension system and semi-active suspension system, which is based on operational mode to increase the safety of the vehicle, ride comfort and overall performance of the vehicle.

Earlier the passive suspension systems were used in the automobile which lacks feedback mechanism in the damper with no controlled springs. On the contrary, active suspension system has a wide range of operation as adjust the damping force with the help of actuator and thus adapting variations of roads. Active systems were studied in 1960 and many further models were proposed. The modification in active suspension is necessary due to the fact that it required a high supply of current. Then came semi-active suspension system in the 1970s and gained popularity because of output damping force is highly adaptive with very lesser power consumption than active suspension system. Therefore, semi-active suspension system usually falls in between the active suspension system and semi-active suspension system. The fluid used in the semi-active systems is electro-rheological (ER) fluid and magneto-rheological (MR), these types of fluids have micron size particles of iron suspended in the oil. These vary the stiffness of the fluid when the voltage is applied to the fluid as the iron particles align themselves in the externally applied magnetic field. The semi-active control is studied by Lai et al. [1] and Yao et al. [2] In this paper Bouc-Wen model of the MR fluid is implemented which was used by Spencer et al. [3].

IOP Conf. Series: Materials Science and Engineering **404** (2018) 012039 doi:10.1088/1757-899X/404/1/012039

Daniel Fischer et al [5] created a mathematical model using variable damper system for active suspension systems as well as designed active parts for fault detection. Eski et al. [6] designed a robust control, based on neural networking, for the full suspension system and compared it with standard PID. G. Priyandoko et al. [7] implemented a hybrid control technique for the control of active suspension system using adaptive neuro active force and skyhook technique. M. S. Kumar [8] made an active suspension system for quarter car model using PID control strategy. P. E. Uys et al. [9] investigated spring and damper settings for optimal ride comfort studied at different road profiles. Daniel A. Mantaras et al. [10] studied the kinetic behaviour of McPherson type steering suspension and determined caster steer and camber angle.
In this paper, a Proportional-Integral-Derivative (PID) controller is used which controls the error between the damping force and the force from the road. The simulation model of the quarter suspension system of an automobile is done in MATLAB Simulink environment with the application of Bouc-Wen fluid model with a controller.

This paper tries to study the performance of semi-active active suspension control in a two degree of freedom model of quarter car.

2. System Modelling

The mathematical model of semi-active suspension is based on following observations

- The suspension system is assumed to be linear or approximately linear for quarter car considering two degrees of freedom.
- The minor forces are neglected such as flex in the vehicle joint and gear system etc. thereby reducing the complexity of the vehicle.
- The damping and stiffness property is also possessed by the tyre material which is to consider.

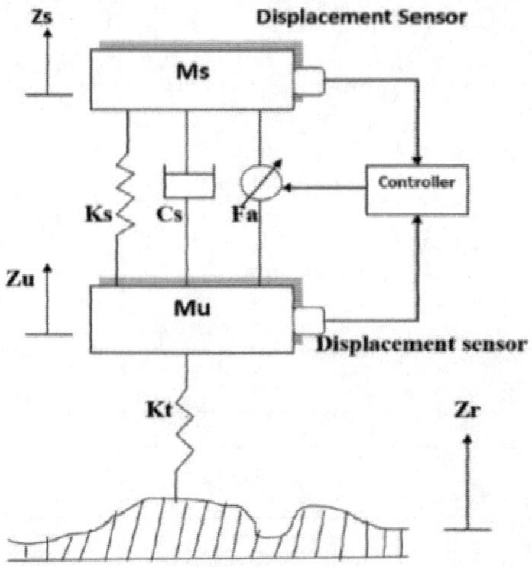

Fig 1. Quarter Semi-active suspension system [4]

From fig 1. following equation has concluded

$$m_s\ddot{z}_s + k_s(z_s-z_u) + c_s(\dot{z}_s-\dot{z}_u) + U(t)=0 \quad (1)$$

$$m_u\ddot{z}_u + k_s(z_u-z_s) + c_s(\dot{z}_u-\dot{z}_s) + k_t(z_u-z_r)= U(t) \quad (2)$$

Bouc-Wen equation is as follows:

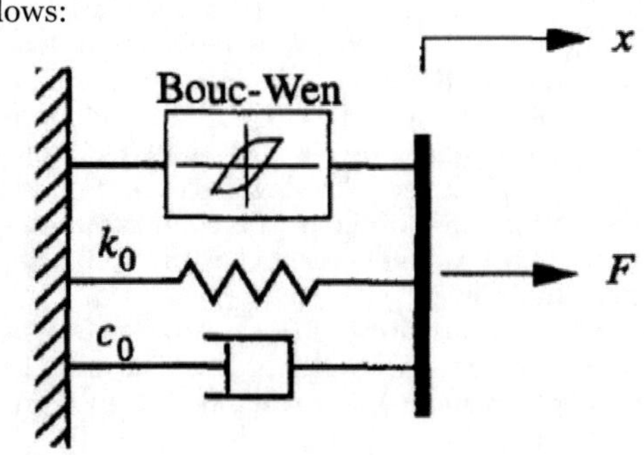

Fig 2. Bouc-Wen model for MR dampers [3]

From the fig 2. the following equations have been derived

$$F = c_o\dot{x} + k_o(x - x_c) + \alpha z \tag{3}$$

$$\dot{z} = -\gamma|\dot{x}|z|z|^{n-1} - \beta\dot{x}|z|^n + A\dot{x} \tag{4}$$

where,
F=Damping force
x=Initial displacement of the piston
k_o=stiffness of fluid
c_o=damping coefficient of the fluid
By adjusting parameters γ, β, and A of the above equations (3) and (4) one can control the smoothness of pre-yield and post-yield region of the MR fluid.

z is an evolutionary variable that encompasses the history of the dependence of response. The parameters depend upon the voltage with is given by equation as:

$$\alpha = \alpha_a + \alpha_b u \tag{5}$$

$$c_o = c_{oa} + c_{ob} u \tag{6}$$

where u is the output obtained from the first-order filter and given relation in terms of voltage as:

$$\dot{u} = -\eta(u - v) \tag{7}$$

For this purpose, the following values are taken

Sr. No.	Parameters	Values
1	γ	1 cm^{-2}
2	β	-1 cm^{-2}
3	A	1.5
4	n	2
5	K_o	300 N/cm
6	c_{oa}	4400 N-s/cm
7	c_{ob}	442 N-s/cm
8	η	190
9	α_a	10872 N/cm
10	α_b	49616 N/cm

Table 1. Parameters used for Simulink model of Bouc-Wen

Parameter values for equation (5), (6) and (7) are taken as from Table 1. The control implementation of the Bouc-Wen in suspension is done with the help of error measurement. The error is continuously taken as the difference between the force experienced on the sprung mass due to road excitation and the damping force generated by the simulated Bouc-Wen model. This error is further reduced by the PID controller.

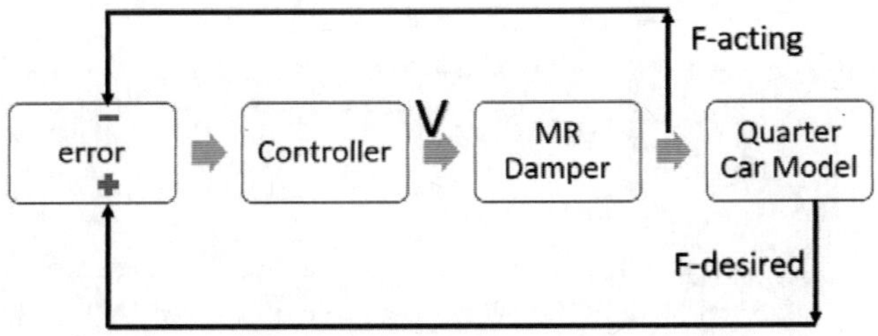

Fig 3. Flowchart of the control system

3. Simulation under MATLAB SIMULINK

The mathematical model of the quarter semi-active suspension system is made in the simulation block by using the equations. The modelling in Simulink is based on the equations (1) and (2). The model is made based on Fig 3.

Sr. No.	Parameter	Symbol	Quantities
1	Mass of vehicle body	M_s	2500 kg
2	Mass of the tyre and suspension	M_u	320 kg
3	The coefficient of suspension spring	K_s	80000 N/cm
4	The coefficient of tyre material	K_t	50000 N/cm
5	Damping coefficient of the dampers	C_s	15020 N-s/m

Table 2. Parameters used in the system simulation. Adapted from [4].

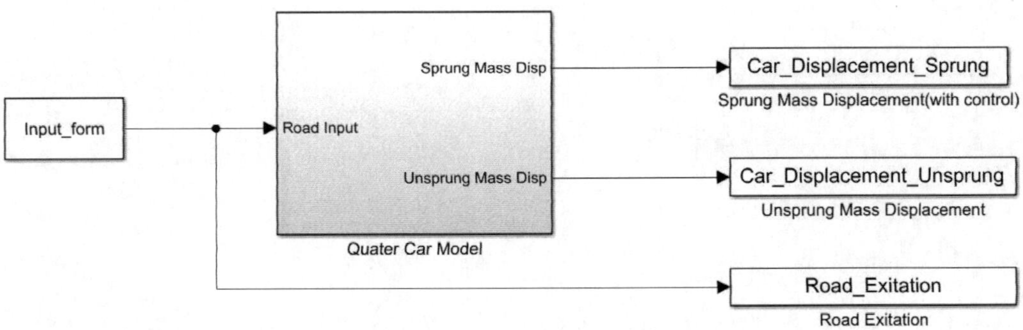

Fig 4. Simulation Model of Quarter Semi-Active Suspension System

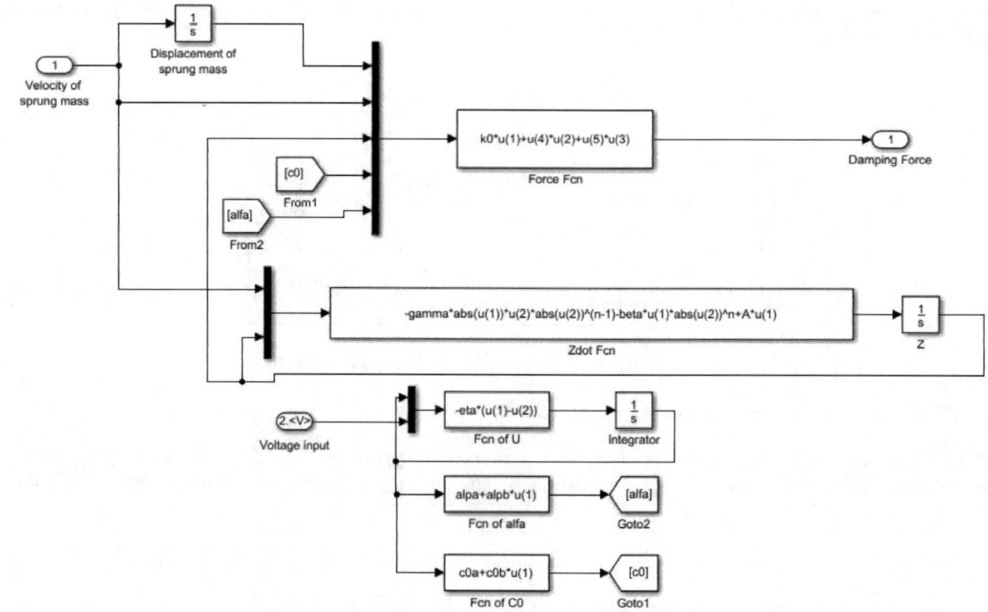

Fig 5. Simulink Model of Bouc-Wen

In Fig 4. Simulation model of sprung mass and un-sprung mass have been made under quarter car model mask. The input forms are sine excitation and random excitation. The sprung mass is experiencing the force excitation generated from the road and that force is generating a specific damping force via a voltage command. The error of force transmitted to sprung mass and damping force produced is reduced by the controller. Thus, gives the desired suspension affect.

In Fig 5. the simulation model of Bouc-Wen fluid model is applied to the suspension. The Bouc-Wen model covers the non-linear hysteric behaviour of the MR fluid and velocity of piston – damping force graph is shown below.

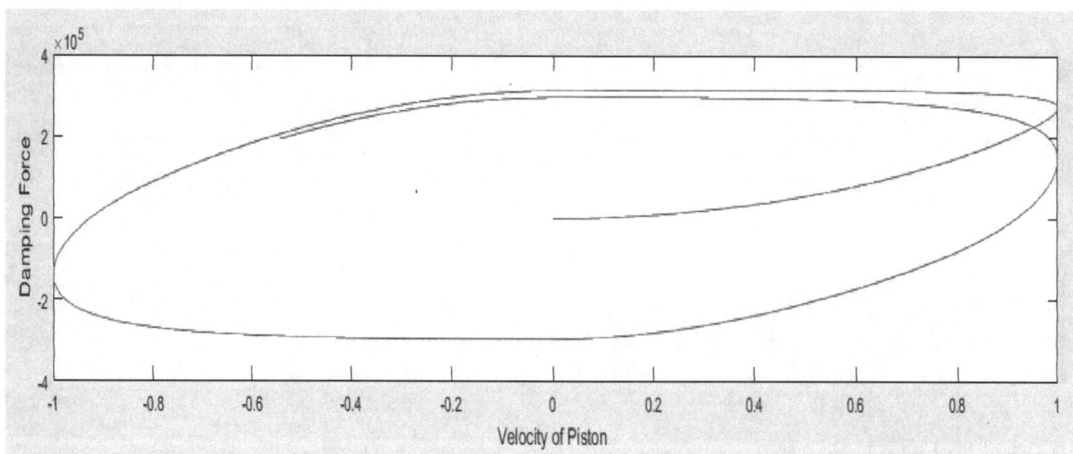

Fig 6. The hysteric behaviour of MR fluid

IOP Conf. Series: Materials Science and Engineering **404** (2018) 012039 doi:10.1088/1757-899X/404/1/012039

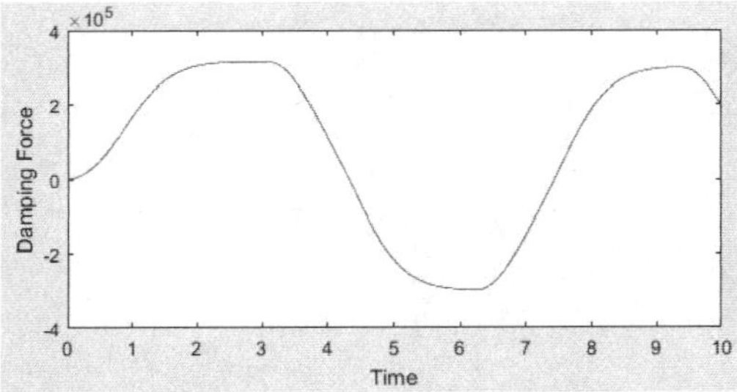

Fig 7. Damping Force v/s time graph

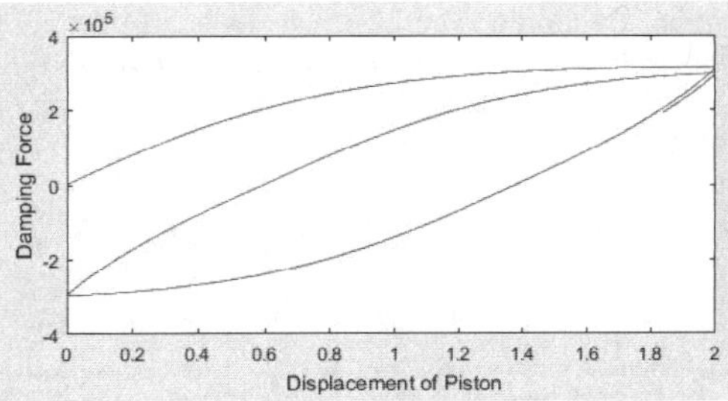

Fig 8. Damping Force v/s Displacement of piston

The Fig. 6 gives the hysteric behaviour which is observed in the MR fluid and how the damping force varies with the rate of change of piston velocity. Similarly, in Fig 7 and Fig 8 the damping force with respect to time and displacement of the piston is studied.

4. Simulation Results

The error between the external force from the road and the available damping force is measured by the PID controller. The damping force is produced by the Bouc-Wen model in accordance with the applied force. For tuning the controller, the following values are taken [4]

K_p=800
K_d=5000
K_i=3500

These values are taken by adjusting manually in the Simulink considering settling time and peak overshoot.
The road profile in the suspension system is to be taken as first the sine input and then the random input on the output is observed through the simulation.

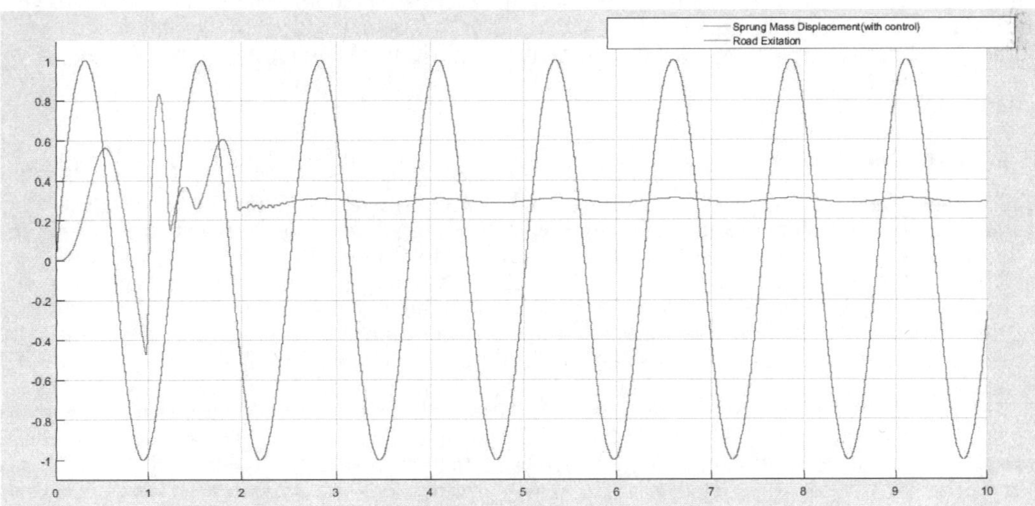

Fig 9. Sprung mass displacement when road profile is a sine input

Fig 9 gives us the displacement of sprung mass when the road input profile is in form of sine wave. With initial overshooting and undershooting, the sprung mass becomes stable. With initial overshooting and undershooting, the sprung mass becomes stable. The sprung mass becomes stable in almost 2.5 sec with the sprung displacement of 0.8mm.

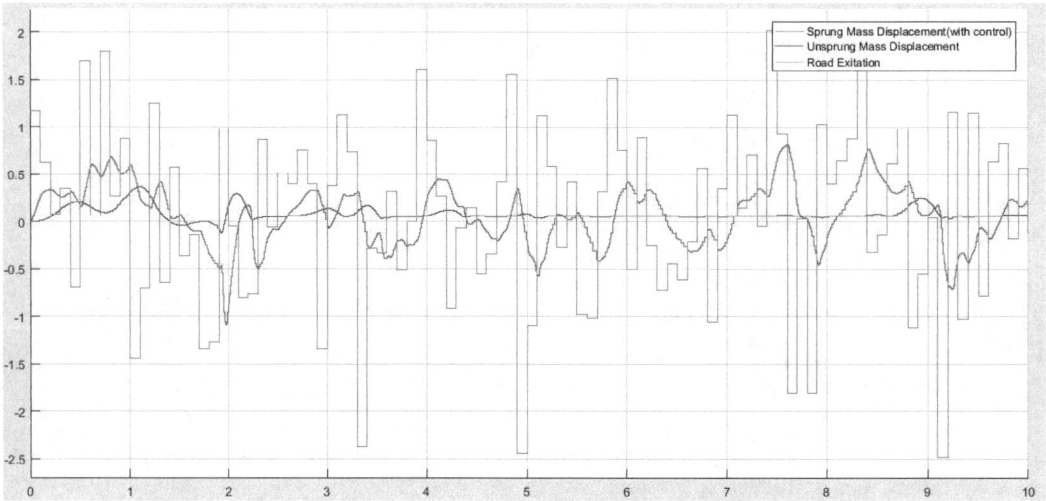

Fig 10. Sprung and un-sprung mass displacement when road profile is random input

Fig 10 gives us the displacement of sprung mass when the road profile is in form of random input. Here the case is due to random motion the sprung mass comes in good stability after some time. At the start, the oscillations are significantly high but after 3.5 seconds the sprung mass becomes stable at random road excitation with a displacement of 0.1mm from the initial position.

On the other hand, the un-sprung mass displacement is comparatively high but it is not our concerned. Thus, we can observe PID is doing controlling the damping force in accordance with road profile. The linearizing in case of sine road profile is comparatively easier than random input profile as the mean of sine is zero, therefore, it is linear in less time. Other values of proportional, derivative and integral gain can also yield desired results. MATLAB coding interface help in obtaining the hysteresis curve and displacement-time and displacement-velocity graphs. By changing values of α, β, γ, η a smoother hysteric can be obtained.

5. Conclusion

In this present study, every aspect of quarter car model is considered including the tyre stiffness and its damping also controller system along with the complex MR fluid model has been presented. The hysteresis curve is obtained between the damping force and velocity of the suspension piston as this model studies the non-linear hysteric behaviour of the MR fluid. Thus, considering all the fluid parameters the control is efficient thereby, increasing the stability of the vehicle. It also demonstrated the random excitation and also with periodic excitation and after few seconds the sprung mass is becoming stable which account for better stability for the semi-active system.

Future work may include a better control technique for linearizing the sprung mass displacement in less time such as a LQG (Linear Quadratic Gaussian) control technique can be best applied and used in high-speed cars and trains also a robust method in order to make cost-effective semi-active suspension system. The Bouc-Wen MR model gives a vast variety of applications such as from cars to trains. This is a technology in which extensive researches are being carried out carried out worldwide.

References

[1]. Lai, C.Y. and Liao, W.H. (2002) Vibration control of a suspension system via a magneto rheological fluid damper, Journal of Vibration and Control, 8(4), 527-547.

[2]. Yao, G.H., Yap,F.F., Chen,G., Li,W.H. and Yeo,S.H.(2002), MR damper and its application for semi active control of vehicle suspension system, Mechatronics,12 (7),963-973.

[3]. Spencer Jr, B. F., S. J. Dyke, M. K. Sain, and JDf Carlson. "Phenomenological model for magnetorheological dampers." Journal of engineering mechanics 123, no. 3 (1997): 230-238.

[4].K.Dhananjay Rao. Modeling, "Simulation and Control of Semi Active Suspension System for Automobiles under MATLAB Simulink using PID Controller" Third International Conference on Advances in Control and Optimization of Dynamical Systems.

[5]. D. Fischer, R. Isermann, Mechatronics semi-active and active vehicle suspensions, Control Engineering Practice.12 (2004) 1353-1367.

[6]. I. Eski, S. Yildirim, Vibration control of vehicle active suspension system using a new robust neural network control system, Simulation Modeling Practice and Theory. 17 (2007) 778-793.

[7]. G. Priyandoko, M.Mailah, H.Jamaluddin, Vehicle active suspension system using skyhook adaptive neuro active force control, Mechanical Systems and Signal Processing. 23(2009)855-868.

[8]. M. S. Kumar, Development of active suspension system for automobiles using PID controller, Proceedings of the World Congress onEngineering. 2(2008).

[9]. P.E. Uys, P.S. Els, M. Thoresson, Suspension settings for optimal ride comfort of off-road vehicles travelling on roads with different roughness and speeds,Journal of Terramechanics. 44(2007) 163–175

[10]. D. A. Mantaras, P. Luque, C. Vera, Development and validation of a three-dimensional kinematic model for the McPherson steering and suspension mechanisms, Mechanism and Machine Theory. 39(2004) 603–619.

IOP Conf. Series: Materials Science and Engineering **404** (2018) 012040 doi:10.1088/1757-899X/404/1/012040

Optimization of Multi-Gate Systems in Casting Process: Experimental and Simulation Studies

Mohd. Bilal Naim Shaikh[1], Shazeb Ahmad[2], Arfeen Khan[3], Mohammed Ali[4]

[1] Research Scholar, Department of Mechanical Engineering, Aligarh Muslim University, Aligarh, India
[2, 3] UG student, Department of Mechanical Engineering, Aligarh Muslim University, Aligarh, India
[4] Professor, Department of Mechanical Engineering, Aligarh Muslim University, Aligarh, India

E-mail: mohammed_bilalnaimshaikh@zhcet.ac.in

Abstract. This paper proposes an improved gating system design for casting process using experimental and simulation studies. The principle motivation behind optimization of gating system is to overcome the defects like shrinkage, gas porosity, slag, inclusion, cold shuts and misruns etc. Four different designs having different sprue location and runner cross-sectional area are taken into considerations for study. Discharge through multiple gates which are connected to horizontal runner has been evaluated as a performance parameter for comparing the designs. Experiments are conducted by tap and collect method using water and the volume of flow through each gate is observed for all designs. A simulation-based study is carried out for all four designs on finite-volume commercial code FLUENT. Boundary and initial conditions, such as velocity, pressure, temperature, etc., was reasonably set. It is found that the flow rate, as well as average velocity through the farthest gate, is higher for the parallel runner while this difference gets reduced with a tapered runner. Furthermore, design with a central position of sprue also minimizes the difference in flow rates as well as velocity. Finally, a design with a tapered cross-sectional area and justified position, as well as orientation of different gates, is finding out to be optimum having balance flow rate through the four gates with less filling time.

Keywords: Casting, runner, gates, flow velocity, flow rate.

1. Introduction

In casting process, gating system plays vital role to produce a good quality casting. A gating system controls mould filling process as well as flow pattern in the gating system. Multiple gates are require to ensure proper distribution of flow along with sufficient temperature. In general, different sections of a casting are required to fill at approximately the same flow rate and fill time. This will lead to similar conditions for solidification, and thereby less variation in mechanical properties. The main function of gating system is to lead clean molten metal from ladle to the casting cavity ensuring uniform, smooth as well as complete filling of mould. The design of the gating system depends on the number, size as well as shape of the casting, to ensure uniform metal flows to all sections while maximizing the casting yield. The rate of flow of metal is regulated by the smallest cross section referred to as choke, which is typically at the bottom of sprue or the gates. The ratio of cross-section area of sprue bottom, runner, and gate, which is called the gating ratio, affects the rate of molten metal flow in various sections of gating system.

To study the fluid flow through gating systems and to evaluate their design, three approaches have been adopted in casting industry. These are: (1) empirical relations derived from experimental observations, (2) hydraulics based analysis involving Bernoulli's and continuity equation, and (3) numerical simulation involving solution of the mass and momentum conservation relations. Two basic assumptions have to be made before hydraulics can be applied to gating problems. The first is that all molten metal poured remains liquid until filling of mold cavity is completed and second one is that the molten metal behaves like true liquid. With this background a systematic study is taken up in which physical models of gating system are made up of acrylic and water is used in place of molten metal].

Complete similitude between real time casting and water model experiments is met so long as certain non-dimensional numbers are same in the both case. These numbers are Weber number, Reynolds number, Froude number and Euler number. Experiments have revealed that Weber number is very insignificant in gravity casting since surface tension plays very much less role. While Euler number is predominantly used in pressure die casting where pressure difference is imposed upon metal by some external element. In gravity casting, flow is assisted by gravity. Hence such situation is not arising. The relevant non-dimensional numbers in water model experiments are found to be Reynolds number and Froude number. If the both number are equal, then condition of physical similarity is met and maximum similitude is achieved. The conditions to ensure analogy are:

$$Re_1 = Re_2 \quad \text{and} \quad Fr_1 = Fr_2$$

From the above two equations, we can conclude that the similitude is greatly influence by the kinematic viscosities of different fluid used in model and the real casting.

Table 1: Properties of aluminium and water

Liquid	Temperature(°C)	Density (gm/cm^3)	Dynamic viscosity(centipoises)	Kinematic viscosity(centistokes)
Water	20	1	1	1
Aluminium	700	2.37	3	1.27

The kinematic viscosities of aluminium is similar to that of water, and full scale water models of gating systems and cavities for Al should therefore give a good indication of flow conditions in the actual castings.

The literature review is mainly focusing on design and optimization technique based on casting related defects and their research and outcomes. This includes the development of mathematical models of flow in gating systems, based on experimental observations. In one of the earliest such studies, it was concluded that the farthest gate from the sprue exit had the largest flow of molten metal. It was reported that the flow in runner is largely independent of the length of runner between the gates [1]. Martin [1]reviewed the work of different researchers on gating ratios and concluded that the selection of the most appropriate gating ratio is influenced by the formation of dross and metal impingement against mold walls and cores. Other researchers [2], [3] , [4] proposed methods to determine the relative flows from different gates by determining the losses in the channel flow. In general, it has been observed that the simulated filling sequence is somewhat different from experimental observations [5]. In one study, while water model experiments were observed to be useful for comparing simulation results, it was concluded that software simulation cannot entirely replace actual experimental work [6]. In another work, comparison between simulation and experiments, it was found that due to wall friction and surface tension the shape of liquid front was different between the both [7]. F. J. Bradley [8] derived an energy balance equation for each path and a continuity equation at each node were applied for the whole system and solved for simultaneous algebraic equations to predict flow rates in each segment. Ravi and Joshi [9] worked on Auto-CAST software and they describe how it assists in designing, simulating, modelling, analysing and improving cast products properties. Bhatt and Barot [10]reported that the design optimization of gating system and simulation reduced the casting defects. Masoumi et al. [11] reported the effect of gating design on mold filling for casting processes. The validation results showed that the design as well as shape of the gate and the gating ratio have a great influence on the pattern of mold filling. Sutaria [12] worked on a new idea where optimization of casting feeding is done with the help of feed- paths. Literature on optimization of gating system recommends minimizing the in-gate velocity of melt, maximizing the yield, minimizing warpage and optimizing location of in-gate. This work focuses on the maximizing the filling rate and minimum difference in flow rate as well as velocity through in-gates. Higher filling rate is useful to increase the production rate of castings. Higher filling rate is also required in thin and long castings which lose heat very rapidly.

2. Methodology

2.1. Water Modelling

Many researchers have studied fluid flow in gating systems and mould cavities with the aid of transparent models usually made of Perspex or other plastics, in which the flow of metal is simulated by that of water. This method has the substantial advantage that the pattern of flow is clearly visible in all parts of the gating system and the cavity, but its validity, depends entirely upon the correctness of the assumption that the flow of water is closely analogous to that of the molten metal.

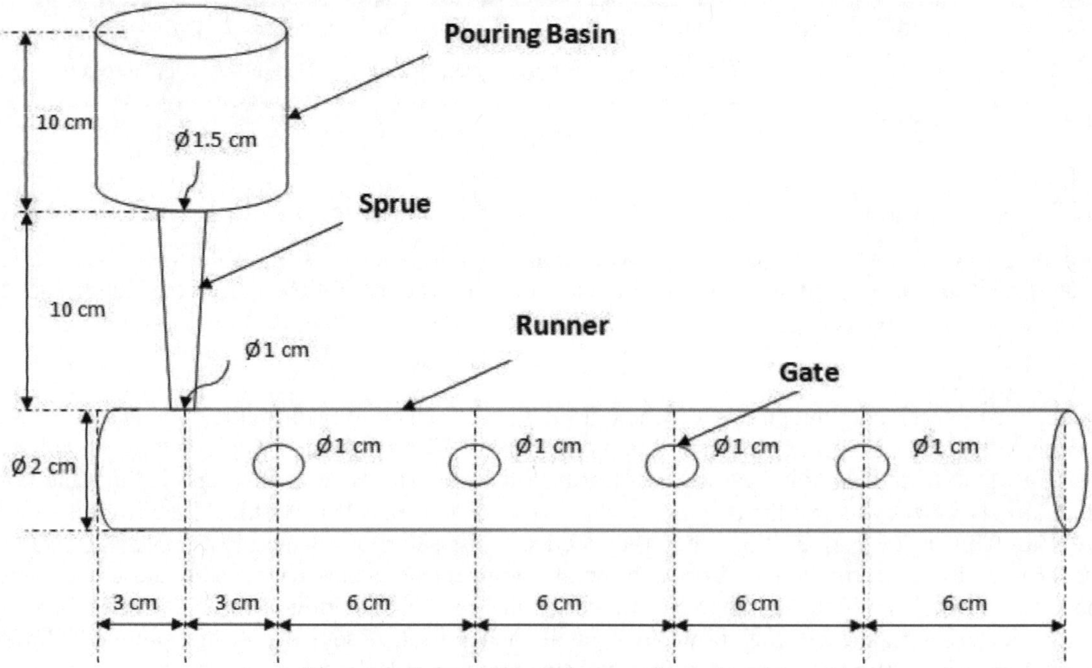

Figure 1: Schematic of the considered gating system

Table 2: Detailed specification of considered four gating system designs

	Runner	Runner's Dimensions	Sprue Position	Gates Position w.r.t. sprue
Design 1	Parallel	Ø 2 cm throughout length	At end	All four on one side
Design 2	Parallel	Ø 2 cm throughout length	Central	Two-two on both side
Design 3	Tapered in one direction from one end to other	Ø 2 cm at one end & Ø 1 cm at the other end	At end	All four on one side
Design 4	Tapered in both direction from centre	Ø 2 cm at centre & Ø 1.3 cm at both end	Central	Two-two on both side

According to the theory, a water model of gating system and cavity should accurately reproduce the character of (steady-state) flow of a metal in the corresponding sand-moulded gating system and cavity, provided that the Reynolds number in the two cases are about equal. Reynolds number is dependent on the diameter of the flow channels and on the kinematic viscosity of the liquid, and is therefore of interest to compare the kinematic viscosities of liquid metals with that of water and other fluids.

The dimensions of gating systems used in water model experiments are calculated using simple relationships. The design of a gating system begins with the selection of the appropriate gating ratio and the alloy for which the study is proposed to be carried out. Popular gating ratio for an alloy of aluminium is 1:4:4. So gating ratio of 1:4:4 is chosen to illustrate the design procedure.

2.2. Simulation Study

The whole filling process for each considered design were simulated using ANSYS 16 fluent to determine the optimum one. It was based on a transient 3D situation. Used material for simulation was aluminium and used properties for simulation was listed below. Before initializing the whole model, required conditions were being applied. Since, we only wants to focus on the flow behaviour of the molten metal, therefore the energy and viscous model were used to simulate the flow. Simulation was run for 250 iterations with 10000 mesh size for each design concept under the same conditions. Whole filling process was considered to be as gravity casting. All the properties used were listed as bellow.

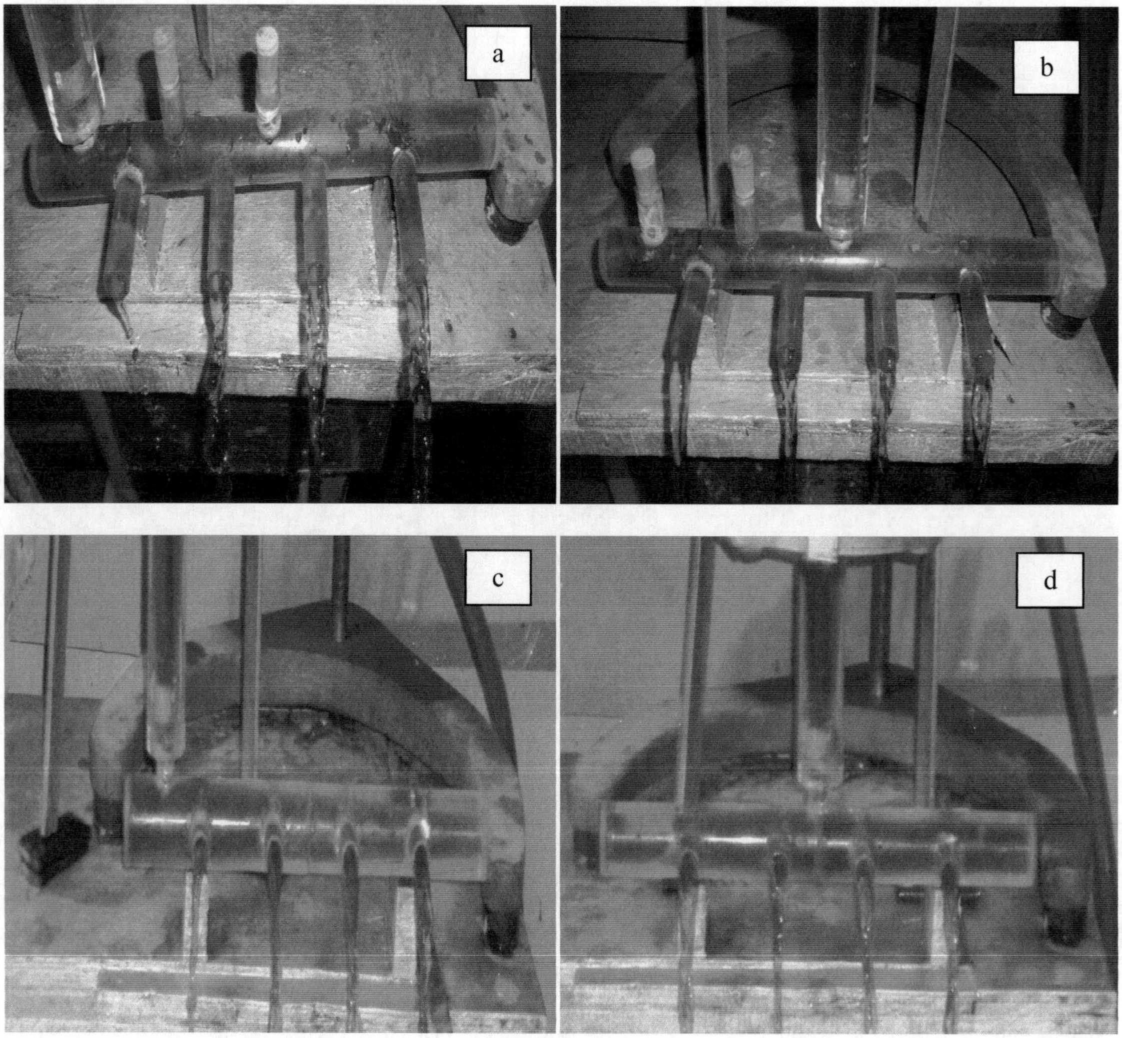

Figure 2: Acrylic model for water modelling of considered four Gating Systems.

Table 3: Parameters taken for simulation study

	Mould's Material	Metal Used
Material	Green sand	Aluminium
Density	1.5 g/cm^3	2.375 g/cm^3

Table 4: Aluminium properties

Pouring temperature	900 ° C
Melting temperature	660 ° C
Viscosity(stokes)	~ 0.0012

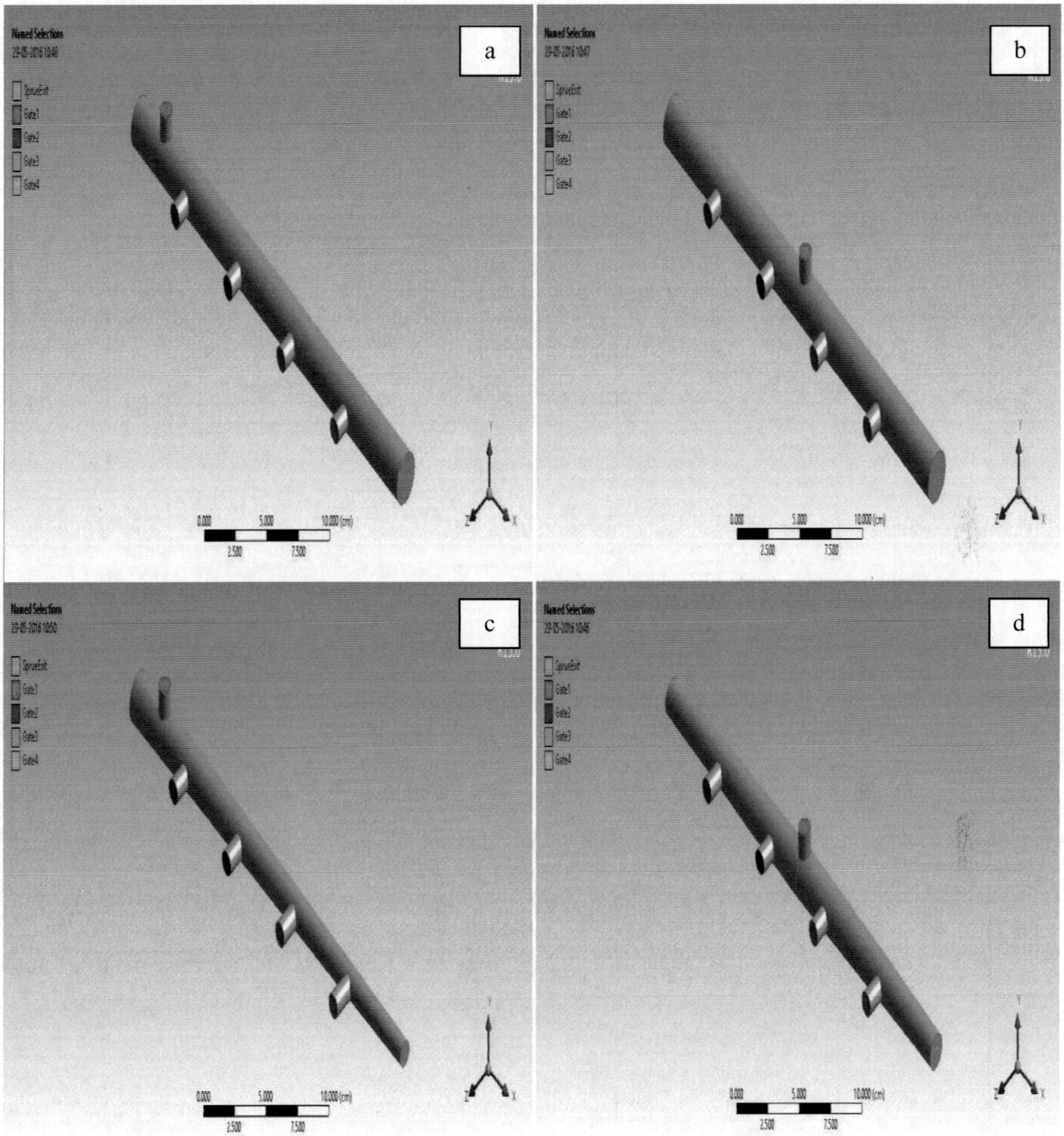

Figure 3: 3D ANSYS model for simulation of considered four Gating Systems.

3. Results and Discussion

3.1. Water Model

IOP Conf. Series: Materials Science and Engineering **404** (2018) 012040 doi:10.1088/1757-899X/404/1/012040

Flow rates and flow velocities through different gates were calculated using tap and collect method. Central position of sprue minimizes the difference in velocities as well flow rate from the gates which is required for a good casting. Tapering of runner further minimize the difference. Hence we can observe that design 4 would be the best one for optimum casting conditions.

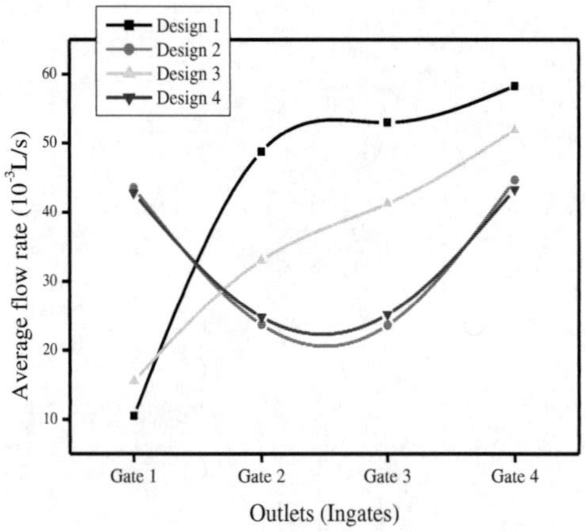

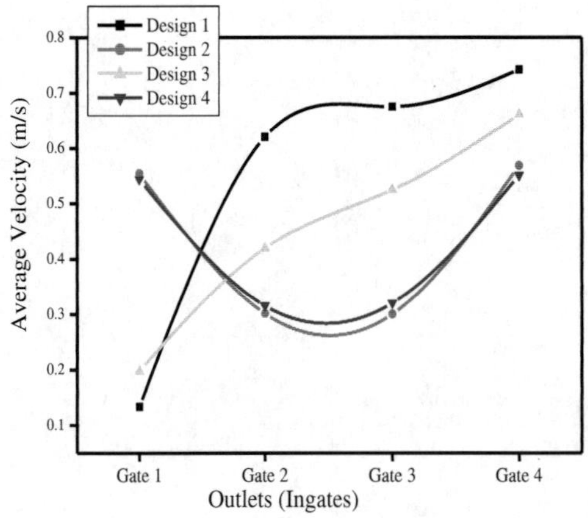

(a) (b)

Figure 4: (a) Flow rate and (b) flow velocity through different gates in water model.

3.2. *Simulation Model*

Results from the simulation studies show the similar trend as observed in water model experiments. Design 4 comes out to be the optimum one. Molten Aluminium behaves same as water.

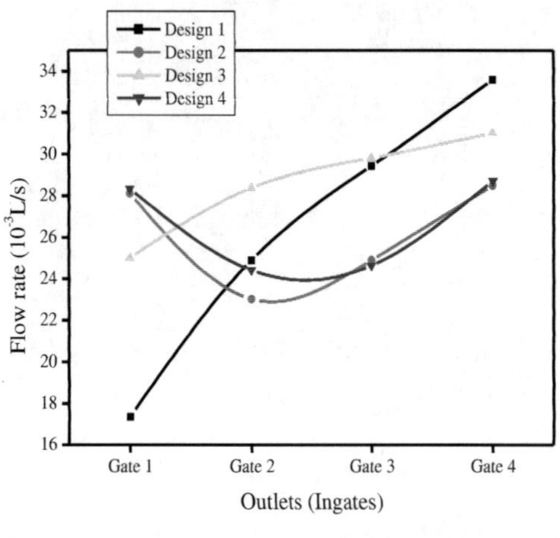

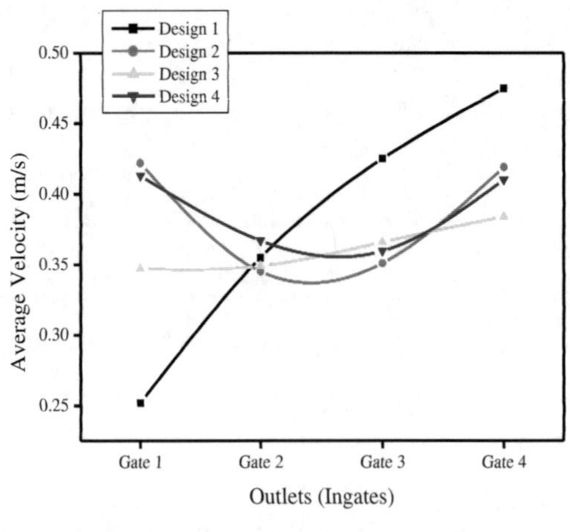

(a) (b)

Figure 5: (a) Average flow rate and (b) average flow velocity through different gates in water model.

3.3. *Modification of design*

To further optimize the gating system design, some modification has been done on design 4. Sharp bends had been replaced by acute angle joint to facilitate streamlined flow. Due to this modification, losses due to bend has been minimized and streamlined as well as uniform flow of liquid has been achieved.

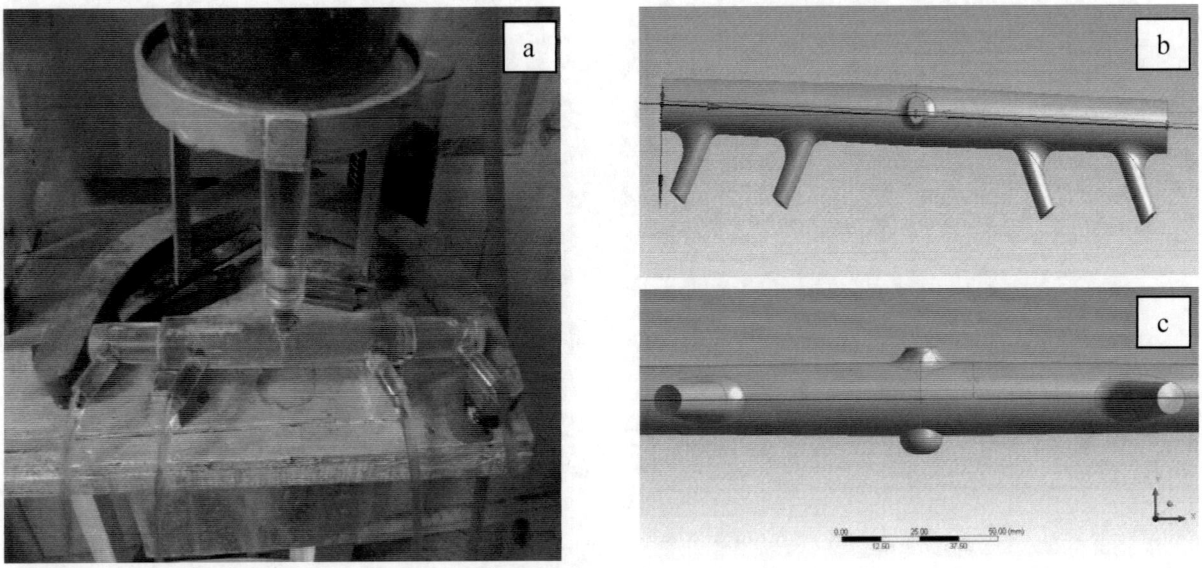

Figure 6: (a) Modified acrylic model, (b) & (c) 3D ANSYS model of modified design used for simulation.

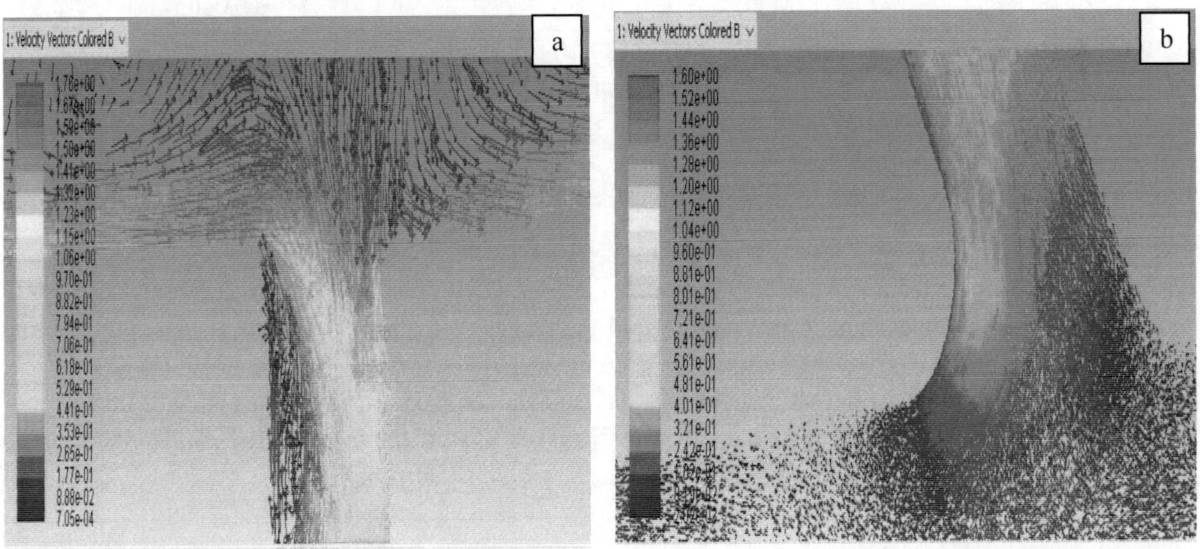

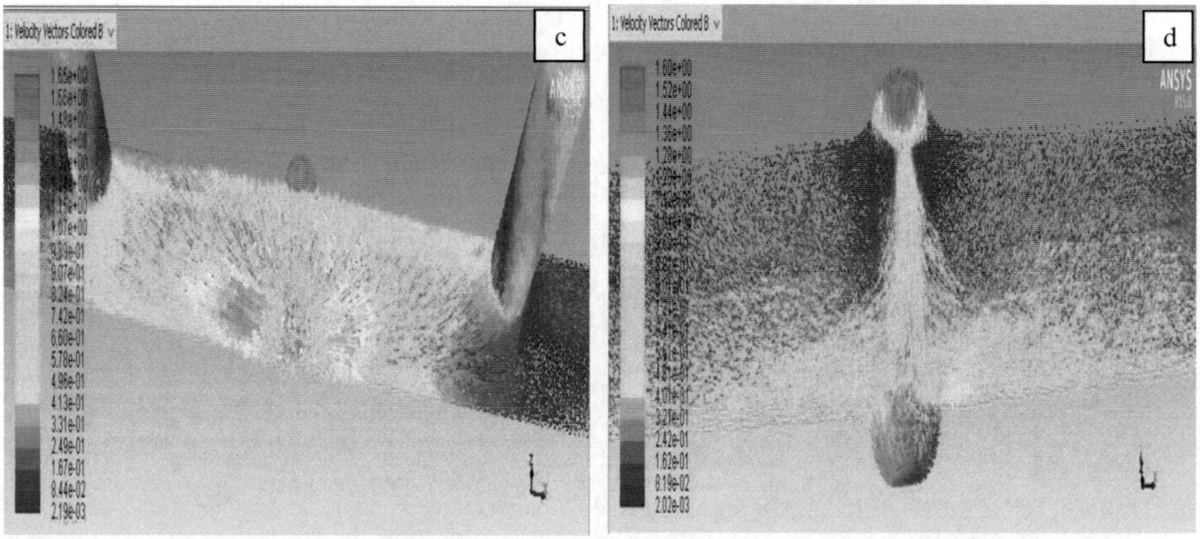

Figure 7: Velocity profile at bends (a) before (b) after modification and at sprue exit (c) before (d) after modification

Table 5: Flow parameters for the modified design through water model

Modified Design	Average flow rate (mL/s)	Average flow velocity (m/s)
Gate 1	42.23	0.537
Gate 2	39.20	0.494
Gate 3	39.18	0.498
Gate 4	41.80	0.532

4. Conclusion

From the above study, we can conclude following:

1. The location of gate in multi-gate system as well as junction nature i.e. how the gate is attached with runner (whether a sharp bend is there or a smooth bend) beside of runner cross-sectional area also affect the flow parameters.
2. From the flow measurement experiment, we find that to maintain constant pressure throughout the runner, we have to reduce the cross sectional area in direction of flow. So by locating the last gate at the end of runner would increase the casting yield.
3. The flow rate from the four proposed designs of gating systems, design 1 and 3 are highly varying whereas design concept 2 and 4 gives somewhat uniform flow rate hence on the basis of flow rate criteria design 1 and 3 comes out to be worst while 2 and 4 gives more satisfactory results.
4. Hence design concept 4 fits best in both the deciding criteria i.e. flow rate and velocity at ingate. Further some modification have been done in the design concept 3 to avoid the high erosion at the bottom of sprue and negative pressure at sharp bends.

5. Simulation studies also shows very much similarity in the trend of flow rate as well as flow velocity with water model. In addition, variation between the both can be attributed to the different properties of considered fluids.
6. Simulation of this modified version of design concept 4 gives better results as shown below.

References

[1] G. Martin, "Principles of gating: A review," *Foundry Trade Journal ,* pp. 63-69, 1953.

[2] M. C. Berger, "A theoretical basis for the design of gates," *Foundry,* pp. 112-117 and 237-242, 1951.

[3] R. Ruddle, "The running and gating of sand castings: a review of the literature," *Institute of Metals Monograph; Institute of British Foundrymen,* 1956.

[4] M. Srinivasan, *Applied hydraulics to gating systems,* The University of Birmingham, 1962.

[5] B. Sirrell, M. Holliday and J. B. Campbell, "Benchmark testing the flow and solidification modeling of Al castings," *The Journal of The Minerals, Metals & Materials Society,* vol. 48, pp. 20-23, 1996.

[6] J. Dumaillet and G. Wilson, "Comparison of flow modification through foundry filters using both water modeling and simulation software," *Transactions of American Foundry Society,* vol. 110, pp. 1-12, 2002.

[7] F. Juretzko and D. Stefanescu, "Comparison of mold filling simulation with high speed video recording of real-time mold fillin," *AFS Transactions,* vol. 113, pp. 1-11, 2005.

[8] F. Bradley and S. Heinemann, "A hydraulics based optimization methodology for gating design," *Applied Mathemaatical Modelling,* vol. 17, pp. 406-414, 1993.

[9] B. J. D. Ravi, "Feedability analysis and optimization driven by casting simulation," *Indian Foundry J.,* vol. 53, no. 6, pp. 71-78, 2007.

[10] H. B. R. Bhatt, "A design optimization of feeding sys- tem and solidification simulation for cast iron," in *Proceeding on 2nd International Conference on Innovations in Automation and Mechatronics Engineering*, 2014.

[11] M. H. J. H. M. B. A. Masoumi, "Effect of gating design on mould filling paper," *AFS Transactions,* Vols. 05-152, no. 2, pp. 1-12, 2005.

[12] M. Sutaria, "Casting simulation case study: shaft pin (cast iron- green sand casting)," *Indian Foundry J.,* vol. 56, no. 12, p. 5, 2010.

IOP Conf. Series: Materials Science and Engineering **404** (2018) 012041 doi:10.1088/1757-899X/404/1/012041

Advanced Technology in Desiccant Wheel Air Conditioning System: *A Review*

Amarjeet Kumar Prasad[1] and Laxmikant Yadav[2*]
Mechanical Engineering Department
MMMUT, Gorakhpur[1,2*]
Corresponding author e-mail: laxmikant.yadav2149@gmail.com

Abstract. Desiccant wheel air conditioning have been considered as one of the effective process to control humidity of ambient air. Desiccant wheel air conditioning system do not use any ozone-depleting refrigerant and also intake lesser energy compared to vapour compression system. All the types and kind of desiccants have been discussed here. Such as solid desiccant system, liquid system have been reviewed briefly. Finally, a summary of economic prospect, future scope, environmental friendly technology and future scope of desiccant wheel air conditioning system have been discussed briefly.

Keywords: *Desiccant,* solid *desiccant wheel air conditioning system, liquid desiccant wheel air conditioning.*

1. Introduction.

The energy used by the air-conditioning system in the homes and in the commercials places now accounts for nearly 45%; the International Institute of Refrigeration in Paris conducted this research. Of the world total energy consumption, almost 15% of the energy is used by the

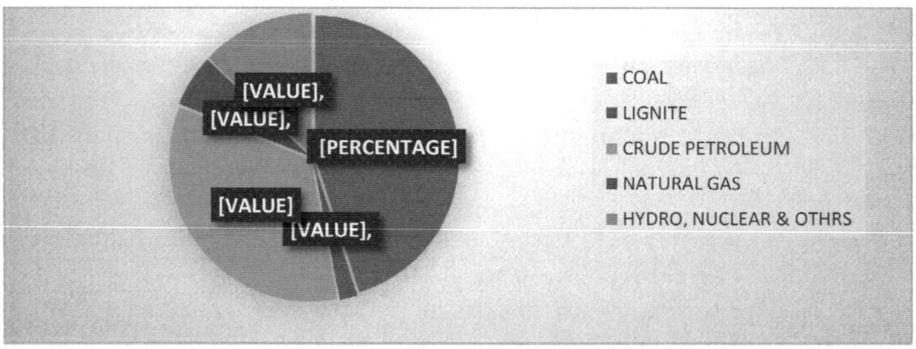

Fig1: Source Consumption of Energy during 2016-17 (3)

air conditioning system (1). As we know, there is continuous rise in the world population and with advancement of new technology; there is increase in the per capita energy demand, this lead to the increase in the energy consumption worldwide. In spite of great progress in technology, we still rely upon fossil fuels as the primary source of energy to fulfil our demands. Most air conditioning system uses vapour compression based system and uses coolant like chlorofluorocarbons, hydro chlorofluorocarbons which are major gases for ozone layer depletion. Other causes of ozone depleting involve refrigeration, coolant. Fire protection system etc. which are being used today. Some refrigerant such as CFC, HCFC have been in there cooling system for 70 years (4).

The major concern of engineers while designing a building is to regulate the balance between thermal comfort, inside space air quality and energy utilization. Air-conditioning systems are designed in such a way to give maximum human comfort. In the space provided best inside air quality is given. An operating parameter of temperature varying between 18.8° C to 26.6°C and humidity between 40 to 70% is commonly considered acceptable for human comfort (5). Additionally, the physical activity and clothing also determines the thermal comfort of human

IOP Conf. Series: Materials Science and Engineering **404** (2018) 012041 doi:10.1088/1757-899X/404/1/012041

being and ventilation plays major role. High percentage of moisture in the interior space can also give rise to various problems, such as various health problem decay of wooden furniture

2. DESICCANT

A desiccant is a substance or chemical that absorbs or attracts moisture from the air, causing a state of dryness (desiccation) in its vicinity. This chemical used to attract and remove moisture from air or gas and keep the products dry which, in turn, inhibits the corrosion process. These chemicals are used in product containers during shipping to protect against moisture damage. Descants also prevent condensation of moisture between window panes. Desiccants are available in various forms and are widely used in many industries, such as:

- Packing
- Electronics
- Manufacturing

There are two major categories for types of desiccants, each with their own set of chemicalproperties and functional requirements. These categories include

- solid desiccants and
- liquid desiccant

3. DESICCANT SYSTEMS:

Desiccant cooling systems include rotary desiccant wheel, processed air, evaporative cooling system, hot air for regeneration of desiccant. Desiccant wheel air conditioning systems can be divided on the basis of type of desiccant material used:
(1) Liquid desiccant air conditioning systems,
(2)Solid desiccant air conditioning systems

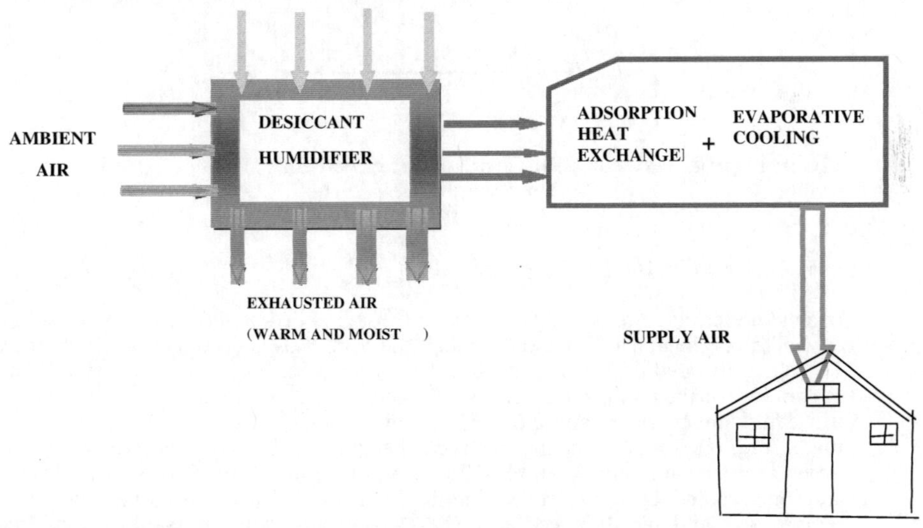

Fig2: Principle of Desiccant Dehumidification and Air Conditioning

A. *Liquid Desiccant Air Conditioning Systems*

Figure below shows simple liquid desiccant systems containing a conditioner and a regenerator. Water vapour from the incoming ambient air is removed in the conditioner. Where desiccant material soaks the humidity from the incoming air. Due to the difference in vapour pressure there mass transfer takes place and heat is liberated due to the condensation of water molecule present in the processed air, this air is sent to the evaporative cooler to bring down the temperature of air and further the desiccant is pumped back to the regenerator to further absorb the moisture from the incoming air. Before the desiccant enter to the regenerator its moisture is removed by sending it to the liquid –liquid sensible heat chamber. Where temperature of the desiccant is increased and moisture is removed. This liquid desiccant is further send to the regenerator to absorb the moisture from the incoming air.

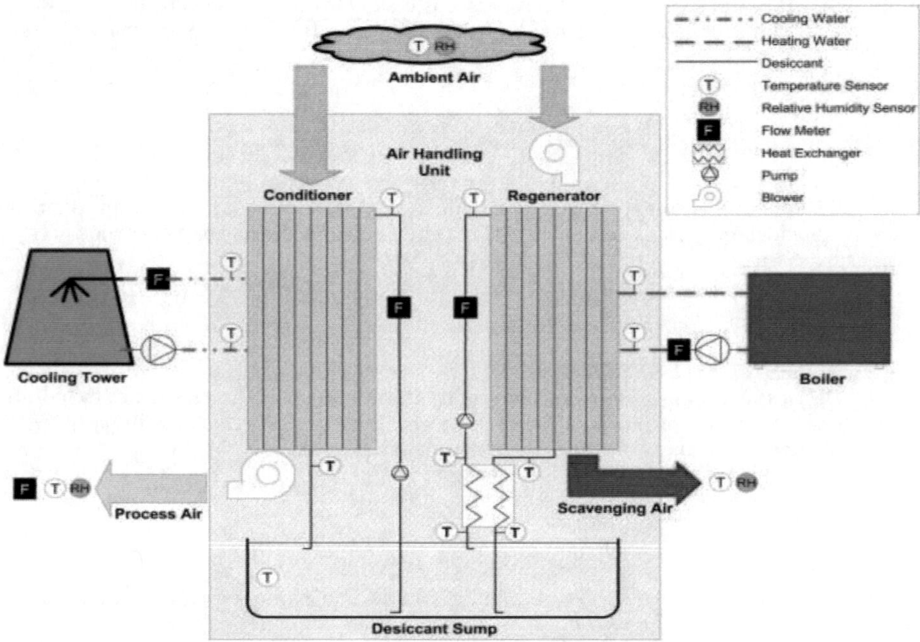

FIG 3: Liquid Desiccant Air Conditioning System (7)

B. Solid Desiccant Air Conditioning Systems

In solid desiccant systems, air is circulated through a bed of absorptive material like silica gel or zeolite. As the moist air passes through the bed, it gives up water vapour to the desiccant. Then the saturated desiccant is heated. The basic process of solar assisted desiccant cooling system is describe on the figure below. Hot and humid air enters the slowly rotating desiccant wheel and moisture is removed by adsorption of water (1-2). Since the air is get heated up by the adsorption heat, this heated air is further passed through a heat recovery wheel (2-3), which shows drop in temperature of the incoming ambient air. Further, the air is humidified and thus cooled by a controlled humidifier (3-4) according to the supply value of air temperature and moisture content. The exhaust air, which is coming from the room, is humidified (6-7) near to the saturation point in order to show an effective heat recovery (7-8). Finally, the desiccant wheel has to be regenerated (9-10) by applying hot air in a comparatively low temperature range from 50°C to 75°C and to allow continuous process of the dehumidification of air.

Flat-plate solar thermal collectors or evacuated solar thermal collector are normally used as heating system in solar assisted desiccant cooling systems. Typically desiccant material is loaded in to the rotating wheel or attached in to the honey-comb wheel, wheel rotates slowly between the processed air and heated air flow (2).

IOP Conf. Series: Materials Science and Engineering **404** (2018) 012041 doi:10.1088/1757-899X/404/1/012041

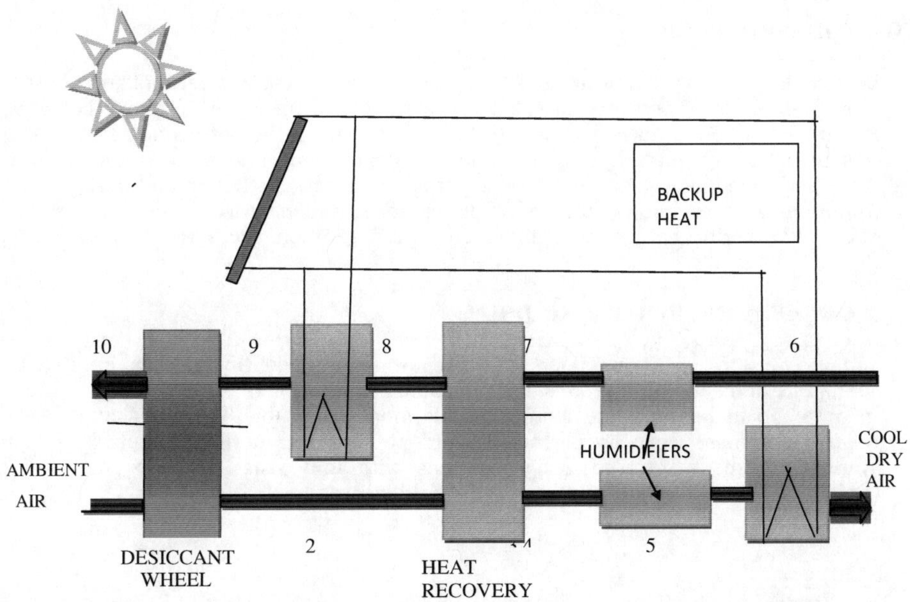

Fig4: Diagram of Solid Desiccant Wheel Air Conditioning System

4. ADVANTAGE

Desiccant cooling systems can decrease the moisture levels in the desired spaces, saving hundreds of rupees in unnecessary repairs.
Desiccant systems also enhance indoor air, rate of air ventilation is improved, and clear away air pollutants and odours.
Similarly, restaurants, schools, and hospitals and other commercial places can use desiccants cooling system to remove air moisture, smell, and pollutants for a healthier environment and productive environment.
Improved perseveration of frozen foods and refrigerated products.

5. ENVIRONMENTAL FRIENDLY TECHNOLOGY

Thevapour compression cycle are old and traditional system, which heavily causes serious environmental effect like greenhouse effect, ozone layer depletion and global warming. In order to decrease these effects traditional cooling system need to be replaced by desiccant cooling system. Desiccant wheel air conditioning cools and dehumidify to produce fresh air whereas VC system uses some refrigerant to cool the air.
Desiccant wheel air conditioning system is free from refrigerant in comparison to VC system thus produces no harmful effect. Desiccant wheel air conditioning system facilitate zero ozone depletion potential and zero global warming potential, these is due to the working fluid used is natural in nature .Therefore desiccant wheel air conditioning system can be reflected as eco-friendly air conditioning system. The above statement produces strong view that desiccant wheel air conditioning system can be considered as economical and eco-friendly technology which reduces serious environmental problem and ensure better air quality in cooling which ensure the better health quality of the users. In comparison to vapour compression system and desiccant wheel air conditioning system confirms that the three is decrease in the carbon dioxide up to 13%. The major advantage of the desiccant wheel air conditioning is principally heat driven cycle. Therefore it allow to use low grade waste heat energy and many other renewable energy like solar energy geothermal energy etc. when we impregnate waste heat in desiccant cooling system it reduces the cost or save the cost of the system up to the 87% (8).

IOP Conf. Series: Materials Science and Engineering **404** (2018) 012041 doi:10.1088/1757-899X/404/1/012041

6. ECONOMIC PROSPECT

Desiccant wheel air conditioning system with sensible cooling technologies have already started showing its presence in the market. Countries mainly which have hot and humid climatic conditioning have already started using other means of cooling system which are economical and eco-friendly and in comparison to the conventional VC system. A comparison study and economic evaluation between vapour compression system and desiccant wheel air conditioning system shows that the hybrid desiccant system performance increased by the use of the solar thermal energy and ambient air for refrigeration. Analysis indicated that replacing the.

7. GENERATING FUTURE SCOPE

Today's need of the hour is to generate awareness, identify various hurdles to wider acceptance and deriving solutions through commercialization of the technology in the country. In order To make desiccant coolingsystems more acceptable, the industry along with the government must work as partners to Adopt advanced desiccant materials, equipment available in the market to design systems with low costs and improved performance (1).Demonstrate desiccant systems in real-world environment.

8. CONCLUSIONS

Desiccant wheel air conditioning was reviewed in this work. Both liquid and solid desiccant air conditioning system were strongly discussed in this paper. Conclusions are as follows:

- Desiccant wheel air conditioning system do not use any ozone-depleting refrigerants.
- Desiccant wheel air conditioning system is mainly suitable for removing moisture from the air. If cooling is needed than a hybrid, system is to be installed

9. REFERENCES

[1]. Minaal Sahlot* and Saffa B. Riffat. Department of Architecture and Built Environment, University of Nottingham 1 October 2014. Desiccant cooling systems: a review 2016: 0: 1–17

[2]. Dr. U. V. Kongre, D. P. Mahure, P. et al. International Journal of Engineering Research 01 April 2014; A Review of Solar Desiccant Air Conditioner ISSN:2319-6890

[3]. Remme U, Trudeau N, Graczyk D, et al. Technology Development Prospects for the Indian Power Sector. International Energy Agency [IEA], 2016.

[4]. A fonso CFA. Recent advances in building air conditioning systems. Appl Therm Eng 2006;26:1961–71.

[5]. Armstrong J. CIBSE Conscise Handbook. Norfolk: Page Bros (Norwich)Ltd.Norfolk, 2008.

[6]. Chen Y, Yin YG, Zhang XS. Performance analysis of a hybrid air conditioning system dehumidified by liquid desiccant with low temperature and low concentration. Energy Build 2014;77:91–102..

[7]. Ahmed CSK, Gandhidasan P, Al Farayedhi AA. Simulation of a hybrid liquid desiccant based air-conditioning system. Appl Therm Eng 1997;17:125–34.

[8]. Sarbu I, Sebarchievici C. Review of solar refrigeration and cooling systems. Energy Build2013;67:286–9.

Water heating system using concrete slab solar collector with phase change material: - a review

Vikas Yadav[1] Prashant Saini[2]

[1]M.tech Student, MED, MMMUT, Gorakhpur, U.P. India

[2]Assistant Professor, MED, MMMUT, Gorakhpur, U.P. India

Corresponding author e-mail: prashant91saini@gmail.com

Abstract: Keeping present scenario of energy crisis in mind we need to move towards renewable sources of energy in which solar energy is one form of available renewable energy on earth and it is environment friendly and free in nature. Hence we have to move towards the efficient utilization of solar energy. It has a main drawback as it is only available in day time or limited period of time, hence using phase change materials in different solar energy storage setup is a way to store solar energy and utilize that in day time as well as night also. In concrete slab solar collector, solar energy is absorbed and transferred into the medium, space and fluid for domestic as well as industrial heating applications. In this paper we would be doing a critical review of the water heating system using concrete slab solar collector with phase change material (PCM). Further various performance parameters and setbacks would be discussed pertaining to solar heating systems.

Keywords: concrete slab solar collector, PCM, latent heat and glazing.

1. Introduction

The permanent rise in emission of greenhouse gases into atmosphere and the continuous hike in fossil fuel prices are the cause behind efforts to efficiently use of various sources of renewable energy [1]. A renowned professor Richard E. Smalley talked about humanity's top problems for next fifty year. According to him depletion of non-renewable energy is the biggest problem for humanity. In many countries, direct solar energy is taken as future source of energy. Efficient accumulation of renewable energy reduces the consumption of non-renewable sources and meet the requirements. It also increase the performance and dependability of renewable energy systems. In solar collectors, it is needed for an effective storage of energy in which the most part of the heat collected during periods when sun is over headed, and later released for utilization during the night [2]. Solar energy utilization is categorized in two form active and passive solar heating. Active solar heating have higher efficiency compare to passive. Active heating is used in industrial purpose where high load and demand is needed [3].One of the efficient techniques for storing thermal energy in concrete slab solar collector with the use of phase change materials (PCM). An efficient thermal storage system is required for solar energy collector. Thus, the successful function of solar energy depends, to a point, on the method of energy storage used. So using phase change material is best for enhancement of passive heating system. The latent heat of PCM energy that needs to be absorbed or released when a material changes phase from solid state to liquid state or vice versa. Following is the list of energy storage method:

1.1 Energy storage methods:

The energy is stored in different manner like thermal, electrical and mechanical etc.

1.1.1 *Mechanical energy storage:* Flywheel is a mechanical energy storage device which store intermediate energy and releases for smoothing the power output of an energy source.

1.1.2 *Electrical energy storage*: Battery is an example of electrical energy storage. It is connected to direct current uses direct electricity but when disconnected it uses stored chemical energy and change into electrical. The familiar type of batteries is the lead acid and Ni–Cd.

1.1.3 Thermal energy storage: The thermal heat storage is mainly of two types (i) Sensible or specific heat type and (ii) latent heat type, in former type of thermal storage heat is absorbed or released in direct contact as hot water absorbs heat when come in contact with rocks. In latent heat type, storage uses the material which can change phase when absorbed or releases heat as their heat of fusion.

1.1.3.1 Sensible heat storage: In sensible heat storage, storing of energy is cause of temperature rise. Sensible heat storage system uses the heat capacity for the rise and fall in temperature of the substance during the process of absorbing and releasing. Sensible heating is two type (a) direct contact (b) indirect contact in direct contact storage system both substance in contact and come under thermal equilibrium in indirect contact a thin wall separate the absorbing and releasing medium.

1.1.3.2 Latent heat storage; It is depend on storing of energy when material goes under phase change from liquid to solid or solid to liquid. The change of phase of liquid to the vapor at constant temperature and pressure is latent heat and storage of energy in heat fusion is latent heat storage. In latent heat transition of solid-solid, heat is absorbed as the material is changed from one crystalline to another [4].

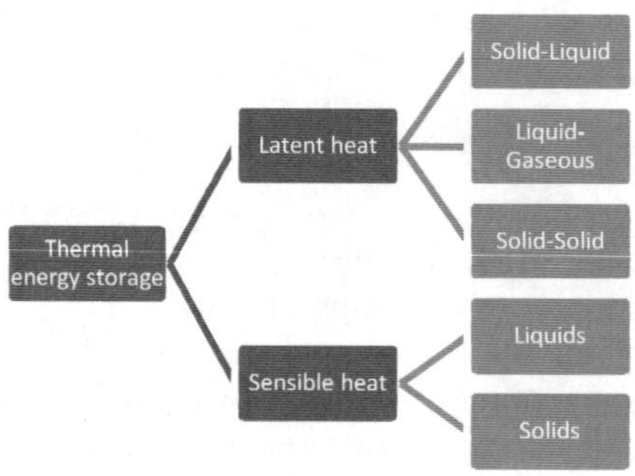

Figure 1: Type of thermal energy storage

2. Types of solar collector

A solar collector is a type of heat exchanger that absorbed heat from solar energy. A conventional type heat exchanger differs in many manner from solar collector. Solar collector is different type it is concentrating and non-concentrating type solar collector in non-concentrating type solar collector absorber and collector are is same but in non-concentrating type solar collector storage is bigger than the absorber. Different type of solar collectors have been discussed as following:

2.1 Flat plate collector

Flat plate collector is discovered by Hottel and Whillier in 1950. It has several components in the set-up, there is a material surface known as absorber and further solar insolation falls on it. For reducing heat loss by the absorber a transparent cover is used which trap the most of the insolation between the absorber and

IOP Conf. Series: Materials Science and Engineering **404** (2018) 012042 doi:10.1088/1757-899X/404/1/012042

transparent surfaces and also prevent absorber plate from dust [5]. For better absorption of radiation absorber have selective coatings. After serpentine design of pipe, Rittidech and Wannapakne [6] proposed new design in 2006. A flat plate solar collector operating in association with a closed-end fluctuating heat pipe was discovered. This system consisting closed end pipe have reasonable cost effective alternative method compared to other solar collector that utilizes heat pipe. This closed-end system could attain 62% efficiency.

2.2 Evacuated tube collector

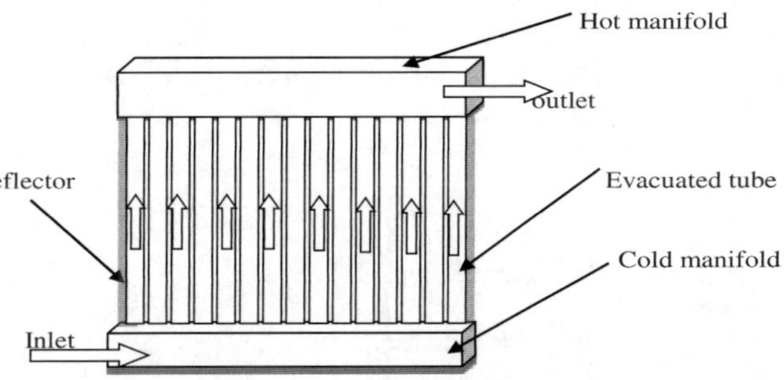

Figure 2: Evacuated tube collector

After combining multiple glass tube evacuated heat pipe tubes is composed, each tube having an absorber plate attached to a heat pipe. Heat is transferred to the circulating water or any antifreeze mix for water heating in a heat exchanger called a manifold. Manifold is covered with insulating material for reducing heat loss [7]. It is works better in colder condition compare to flat plate collector. Evacuated tube has disadvantage in cold environment as snow and frost may affect the performance and getting stuck in between pipes gaps. Presently the main component in solar thermal utilization is glass tube collector because of they have high heat absorbing efficiencies and comparatively less heat loss to flat plate collector.

2.3. Compound parabolic concentrator

Evacuated tube collector and flat plate collectors are broadly used in solar thermal storage system mainly to provide temperature in low to mid-range like 20 °C to 120 °C. For maximizing the thermal storage type of solar collector is used which have parabolic concentrator are non-imaging type and having the ability to reflect most of the incident insolation to the absorber [5]. The sunlight which is collected using a tracking receiver is concentrated at the focal line of a spherical reflector. Receiver is containing pipes which carry fluid for thermal changes. Compound parabolic concentrator is also termed as solar bowl.

In past years, there was many type of non-imaging concentrator. Kaiyan et al. [8] have introduced different types of parabolic concentrator which contain many type of dimpled surface that could deliver forward reflective and imaging light beam. Figure-2 shows system which consist a parabolic and a flat contour.

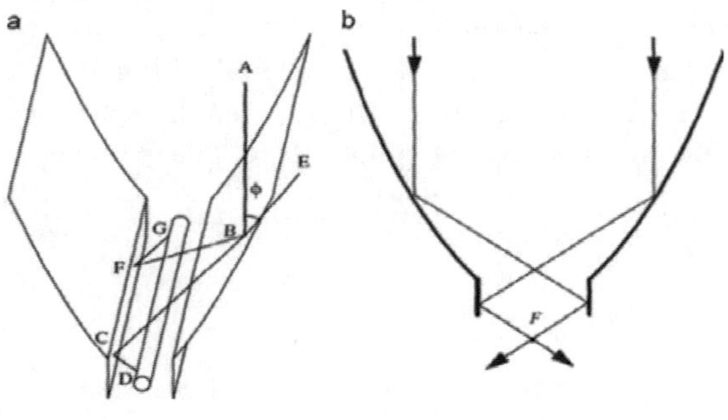

Figure 3: Figure of the different curved surface concentrating collector (a) Light transmission pattern; (b) concentric model and the passed light *[8]*

IOP Conf. Series: Materials Science and Engineering **404** (2018) 012042 doi:10.1088/1757-899X/404/1/012042

2.4. Latent heat storage materials

Phase change materials (PCM) are used as latent energy saving materials because of latent heat absorption, it can largely increase the thermal behavior of buildings compared to conventional building materials [9]. The thermal energy transfer occurs in latent heat storage material when a material changes phase from solid to liquid, or vice-versa [10]. It is unlike conventional storage material phase change material absorbs and release heat at nearly constant temperature. They absorb more heat per unit volume than sensible heat storage material like water or rock. Latent heat storage is most accepted because of high releasing and storing capacity of PCM at a constant temperature and equalizes energy imbalance between production and demand [11]. Here is many type of phase change material like inorganic compounds, organic compounds, salt that absorb and release heat at a range of temperature [12]. Phase change material is used in solar thermal storage, energy efficient buildings etc.

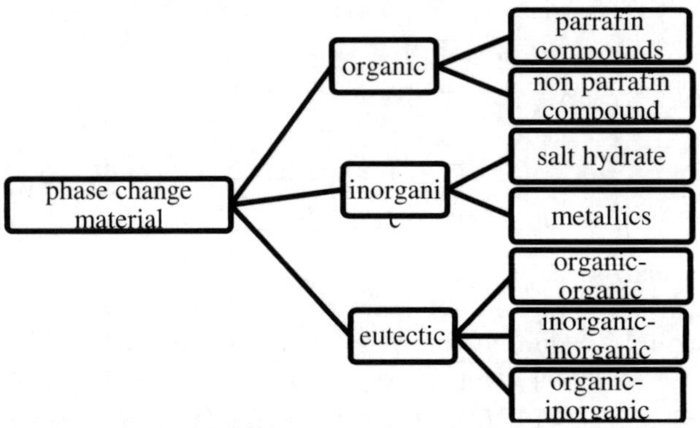

Figure 4: Classification of PCM conventionally applied for latent heat storage [9]

3. Water heating system

Water heating system is basically tow type

(i) Sensible or active heating
(ii) Latent or passive heating

In the sensible heat storage the temperature of the fluid changes at charging or discharging of storage but in the latent heat storage type the temperature of fluid is changes when it goes to change in phase or state. Latent heat storage system have an advantages of less heat loss and high storage capacity as compared to sensible heat storage.

P.B.L Chaurasiya [1] In day time during winter hot water can be obtained at medium temperature range 35 °C to 55 °C by using concrete slab of $2m^2$ area as heat storage. Slight change in the structure of roof or modifying the roof and embedding the network of aluminum pipe which gives a low cost solar storage in terms of water heating. This water heating technique is very easy to construct in buildings. This is the sensible heating process in which solar energy directly used as water heating system.

Atul Sharma et al. [12] Reviewed the availability of solar energy with phase change material with multiple application which is very useful for human being and for energy conservation. The reviewer presents the current research in thermal energy storage with focus on thermodynamic properties of phase change material. The thermal storage used as a part of solar air heating, water heating, cooking and space heating for buildings in night or when solar energy is not available.

Shukla [13] discussed latent heating with PCM at two different type of solar water heater both have different design one has tank type storage and other has joined type of storage using a reflector. They were found to be 45% and 60% efficiency respectively. Both the collector uses paraffin wax as latent energy storage.

Suat et al. [14] Introduced a traditional passive solar water heating system using phase change material sodium thiosulfate penta hydrate ($H_{10}Na_2O_8S_2$) were experimentally examined in November and enhancement of solar water heating system performance with same heating system without phase change material was investigated. It is observed experimentally that total heat absorbed by the thermal energy storage system using phase change material is approximately 2.6 to 3.5 times of traditional sensible heat storage system in this also found that in hydrated salt disodium hydrogen phosphate have higher thermal energy storage.

Cabeza et al. [15] Constructed a plant to observe the behavior of PCM in real life conditions. Several test is done with two, four, six phase change material component were carried out in real setup. 90% volume of sodium acetate tri-hydrate and 10% volume of graphite was picked as the PCM. Author finished that the use of PCM in water tank for heating of water is very worthy technology. Different shape and dimension of concrete slab and pipe embedding ratio is used and efficiency is measured.

4. Problems of solar water heating systems

The use of solar energy for water heating is discovered in late 1800 when open water tank is left in open for warming of water [16]. The key use of solar energy is come under consideration in 20th century. The prospective of solar water heating system is mainly depend on accessibility and availability of solar insolation and cost that is affordable for households [17]. After studying it is seen that common water heating system is not able to generate temperature above 100 °C and this setup is mostly used as preheating of water for the devices where water temperature is above 100 °C.

4.1 Problem

In previous decade solar water heating system gained most popularity due to high energy cost and to avoid harmful environment and developed environment friendly energy sources. Even though people is aware about clean energy its usage and acceptance are still inadequate. The main problem for less acceptability is economic and technical it is different in every region.

4.2 Economic problem

Several economic problem obstructed uses of solar storage system this is cause of awareness issue, lack of energy policy, high initial cost, low level of affordability this problem occur in developing countries.in developed countries poor promotion and most of customer did not know the difference between PV cell system and solar storage system. For solving this economic problem government must take a step and make policies for better regulation, promotion and provide subsidy. To increase awareness about renewable sources benefits promotional event should organized.

4.3 Technical problem

Technical problem which occur in solar water heating system is higher energy loss during night by energy conversion system and the different shape of surfaces is necessary for better absorptivity of solar insolation to avoid this technical problem there must be a reflector or transparent surface which reduce the heat loss during night [18]. Presently solar water heating system have very low efficiency to increase this many research work is needed.

5. Conclusion

This paper has been carried out a critical review of thermal energy storage systems and various design of concrete slab solar collector with uses of latent heat mechanism. These systems are very much beneficial for society and environment also. In this paper advancement of solar collector and phase change materials has been classified for enhancement of efficiency of heating systems. Further technical and economic problems and their outcomes have also been discussed pertaining to solar energy storage systems. The application of such systems are in domestic as well as industrial heating. Still the implementation of concrete slab solar collector has not commercialized due to less awareness. Finally this review recommend the installation of concrete slab solar collector because of its less cost and better efficiency.

References

[1] P. Chaurasia, "Solar water heaters based on concrete collectors," *Energy,* **vol. 25**, pp. 703-716, 2000.

[2] W. A. B. John A. Duffie, Solar Engineering of Thermal Processes, John Wiley & Sons, Inc., Hoboken, New Jersey, 2013.

[3] E. E. A. a. O. V. E. N. V. Ogueke, "A review of solar water heating systems," *renewable and sustainable energy,* 2009.

IOP Conf. Series: Materials Science and Engineering **404** (2018) 012042 doi:10.1088/1757-899X/404/1/012042

[4] J. P. H. P. G. a. G. DATTA, "A SOLAR WATER HEATER WITH A BUILT-IN LATENT HEAT STORAGE," *energy conversion management,* **vol. 25**, pp. 51-56, 1985.

[5] K. n. P. J. Ruchi Shukla, "Recent advancesinthesolarwaterheatingsystems:Areview," *SciVerse ScienceDirect,* **vol. 19**, no. 173-190, 2012.

[6] S. W. S. Rittidech *, "Experimental study of the performance of a solar collectorby closed-end oscillating heat pipe (CEOHP)," *Applied Thermal Engineering,* **vol. 2**, no. 1978–1985, p. 27, 2007.

[7] M. M. F. Amar M. Khudhair, "A review on energy conservation in building applications with thermal storage by latent heat using phase change materials," *Energy Conversion and Management,* **vol. 45**, pp. 263-275, 2004.

[8] H. KaiyanH, "An ovel multiple curved surfaces compound," *SolarEnergy,* **vol. 9**, no. 523, p. 85, 2011.

[9] J. V. J. N. Fre´de´ric Kuznik, "Optimization of a phase change material wallboard for building use," *science direct,* **vol. 28**, pp. 1291-1298, 2008.

[10] M. S. L. a. Y. R. Ahmed Hassan, "Micro encapsulated phase chyange material a Review of Encapsulation, Safety and Thermal Characteristics," *sustainability,* 2016.

[11] A. S. C. C. ,. T.-N. W. Atul Sharma, "Development of phase change materials (PCMs) for low temperature energy storage applications," *Sustainable Energy Technologies and Assessments,* 2014.

[12] V. T. C. C. D. B. Atul Sharma, "Review on thermal energy storage with phase change materials and applications," *Renewable and Sustainable Energy Reviews,* **vol. 13**, pp. 318-345, 2009.

[13] D. B. ,. R. S. Anant Shukla, "Solar water heaters with phase change material thermal energy storage medium: A review," *Renewable and Sustainable Energy Reviews,* **vol. 13**, p. 2119–2125, 2009.

[14] S. A. E. A. G. A. Y. A. F. Canbazoglu S, "Enhancement of solar thermal energy storage performance using sodium thiosulfate pentahydrate of a conventional solar water-heating system," *energy and building ,* **vol. 37**, no. 3, pp. 235-242, 2005.

[15] I. M. S. C. R. J. N. M. Cabeza LF, "Experimentation with a water tank including a PCM module," *solar energy materials and solar cells,* **vol. 90**, no. 9, pp. 1273-1282, 2006.

[16] P. C. E. a. B. N. M. Smyth, "Integrated collector storage solar water heaters," *Renewable and Sustainable Energy Reviews,* **vol. 10**, pp. 503-538, 2006.

[17] A. M. Pallav Purohit, "CDM potential of solar water heating systems in India," *Solar Energy,* **vol. 82**, pp. 799-811, 2008.

[18] L. H. A. S. J. E. G. N. Dukhan, "Initial Analysis of PCM Integrated Solar Collectors," *Solar Energy Engineering,* **vol. 128**, pp. 173-177, 2006.

Energy absorption, deformation and crushing behavior of bi-metallic tubes with different cross-sectional shapes under axial loading

Afaque Umer[1], Bakhtawar Hasan Khan[2], Salman Khan[3], Azaz Alam[4] and Mohd. Reyaz-Ur-Rahim[5]
[1,2,3,4,5] Department of Mechanical Engineering, Integral University, Lucknow, Uttar Pradesh, India.

Abstract- Thin walled tubes because of their lightweight, economic manufacturing and efficient energy absorption characteristics are widely used in safety applications contemporaneously many researchers have investigated the behavior of thin-walled metallic tubes with the different cross section under identical test conditions for assessing its effectiveness in crashworthiness. This paper presents the deformation and energy absorption characteristics of bi-tubular tubes along with several different cross-sectional shapes whose base was set as the boundary condition whereas in reference to the axis of the tubes the load was applied at the opposite end with the help of a rigid wall. The analogy of double-walled tubes with different cross-section alters the deformation modes substantially and resulted in significant increase in the energy absorption capacity over bi- metallic tubes as an effect of stress wave propagation. The overall results divulge that the proposed geometry can be a good alternative due to stabilized deformation mode in safety applications such as a locomotive, automotive industries.

Keyword: Bi-metallic tubes; crashworthiness; cross-section; energy absorption

1. Introduction

The preferment in technology has led to automobiles being able to hit the 300mph milestone with ease contemporaneously increasing the probability of casualty. Keeping in mind for the occupant safety in recent decades a lot of research has been done to improve the crashworthiness characteristics of structures. Hollow tubes are the conventional tubes for energy absorption systems under various loading conditions. Quasi-static axial loading is preferred for finding the effectiveness of any structure. One of the methods of improving the crushing behavior of metallic tubes is by changing their cross-section as energy absorption of bodies depends on their shapes. It is not always preferable to increase the amount of energy absorbed rather a reduction in peak force is also desired for the safety of both the occupant and vehicles. The rendition of the tubular structure is exclusively dependent on either material or the cross-sectional shape. The behavior of metallic tubes under axial loading with different cross-sectional shapes such as circular, rectangular, multi-corner metal columns, hexagonal, polygonal and star-shaped had been studied both theoretically and experimentally for finding the behavior in terms of peak forces modes of buckling and energy absorbing characteristics [1-2]. Z. Fan concluded both experimentally and theoretically that the value of peak force changes accordingly with the variation in a number of corners [3]. Previous research shows that introduction of corrugation perpendicular to the axis of tube results in significant increase in buckling load along with the energy absorption capacity and crushing force efficiency [4-6]. It was also found that the crushing behavior can be stabilized by combining the geometries of the tubular structures [7]. Goel MD compared the behavior of concentric bi-tubular and tri-tubular empty and foam filled deformations in terms of deformation modes and energy absorptions. It was found that bi-tubular and tri -tubular structure was far ahead in terms of energy absorption and crushing behavior pattern [8]. In this research, the effectiveness of the bi-tubular structures under quasi-static axial loading in terms of energy absorption and peak force with different cross-sections, namely circular, rectangular, square, hexagonal and octagonal shapes is presented. The analysis is simulated using non-linear finite element analysis

IOP Conf. Series: Materials Science and Engineering **404** (2018) 012043 doi:10.1088/1757-899X/404/1/012043

Nomenclature

C	Circular	E	Octagonal
S	Square	N	None
R	Rectangular	O	Outer periphery
H	Hexagonal	I	Inner periphery

software ABAQUS/Explicit. A comparative analysis is carried out to demonstrate the optimal design of bi tubular structures with different cross-sectional shapes.

2. Numerical modeling

2.1 Material model

Low carbon steel ASTM A36 and aluminum alloy AA6060 were used for modeling single, bi-walled tubes. The material properties for A36 steel were taken from the tensile test data performed by Z. Fan et al [3] whilst aluminum properties are taken from the experimental testing data provided by Kumar AP et al [7].

Table 1. Parameters for A36 steel and AA6060 alloy.

Specimen	Density (kg/m^3)	E (GPa)	Poisson's Ratio
ASTM A36	7800	210	0.3
AA6060 T4	2700	71	0.33

2.2 Finite element model

The finite element simulation was carried out using ABAQUS/Explicit non-linear element code. The C3D8R element was used to model the tubes whilst the two restraining rigid plates were modeled using a 4 -node 3-D bilinear rigid quadrilateral elements. A total of ten tubular configurations were modeled with single and bi-tubular structures with different cross-sectional shapes combinations were simulated. Primarily the outer tube was kept as circular and diversification was racked up by altering the inner tube and then the antipodal profiles were made by keeping the inner tube circular and varying outer tubular cross section. Throughout the simulation, in all the alteration the mesh size was kept constant. All degree of freedom of the bottom plate was fixed. Quasi-static axial displacement was given to the upper plate to crush the tubes to half of its length. All the analysis was performed by keeping the step time stable at 70 ms. For simulating the contact interaction between the bi-metallic tubes and the two plates general contact algorithm was used with a coefficient of friction of value 0.2.

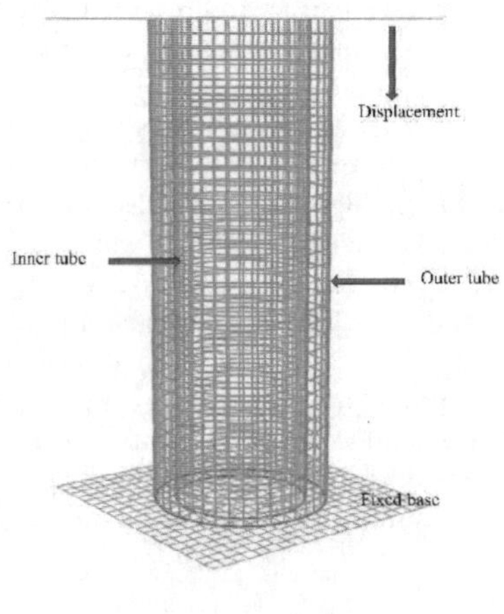

Figure 1 Finite element model of specimen COCI

IOP Conf. Series: Materials Science and Engineering **404** (2018) 012043 doi:10.1088/1757-899X/404/1/012043

Table 2. Dimensions and profile of tubes under study.

Specimen ID	Perimeter(mm) Outer tube	Perimeter(mm) Inner tube	Length (mm)	ThickNess(mm) Outer tube	ThickNess(mm) Inner tube	Profile
CONI	400	-	350	1	-	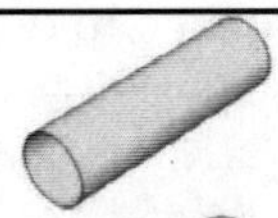
COCI	400	300	350	1.00	1.00	
CORI	400	300	350	1.00	1.00	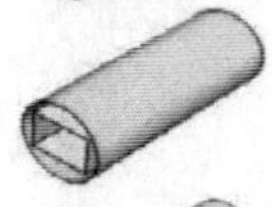
COSI	400	300	350	1.00	1.00	
COHI	400	300	350	1.00	1.00	
COEI	400	300	350	1.00	1.00	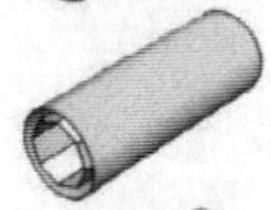
ROCI	400	300	350	1.00	1.00	
SOCI	400	300	350	1.00	1.00	
HOCI	400	300	350	1.00	1.00	
EOCI	400	300	350	1.00	1.00	

3. Results and discussion

3.1 CONI:

In the specimen testing of the CONI model, The values recorded were the lowest as compared to any other geometry probably because of the fact that it is a single-celledthin-walledmember while the other specimens were multi-walled. Due to their multi-wallednature, they have shown appreciable results in terms of the testing parameters. The peak force was recorded at about 10 mm deformation with a value of 104.706 kN.The average crush force was found to be at 56.030 kN and the energy absorbed was about 11.765 kJ, which was least among all the geometries.

3.2 COCI:

IOP Conf. Series: Materials Science and Engineering **404** (2018) 012043 doi:10.1088/1757-899X/404/1/012043

The COCI specimen performed much better than the simple circular tube model and showed a 28.28% increase in the value of Peak force which was found around 125mm of deformation. The value of peak force was recorded to be about 134.916 kN. A similar rising trend was found in case of the values of average crushing force and energy absorbed values which were found to be 79.109 kN and 16.611 kJ respectively. The energy absorption capacity increased by about 1.4 times when compared to the CONI model.

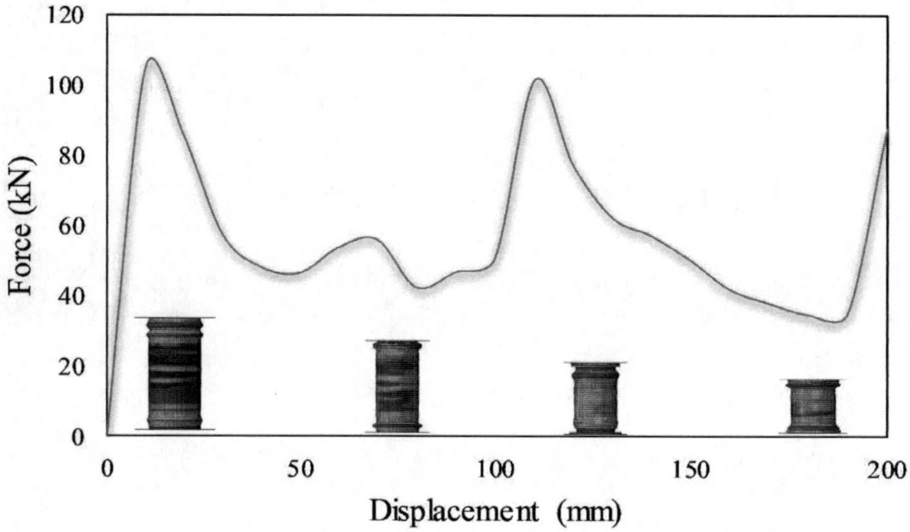

Figure 2. Force vs. displacement curve for specimen CONI

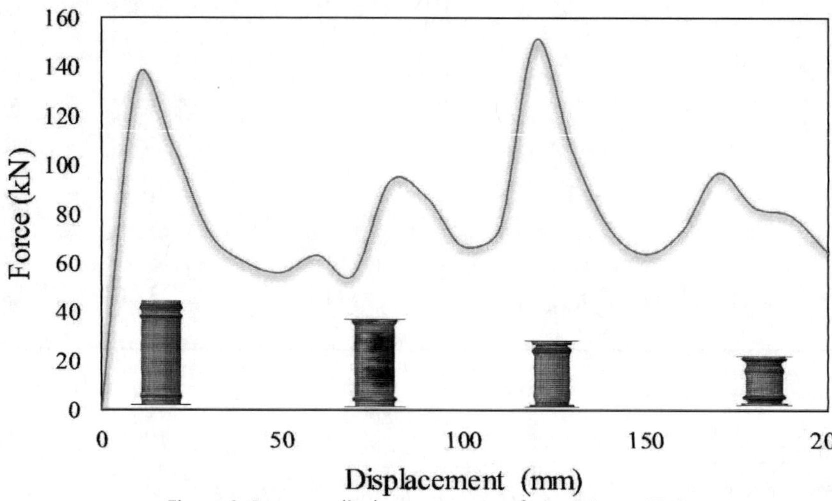

Figure 3. Force vs. displacement curve for specimen COCI

3.3CORI:

The results of the CORI and COCI models were quite similar and comparable but in referenceto CONI models, there was a remarkable difference. The CORI models gave better results for the average crush values and the and energy absorbed values than the COCI models while the value of peak force decline negligibly when compared to the COCI section. The energy absorbed values were about 1.45 times the energy absorbed values in the CONI models also this value of 17.170 kJ was the highest among the specimens having a circle as the outer shell. The value of average crush force was about 81.77 kN. The value of peak force was recorded after 110 mm of deformation.

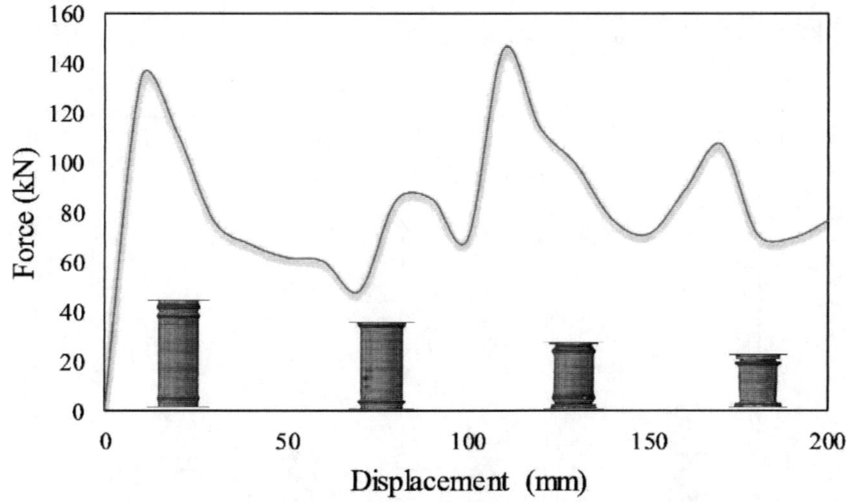

Figure 4.Force vs. displacement curve for specimen CORI

3.4*COSI*:

The CORI and the COSI models were quite contemporary in their performances the values of the Peak force, average crush force and energy absorbed of the above mention model were having hardly a difference of 0.16%. Thus, it can be said that the results obtained by the variation of inner wall as rectangle and square revealed similar results. But In accordance with the CONI models, the peak force increased by about 27.90% having a value of 133.921 kN. The values of average crush force and the energy absorbed also got increased in similar proportion.

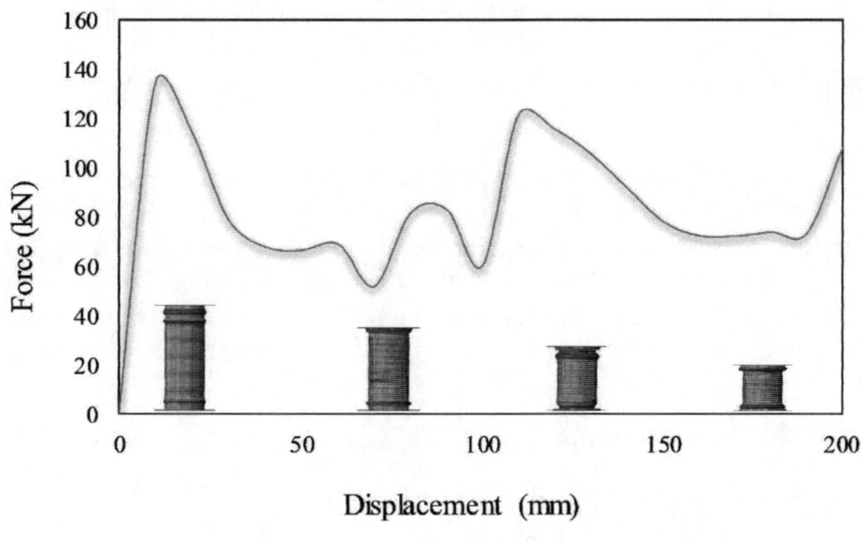

Figure 5. Force vs. displacement curve for specimen COSI

3.5*COHI*:

Among all the models that had the circular section at the outer core the value of peak force was highest for the COHI model and showed a 1.30-fold increase in comparison to CONI model. But the values of average crush force and the energy absorbed were lower when compared to the CORI, COSI and COCI models. The value of average crush force was noted as 78.170 kN, and similarly, the value of energy absorbed was found to be 16.414 kJ. The value of peak force was found to be around 120mm of deformation of the model.

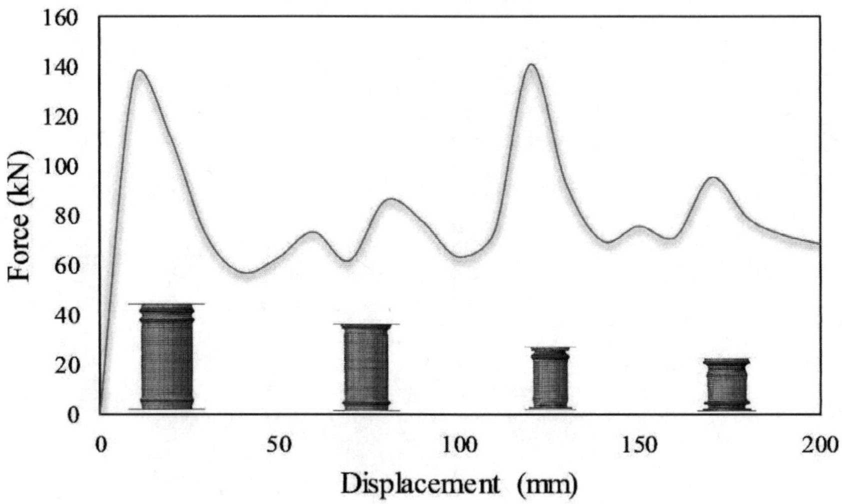

Figure 6. Force vs. displacement curve for specimen COHI

3.6*COEI*:

In this module, the value of peak force was greater than the value of CONI but it was lower than any other value, alternatively, it can be stated that the second highest value of peak force was recorded in the COHI section. A similar situation was observed with the average crush values and the energy absorbed values, the values were greater than the CONI model results and were lower when compared to any other specimen.135.196 kN, 78.170 kN, 16.414 kJ, were the value noted for the peak energy absorption, average crush force and energy absorbed respectively.

3.7*ROCI*:

The peak force for the ROCI was found at about 75 mm of deformation and the value so observed as 138.344 kN. It showed a 32 % increase in the value of peak force when compared to the CONI model. The values of crush force and the energy absorbed showed even better results by registering an 85% increase in the above-mentioned values. The value of average crush force and the energy absorbed were recorded as 103.892 kN and 21.815 kJ respectively. Thus it can be said that the geometries with a polygon as outer shell gave much better values than the specimens having a circle as their outer periphery.

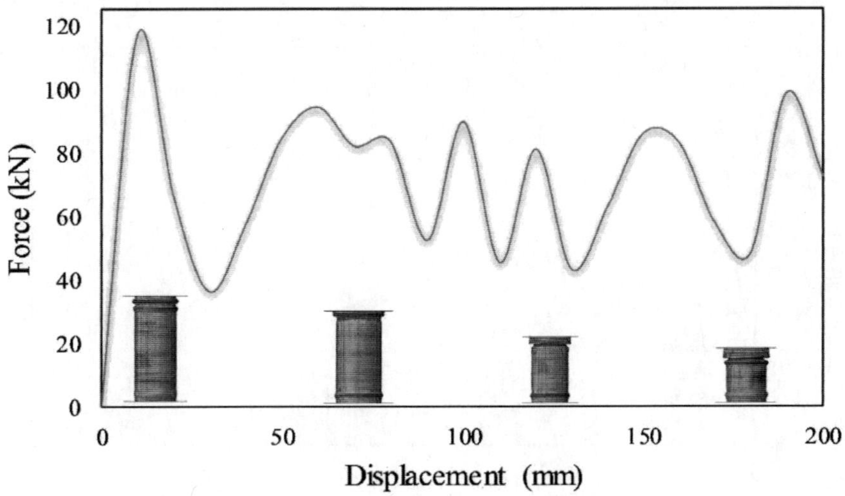

Figure 7. Force vs. displacement curve for specimen COEI

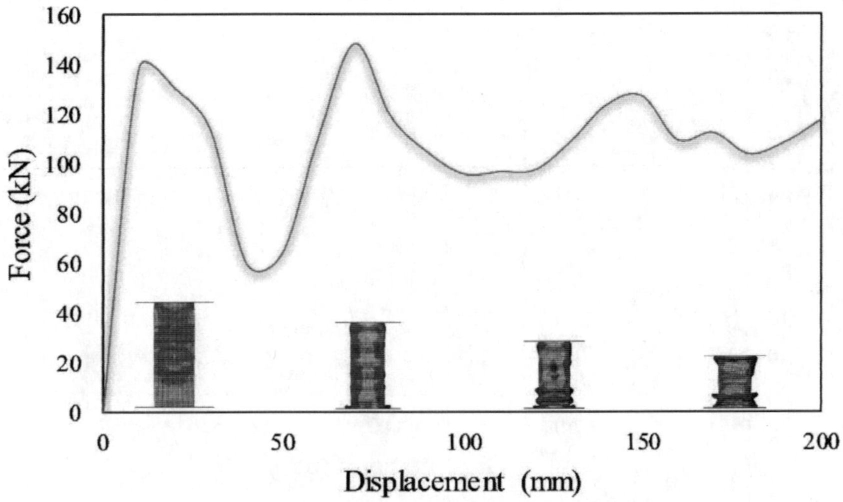

Figure 8. Force vs. displacement curve for specimen ROCI

3.8 SOCI:

Among all the above numerical simulation specimen the best results were obtained in the SOCI geometry the values of peak force, average crush force and the energy absorbed was recorded in this specimen the values of crush force and the energy absorbed increased by 94% in reference to the CONI model and the value of peak force showed 1.32 times increase in comparison to the CONI model. The value of peak force was recorded at instance before the 200mm deformation of the specimen.

3.9 *HOCI*:

The values of the HOCI model was slightly less than the SOCI model but it was quite comparable with the ROCI. The peak force was recorded as 138.570 kN, the crush force as the 100.975 kN while the value of energy absorbed was 21.203 kJ. The values may be slightly lower than the SOCI model but it appreciably good when we consider the geometries with outer shell as a circle.

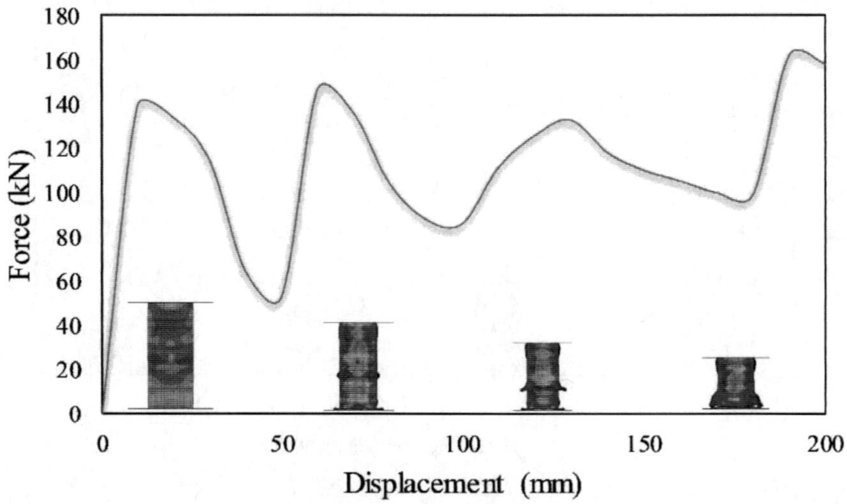

Figure 9. Force vs. displacement curve for specimen SOCI

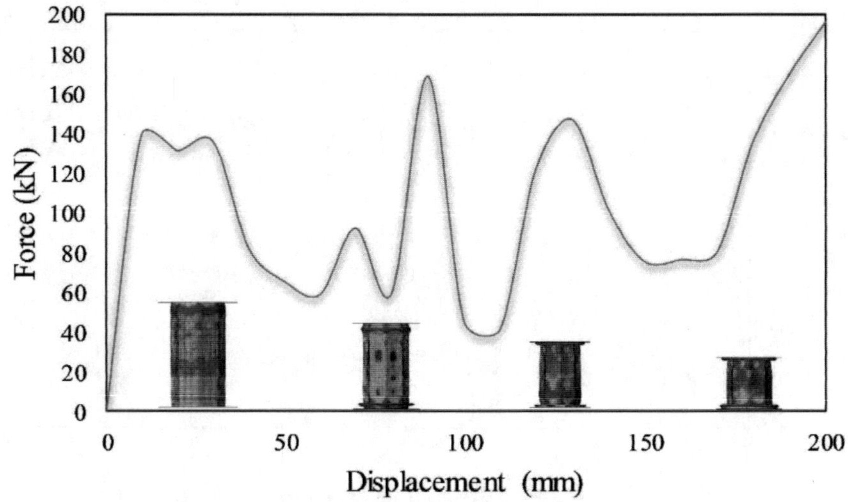

Figure 10. Force vs. displacement curve for specimen HOCI

3.10 *EOCI*:

The values of the EOCI were quite similar to the values of the SOCI, that is to say, that the values so noted were having appreciably high values in terms of peak force, average crush force and energy absorbed. This model outperformed the other models.

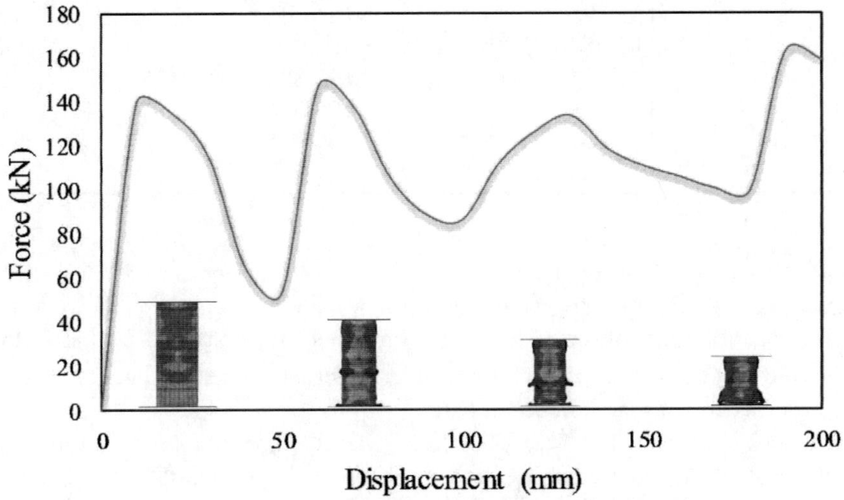

Figure 11. Force vs. displacement curve for specimen EOCI

A comparison of peak force values of the various specimen is presented in Figure 12.

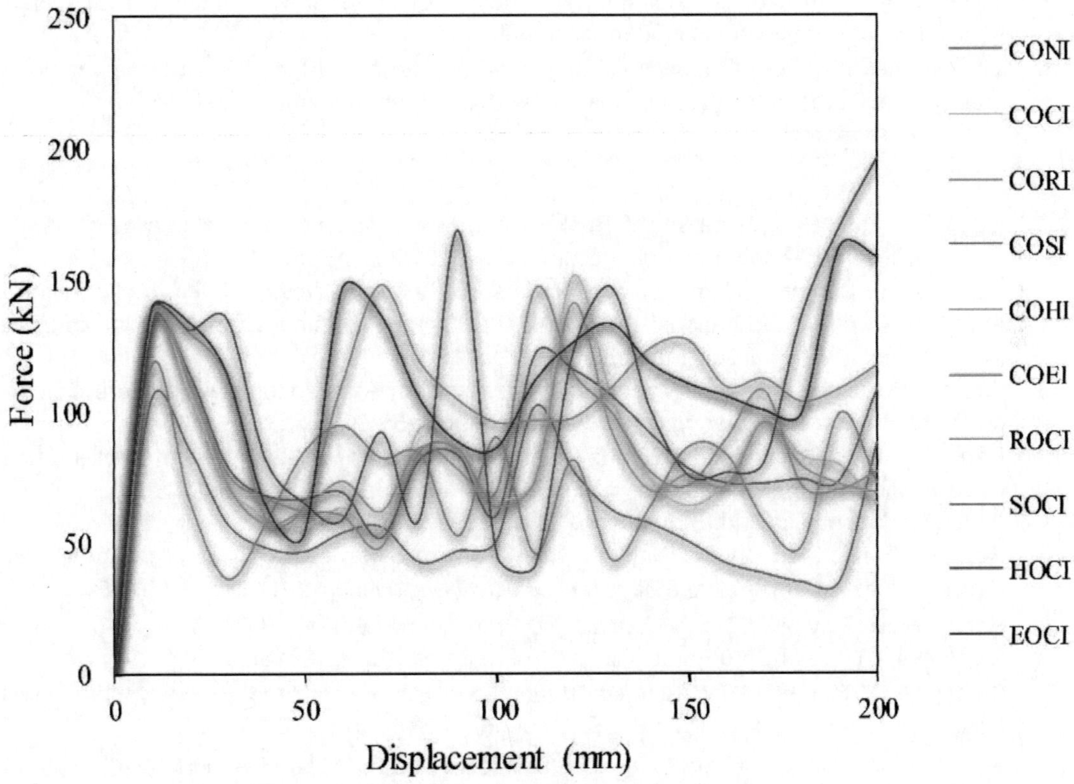

Figure 12. Force vs. displacement curve for all specimens

Table 3. Summary of results of different profiles (stroke: 200 mm)

Specimen	Peak force (kN)	Average crush force (kN)	Energy (kJ)
CONI	104.706	56.030	11.765
COCI	134.916	79.109	16.611
CORI	133.630	81.77	17.170
COSI	133.921	81.651	17.145

COHI	135.196	78.170	16.414
COEI	117.052	68.591	14.402
ROCI	138.344	103.892	21.815
SOCI	138.938	108.876	22.862
HOCI	138.570	100.975	21.203
EOCI	138.938	108.876	22.862

4. Conclusion

The main aim of the paper was to analyze the different cross sections of the bimetallic tubes on the basis of the parameters like Energy absorption, deformation and crushingbehavior, On the basis of the numerical simulations the results of the mentioned geometries can be summarized as follows:

- It can be said that the values of peak force, average crush force, and energy absorption does vary in accordance with the number of sides.
- In some cases, like HOCI, COCI, ROCI, EOCI, SOCI, etc. The peak force values are not found at the starting of the crushing, rather it happens at a much later stage, this phenomenon happens due to the densification of the material.
- Geometries having polygon as the outer shell revealed better results than the geometries having the circular shells as their outer shell. In reference to the test parameters, i.e. peak force, average crush values, and energy absorbed.
- The minimum values of the crushing force and energy absorbed were found in the CONI model, because of the absence of an inner wall.
- The maximum Values of the simulation results were found for the geometries having polygon cross section at the outer shell and that too with a maximum number of sides.

References

1. Liu M, Zhang L, Wang P, Chang Y. Buckling behaviors of section aluminum alloy columns under axial compression. Engineering Structures. 2015 Jul 15;95:127-37.
2. Tarlochan F, Samer F, Hamouda AM, Ramesh S, Khalid K. Design of thin wall structures for energy absorption applications: Enhancement of crashworthiness due to axial and oblique impact forces. Thin-Walled Structures. 2013 Oct 1;71:7-17.
3. Fan Z, Lu G, Liu K. Quasi-static axial compression of thin-walled tubes with different cross-sectional shapes. *Engineering Structures.* 2013 Oct 1;**55**:80-9.
4. Rahim, M. R. U., Akhtar, S., & Bharti, P. K. (2017). Comparative Analysis of Buckling Load of Circular and Corrugated Tubes by Utilizing Key Performance Indicators. International Journal of Applied Mechanics and Engineering, **22**(3), 789-797.
5. Reyaz-Ur-Rahim M, Bharti PK, Umer A. Axial Crushing Behaviors of Thin-Walled Corrugated and Circular Tubes-A Comparative Study. *Technological Engineering.* 2017 Oct 26;**14**(1):5-10.
6. Rahim, Mohd Reyaz Ur, Shagil Akhtar, and Prem Kumar Bharti. "Finite Element Analysis for the Buckling Load of Corrugated Tubes." Vol-**2**, Issue-7, July (2016): 935-939.
7. Kumar AP, Mohamed MN. Crush performance analysis of combined geometry tubes under axial compressive loading. Procedia Engineering. 2017 Jan 1;173:1415-22.
8. Goel MD. Deformation, energy absorption and crushing behavior of single-, double-and multi-wall foam filled square and circular tubes. Thin-Walled Structures. 2015 May 1;90:1-1

IOP Conf. Series: Materials Science and Engineering **404** (2018) 012044 doi:10.1088/1757-899X/404/1/012044

Interaction of water soluble polyacrylic acid with mild steel / hydrochloric acid interface

Abhishek Dwivedi[1], P.K. Bharti[1], Sudhish Kumar Shukla[2*]

[1] Department of Mechanical Engineering, Integral University, Lucknow-226026 India
[2] Department of Chemistry, Manav Rachna University Faridabad-121004 India

* Corrsponding Author

E-mail: sudhish.shukla@gmail.com[2]; abhishek@iul.ac.in[1]

Abstract: Many of the polymers such as polyanilines, polyanthranilic acids have been reported as one of the efficient corrosion inhibitors for mild steel in acidic media. In view of the major limitation of insolubility of polymers, we have taken water soluble polyacrylic acid for the corrosion inhibition process. The corrosion inhibition property of polyacrylic acid is demonstrated for mild steel in 0.5 M hydrochloric acid using electrochemical impedance spectroscopy, Tafel polarization and weight loss methods. All the results are found to be in well correlation and the inhibition efficiency shows upto 94% in 0.5 M hydrochloric acid for 3hrs duration. Surface studies are also done by scanning electron microscopy

Keywords: Polymer, Weight loss, Electrochemical calculation, Inhibition

1. Introduction

Deterioration and degradation of metal due to environmental circumstances is known as corrosion. Main cause of corrosion is environmental moisture which leads to the oxidation of metal surface. Metals in ores occur in mixed form of oxides, sulphates; carbonates etc. from those pure metals are extracted. These pure metals are highly unstable and react easily with oxygen and water contents in presence of carbon dioxide.

To reduce the effect of corrosive environment on metals corrosion inhibitors are used. These inhibitors are coated over the metal surface to inhibit the contact of metal surface and moisture. These inhibitors are working on the phenomenon of adsorption and can be explained by mainly Langmuir adsorption isotherm [1]. There are different types of inhibitors used, such as heterocyclic organic compounds, organic dyes, drugs natural products etc. but most efficient corrosion inhibitors are soluble polymers.

Acids are widely used as pickling, descaling, cleaning and for other purposes in different industries [2-6]. Hydrochloric acid is highly corrosive to the metals or mild steel (alloy of pure metals). So, the efficiency of water soluble polymers as corrosion inhibitors is checked in the same solution of 0.5 M hydrochloric acid. Efficiency can be calculated by gravimetric analysis, potentiodynamic polarization and electrochemical impedance spectroscopy. This is also supported by the surface studies done by Atomic force microscopy.

In recent years, studies on corrosion inhibitors focused on different polymers as inhibitor such as polyaniline, polyanthranilic acid, poly-diphenyl amine etc. [7-9]. Most of the polymeric materials are insoluble in nature. This is the major drawback of polymeric material to be used as corrosion inhibitors. For this purpose polymeric materials can be sulphonated, which increases the solubility of polymeric material in acidic medium. Polymeric compounds interact with the metal surface through the pi bond present in moiety [10].

Various factors like steric hindrance, aromaticity and electronic structure of inhibitor molecule affect the adsorption of corrosion inhibitor compounds. In addition of those functional groups;

electron density at donor site, size of the molecule and molecular weight also affect the corrosion inhibition process.

Polymers are excellent corrosion inhibitors for metallic artifacts in acid environment [16-21]. Presence of a small quantity of polymers may be effective in inhibiting the corrosion of metals in aqueous acidic environment [22-23]. The highly insoluble nature of most of the polymers in aqueous medium is the major limitation in the application of corrosion inhibition. Mainly polymers were used as coating material instead of inhibitors. Recently, several attempts have been made to solubilize the polymers for the use of corrosion inhibitor [24-26].

In this article we present the inhibitive behavior of water soluble polyacrylic acid (PAA) on mild steel sample in 0.5M hydrochloric acid solution. The surface coverage of PAA over mild steel surface in corrosive environment was studied using Atomic force microscopic techniques.

2. Experimental:

Polyacrylic acid (PAA) was purchased from Merck India and sulphonated it in laboratory to increase solubility in water.

Weight loss and electrochemical studies were done on mild steel strips of composition reported elsewhere [1]. Hydrochloric acid (AR grade) procured from Merck and double distilled water is used for the dilution of aggressive medium in present study.

2.1. Weight loss Studies

Weight loss measurements were done on the $10cm^2$ surface area of mild steel sample for 3hrs duration and all the experiments were conducted in 0.5M HCl at 35^oC for 3 hrs duration. The complete studies were done in the conical flask having glass stopper with different concentration of PAA inhibitor to obtain the optimum inhibitor concentration. Such studies were conducted for optimum concentration of PAA on various temperature ranges, time interval, and concentration of acid. Inhibition efficiency (%) and surface coverage (*θ*) of the PAA on mild steel were obtained by: [27,28]

$$I.E.(\%) = \frac{W_o - W_i}{W_o} \times 100 \qquad \text{(1)}$$

$$\theta = \frac{W_o - W_i}{W_o} \qquad \text{(2)}$$

W_o and W_i represent the weight loss of metal in the absence PAA inhibitor and in the presence of PAA inhibitor concentration respectively.

2.2. Electrochemical Studies

Three electrode cell assembly was used to conduct all the electrochemical studies including tafel polarization, linear polarization and electrochemical impedance spectroscopy [29, 30] at 35°C. Mild steel sample will be used as working electrode. 1 cm^2 area of the mild steel was exposed and the rest of the strip covered as reported elsewhere [1]. Standard calomel electrode is used as reference electrode whereas Counter electrode is made up platinum. All the experiments were conducted on Potenentiostat / galvanostat of Gamry instruments Inc. USA, (model G-300) and EIS software. All the electrochemical measurements were carried out after stabilization of the open circuit potential in the aerated enveronment.

Tafel polarization, linear polarization resistance and electrochemical impedance spectroscopy experiments were carried out on the detail parameters reported elsewhere [1]

3. Result and discussion:

3.1. Weight-loss method

3.1.1. Effect of inhibitor concentration

Weight loss techniques were done for the various inhibitor concentration of PAA. The weight loss in mg, % inhibition efficiency and corrosion rate in mmpy were listed in table 1. Trend of the inhibition efficiency with respect to increase in PAA concentration graph is plotted in figure1. Tendency of increment of % inhibitor efficiency with the increase in PAA concentration in 0.5 M hydrochloric acid solution is already revealed from figure1. From table 1, it is clearly visible that the % inhibition efficiency increases and corrosion rate decreases with the increment of concentration of PAA. Maximum inhibition efficiency shown at 100 ppm concentration of PAA used i.e. upto 94.6%. The graphical representations of error bar have also been shown in figure1 represent the variability of data used to indicate the error or uncertainty in a reported measurement. It gives a general idea of mean, standard deviation and precision of measurement during triplicating observation.

Table 1. Corrosion inhibition data from weight loss studies of mild steel of surface area $10cm^2$ for 3hrs at 35°C in aqueous acidic solution of 0.5N hydrochloric acid in absence of inhibitor and in presence of various concentrations of Polyacrylic acid.

S.No	Inhibitor concentration (ppm)	Weight loss (mg)	*IE* (%)	CR (mmpy)
1.	Nil	81.05	-	30.18
2.	15	41.82	48.4	15,57
3.	25	27.95	65.5	10.41
4.	50	16.85	79.2	6.27
5.	75	8.59	89.4	3.20
6	100	4.37	94.6	1.63

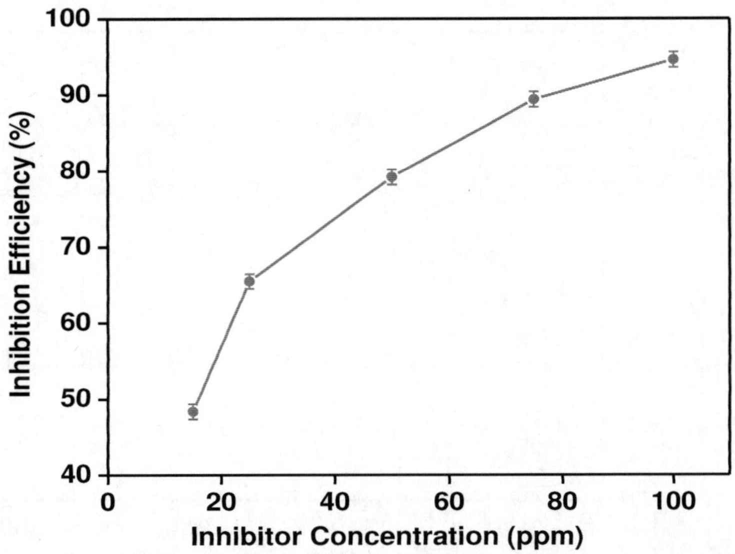

Figure 1. Change in % inhibition efficiency with the varying concentration of Polyacrylic acid in 0.5M HCl on the surface of mild steel at the temperature 35°C in 3hrs immersion time

3.1.2. Effect of Immersion time

Figure 2 shows the % inhibition efficiency of PAA against the variation in immersion time. From the figure2, it is clearly revealed that the inhibition efficiency slowly decreases with increase in time interval. This result is suggesting that the PAA inhibitor is adsorbed firmly on the surface of mild steel and the very slow rate of desorption of inhibitor molecule [27].

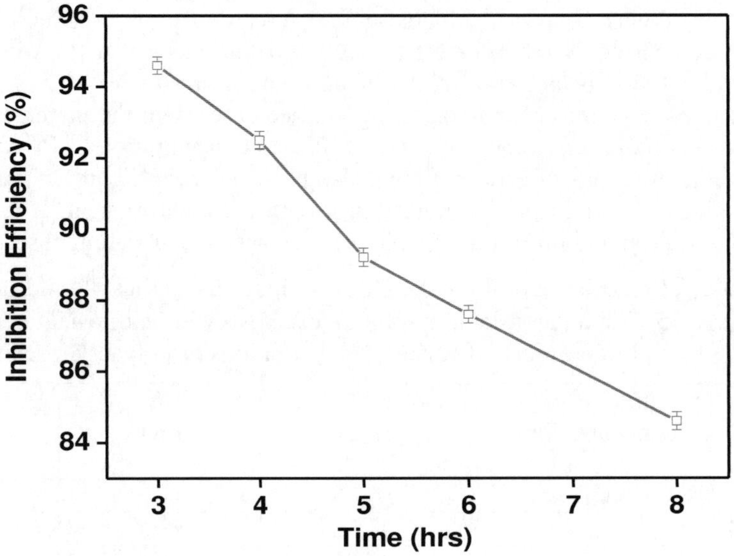

Figure 2. Change in % inhibition efficiency in 0.5M hydrochloric acid on the surface of mild steel of $10cm^2$ of exposed area with different time intervals at 100 ppm concentration of PAA at 35^oC temperature

3.1.3. Effect of solution temperature

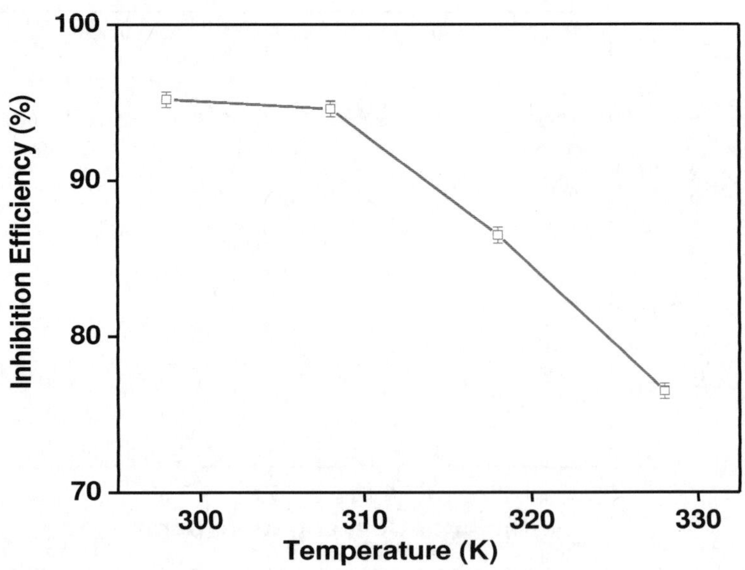

Figure 3. Trend in % inhibition efficiency in 0.5M HCl on the surface of mild steel of $10cm^2$ exposed areas at varying temperature range for 3hrs immersion time at optimum concentration of PAA inhibitor

Effect of temperature on % inhibition efficiency graph is shown in figure 3. This was studied at the 100 ppm of the concentration of PAA at wide temperature range of 298-328K. It is evident from graph that the % inhibition efficiency is lowering with increase in the temperature of the aquous acid solution of hydrochloric acid. Decrease in the % inhibition efficiency with the rise of solution temperature may due to the desorption of the molecule of PAA inhibitor from the surface of mild steel [1]

3.1.4. Effect of acid concentration

Figure 4 clearly shows the trend of % inhibition efficiency with the variation of the acid concentration. It is clearly evident from the figure4 that the inhibition efficiency is decreasing with the increase in the hydrochloric acid concentration. This clearly indicates that the PAA is a very good corrosion inhibitor for the less acidic medium. It gives best result in 0.5M hydrochloric acid solution.

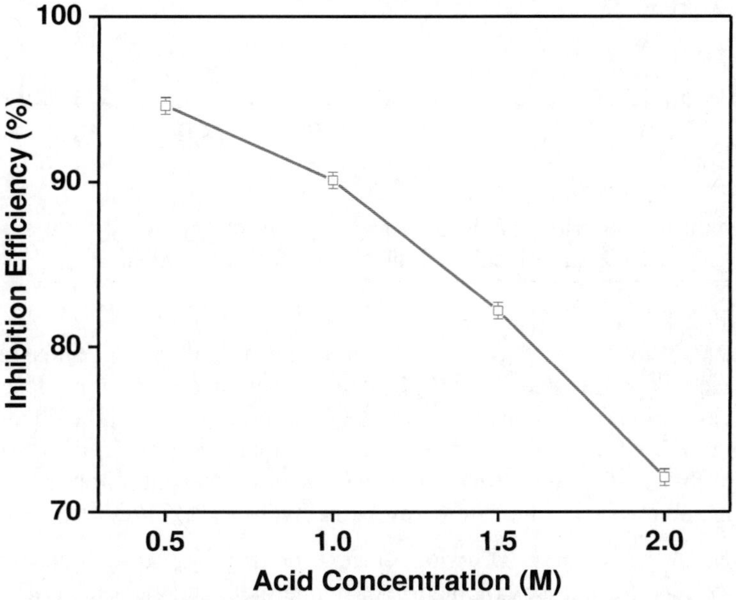

Figure 4. Trend in % inhibition efficiency on surface of mild steel of exposed area 10cm^2 at 35°C at 100 ppm concentration of PAA immersed for 3 hrs with different acid concentration

3.1.5. Adsorption isotherm

"Adsorption isotherm study is the tool to explain the behavior of adsorption of PAA molecule on the surface of the metal to know the mechanism of adsorption. Langmuir, Temkin, and Frumkin are the most widely used isotherms to explain the mechanism of adsorption" [9]. The surface coverage values (θ) required for the adsorption isotherm calculation was deduced from the weight loss parameters given in table 1.

In all the three isotherms were tested and Langmuir adsorption isotherm were found most appropriate adsorption isotherm to explain. Langmuir adsorption isotherm is expressed by:

$$\frac{C_{inh}}{\theta} = \frac{1}{K_{ads}} + C_{inh} \qquad \text{.................................... (6)}$$

where, concentration of PAA corrosion inhibitor represented by C_{inh} and equilibrium constant of adsorption is represented by K_{ads} in acid medium. Straight trend line with the rigration coefficient (R) is almost unity which is obtained by C_{inh} vs C_{inh}/θ graph (Figure 5). Unity value of R represent the best description of adsorption behavior through Langmuir adsorption isotherm [31,32].

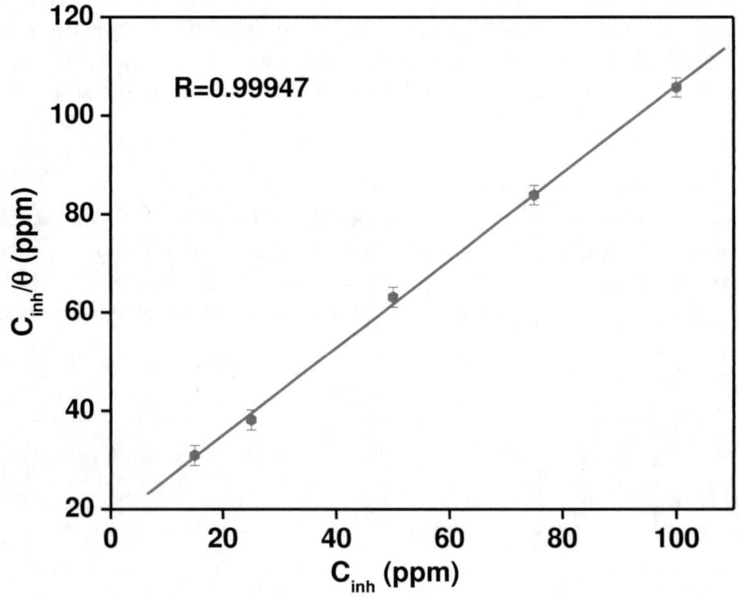

Figure 5. Langmuir's adsorption graph for the adsorption of PAA inhibitor molecules in 0.5M Hydrochloric acid solution on surface of mild steel

3.2. Tafel polarization study:

Figure 6 represents the effect of PAA corrosion inhibitor on both anodic and cathodic polarization curves on the surface of mild steel in 0.5M hydrochloric acid solution. The potentiodynamic polarization study was carried out to understand kinetics of both anodic and cathodic reactions. After examination of figure 6, both the cathodic and anodic reactions were minimising with the addition of the PAA inhibitor in this study. This result gives the inference that inhibitor reduced the reaction of anodic dissolution and on the same time it suppresses the hydrogen evolution reactions on cathode.

Corrosion potential (E_{corr}) and corrosion current density (I_{corr}) like electrochemical kinetic parameters can be obtained from the extrapolation of both anodic and cathodic polarization curves and the values are reported in Table 2. Decrease in the corrosion current density (I_{corr}) by addition of inhibitor molecule. Variation in more than 85mV corrosion potential with respect to blank solution, results the cathodic or anodic type of inhibitor [33, 34]. In this study, change in corrosion potential between tafel plot uninhibited and inhibited sample is only 02mV which clearly indicates that the PAA is a mixed type of inhibitor. % inhibition efficiency calculated from tafel polarization is in good agreement with the weight loss results.

Table 2. Tafel polarization and linear polarization data for mild steel corrosion without and with PAA inhibitor in 0.5M aqueous hydrochloric acid solution

Inhibitor Conc. (PPM)	Tafel data					Linear Polarization data	
	$-E_{corr}$ (mV vs SCE)	b_a (mV dec^{-1})	b_c (mV dec^{-1})	I_{corr} ($\mu A\ cm^{-2}$)	IE (%)	R_p ($\Omega\ cm^2$)	IE (%)
Nil	469	75	150	2140	-	20.1	-
100	467	69	158	148	93.1	395.5	94.9

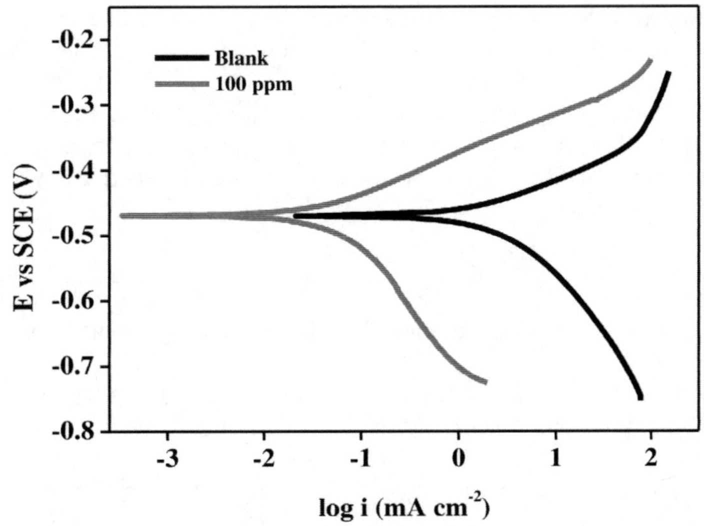

Figure 6. Tafel polarization curve in absence of inhibitor and presence of optimum concentration PAA inhibitor molecule on the surface of mild steel in 0.5M hydrochloric acid solution.

3.3. *Linear polarization Study:*

Polarization resistance parameters and % inhibition efficiency were calculated and reported in table 2. Calculation of Polarization resistance from the slope of the potential current lines by:

$$R_P = A\frac{dE}{di} \qquad \text{...(7)}$$

In above reaction A is area of the working electrode surface, dE and *di* is change in potential, and change in current respectively. It is clearly revealed from the data listed in table2 that the inhibition efficiency increases with the addition of corrosion inhibitor.

3.4. *Electrochemical impedance spectroscopy:*

Electrochemical impedance spectroscopy measurements were carried out to investigate the corrosion behavior of mild steel surface in 0.5 M aqueous hydrochloric acid with and without PAA inhibitor. A Nyquist plot of the impedance spectra of the corrosion of mild steel is shown in figure 7. In the figure 7, at high frequency a single semicircle has clearly been shown which attributed to charge transfer of the process of corrosion and the diameter of semicircle was increased with addition of inhibitor. In the figure 8 impedance spectra shows depressed semicircle under real axis and look like a depressed capacitive loop. This is due to the heterogeneity on the mild steel surface. This heterogeneity may arise by the surface roughness, distribution of the active sites dislocations or adsorption of the inhibitor molecules [35-37]. Electrochemical impedance data must be correlated with various theoretical circuit proposed by F. Mansfeld. The equivalent circuit having constant phase angle was introduced to explain the impedance data which is applicable to explain the iron / acid interface model [38]. From the fitting of equivalent circuit of electrochemical impedance parameters (R_s, R_t Y_o and n) were obtained and reported in Table3. Constant phase angle calculation was done as reported by J.R. McDonold [39]. In the iron/acid interface, roughness or uneven current distributions on the electrode surface ideal capacitor behavior is not observed which results the dispersion in frequency [36,40,41]. C_{dl} values were derived and listed in table3 from the Constant phase element parameters calculated by using equation [42]:

$$C_{dl} = (Y_{o.}\, R_t^{1-n})^{1/n} \qquad \text{...(9)}$$

It is clearly evident from the data represented in table 3 that the R_t values increases with the addition of optimum concentration of PAA corrosion inhibitor. Increase in R_t value is attributed that inhibitor molecule from a protective film on the interface of metal and solution. The higher value of n in the inhibited sample in comparison to the uninhibited sample is due to the decrease in surface heterogeneity. Surface heterogeneity may decrease due to adsorption of PAA inhibitor molecules on the active adsorption sites [43]. Double layer capacitance (C_{dl}) values were decreased with addition of PAA inhibitor. Following equation shows the relation between Protective layer thicknesses (d) and double layer capacitance C_{dl} [44]:

$$C_{dl} = \frac{\varepsilon \varepsilon_o}{d} \quad \text{..(10)}$$

In above equation ε is dielectric constant of protective layer and ε_o is permittivity of free space.

Table 3: Electrochemical impedance parameters calculated without inhibitor and with the optimum concentration of PAA inhibitor molecule for the corrosion of mild steel in 0.5M hydrochloric acid solution

Inhibitor concentration (PPM)	R_s ($\Omega\ cm^2$)	R_p ($\Omega\ cm^2$)	Y_o ($\mu F\ cm^{-2}$)	N	C_{dl} ($\mu F\ cm^{-2}$)	IE (%)
Nil	1.23	8.3	1181.5	0.858	55	-
100	1.19	260.2	69.3	0.812	27	96.8

Inhibition efficiency data reported in Table 3 is in quite good agreement with the other techniques used in the studies.

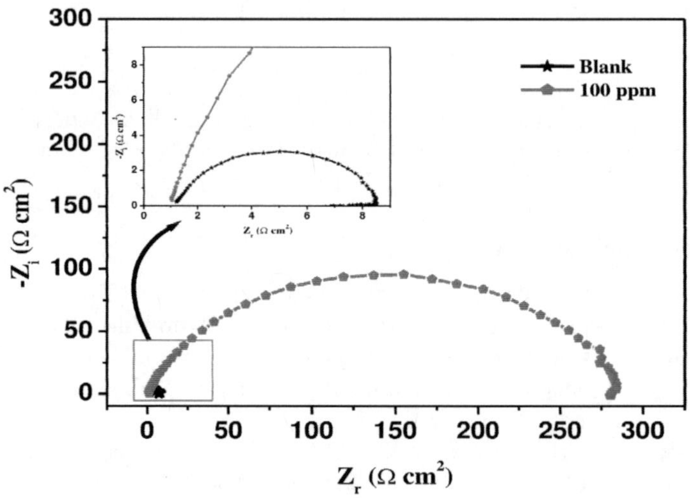

Figure 7: Nyquist plot of corrosion of mild steel in absence and presence of optimum concentration of PAA inhibitor in 0.5M hydrochloric acid solution

3.5. *Surface Morphology of Specimen*

Surface morphology of the mild steel sample was studied before and after corrosion in presence and absence of the inhibitors. The surface morphology supported the formation of adsorbed polymer film over the steel samples. In the figure 8 (b) it is clearly shown in that the steel samples are getting cracks due to the acid corrosion. However, the presence of polyacrylic acid as inhibitor retard the corrosion and surface showed more smoothness in figure 8 (c) which is less smooth than figure 8 (a) which is more smooth as the polished one.

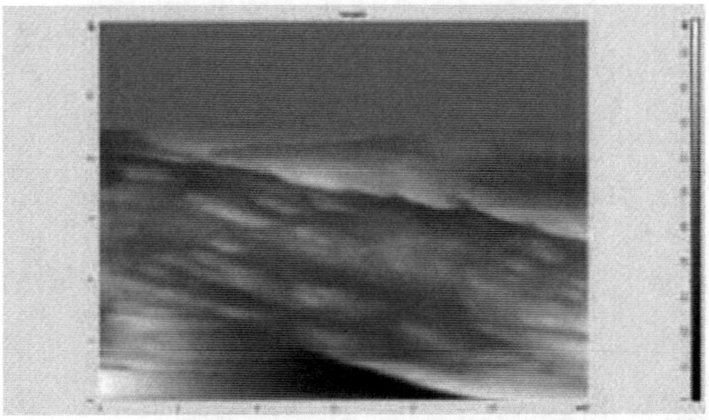

(a) Polished sample

(b) sample immersed for 3 hrs in 0.5 M HCl without inhibitor

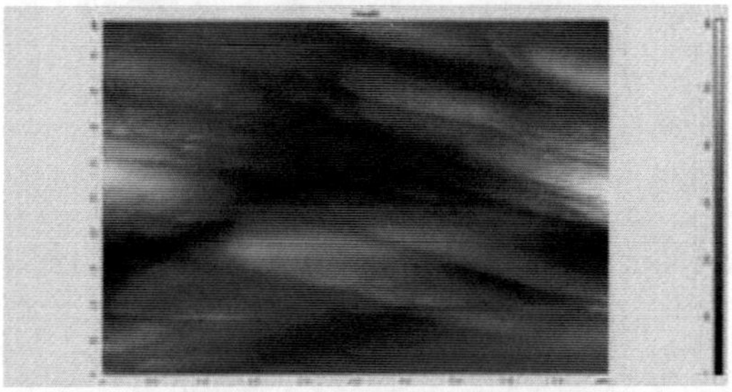

(c) sample immersed for 3 hrs in 0.5 M HCl with 100 ppm PAA inhibitor

Figure 8. Surface morphological analysis through atomic force microscopy of mild steel sample

4. Conclusion:

- PAA exhibit good inhibition towards corrosion on the surface of mild steel in the medium of aqueous hydrochloric acid.
- From the potentiodynamic polarization techniques it is clear that the used PAA inhibitor exhibit mixed-type character and alter both the anodic as well as cathodic type of reactions.
- Increase in the PAA inhibitor concentration up to the optimum concentration of 100 ppm results the increase in inhibition efficiency.
- These studies suggest that the polyacrilic acid exhibit good corrosion inhibition property towards the mild steel surface in the lower concentration of aqueous hydrochloric acid solution.
- Good inhibition property of PAA against the mild steel corrosion was also supported by the atomic force microscopic characterization.

5. Acknowledgement: The author (Abhishek Dwivedi) gratefully acknowledges Dr. Minaxi B. Lohani, Associate Professor, Department of Chemistry, Integral University, for her valuable suggestions. The manuscript communication number: - IU/R&D/2018-MCN000416 has been provided by the Research and Development Cell, Integral University, Lucknow to this article.

6. References:

1. S.K. Shukla, Corrosion inhibitive effect of N-(6-methoxyquinolin-8-yl)-pentane-1,4-diamine on mild steel / sulphuric acid solution interface; *J. Mater. Environ. Sci.* 6(2015) 1779-1786.
2. M. Lagrenee, B. Mernari, M. Bouanis, M. Traisnel, F. Bentiss, Studies of the mechanism and inhibiting efficiency of 3,5-bis(4-methylthiophenyl)-4H-1,2,4-triazole on mild steel corrosion in acidic media *Corros. Sci.* 44 (2002) 573-588.
3. M.A. Quraishi, R, Sardar, Aromatic triazoles as corrosion inhibitors for MS in acidic environment, *Corrosion* 58 (2002)747-756.

4. M.A. Quraishi, M. Athar, H. Ali, Corrosion inhibition of carbon steel in hydrochloric acid by organic compounds containing heteroatoms; *Brit. Corros. J.* 37 (2002) 155-158.
5. M.A. Quraishi, F.A. Ansari, Corrosion inhibition by fatty acid triazoles for mild steel in formic acid; *J. Appl. Electrochem.* 33 (2003) 233-238.
6. M.A. Quraishi, S. Khan, Inhibition of mild steel corrosion in sulfuric acid solution by thiadiazoles; *J. Appl. Electrochem.* 36 (2006) 539-544.
7. M. Abdeli, N.P. Ahmadi, R.A. Khosroshahi, Nile Blue and Indigo Carmine organic dyes as corrosion inhibitor of mild steel in hydrochloric acid; *J. Solid State Electrochem.* 14 (2010) 1317-1324
8. A.S. Fouda, Ali H. El-Azaly, R.S. Awad, A.M. Ahmed, New benzonitrile azo dyes as corrosion inhibitors for carbon steel in hydrochloric acid solutions; *Int.J. Electrochem. Sci.* 9 (1014) 1117-1131
9. S.K. Shukla, A.K. Singh, L.C. Murulana, M.M. Kabanda, E.E. Ebenso Inhibitive effect of Azorubine dye on the corrosion of mild steel in hydrochloric acid medium and synergistic iodide additive; *Int.J. Electrochem. Sci.* 7, (2012) 5057-5068
10. G. Schmitt, Application of inhibitors for acid media; *Britt. Corros. J.* 19(1984) 165-176.
11. F. Bentiss, M. Lagrenee, M. Traisnel, J.C. Hornez, The Corrosion Inhibition of Mild Steel in Acidic Media by a New Triazole Derivative; *Corros. Sci.* 41 (1999) 789–803.
12. E. McCafferty, V. Pravdic, A.C. Zettlemoyer, Dielectric behavior of adsorbed water films on the α-Fe_2O_3 surface, *Trans. Faraday Soc.* 66 (1970) 1720-1731.
13. F.B. Growcok, W.W. Frenier, P.A. Andrezzi, Inhibition of Steel Corrosion in HCl by Derivatives of Cinnamaldehyde: *Part II.* Structure–Activity Correlations, *Corrosion* 45 (1989) 1007
14. I. Lukovits, E. Kalman, G. Palinkas, Nonlinear Group-Contribution Models of Corrosion Inhibition, *Corrosion* 51 (1995) 201–207.
15. R.C. Ayers Jr., N. Hackerman, Corrosion inhibition in HCl using methyl pyridines, *J. Electrochem. Soc.* 110 (1963) 507–513.
16. M. Rohwerder, A. Michalik, Conducting polymers for corrosion protection: What makes the difference between failure and success?, *Electrochim Acta* 53 (2007) 1300–1313.
17. U. Riaz, S.M. Ashraf, S. Ahmad, High performance corrosion protective DGEBA/polypyrrole composite coatings, *Prog. Org. Coat.* 59 (2007) 138–145.
18. C. Jeyaprabha, S. Sathiyanarayanan, G. Venkatachari, Effect of cerium ions on corrosion inhibition of PANI for iron in 0.5M H_2SO_4, *Appl. Surf. Sci.* 253 (2006) 432–438.
19. S. de Souza, Smart coating based on polyaniline acrylic blend for corrosion protection of different metals, *Surf. Coat. Technol.* 201 (2007) 7574–7581.
20. E. Armelin, R. Oliver, F. Liesa, J.I. Iribarren, F. Estrany, C. Alemán, Marine paint formulations: Conducting polymers as anticorrosive additives, *Prog. Org. Coat.* 59 (2007) 46–52.
21. J.E.P. da Silva, S.I.C. de Torresi, R.M. Torresi, Polyaniline / poly(methylmethacrylate) blends for corrosion protection: The effect of passivating dopants on different metals, *Prog. Org. Coat.* 58 (2007) 33–39.
22. R.C. Patil, S. Radhakrishnan, Conducting polymer based hybrid nano-composite for enhanced corrosion protective coatings, *Prog. Org. Coat.* 57 (2006) 332–336.
23. S. Sathiyanarayanan, S. Syed Azim, G. Venkatachari, Preparation of polyaniline-TiO_2 composite and its comparative corrosion protection performance with polyaniline, *Synthetic Met.* 157 (2007) 205–213.
24. C. Jayaprabha, S. Sathiyanarayanan, G. Venkatachari, Polyaniline as corrosion inhibitor for iron in acid solutions, *J. Appl. Poly. Sci.* 101 (2006) 2144–2153.
25. N.M. Martyak, P. McAndrew, Corrosion performance of steel coated with co-polyamides and polyaniline, *Corros. Sci.* 49 (2007) 3826–3837.

26. G.A. Moraga, G.G. Silva, T. Matencio, R.M. Paniago, Poly(2,5-dimethoxy aniline)/ fluoropolymer blend coatings to corrosion inhibition on stainless steel, *Synthetic Met.* 156 (2006) 1036–1042.
27. S.K. Shukla, M.A. Quraishi, R. Prakash, A self-doped conducting polymer "polyanthranilic acid": An efficient corrosion inhibitor for mild steel in acidic solution, *Corros. Sci.* 50 (2008) 2867-2872.
28. G. Ji, S.K. Shukla, P. Dwivedi, S. Sundaram, R. Prakash, Inhibitive effect of Argemone mexicana plant extract on acid corrosion of mild steel, *Ind. Engg. Chem Res.* 50 (2011) 11954-11959
29. P.C. Pandey, R. Prakash, Characterization of Electropolymerized Polyindole; Application in the construction of a solid-state, ion-selective electrode. *J. Electrochem. Soc.* 145 (1998) 4301–4307.
30. J. Mohan, S. Joshi, R. Prakash, R.C. Srivastava, Novel Ni(II) Mixed Ligand Complex Modified Electrode: Catalytic Effect on Anodic Oxidation of Phenol, *Electroanalysis* 16 (2004) 572–576.
31. S.K. Shukla, E.E. Ebenso, Corrosion inhibition, adsorption behaviour and thermodynamic properties of streptomycin on mild steel in hydrochloric acid medium *Int. J. Electrochem. Sci.,* 6 (2011) 3277.
32. S.K. Shukla, M.A. Quraishi, Effect of some substituted aniline-formaldehyde polymers on mild steel corrosion in hydrochloric acid medium, *J. Appl. Poly. Sci.* 124 (2012) 5130-5137.
33. E.S. Ferreira, C. Giancomelli, F.C. Giacomelli, A. Spinelli, Evaluation of the inhibitor effect of l-ascorbic acid on the corrosion of mild steel, *Mater. Chem. Phys.* 83 (2004) 129-134.
34. W.H. Li, Q. He, C.L. Pei, B.R. Hou, Some new triazole derivatives as inhibitors for mild steel corrosion in acidic medium, *J. Appl. Electrochem.* 38 (2008) 289-295.
35. W.R. Fawcett, Z. Kovacova, A.J. Motheo, C.A. Foss Jr, Application of the ac admittance technique to double-layer studies on polycrystalline gold electrodes, *J. Electroanal. Chem.* 326 (1992) 91-103.
36. K. Juttner, Electrochemical impedance spectroscopy (EIS) of corrosion processes on inhomogeneous surfaces, *Electrochim. Acta.* 35 (1990) 1501-1508.
37. A.K. Singh, S.K. Shukla, E.E. Ebenso, Cefacetrile as corrosion inhibitor for mild steel in acidic media, *Int. J. Electrochem. Sci.* 6 (2011) 5689-5700.
38. F. Mansfeld, Recording and Analysis of AC Impedance Data for Corrosion Studies, *Corrosion.* 37 (1981) 301-307.
39. J.R. Macdonald, Impedance spectroscopy and its use in analysing the steady state AC response of solid and liquid electrolytes, *J. Electroanal. Chem. and Interfacial Electrochem.* 223 (1987) 25-50.
40. U. Rammelt, G. Reinhard, The influence of surface roughness on the impedance data for iron electrodes in acid solutions, *Corros. Sci.* 27 (1987) 373-382.
41. J. Pang, A. Briceno, S. Chander, A Study of Pyrite/Solution Interface by Impedance Spectroscopy, *J. Electrochem. Soc.* 137 (1990) 3447-3455.
42. A. K. Singh, S.K. Shukla, M. Singh, M.A. Quraishi, Inhibitive effect of ceftazidime on the corrosion of mild steel in hydrochloric acid solution, *Mater. Chem. Phys.* 129 (2011) 68.
43. M. Lebrini, M. Lagrenee, H. Vezin, M. Traisnel, F. Bentiss, Experimental and theoretical study for corrosion inhibition of mild steel in normal hydrochloric acid solution by some new macrocyclic polyether compounds, *Corros. Sci.* 49 (2007) 2254-2269.
44. H.H. Hassan, Perchlorate and oxygen reduction during Zn corrosion in a neutral medium, *Electrochim. Acta.* 51 (2006) 5966-5972.

IOP Conf. Series: Materials Science and Engineering **404** (2018) 012045 doi:10.1088/1757-899X/404/1/012045

Rolling Barriers: Emerging Concept to Reduce Road Accidents: *An Indian Perspective*

Muhammad Farhan[1], Mohd. Anas[2], Mohammad Azeem[3]

[1,2]Department of Mechanical Engineering, Integral University, Lucknow
[3,] Department of Mechanical Engineering, AMU, Aligarh

E-mail: mufarhan@iul.ac.in, anas@iul.ac.in

Abstract. The latest emerging technologies for road safety focuses on finding ways to avoid or minimize road accidents to road users with special concern by reducing the causes of road accidents. As depicted by data of certain advanced countries like Korea, Malaysia, Australia, United States of America, the major number of accidents causing death was very high during a previous couple of years due to the increased number of vehicles on road, which is getting unmanageable. However, Urethane Rollers invented in Korea has served to re-direct the uncontrolled moving vehicles and to balance it again causing reduction of accidents. In this paper, a study is carried out to explain its need in India perspective for using "Rolling Barriers" (RB) which has minimized the accidents rate in the above-mentioned countries. Rolling barriers provides cushioning effect during a crash, reduces the high-speed effect, constitutes material resilience with stiffness and have other performance characteristics that reduce injury to occupants and damage to the vehicle. The roller barriers are extremely effective and its implementation has given signified results in reducing the road accidents at flat roads, curved roads sections, ramps, medians, entrances/exit ramps in the parking garage etc. steep curved roads as in the mountainous terrain.

Keywords: Accidents, Urethane, Rolling Barriers, Stiffness, Resilience, Crash cushioning.

1. Introduction

The Rolling Barriers are the barrier that absorbs impact energy and converts that impact energy into rotational energy and directs the vehicles forward rather than potentially breaking through an immovable barrier.[1] In India, the year 2016 has recorded the highest number of road accidents around 4,80,652 which comprised 1,50,785 deaths. The Indian Express, dated 11 Sept. 2017 has claimed in the report by Transport Research Wing under Ministry of Road Transport and Highways, Government of India that there are 17 deaths on road every hour in India. As per the records, 34.5 % accident deaths occurred on National Highways, while 27.9 % accident deaths took place on state highways. The report says that over speeding is the biggest cause of casualties i.e. 66.5 % of all road accident with 61% of death due to over speeding. Other factors constitute the unchecked surge in motorization and human errors. Cities with large population saw majority of accidents. Chennai leads with 7486 accidents followed by Delhi 7,375 accidents. [2].

Even though efforts are being made by the Govt. to reduce the severity of road accidents but the results have not been significantly fruitful. Primarily it is based on 4 E's (i.e.) Education, Engineering, Enforcement and Emergency care. Nearly 700 Accidents prone spot have been found out by Ministry of Road Transport and Highways (MRTH) [3]. Safety road roller are also being used as a barrier for safety of the road users in certain countries like Malaysia Road Accessory (RB) prevents driver and passenger from fatal and destructive accident by not only absorbing shock energy but also converting shock energy into rotational energy.[4] Safety roller can be recycled and eco-friendly. Safety roller barriers are easily adjustable by adding or removing a roller.

Figure 1: Rolling Barriers at the Roadside and vehicle collision.[5]

Figure 2. Components
1. Roller absorbs collision shock (shock energy to rotational energy)
2. Front rail absorbs 2nd shock
3. Back rail absorbs 3rd shock
4. Metal pipe inserted in to strengthen post.[4]

Kim et al (2004) in a study concluded that the use of longitudinal barriers results in reduction of the accidents by more than 50 percent in a year. In a strength performance test of the 8-ton truck, and in a passenger protection test with 1.3-ton passenger car, the RB satisfies the MOCT's Guidelines for Installation and Management of Road Safety Facilities.[6]. Wadekar et al (2017) published that the accidents that occurs on major roadways like expressway has observed 14,500 accidents resulting to 1,400 deaths since its inception. India, being on the threshold of fetching a the place among the developed country needs to make its efforts that withstand on Shock Absorbing Rolling Barrier. RB can help to reduce accidents and fatalities. [7]

Rao. et al in 2017 his analysis of road accident data 2015 exposes that about 1,374 accidents and 400 deaths takes place every day on Indian roads which further interprets that 57 accidents and loss of 17 lives on an average every hour in the country. He concluded, the RB safes life and also prevents maximum damage level of the vehicles and established that the rolling barrier systems are the future technology in transportation engineering.[5] Reddy et al (2017) perceived that every year approximately 1.25 million people die over the world due to road traffic crash. The federal highway administration suggested that the guardrail installed in the road way can minimize the large number of accidents. A Korea based company ETI (evaluation in traffic innovation) is involved in design rolling barriers system. [8]

2. Barriers

Rolling Barriers are also known as guardrails, longitudinal barriers. RB retain vehicles on their path and check vehicles from hitting with hazardous obstacles and roadway such as rocks, traffic sign posts, bridge supports, trees and construction walls. Rolling Barriers are characterized in two types, by the

IOP Conf. Series: Materials Science and Engineering **404** (2018) 012045 doi:10.1088/1757-899X/404/1/012045

functions they serve and conferring to the degree that how much they can deflect a vehicle when it hits into the RB. [9]

2.1 Types of barriers according to its functions
2.1.1 Roadside barriers: Its primary function is to protect traffic from outside obstacle and hazards. The hazard could be such as steep slope, heavy rigid body (eg. stone) and accumulated water body.
2.1.2 Median barriers: Its function is to prevent vehicles to cross over a median to avoid head-on collision.
2.1.3 Bridge barriers: it is generally having greater dimension in comparison to road side barriers. This barriers prevent vehicles from crashing off the side of the bridge and jumping onto the river, railways and roadways below.
2.1.4 Work zone barriers: they are temporary barriers and can be reallocated depending upon the work zone. It prevents the traffic from work zone hazards.[9]

2.2 Types of Barriers according to stiffness
2.2.1 Flexible barriers: it consists of weak post corrugated guide rail and cable line barriers.
2.2.2 Semi rigid barriers: They Includes heavy post blocked out corrugated guiderail with a line box beam guiderail and thrie-beam rail. They deflect 3-6feet (0.91 to 1.83m).
2.2.3 Rigid barriers: they are simple rigid body and made with cement concrete. The deflection of rigid barrier is almost zero. [9]

Table 1. Percentage of collision of vehicles on various objects/obstacles

S. no	Objects of impact of vehicles	Total percentage
1.	Concrete Barriers	23%
2.	Guard rail	19%
3.	Flowerpots and curb stones	15%
4.	Bridge wall	13%
5.	Overhead bridge pillar/tunnel wall	13%
6.	Trees	5%

3. Working mechanism of Rolling Barriers

They consist of both flexible property and semi rigid property of barrier stiffness. RB are different in mechanisms than the other types of normal barriers. It also reduces the hazard and road accidents. Urethane has become the material of choice in so many of today's performance driven applications because it illustrate extraordinary physical and mechanical properties that other material simply can't match. [10]

Roller Barriers absorbing the collision shock in three points, the barrel, the buffering bracket, and the liquidity of the rail shocking mechanism. The shock energy absorb from a curved crash is absorbs then the shock is converting in a rotational energy. Shock energy absorbs the gain by its upper and lower safety guard rail. Shock energy will be again by the back side of safety rails.

Finally the safety roller barriers will leave the cover back to the lane. Thus, this decreases the car damage and ensures the safety of drivers and passenger in the vehicles. Besides strong resilience and shock energy absorbing capacity of RB the cost and maintenance of repair significantly less, this is another economic advantage of this technology. [11]

Table 2: Percentage of type of accidents

S. No	Types of accidents	Constituting percentage
1.	Run off road to left	32%
2.	Run off road to right	23%
3.	Collision with another vehicle moving ahead or waiting	21%
4.	Collision with another vehicle which starts, stops or Stationary	10%
5.	Collision with another vehicle moving laterally in same direction	5%
6.	Collision with another oncoming vehicle	2%

Figure 3. Rolling Barriers at the midroad and vehicle collision with diversion.[4]

Figure 4. Rolling Barriers at the U-turn.[8]

Rolling Barriers can be for two way traffic by installing them at the midroad and used as dividers (figure 3). Roller Barriers can be used at turning of roads to avoid accidents at turnings. The figure 4 shows is the RB installed at the U turn, to reduce fatal damage during the turning. U turns are much prone to accidents.

4. Test performance comparison during crushing between conventional barriers and RB

Performance test has been carried out to compare normal and RB to measure the degree of damage imparted to the barrier during a vehicle impact. The figure 5 shows the representations of the crash test (test 1). It was observed that the conventional barriers experienced more damage in comparison to RB. In a similar crash test (test 2) a comparison is done between passenger car and heavy vehicle impact to a RB, as shown in figure 6. It was seen that impact of passenger car delivered no damage to the RB, while slight damage is recorded in case of heavy vehicle. [4]

Findings of crash test

4.1 Passenger Safety Performance:
Theoretical Head Impact Velocity (THIV): 32.4kmh (Below 33kmh)
Post Impact Head Deceleration (PHD): 9.9 gs (Below 20g's).
4.2 Scatter Prevention Performance:
No scatters of the safety barrier.
4.3 Test Vehicle Behavior Performance:
Not overthrown or a sudden stop after collision.
76.9% (Exit speed: 78.4kmh)

IOP Conf. Series: Materials Science and Engineering **404** (2018) 012045 doi:10.1088/1757-899X/404/1/012045

43.7% (Exit angle: 8.74degree)
Result: Satisfied with Criteria. [4]

Test 1: SB4 Crash test

Figure 5. Crash test comparison between conventional barriers and RB (KSI Korea).[8]

Test 2: SB5 Crash test

Figure 6. Crash test for car and truck collision at RB (KSI Korea).[8]

5. *Future works*

In the domain of material advancements in the last few decades we have developed exciting materials in engineering science. For materials election or fabrication of RB and cost factors will play a significant role.

Materials: While selecting material for RB we need to see the properties required in the installed component.one of the most important properties in this regard is resilience and shock absorbing capacity, as this is the primary feature of the RB in absorbing the impact Energy imparted by the decelerating vehicle .Crash cushioning property plays an important role in fulfilling the main purpose of the RB. Another property which can add to the Functionality of the RB is thermal resistance, as large amount of heat is generated during the impact event. Using materials that are thermal insulators will ensure the proper working of the RB.

Cost optimisation: Cost is another factor to decide the material of RB, as large number of rollers is to be installed at the side and the middle of the road.

6. Implementation of RB

The efficient implementation of the RB will cater multiple objectives as discussed above, however RB can be applied resourcefully at following sites

- National Highways and major roadways require its competent use.
- Other accident prone sites like in curved road sections, U turns etc.
- Gradients and slopes in the urban or state or national road arteries.
- Inclines in parking lots and garages.

7. Conclusion

India is among those developing economies in the world which will soon take up the band of developed nations. The high growing economy not only requires monetary growth but also the life safety and high status of living of the nationals. Due to the high number of road fatalities it is imperative for India to take measures towards the safeguard of human life by eradicating or by minimizing road accidents. Rolling barriers in this context will serve a potential role for the achievement of higher degree of safety on road. RB not only absorbs the vehicle impact but also

IOP Conf. Series: Materials Science and Engineering **404** (2018) 012045 doi:10.1088/1757-899X/404/1/012045

safeguards the vehicle damage and secondarily diverts the wrong moving vehicle. By implementing RB in the cities India will come in that group of countries who have already using this technology. The Indian highways records large number of accidents and this technology will definitely elucidate this problem in a highly efficient manner. RB will safeguard the vehicle & human life and also increase the overall efficiency of the traffic flow. As the undue traffic jams due to accidents will be decreased.

8. References

[1] https://interestingengineering.com>korea.

[2] www.indianexpress.com, Transport research wing under Ministry of Road Transport and Highways, Government of India.

[3] 2016, Ministry of road Transport and highways, *Geography and You*, **16**

[4] ksi@ks0486.com/wp/product/highway-roller-barriers-system.

[5] Nagadarshan Rao, Nagaraja bhagav, Naimish G Jagani, Jaydeep N Maravia 2017 An Alternative Method for Barriers: Rolling Barrier system, *International Journal of Latest Engineering And Management research* **02** 37

[6] Kyung-Whan Kim and Bu-Yong Shin 2004 A study on the characteristics of rolling barriers, *KSCE Journal of Civil Engineering*, **8**, 135

[7] Akshay Wadekar, Neeraj Tilekar, Chinmay sawalkar 2017 Introduction of a New Road Barrier System for Safety Management and Efficiency on the Mumbai-Pune Expressway, *International Journal of Engineering Research in Mechanical and civil Engineer(IJLEMCE),* 48

[8] B. Mahendra Reddy, M. Sri Priya, 2017, New Roller Barrier System Management And Efficiency on Highways, *International Journal of Engineering Science and Computing* **07**- 14415-14418,

[9] http://en.m.wikipedia.org/wiki/traffic_barrier.

[10] www.mearthane.com/about-mearthane.

[11] admin@monrio.com.ph.

IOP Conf. Series: Materials Science and Engineering **404** (2018) 012046 doi:10.1088/1757-899X/404/1/012046

Role of Renewable Energy in Indian Economy

Aaditya Ranjan Srivastava[1], Manavvar Khan[1], Faraz Yusuf Khan[2] & Shrish Bajpai[3]

[1] UG Student, Mechanical Engineering, Faculty of Engineering, Integral University, Lucknow, India.
[2] UG Student, ECE Department, Faculty of Engineering, Integral University, Lucknow, India.
[3] Assistant Professor, ECE Department, Faculty of Engineering, Integral University, Lucknow, India.

saadityaranjan@gmail.com, manavvark21@gmail.com , ferez1905@outlook.com, shrishbajpai@gmail.com

Abstract. Renewable energy sources are non-conventional energy sources which are the need of the hour. Renewable energy sources contribute a lot in the reduction of green house gases emission along with having several socio-economic advantages when compared to carbon-based fossil fuels. Renewable energy sources are the answer to India's energy demands which are growing and are expected to grow exponentially in the future. Through the "Make in India" initiative of the Indian Government, Renewable Energy will account for significant job creation in urban areas as well as rural areas along with several other renewable energy policies which cover Solar Energy, Wind Energy among other renewable energy sources. Renewable energy is responsible for sustainable jobs resulting in a boon for the youth of India. This paper highlights the significance of renewable energy, impact of renewable energy on the social and economic landscape of India and the government initiatives for the promotion and expansion of renewable energy sources in the country.

Keywords: renewable energy, fossil fuels, Indian economy, Government effort.

1. INTRODUCTION:

Energy is the basic exigency in any work field. Energy is preponderating of two types that is: Renewable and Non-Renewable Energy. Our main focus and pursuit is on Renewable Energy.

Renewable Energy has its rootlet in nature and it can be replenished time to time. Renewable can be used in any field, like power generation, transportation, etc. In today`s epoch there are several technologies developed just like Solar panel, Wind Mill, Wind Turbine etc. for generation of renewable energy. Renewable Energy can be proliferating from sunlight, wind, tidal, waves, biomass etc. Today the world is moving towards the renewable energy due to its easy and infinitely presence in nature.

IOP Conf. Series: Materials Science and Engineering **404** (2018) 012046 doi:10.1088/1757-899X/404/1/012046

In 19th century, coal was used as an energy evacuation but in last few years renewable energy is superseding the coal, petroleum as well as fossil fuel energy. As this field is accenting diurnally the jobs are also skyrocketing [1].

2. ENERGY DEMAND OF INDIA DUE TO POPULATION:

It is a well-known fact that India has the second largest population in the world and is also the fastest growing population in the world. It is estimated that India's population will overtake China's by the year 2025. India has done well to provide adequate energy for its growing population being the third biggest electricity consumer in the world, however India has the lowest per capita electricity consumption (Unit) among the developed and developing countries in the world. India's median age is 27 years, having a considerable amount of youth population it will need to satisfy their growing energy needs. One must also remember that India is an active participant of the Paris Climate Agreement hence through renewable energy India is looking forward to bringing down its greenhouse gasses emissions ushering socio-economic change.

Thermal energy provides for 61% of total installed capacity, considering the Indian Government's commitment to renewable energy policies such as Jawaharlal Nehru National Solar Mission and Faster Adoption and Manufacturing of (Hybrid &) Electric Vehicles in India (FAME), the involvement of Thermal Energy is set to decrease.

Not only will the investment in renewable energy pay off it will also decrease the burden on India's foreign reserves most of which are spent on purchasing fossil fuels, in addition to creating a massive amount of sustainable jobs for the exponentially growing youth population of India.

India's geographical position is a blessing for the utilization of renewable energy as it has almost 300 sunny days which are a boon for the solar energy industry in India and a coastline of 7500KM for the installation of wind energy farms. India's gargantuan rural population will also contribute to its energy needs through Bio-Gas plants. With the even distribution of production of energy in India which has been made possible through renewable energy, India's per capita electricity consumption (Unit) will also increase signalling a positive socio-economic trend.

3. RENEWABLE ENERGY IN INDIA:

Taking lessons from the 1979 energy crisis, several countries began exploring the possibilities of Renewable Energy with an increased zeal. India formally began its journey into the world of Renewable Energy in 1981 when The Commission on Additional Sources of Energy (CASE) was established additionally the Department of Non-conventional Energy Sources (DNES) set up in 1982. Financial aspects of the Indian renewable energy sector were to be the responsibility of IREDA

(Indian Renewable Energy Development Agency) which was founded 1987. The Ministry of Non Conventional Energy Sources (MNES) which later became the Ministry of Non-Conventional Energy Sources were initiated in 1992 as a separate ministry with a Minister of State in charge.

India stands 3rd in the list of countries that produce the most electricity as of 2016; it's the 5th largest wind energy producer having wind energy potential of 102.8 GW. It has Hydro energy potential of 19.7 GW, Bio power potential of 22.5 GW, Solar power potential of 6 GW and it is expected that 22,000 MW of solar power would be generated by 2022 [3]

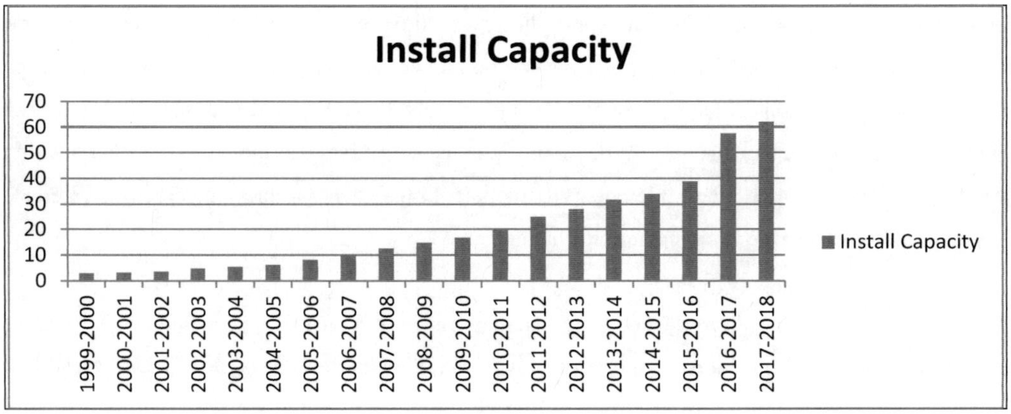

Fig 1: Renewable Energy Installed Capacity Growth for 18 Year to Year (in GW) [4]

4. GOVERNMENT EFFORTS TO REDUCE POLLUTION AND TO PROMOTE THE CONCEPT OF RENEWABLE ENERGY:

In last few years, India is working on making environment pollution free through use of renewable energy. Government of India has taken umpteen steps towards it; they have started with automobile sector which is a major part responsible for pollution. Battery Operated Vehicle is one of the government initiatives to slacken the defile environment.

Jawaharlal Nehru Solar Mission [5], Faster Adoption and Manufacturing of Electric Vehicles (FAME) etc. are the epochal prelude in renewable energy field. Whereof increasing focus on renewable resources in order to reduce reliance on conventional fuels, coal in particular, for the long-term [6].

Presently, there are huge investments opportunities in the Indian RE sector. Infrastructure growth has been made a national priority by our new Union Government, which is led by Prime Minister Narendra Modi. This sector is expected, including the manufacturing, to drive the economy in the next five years. Keeping up with this vision, MNRE has scaled up renewable targets manifold to about 175 GW by 2022 which accounts to 90 per cent of this volume dedicated to solar and wind-based power.

5. JOB CREATION:

Due to the neoteric underlay of Renewable energy field, there are agglomeration jobs in this sector which not only lead to power evulsions but also a good realm for deputation of Indian youth in energy sector. Energy is the initial thing which all industries requires, no any machine or doohickey can run without energy (except manual machines). So, for the generation of energy companies and government have to employ the roustabout. Government has taken umpteen initiatives for development of renewable energy in India through "MAKE IN INDIA" project. Tata Power Solar Systems, Waaree Solar etc. are the major companies who are outbidding the solar energy sector in India. Vestas India, Inox Wind etc. are the major companies who are superinducing the wind energy sector in India. Institutional & Domestic Biogas Plant, FRD Biotech etc. are the major companies who are developing the biogas energy sector in India.

6. CONCLUSION:

Indian government is interested to generate the employment in the power sector which is unstrapped till now. Still India depends upon foreign oil export to run automotives and industries, which ate a large portion of foreign currency earn by the India through exports. Since sun is shining all 12 months in major portion of India, Solar energy is a good alternative of generation of energy. India has a coast line of 7500 KM, which is a good place to generate the wind energy. According to census 2011 around 69% of Indian citizens lives in rural area and involved in husbandry and agriculture.

Above three renewable energies are new to Indian market and can generate a lot of employment in manufacturing, installation and maintenance of renewable products at the dedicated site. Indian Government is pushing this sector because energy generation through this does not create green house gases, creates clean environment and are infinitely available.

References

[1] Narula, K., Reddy, B. S., Pachauri, S., & Dev, S. M. (2017). Sustainable energy security for India: An assessment of the energy supply sub-system. Energy Policy, 103, 127-144.

[2] PLANNING COMMISSION of India, 2002, Indian Goverment Report.

[3] Ramachandra, T. V., Jain, R., & Krishnadas, G. (2011). Hotspots of solar potential in India. Renewable and Sustainable Energy Reviews, 15(6), 3178-3186.

[4] Ministry of New and Renewable Energy Report on Renewable Energy Installed Capacity Growth (2015), http://mnre.gov.in/file-manager/annual-report/2014-2015/EN/Chapter%201/chapter_1.htm.

[5] Tripathi, L., Mishra, A. K., Dubey, A. K., Tripathi, C. B., & Baredar, P. (2016). Renewable energy: an overview on its contribution in current energy scenario of India. Renewable and Sustainable Energy Reviews, 60, 226-233.

[6] Bajpai, S., & Kidwai, N. R. (2017). Renewable Energy Education in India. Comparative Professional Pedagogy, 7(4), 103-113.

[7] Krishnamurthy, S., Joseph, S., Pradhan, V., & Rao, P. (2017). Empowering Women of Rural India for Renewable Energy Adoption–An Exploratory Factor Analysis. Indian Journal of Science and Technology, 10(37).

[8] Rehman, S., & Hussain, Z. (2017). Renewable energy governance in India: challenges and prospects for achieving the 2022 energy goals.

[9] Sindhu, S., Nehra, V., & Luthra, S. (2017). Solar energy deployment for sustainable future of India: Hybrid SWOC-AHP analysis. Renewable and Sustainable Energy Reviews, 72, 1138-1151.

[10] Sawle, Y., Gupta, S. C., & Bohre, A. K. (2018). Socio-techno-economic design of hybrid

renewable energy system using optimization techniques. Renewable Energy, 119, 459-472.

IOP Conf. Series: Materials Science and Engineering **404** (2018) 012047 doi:10.1088/1757-899X/404/1/012047

Effect of sustainability-related activity on behaviour: consumer vs. designer

Bijendra Kumar* Prabir Sarkar

Department of Mechanical Engineering, IIT Ropar, Punjab-140001, India

*E-mail: bijendra.kumar@iitrpr.ac.in

Abstract. The lack of awareness of the possible consequences of natural resource depletion is the major reason behind human carelessness toward nature. In this regard, designers who have a greater understanding of using materials for different applications, are to be more concern about its scarcity. It has been hypothesized in this work that compared to end users, designers who have direct experience with the consequences of material depletion would be more inclined to take responsibility to preserve the minerals. We made a survey questionnaire and collected the data from participants (N=100) across India to understand the relation between the designer (N=50) and consumer (N=50) perceptions of material use and standards related to the sustainable product. We found that designers are more concerned about the appropriate use of natural resources. They are keen to act effectively with a mind-set to preserve natural mineral resources. We can say that these perceptual differences are an indication of enthusiasm to save the climate. Therefore, highlighting the relationship between consumer and designer perception about material consumption and development of related policies may be a useful strategy for increasing the awareness toward conservation of minerals.

Keywords: Sustainable Product; Minerals; Consumer; Designer, Perception, Sustainable Behaviour

1. Introduction

The commonly used definition of sustainable development was given by Brundtland [1], which states "Sustainable development is the development that meets the need of present without compromising the ability of future generation to meet their own". However, the sustainable product can be defined as "A product which has the ability to continue a well-defined social, economic and environmental performance forever". It is known that nature is the source of all mineral required survival of all living beings. Thus, if we continue to consume natural resources without conservation, the survival will be difficult for all living beings [2]. To save the natural resources, the concepts such as designing products with 3R [3] principles and use of sustainable product [4,5] have already been introduced in many developed and developing nations. However, these targets cannot be achieved without major changes in societal structures that will necessarily require a commitment of the wider public to use the sustainable product and save minerals [6]. Many surveys have been conducted on public perception of the sustainable product [4,7–10] and many people are concern about the sustainable product. However, they are not adopting sustainable behaviour [4] may be due to the perception that it is a distant issue i.e. sustainable product are not available thus people are not expected to use it [11].

Since during above-mentioned survey, people have indicated a lack of availability of sustainable products. Thus, the role of a designer comes into the picture. Designers are expected to design the sustainable product by taking care of conservation of minerals as well as public demand. Hence, designers are under more pressure to fulfil public demand by designing more number of the sustainable product. Designers are using different sustainability criteria to design sustainable product [12,13]. Additionally, design researcher also trying to get a number of design methods to assist the designer in various design stages to design a sustainable product [14]. However, selection of material is one of the criteria, as a wrong selection or wastage of material can have larger associated impacts on the environment[7,8]. For example, researchers are using the minimum amount of lubricants for

IOP Conf. Series: Materials Science and Engineering **404** (2018) 012047 doi:10.1088/1757-899X/404/1/012047

machining [15], so that the wastage of lubricants, as well as associated environmental impacts, can be reduced. Thus, we can save the minerals directly as well as we can initiate various activities to save minerals indirectly [16–18], such as minimizing the use of technology [19] by understanding the use suitable technology for the specific application.

We observe, that the lack of awareness about the possible consequences of resource depletion might be one of the possible reason that people do not care about conservation of minerals. We conducted an online survey to understand the awareness about the importance of material from the perspective of the designer (who probability more aware of materials consumption and conservation) and consumer (who are the end users of the product). We believe that highlighting of studies related to the perceptual difference of people would be beneficial for the public so that they can take appropriate action to improve their behaviour towards the environment.

2. Methodology

2.1. Formulation of survey

We prepared an online survey questionnaire to observe designer and consumer perceptions of materials and standards related to sustainable products. Questions were prepared by a penal of two academic researchers and further refined after drawing feedback from a pilot study. Two designers and two consumers participated in the pilot study. Questions assessing individual's action and the importance of material and standards relating to the sustainable product are given in Table 1. Participants were asked to indicate their response on a rating scale of one to three, where one represents the least preferred choice and three represents the highly preferred choice. Although the questions were not straightforward, however, we intended to understand their perception through closely related activity happening around them. We have sent the online survey link to more than 100 people until we get the required sample size of informed participants. The sample size was decided to be 100 (50 designers and 50 consumers) consisting of adult participants from different regions of India.

Table 1: Questions assessing personal activities and behaviour

Concept	Questions	Option	Designer Mean (SD)	Consumer Mean (SD)
Profession	What is the best way to describe your profession?	Designer / Consumer		
Perception regarding the involvement in different activity of using material and standards	Recyclable material	Three-point scale (No-Less-High)	2.80 (0.45)	2.70 (0.58)
	Long lasting product		2.76 (0.48)	2.10 (0.81)
	Standard compliant product		2.54 (0.54)	2.12 (0.72)
	Life cycle of the product		2.68 (0.55)	2.16 (0.79)
	Optimum material		2.68 (0.51)	2.02 (0.77)
Understanding one's responsibility for conservation of natural resources through regular habits and activities	Use biodegradable utensil	Three-point scale (No-Rarely-Often)	2.06 (0.79)	1.54 (0.84)
	Energy Conservation at home		2.82 (0.44)	2.22 (0.89)
	Energy Conservation at Office		2.60 (0.64)	1.98 (0.91)
	Take own bags for shopping		2.62 (0.57)	1.94 (0.93)
	Use separate bin for waste disposal		2.34 (0.77)	1.84 (0.77)

2.2. Analysis of survey data

We collected the responses using Google® survey followed by their compilation in a Microsoft Excel® worksheet. Further, mean and standard deviation of responses were evaluated. This was followed by an ANOVA single factor analysis with $\alpha<0.01$ and 99% confidence level to examine the difference in perception and personal activities of designer and consumer. The variables such as age, gender, educational background were considered separately within the analysis, to ensure the results free from their effect.

3. Results and Discussion

3.1. Perception regarding the involvement in different activity of using material and standards

We calculated the mean and standard deviation of responses on a scale of one to three. Analysis of responses reveals that the designer, as well as consumer both, are having a similar concern about the importance of using the recyclable material for the sustainable product. However, on the importance of other related aspects (i.e. the importance of long lasting product, standard compliant product, the life cycle of product and use of optimum material) in developing a sustainable product, designers are more concerned than consumers as shown in Fig 1. With ANOVA single factor analysis of mean values from Table 1, we found that the perceptual difference of various aspects is significantly different between designer and consumer. We also found the significant difference between their activities. For example, we can interpret the results that consumers are the least concern about the use of optimum material with a mean rating of 2.02 and highly concern of recyclable material with a mean rating of 2.70, which differ significantly.

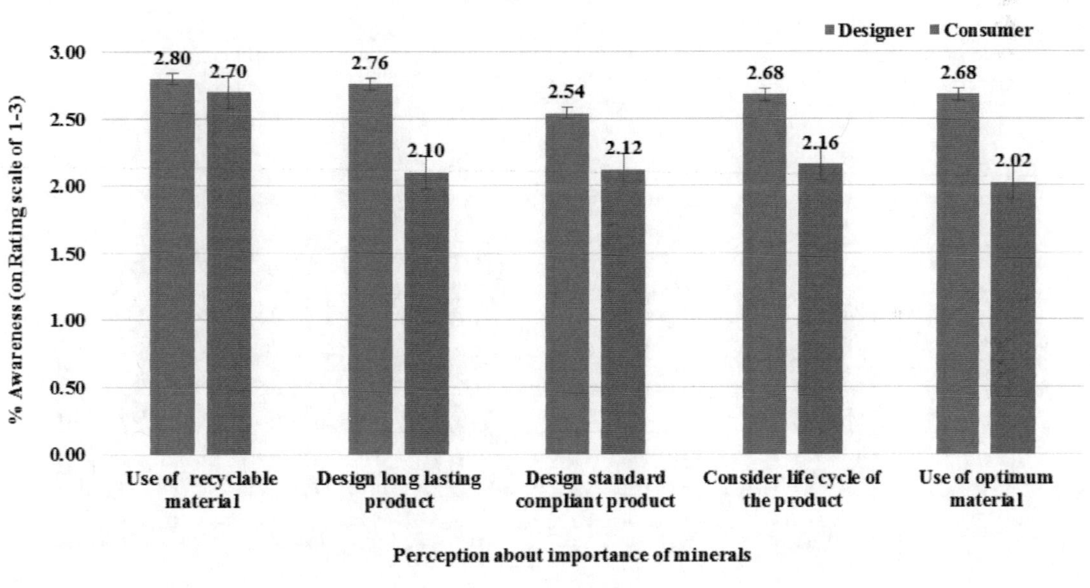

Figure 1. Perception regarding the involvement in different activity of using material and standards

The significant difference between a designer and consumer perception support our hypothesis, that, designers who are directly involved with designing products, should be more concern of the matter in comparison to the consumer. Although consumers' awareness of recyclable material might have increased in recent years, they are still unaware of using the optimum material for sustainable product development. Therefore, it is the responsibility of the manufacturers to share material and energy consumption details with consumers. In addition, the current status of the material available in nature and possible consequences of its scarcity should also be mentioned. This may help in changing the consumer attitude towards the sustainable use of the product.

It was observed that designers also had a difference of opinions among themselves. This may be attributed to their individual preferences or to their unawareness towards possible ways to introduce sustainability into the product. For example, within designers, we can interpret that, the designer has

IOP Conf. Series: Materials Science and Engineering **404** (2018) 012047 doi:10.1088/1757-899X/404/1/012047

shown least concern of producing the standard compliant product. However, it is unclear from the survey that whether the designers are aware of sufficient standards or there is non-availability of a standard related to sustainable products. Thus, it should be assessed that whether there is a need for an increase in design standards for sustainable products or not. In addition, we should focus on increasing the methods to support the designer in various design stages of a sustainable product.

3.2. Understanding one's responsibility for conservation of natural resources in regular activities

The responses related to individual's responsibility toward nature we found that cost plays a major role in adopting a sustainability habit. For example, energy saving at home is highly preferred, whereas, use of biodegradable plates and utensils is a least preferred habit. In addition, the popularity of the energy saving campaign in the country over activities such as the use of biodegradable products may also be a major reason for the disparity. Furthermore, it is interesting to notice that people are adopting the habits such as energy saving at the office, use of reusable carry bags and use of distinct bins for a typical garbage.

A deeper interpretation of results shows a significant difference between the mindset, individual habits, and practices of a designer and a consumer. For example, it was observed that designer feel significantly more responsible for their actions in comparison to the consumer as shown in Fig 2. This is a further support to our hypothesis that designer is more inclined to take responsibility towards conservation of natural resources.

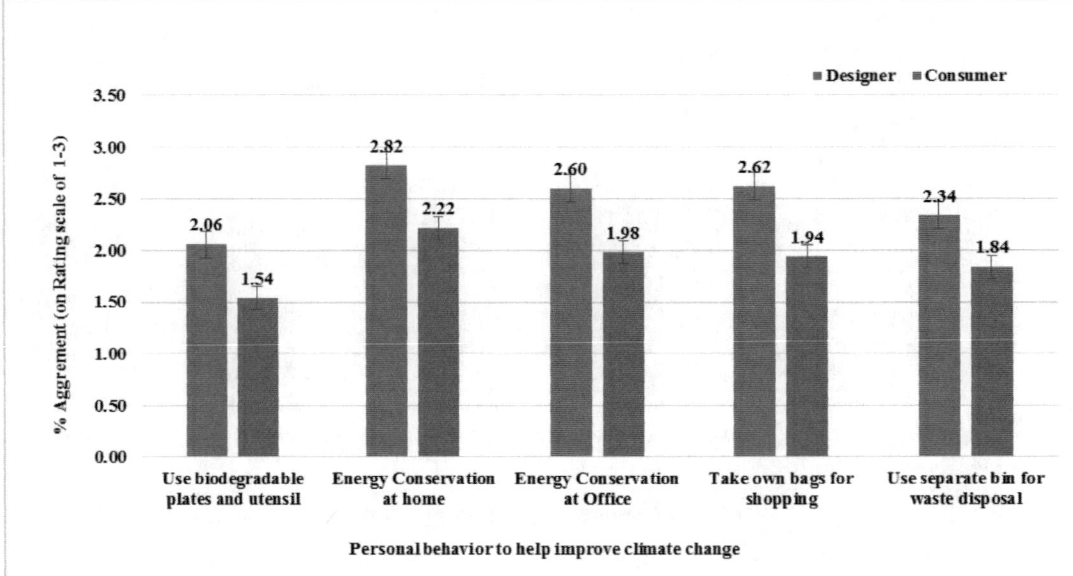

Figure 2. Personal behaviour to help improve climate change.

3.3. Understanding relation between perception and personal activities

To understand the relation between perception regarding the importance of material and one's responsibility for conservation of natural resources, we observed the variation of rating as shown in Table 2.

Table 2. The relation between perception and personal activities

	Designer (Mean)	Consumer (Mean)
Awareness	Varies between 2 to 3	Varies between 2 to 3
Activities	Varies between 2 to 3	Varies between 1 to 2

From Table 2, it is clear that average ratings of awareness toward sustainability and environment conservation are for both designer and consumer are on the higher side (i.e. lies between 2 and 3). Furthermore, designers rating for sustainable practices also varies from 2-3, however, consumers rating for sustainable practices are at the lower side (i.e. lies between 1 and 3). It is shown that except the practice of saving energy at home, consumers always lag behind designers while adopting sustainability practices in every aspect of individual habits and practices, designers as a whole designer are always more concerned, hence awareness plays an important role to adopt sustainable practices. Another possibility of consumer energy saving activity is, that they really feel the absence of electricity, thus they care more for energy saving. Subsequently, consumers doing other activities rarely, either they are not aware or the required product to perform that activity is not there, such as lack of multifunction dustbin.

4. Conclusions

The present study highlights the perceptual difference between a designer and consumer for the importance of materials related to sustainable product development. We also studied the difference in their responsibility toward nature. We found that designer is more concerned about the issues and have a greater interest in adopting sustainable design practices. We did the study with 100 participants, this looks like limitation of our work at first sight, however, we found various studies in literature with around these number of participants such as 86 participants from one country [4], 232 participant from 7 different country on an average of 33 per country [20], 12 participants of one country [21], thus we believe that 100 number of participants is not unusual in the study of consumer perception. Additionally, we found the lack of studies with 50 designer's perception. This is another reason; we have intentionally a limited number of participants to 100 to pertain equal importance to both. However, in future, the detailed study with more number of participants can be carried out for further validation of the findings.

We found that there might be a lack of standards related to the sustainable product that might be the reason, designers have shown least awareness towards a design standard-compliant product. Additionally, development of more policies and increase in awareness regarding mineral conservation would be helpful for achieving sustainable development. We also believe that equal awareness of designers and consumers may ease the process of achieving sustainable development. Thus this study may enhance the knowledge of consumers, as well as designers so that their actions towards natural resource conservation may be improved.

Acknowledgments

Authors are highly thankful to all respondents for spreading their precious time to respond the survey, we also thankful to Google® for support in conducting an online survey. We also acknowledge the comfortable research environment provided by the institute.

References

[1] WCED R 1987 *Report of the World Commission on Environment and Development: Our Common Future*

[2] Goodland R 1995 The Concept of Environmental Sustainability *Annual Review of Ecology and Systematics* **26** 1–24

[3] Mohanty C R C 2011 *Reduce, Reuse and Recycle (the 3Rs) and Resource Efficiency as the basis for Sustainable Waste Management* (UNCRD)

[4] Sarkar P, Kota S and Kumar B 2015 Understanding Consumers' Perceptions of Sustainable Products in India *ICoRD'15–Research into Design Across Boundaries Volume 2* (Springer) pp 99–112

[5] Hoffmann E 2007 Consumer integration in sustainable product development *Business Strategy and the Environment* **16** 322–338

IOP Conf. Series: Materials Science and Engineering **404** (2018) 012047 doi:10.1088/1757-899X/404/1/012047

[6] Spence A and Pidgeon N 2009 Psychology, Climate Change & Sustainable Bahaviour *Environment: Science and Policy for Sustainable Development* **51** 8–18

[7] Karana E, Hekkert P and Kandachar P 2008 Material considerations in product design: A survey on crucial material aspects used by product designers *Materials & Design* **29** 1081–1089

[8] Sharma S, Manepatil S and others 2009 Eco Design: Selection of Material for Important Parts of a Tri-Cycle Through LCA *ICORD 09: Proceedings of the 2nd International Conference on Research into Design, Bangalore, India 07.-09.01. 2009*

[9] Samanta S, Goswami M C, Baidya T K, Mukherjee S and Dey R 2013 Mineralogy and carbothermal reduction behaviour of vanadium-bearing titaniferous magnetite ore in Eastern India *International Journal of Minerals, Metallurgy, and Materials* **20** 917–24

[10] Sharma M and Rani L 2014 Environmentally Sustainable Consumption: A Review and Agenda for Future Research *Global Journal of Finance and Management* **6** 367–374

[11] Lorenzoni I and Pidgeon N F 2006 Public Views on Climate Change: European and USA Perspectives *Climatic Change* **77** 73–95

[12] Cinelli M, Coles S R, Sadik O, Karn B and Kirwan K 2016 A framework of criteria for the sustainability assessment of nanoproducts *Journal of Cleaner Production* **126** 277–87

[13] Singh R K, Murty H R, Gupta S K and Dikshit A K 2009 An overview of sustainability assessment methodologies *Ecological Indicators* **9** 189–212

[14] Kumar B and Sarkar P 2017 A Tool for Generating New and Appropriate Methods for Supporting Various Design Processes *Research into Design for Communities, Volume 2: Proceedings of ICoRD 2017* ed A Chakrabarti and D Chakrabarti (Singapore: Springer Singapore) pp 423–432

[15] Goindi G S, Chavan S N, Mandal D, Sarkar P and Jayal A D 2015 Investigation of Ionic Liquids as Novel Metalworking Fluids during Minimum Quantity Lubrication Machining of a Plain Carbon Steel *Procedia CIRP* **26** 341–5

[16] Jovanovic M, Afgan N and Bakic V 2010 An analytical method for the measurement of energy system sustainability in urban areas *Energy* **35** 3909–20

[17] Larcher D and Tarascon J-M 2014 Towards greener and more sustainable batteries for electrical energy storage *Nature Chemistry* **7** 19–29

[18] Spence A, Poortinga W, Butler C and Pidgeon N F 2011 Perceptions of climate change and willingness to save energy related to flood experience *Nature Climate Change* **1** 46–9

[19] Sarkar P and Kumar B 2014 Identifying tools for remotely located collaborative interactions *Engineering and Computational Sciences (RAECS), 2014 Recent Advances in* (IEEE) pp 1–6

[20] San Jose State University, Lal R C, Yambrach F, San Jose State University, McProud L and San Jose State University 2015 Consumer Perceptions Towards Package Designs: A Cross Cultural Study *Journal of Applied Packaging Research* **7** 61–94

[21] Xia H C and Zhang S Y 2013 A Study on Product Materials Mental Imagery Perception of User Experience Design *Advanced Materials Research* **753–755** 1599–602

IOP Conf. Series: Materials Science and Engineering **404** (2018) 012048 doi:10.1088/1757-899X/404/1/012048

Study of Natural Fibre Reinforced Composites

Mohammad Affan Usmani[1], **Mohd. Anas**[2]
[1]PG Student, Department of Mechanical Engineering Integral University Lucknow
[2]Associate Professor, Department of Mechanical Engineering Integral University Lucknow

E-mail: anas@iul.ac.in, affanusmani62@gmail.com

Abstract. The need of eco friendly material or specific use in automobiles and construction industries has lead to the development of natural fibre composites. Most of the available natural fibres are renewable and posses relatively high strength, stiffness and thermal stability etc. Presently natural fibres are focus of many designers and research scholar due to its low cost and abundant availability in nature. Moreover, processing of these fibres is still going on using conventional production method. Natural fibre composites are being used since ancient Egypt, where straw and clay were mixed together to build walls. The most important type of natural fibre used as composites are flex hemp, jute, straw, bamboo stick etc. which are used due to their desired properties and easy availability. In this paper review of different properties such as Mechanical, Chemical and Thermal properties of natural fibre composite is presented. The main objective of this paper is to study and investigate the existing properties of natural and common composite fibres and to promote the use of natural fibre which is economical and against environmental aggressive material.

1. Introduction

Basically natural fibres got its importance because of their low cost high availability with various desirable properties. Natural fibers are being used from ancient age at different level. Natural fibre can be classified on the basis of their origin. Natural composites are a structural material that consists of two or more constitute that are combined at a microscopic level and are not soluble in each other. Animal fibre Contains wool, silk etc. Avian fibres include sheep's wool, goat hair, horse hair, feathers and feather fibbers etc. Mineral fibres are naturally occurring fibers or slightly modified fiber procured from minerals. These can be further categorized as asbestos, ceramic, metal fibres. Plant fibre is generally comprised of cellulose.

Natural fibre finds its origin since 1000 BC for construction of historical structure in an Egypt. Straw and clay formed first natural composite and were used to build walled structure. During couple of decade composites reinforced with natural fibres have become point of focus for researchers and for industrial application there are different of reinforcement and fillers for different natural fibres like flax, hemp, jute, kenaf, sisal due to their distinguished properties, easier availability and economical usability. Furthermore, they do not have any adverse affect on environment [1].

New composites explored have new application and have contributed in markets for packaging, automotive industries, aviation and shipping sectors etc. Natural fibres reinforced composites are good substitute of metal or ceramic based materials in applications that may be automotive, aerospace, marine, sporting goods and electronic etc areas and their composites. The use of natural fiber as reinforcement in polymer matrix focused the attention towards environmental awareness among all over the world. Various automotive components are already produced natural composites, mainly based on polyester and fibres like flax, hemp, or sisal. Mercedes-Benz used an epoxy matrix with the addition of jute in the door panels in its E-class vehicles, Audi used the door trim panels were made of polyurethane reinforced with a mixed flax/sisal material and Toyota used eco-plastic made from sugar cane and will use it to line the interiors of the cars. Biodegradable bark cloth reinforced green epoxy composites are developed with view of application to automotive instrument panels.

[1] To whom any correspondence should be addressed.

Jute-based natural fibre composites would be suitable for even primary structural applications for example indoor elements in housing, temporary outdoor applications like low-cost housing for rehabilitation and transportation. Wood fibre composites are being used in a large number of applications in decks, docks, window frames. Automobile parts like rear view mirror, visor in two wheeler, billion seat cover and indicator cover etc made with the use of sisal etc natural fibre composites. Recently banana fibre reinforced composites used under-floor protection for passenger cars.

2. Literature review

Ryosuke, et al [2]: Author has experimentally analysed the thermal conductivity and behaviour of natural fire reinforced composites by using manila hemp fibre as a reinforcing fibre because of their high strength and excellent thermal stability. In this experiment a poly lactic acid (PLA) and epoxy resin were used as a matrix material. The samples were mixed with PLA resin and dried for 20 hours at 70^0c in circulation type oven and then pressed by hot pressing machine to complete the fabrication of PLA composites. It was concluded that thermal conductivity decreases by increasing fibre content and thermal conductivity of the composites with fibre content of 76wt% is 0.190 W/(mK).

Braga R. A, et al [3]: In this paper authors have presented the comparative study of mechanical and thermal properties of polyester hybrid composites (i.e. jute and glass fibre) to enhance mechanical properties. Tested Samples were analysed for flexural, thermal, density and impact properties. In this experiment the polyester resin and hardener resin with jute and glass fibre composites were synthesised. Analysis of this paper shows that density of composite with more percentage of jute fibre is a better composite. The composite without glass fibre lost mass as a function of temperature and the composite with more percentage of glass fibre loses less weight with increasing temperature.

MallaSurya Teja et al[4]: In this paper the effect of SiC (as filler) on mechanical and thermal properties of natural sisal fibre composites was investigated. The polyester as bonding material was used. In this experiment hand lay-up technique was adopted in the preparation of unidirectional composites. Clean the mould with shellac NC thinner solution. Apply a thin coating of poly-vinyl alcohol on the interior tile surface and along the edges of the rubber sheet and then dry it for a day. The samples were also post cured at 70 °C for 8 h after removing from the mould. In the same above process composites are also prepared by adding SiC (5%,10%) to resin and proper mixing were done before poured on fibers. Material compositions of samples are as below

Sample code	Polyester	Sisal fiber	silicon carbide (SiC)
A	70	30	0
B	65	30	5
C	60	30	10

Result shows that the tensile strength is increases with the increase of SiC ,that it obtains maximum tensile strength at 10% Sic. The composite with 10 % SiC has the maximum impact strength and it reaches 14.25 N/m^2. Thermal diffusivity of composite decreases with increase of temperature. Temperature at maximum weight loss is nearly constant (350^0C-380^0C).The results have shown that sample C has higher tensile strength (i.e. 72.5 MPa) than sample A (28.6 MPa). Sample B has tensile strength of 45.7 MPa.

Poddar P, et al [5]: The author has examined the thermal and mechanical properties of short areca nut leaf sheath fibre reinforced polypropylene composites. Experimental result shows that a composite with 10% areca nut leaf sheath fibre possesses higher mechanical properties. Increasing the fibre content has reduced mechanical properties. However better fibres matrix adhesion was seen

V. Muthukumar[6]: In this study, the composite materials are reinforced with natural fibres such as Jute, Sisal, Banana, and Palmyra in the Epoxy (LY 556) Matrix. Because of their properties of excellent adhesion, high resistance to chemical and atmospheric attack, High dimensional stability, Free from internal stresses, Excellent mechanical and electrical properties, Odourless, and completely nontoxic. In this experiment the materials used in fabrication process are Epoxy resin-LY556,

Hardner-HY951, Polythene Sheets, ASTM Rubber, Polythene Sheets, Banana, sisal, Palmyra and jute fibres.

Experiment done by hand layup technique which is simplest as well as oldest method of composite fabrication. Hand layup process involve following steps

Step 1: Cut down the fibres as per the required Length.

Step 2: Place the ASTM rubber along the sides.

Step 3: Lay up the mixture of separate parent fibres and hybrid combinations of required proportions over the polythene Sheet. Figure 2 shows process of hand layup method.

Step 4: Pour the polyester resin uniformly and place another polythene sheet over it.

Step 5: Allow a Curing time of about 3 to 6 hours and then separate the composite plate.

3. Discussion

Result shows that the hand layup method can be used for the preparation of composites specimen successfully and The comprehensive mechanical properties are increased to different extents compared with those of parent fibres and hybrid composites; for instance, ultimate breaking strength, Ultimate stress, displacement and Impact strength.

Palmyra individual fibre is better choice than other seven different composites studied because it shows highest ultimate tensile stress under various tests performed. According to the author In future, there is tremendous investments on natural fibre, to bring an great impact on manufacturing sectors.

S. John Paul Devaseelan, et al[7]: Author check the mechanical and thermal properties of various types natural fibre composites like jute, sisal, betel nut polymer because of their higher specific strength at low cost. Sample can be prepared by compression moulding method with 10, 15 wt% sisal and betel nut fibre into polymer mixture. Result shows 10 wt% sisal and betel nut fibre the tensile and compressive properties found to be maximum. The failure morphology is investigated by scanning electron microscope. The natural fibre composites are tested for three properties which are as follows;

(a) *Tensile Strength*:- Various experimental study shows the tensile strength of natural fibre composites is high and it can be improved also

(b) *Young's modulus*:- As shown in various research it is depending on the fibre weight ratio. Young's modulus increased when the fibre weight ratio is increases up to a maximum value after that it will decrease.

(c) *Flexural strength*:- It is also depend upon the fibre loading. If fibre load is increasing flexural strength is increases up to optimal level then decrease.

Table 1. Mechanical Properties of Natural Fiber [8].

Fibre	Density g/cm^3	Elongation	Tensile Strength (Map)	Elastic Modulus (GPa)
Cotton	1.5-1.6	7.02-8.0	400	5.5-12.6
Jute	1.3	1.5-1.8	393-773	26.5
Flex	1.5	2.7-3.2	500-1500	27.6
Hem	1.47	2-4	690	70
Kenaf	1.45	1.6	930	53
Ramie	N/A	3.6-3.8	400-938	61.4-128
Sisal	1.5	2.0-2.5	511-635	9.4-22
Coir	1.2	30	593	4.0-6.0

Chemical Properties:- Chemical properties of natural fiber are those which defined the reactivates with others it show the chemical response when it comes with another material. In various studies basically check the chemical properties of cellulose hemicelluloses legion .Cellulose is an important structural component of natural fiber. It doesn't have any taste and odorless. Hemicelluloses is mixture of several plant polysaccharides of smaller molecular weight than cellulose hemicelluloses are embedded in the cell walls of plant legion is a class of organic polymer. Cellulose and hemicellegion percentage is vary from one to another material, tabulated below.

Table 2. The Chemical Composition of Natural Fibers [9].

Fiber	Cellulose (wt %)	Hemi-cellulose (wt%)	Lignin (wt%)	Waxes (wt%)
Abaca	56-63	20-25	7-9	3
Kenaf	72	20.3	9	--
Jute	61-71	14-20	12-13	0.5
Hemp	68	15	10	0.8
Ramie	68.6-76.2	13-16	0.6-0.7	0.3
Sisal	65	12	9.9	2
Coir	32-43	0.15-0.25	40-45	--
Flax	71	18.6-20.6	2.0	1.5

Thermal Properties:- In thermal properties we can investigated thermal conductivity. Specific heat capacity thermal degradation and thermal diffusivity etc. brief introduction these properties are as below-

(a) Thermal Conductivity: Thermal conductivity is property of material to conduct heat it can be defined as the quantity of heat transmitted through a unit thickness in a direction that normal to a surface of unit area due to a unit temperature gradient under a steady state conditions
(b) Specific Heat Capacity: Differential scanning calorimeter technique using double furnace setup is used for measuring specific heat capacity DSC is a thermo analytical technique in which the difference in the amount of heat required to increase the temperature of sample and reference is measured as a function of temperature.
(c) Thermal Degradation: Thermal degradation by TGA thermo gravimetric analysis is a technique in which the mass and substance is monitored as a function of temperature and time as the sample specimen is subjected to a controlled temperature program in a controlled atmosphere.
TGA measures as a samples weight as it is heated or cold in the Furnace the weight loss for temperature range give the denotes of the sample composition and thermal stability including filler materials and volatiles Thermal diffusivity: although thermal diffusivity can be experimentally measured by flash method in this work it is calculated after knowing thermal conductivity specific heat and density.

4. Conclusion and Scope of Natural Fiber Composites

Natural fiber is renewable raw material which is eco-friendly and renewable material. Natural fiber has good thermal insulation properties. But Strength of the fiber depends upon the fiber loading and young modulus. It also depends on the fiber weight ratio. Natural fiber can be treated by various methods like treatment is most efficient method of chemical modification. It has been used to treat almost all types of natural fiber with successful result. By the combination of the useful properties of two different materials make them as a versatile material in the field of engineering and technology.

This paper gives a summary about the mechanical, thermal and chemical properties of natural fiber with focus on the advantages of natural fiber over all other materials of such categories.

Natural fiber is rapidly growing in both term of their industrial application and fundamental research as they are renewable. They are completely or partially recyclable and biodegradable. It is highly economical as compare to others and suitable for our environment. Partially natural fiber used in automotive industries and bullet proof panels was made from arsines fiber reinforced composition by hand layup process with epoxy as a matrix composition based natural fiber has very good implication in the automotive and transportation industries [10-12].

Further research and investigation are required to overcome obstacles such as moisture absorption toughness and reduced long term stability for outdoor application. Even now natural fiber composition have a potential to be used in many application that do not require very high load bearing capabilities or high temperature working capacity.

References

[1] La Mantia F.P, Morreale M 2011 *Compos A : Applied Science Manufacturing* **42** 579.

[2] Ryosuke Osugi, Hitoshi Talkagi, Ke Liu and Yusuke Gennai 2009 Thermal Conductivity Behaviour of Natural Fiber-Reinforced Composites. Asian Pacific Conference For Material And Mechanics at Yokohama Japan.

[3] Braga R, Silva R. S Magalhaes Jr, P. A. A 2015 Comparative Analysis of The Mechanical And Thermal Properties of Polyester Hybrid Composites By Jute and Glass Fiber. *Journal of Engineering Research and Application* **2** 113

[4] Malla Surya Teja, M V Ramana, D Sriramulu And C J Rao 2016 Experimental Investigation Of Mechanical And Thermal Properties Of Sisal Fiber Reinforced Composites and Effect of SiC Filler Material *IOP Conf. Material Science and Engineering* doi:1088/1757-899x/149/1/012095.

[5] Poddar P, Islam MS, Sultana S, Nur HP and chowdhury AMS. Mechanical and Thermal Properties of Short Areca Nut Leaf Sheath Fibers Reinforced Polypropylene Composites TGA, DSC and SEM analysis *J Material Science Eng* **5** 270.

[6] V. Muthkumar, R. Venkatasamy, A. Sureshbabu, D. Arunkumar 2011 A Study On Mechanical Properties of Natural Fiber Reinforced Laminates of Epoxy Ly 556 Polymer Matrix Composites *International Journal of Production Technology And Management Research* **2** 68

[7] S. Jhon Daul Devaseelan, S Rajkumar.2016 Mechanical and Thermal Properties of Hybrid Reinforcement Polymer Composites. *International Journal of Advances In Engineering Research* **11** 2454.

[8] Santosh S, Bhanuprakash 2017 A Review on Mechanical And Thermal Properties On Natural Fiber Reinforced Hybrid Composites *International Research Journal of Engineering And Technology* **04** 3053.

[9] Kiran Rohit, Savita Dixit 2015 A Review –Future Aspects of Natural Fiber Reinforced Composites. Polymer From Renewable Resources **07** 43.

[10] Anu Gupta, Ajit Kumar.2011 Chemical Properties of Natural Fiber and Mechanism of Chemical Modifications *Asian Journal of Chemistry* **24** 1831

[11] Pradeep P, Edwin Raja Dhas, 2015 Characterization of Chemical And Physical Properties of Palm fibers *Advances In Material Science And Engineering An International Journal* **2** 1831

[12] Singha A. S, Thakur V. K, 2008 Mechanical Properties of Natural Fiber Reinforced Polymer Composites *Bulletin of Material Science* **8** 791

A Novel Design for Highway Windmill through Re-engineering

Faraz Yusuf Khan[1], Sushant Khare[3], Aaditya Ranjan Srivastava[2], Shrish Bajpai[4] and Kamran Rasheed[5]
[1] UG Student, ECE Department, Faculty of Engineering, Integral University, Lucknow.
[2] UG Student, ME Department, Faculty of Engineering Integral University, Lucknow.
[3] Executive, Tata Consultancy Services Limited, Bangalore.
[4] Assistant Professor, ECE Department, Faculty of Engineering, Integral University, Lucknow, India.
[5] Assistant Professor, ME Department, Faculty of Engineering, Integral University, Lucknow, India

Corresponding author E-Mail : ferez1905@outlook.com

Abstract: The purpose of this project is to design a wind turbine which recaptures wind energy from vehicles on the highway. Wind energy is considered as the fastest growing clean energy source; although it is limited by variable natural wind. Highways are good source of wind which left behind wind by the fast-moving vehicle traffic. This wind is useless till it can be used to drive the fan of the wind turbine. This energy is not only unmeasured in potential but also being witnessed going to waste. This energy is unused and could prove to be a boon for power related problems. A preliminary investigation of the characteristics of this VAWT has been done, and a data set has been collected. The objective to design this prototype wind turbine through the re-engineering is to generate power from the wind of the moving vehicle which remains unused. The design of the wind turbine is simple, and it is a low-cost wind turbine. Previously, these wind mills were established in rural areas. The main motive of this project is to install this wind mill in cities for better production of electricity. The wind draft behind the turbine is used as a source, which rotates the blades of wind turbine. This project idea is going to supersede the profanation in atmosphere which was indicted due to scalded fossil fuel.

Keywords: Wind Power, Re-engineering, Highway, Windmill Component.

1. Introduction:

From the advent of 21st century, the need of power rose to levels like never before. As much as new ways were developed still billions of people are living without electricity. According to estimates that number is about a quarter of total population living on the planet which is both shocking and disturbing. In India only 78.7% of population has access to electricity which translates to thousands of villages still living without even a bulb [1]. With ever increasing population and depleting natural resources it is an alarming situation especially for country likes ours, India and other developing nations as well. The reason being our primary sources for generating electricity are fossil fuels [2]. This project or design started as an exploration about ten months back. Here the authors would like to stress on the word – 'reengineering', by reengineering they targeted to make a model which was unlike any made before, using materials which were usable and lying in machines that were not usable. The authors later realized how useful such small wind turbine could be in rural coastal areas.

2. WHY WIND ENERGY AS A SOURCE OF ENERGY:

Wind is a source of energy which is extracted from the blowing wind and is then converted into electricity. This conversion is carried out with the help of wind turbines, wind pumps etc. which are connected with blades which rotates when moving air applies a force on it. The main cause of taking wind as a source for power production is the rapid consumption as well as less availability of the non-renewable resources in present days [3]. This kind of natural fuel has come out to be an excellent alternative fuel and is considered to be extremely beneficial with its super merits which include:

- Wind energy is naturally available and can be easily used at a huge demand with negligible human effort.
- It doesn't pollute nature at all due to zero greenhouse gas emission.
- Round the clock availability everywhere on earth`s surface.
- It can be used in large as well as small scale power production.
- As per return of interest (ROI) the best renewable resource.

Wind is among the best possible source which can be utilized anywhere, anytime and places from high attitude areas to coastal areas through wind turbines. It can be used again and again as many times needed. Wind energy has even helped for employment for the people in manufacturing sector from the constructions, development, installations of the wind turbines to the maintenance criterion [4].

Wind energy has also overtaken many power generation systems like solar panels, gas products etc. with its higher availability, low maintenance and it has even stood high on the scale where the need of petroleum and diesel has been in so much need. Today wind energy is

IOP Conf. Series: Materials Science and Engineering **404** (2018) 012049 doi:10.1088/1757-899X/404/1/012049

answered, emphasized and recommend as the first option in any kind of power generation system by every developed and developing countries.

According to the latest study it's been observed that the wind power generation has increased by 2.5% of the world's electricity in 2011, before it was 0.5% in 1997 which jumped up to 1.5% in the year 2008. The wind energy usage is continuously increased by the new techniques and modern engineering ideas. Many countries like Denmark and Portugal, wind power has contributed around 20% of the total electricity production and it is even increasing with every day [5,6].

The only demerit faced while using wind as a source is the speed at which it blows. The lesser speed of the flowing wind the lesser it the power production or may even pause the production work.

Hence, with the fastest growing mode of power/electricity production and helping to lighting up so many houses and streets, wind is chosen to be the best source in every possible case.

3. DESIGN OF BLADES OF OUR DESIGN:

Modified standard Savonius type blades have been used; the blades have been curved at the bottom and twisted at certain angles. These blades are cut out of used buckets of paint. So,

Fig. 1: Blades of wind mill made by plastic paint cans.

essentially it was a HDPE grade plastic from a raw material perspective. This not only keeps it lightweight but also aligns with the goal of keeping this project green as much as possible. This new modification contributes in getting more drag force.

4. ELECTRICAL GENERATOR USED: DYNAMO

A dynamo is used in this assembly for conversion of mechanical energy to electrical energy. Its rating is 12 V. Dynamos basically are DC generators. Staying committed to making this a reengineering project; the dynamo has been procured from an old model lying dusted. Any alternator, permanent magnet generator with low rpm could have solved this purpose.

Fig. 2: Gear mechanism used in wind mill.

5. GEAR MECHANISM:

Teeth ratio of the VAWT design is set to 3:1. The driving gear has double the number of teeth compared to that of driven gear. The driven gear is then connected to the shaft of dynamo. Gears have been sourced from building block games used for making robots or cars. A higher teeth ratio was intended to be implemented in this VAWT Turbine but due to unavailability of parts the idea was dropped.

IOP Conf. Series: Materials Science and Engineering **404** (2018) 012049 doi:10.1088/1757-899X/404/1/012049

6. THE COMPLETE MODEL

Fig 3: Working model of our wind mill.

In this design, a shaft has been used to support the blades on top. Two bearings are there to rectify any changes in moment of inertia as speed of rotation increases. A custom tripod stand design has been welded to support the complete structure. Cost of the project has been minimized to the highest extent possible by using reusable parts.

7. EXPERIMENTAL ANALYSIS:

This data has been generated by simulating the wind sources. Cars and two-wheelers were used to create a substantial flow of wind. In further testing other artificial sources of wind were used. The results are encouraging. Digital tachometer has been made use for RPM measurement and Digital Multimeter has been used for volts measurement. Results have been shown in the Graph 1 & Table 1

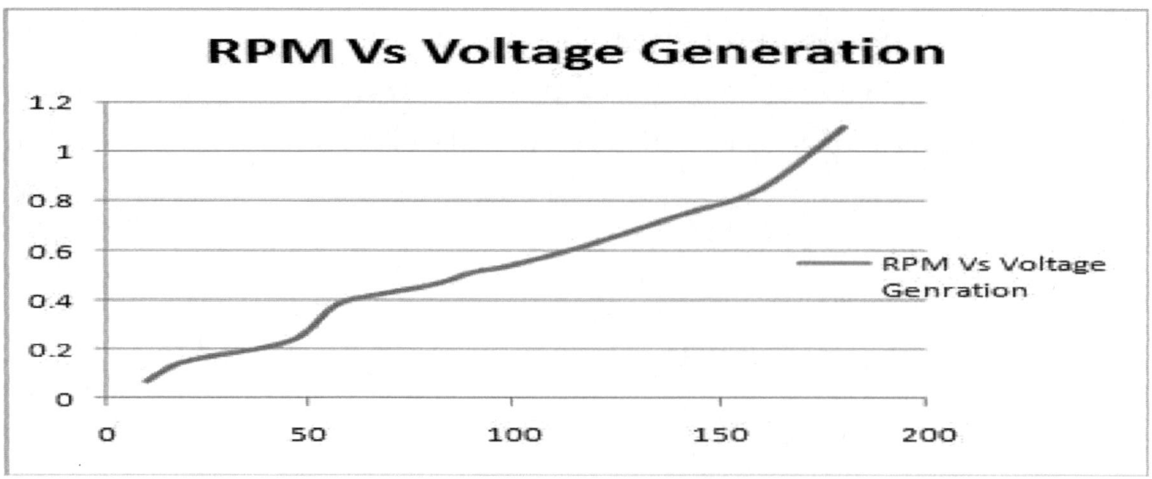

Graph 1: RPM vs Voltage Generation

.

S. No.	RPM	Volt generated (V)
1	10	0.07
2	20	0.15
3	45	0.23
4	55	0.36
5	60	0.4
6	80	0.46
7	90	0.51
8	100	0.54
9	120	0.63
10	140	0.74
11	160	0.85
12	180	1.1

Table 1: RPM Vs. Volt Generated.

8. CONCLUSIONS:

The concerned VAWT model has produced quite satisfactory results. On seeing positive early stage results, attempts were made to enhance those results and fortunately those attempts were successful. At the completion of this project, the authors have learnt something new and are ready to apply this knowledge in future engineering projects as well as in various challenging aspects of life. Similar designs are already being used for testing purposes in many nations around the world. Arizona State University also conducted research in this direction. In Germany too, such designs are seen. They also found a way to look up for maintenance in future by giving it such a design that advertisement hoardings could also be put up on the structure. So, it is safe to say that future looks promising for such designs.

The presented VAWT has yielded quite good results in highway applications. It will be quite successful in areas which are naturally windy such as coastal areas. The model itself can be designed in some modified ways. The authors had a limited knowledge of aerodynamics and blades design. So, may be with more knowledge in future a better design could be come up with. On the applications prospective, wind energy generation is getting much traction in recent times. After solar energy it has the highest throughput among other non-conventional energy resources [7]. The presented wind turbine can also be used in conjunction with solar energy to make it perform better giving it an autonomy period in case wind energy is not available. This model could also be used in metro subways. This was a potential that will be explored since wind generated by metros is quite high compared to other.

References

[1] Pereira, Marcio Giannini, et al. "Evaluation of the impact of access to electricity: A comparative analysis of South Africa, China, India and Brazil." Renewable and Sustainable Energy Reviews 15.3 (2011): 1427-1441.

[2] Abbasi, T., & Abbasi, S. A. (2010). Biomass energy and the environmental impacts associated with its production and utilization. Renewable and Sustainable Energy Reviews, 14(3), 919-937.

[3] Ahmed, S. (2015). Wind energy: theory and practice. PHI Learning Pvt. Ltd.

[4] Abdullah, M. A., Yatim, A. H. M., Tan, C. W., & Saidur, R. (2012). A review of maximum power point tracking algorithms for wind energy systems. Renewable and sustainable energy reviews, 16(5), 3220-3227.

[5] Liu, Yijin, et al. "Comparison study of tidal stream and wave energy technology development between China and some Western Countries." Renewable and Sustainable Energy Reviews 76 (2017): 701-716.

[6] Bajpai, Shrish, and Naimur Rahman Kidwai. "Renewable Energy Education in India." Comparative Professional Pedagogy 7.4 (2017): 103-113.

[7] Mohammed, Y. S., Mustafa, M. W., & Bashir, N. (2014). Hybrid renewable energy systems for off-grid electric power: Review of substantial issues. Renewable and Sustainable Energy Reviews, 35, 527-539.

Analyzing optimum thickness for combination of two thermal insulation materials for building walls

Syed Ali Husain Jafri[1] and Prem Kumar Bharti[1]

[1] Department of Mechanical Engineering, Integral University, Lucknow , India

E-mail: sahjafri@iul.ac.in

Abstract- This paper reports an approach for analyzing optimum thickness for combination of two thermal insulation materials for walls of residential buildings on the basis of life cycle cost. Two thermal insulation materials for achieving contrast purposes can be use as a solution in many applications e.g. if fire resistant material added with Expanded polystyrene (EPS) to improve its fire resistance property, its thermal conductivity increases, where as if we select glass wool as thermal insulation material it has higher thermal conductivity but better fire resistant, thus if we wish to achieve high thermal resistance with better fire resistance at low cost we can use in place of single thermal insulation material a combination of two thermal insulation materials. In order to determine optimum thickness for combination of two thermal insulation materials there is a need to develop an approach, through which optimum insulation thickness for different combination of two thermal insulation materials on the basis of life cycle cost can be analyzed. To study the effect of fraction of one thermal insulation on optimum thermal insulation thickness proposed approach is applied to insulate external wall of a residential building envelope in heating application. Developed total cost objective function analyzed with the help of surface plot. In order to apply the developed approach EPS and glass wool are considered as thermal insulation materials. Through this approach optimum thickness for combination of two thermal insulation materials can be analyzed on the basis of life cycle cost in order to select best combination of thermal insulation materials as per requirement.

Keywords- Combination of two insulation materials; Fraction; Optimum thickness

NOMENCLATURE

$AHLPUA$	annual heating load per unit area of external wall
$ERPUAFH$	energy required per unit area for heating
$EOHS$	efficiency of heating system
H_{loss}	heat loss per unit area of external wall
U	thermal transmittance of composite external wall
T_i	temperature inside the building
T_o	outside temperature
HDD	heating degree day
R_{twt}	total wall thermal resistance without insulation
x	total thickness of thermal insulation materials
f	fraction of total thickness of thermal insulation materials
k	thermal conductivity of first thermal insulation material
k_1	thermal conductivity of second thermal insulation material
$HVOF$	heating value of fuel
$COTIPUA$	total cost of thermal insulation materials per unit are for x thickness
C_i	cost of first thermal insulation material per unit volume
C_{i1}	cost of second thermal insulation material per unit volume
PWF	present worth factor
C_t	total cost of heating and thermal insulation materials per unit area of external wall
C_f	cost of fuel per kg

1. Introduction

Issue of energy crisis and increasing cost of energy along with environment concerned lead to need of energy conservation specially in building sector in the form of heating and cooling [1,2]. As far as energy and environment is concerned building sector is still most inefficient as compared to transportation sector or industries [9]. As an energy conservation measure in envelope load dominated building thermal insulation of components of building envelope is considered important strategy, with increasing need of heating and cooling for maintaining comfort in hostile climatic conditions [28]. By thermal insulating building envelope cost of heating and cooling reduces but at the same time cost of insulation added. On the basis of life cycle cost analysis optimum thickness of thermal insulation for component of building envelop proposed where total cost of energy and insulation becomes minimum [1]. Optimum insulation thickness leads to maximum saving on the basis of life cycle cost analysis [3]. Thus selection of proper thermal insulation material for component of building envelope and determination of its optimum value for a particular climatic condition becomes subject of interest for many researchers across the world [1-25]. A review of different studies related to economically optimum thickness was presented earlier [26]. In subsequent review need to analyze optimum thickness for combination of two thermal insulation materials suggested [27]. Selection of thermal insulation for different components of building envelopes depends upon various parameters like thermal resistance, cost, fire resistance, moisture resistance, compressive strength and weight depending upon application [28]. On the basis of chemical composition thermal insulation materials can be classified as organic, inorganic, combined and new technology materials [29]. In this work a concept of combined thermal insulation material thickness optimization is proposed on the basis of life cycle cost analysis. Using two thermal insulation materials in combination as compared to single thermal insulation material for external wall has number of advantages. Better quality insulation material from fire resistance, water permeability point of view can be used in the external parts whereas low cost material with inferior quality can be used in inner part in such a combination where life cycle total cost becomes minimum.

2. Structure of external composite wall

In order to analyze optimum thickness of combined thermal insulation material, a brick external wall with outer and inner plaster of a residential building is considered which is common in major parts of India.

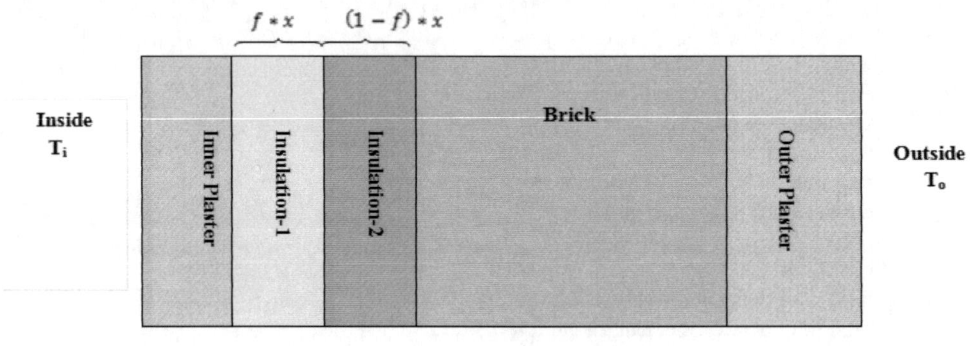

Figure 1. Composite external wall structure

Figure 1, shows the composite external wall considered for present study. In this study inplace of single thermal insulation, two thermal insulation materials are considered. Total thickness of thermal insulation material is divided into two parts i.e, insulation 1 and insulation 2. f is considered as fraction of total thickness of two thermal insulation materials. So out of total x thickness of combined thermal insulation material $f * x$ is the thickness of insulation 1 and $(1 - f) * x$ of second thermal insulation material.

3. Annual heating load because of heat loss from external composite wall

In order to calculate annual heating load degree day model is used that is being used by different researchers [1,3,7,15]. Heating load is analyzed for a unit area of external composite wall for ease of comparing the result of any residential building.

Heat loss per unit area of composite external wall is given by [15]

$$H_{loss} = U * (T_i - T_o) \quad (1)$$

Annual heat loss from external composite wall per unit area is given by

$$AHLPUA = 24 * 60 * 60 * U * HDD \quad (2)$$

Where overall heat transfer coefficient for external composite wall is given by

$$U = 1/\{R_{twt} + (1-f) * \frac{x}{k} + f * x/k_1\} \quad (3)$$

4. Requirement of energy for heating of building to compensate heat loss from external composite wall

In order to maintain constant inside temperature of building, amount of energy required for heating is given by

$$ERPUAFH = \frac{AHLPUA}{EOHS} \quad (4)$$

Cost of heating to compensate heat loss per unit area from composite external wall on annual basis is given by

$$ACHPUA = \frac{ERPUAFH}{HVOF} * C_f \quad (5)$$

$$ACHPUA = \frac{\{24*60*60*U*HDD*C_f\}}{\{R_{twt}+(1-f)*\frac{x}{k}+f*x/k_1\}*EOHS*HVOF} \quad (6)$$

5. Life cycle cost analysis for optimizing thickness of combination of two insulation materials for external wall

Total cost of thermal insulation materials per unit area, on the basis of fraction of thickness considered is given by

$$COTIPUA = C_i * (1-f) * x + C_{i1} * f * x \quad (7)$$

It has been reported as the cost insulation per unit area increases, cost of heating decreases. In order to determine optimum thickness for combination of two thermal insulation materials total cost per unit area of external composite wall represented by equation (9), need to minimize on the basis of life cycle cost as done in previous studies [1,3].

$$C_t = ACHPUA * PWF + COTIPUA \quad (8)$$

$$C_t = \frac{\{24*60*60*U*HDD*C_f\}*PWF}{\{R_{twt}+(1-f)*\frac{x}{k}+f*x/k_1\}*EOHS*HVOF} + C_i * (1-f) * x + C_{i1} * f * x \quad (9)$$

6. Results and discussion

In order to implement the proposed strategy for insulating external wall of residential building data is taken from literature and optimum thickness for combination of two thermal insulation materials is analysed.

Table 1. Parameters used in calculating thickness of combined thermal insulation [21]

Parameter	Value
HDD in °C-days	2614
LPG as fuel for heating	
C_f in Rs per kg $HVOF$ in MJ/kg $EOHS$ in percentage	70 46.04 90
Thermal insulation material 1- Glass wool	
C_i in Rs/m^3	4279 0.038

k in W/mK	
Thermal insulation material 2- Expanded polystyrene	
C_{i1} in Rs/m^3 k_1 W/mK	9421 0.032
Total wall thermal resistance without insulation in m^{2}K/W	0.5858
Present worth factor	9.05

A total cost per unit area of external wall versus combined thickness of two insulation materials with different fraction is plotted. From annual total cost point of view optimum thermal insulation obtained when $f = 1$, i.e. fraction of Glass wool is 1 and that of EPS is zero than optimum insulation thickness is 0.15 m as shown in Figure 2, which is same as obtained by earlier researcher with same design parameters, where single thermal insulation was analysed at a time [21].

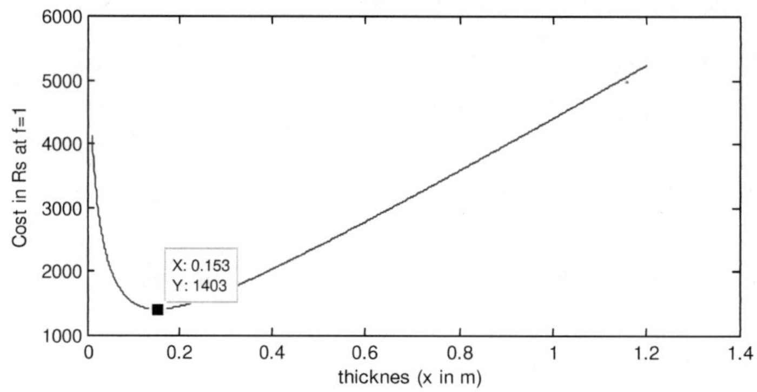

Figure 2. Total annual cost per unit area versus thermal insulations thickness for $f = 1$

Figure 3, shows the variation of optimum thickness of combined thermal insulation materials for different values of f. When $f = 0$ i.e. thickness of Glass wool is zero then although optimum thickness is 0.09 m but annual total cost per unit area is more as compared to total annual cost per unit area of composite external wall for other values of f. Although in the present analysis cost of glass wool is less as compared to cost of EPS but lower value of thermal conductivity leads to better performance on the basis of life cycle total cost.

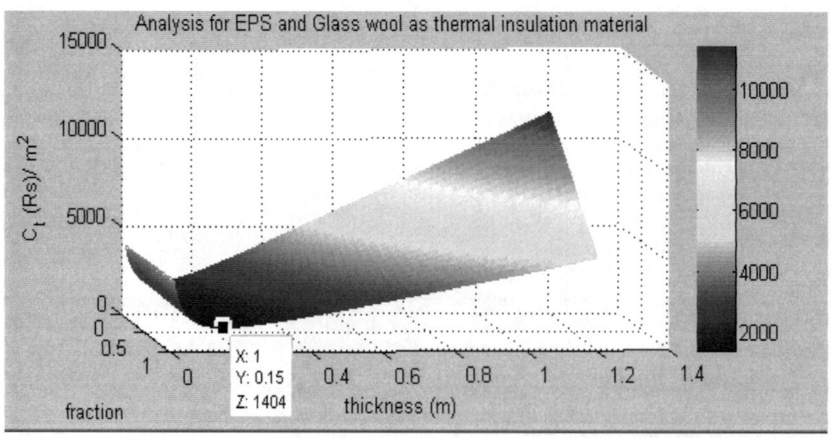

Figure 3. Total annual cost per unit area versus thermal insulations thickness for different fraction f

7. Conclusion

In the present study an approach to determine optimum thickness for a combination of two thermal insulation materials is proposed with the objective to minimize life cycle cost of fuel and thermal insulation for unit area of external wall of a residential building envelope. Developed approach is applied successfully in heating application. Although optimum value of combination of thermal insulation material on global minima is when $f = 1$, which lead to a case of single thermal insulation material. But if space is constrained then designer can use combination of two thermal insulation materials. Thus this approach will be useful in those applications of building thermal insulation where single thermal insulation material not fulfil all the requirements and one property is sacrificed in order to improve the other property, thus combination of thermal insulation materials can be used and their economically best combination can be find out.

Acknowledgement

Manuscript communication number (MCN): IU/R&D/2018-MCN000311 office of Research and Development cell, Integral University, Lucknow.

References

[1] Hasan A 1999 Optimizing insulation thickness for buildings using life cycle cost *Appl Energy* **63** pp115–24

[2] Ahmad EH 2002 Cost analysis and thickness optimization of thermal insulation materials used in residential buildings in Saudi Arabia *The 6th Saudi Engineering Conference* vol. **1** pp. 21–32

[3] Comakli K and Yuksel B 2003 Optimum insulation thickness of external walls for energy saving *Appl Therm Eng* **23** pp473–479

[4] Al-Sallal KA 2003 Comparison between polystyrene and fiberglass roof insulation in warm and cold climates *Renew Energy* **28** pp 603–11

[5] Al-Khawaja MJ 2004 Determination and selecting the optimum thickness of insulation for buildings in hot countries by accounting for solar radiation *Appl Therm Eng* **24** pp 2601–10

[6] Comakli K and Yuksel B 2004 Environmental impact of thermal insulation thickness in buildings *Appl Therm Eng* **24** pp 933–40

[7] Bolatturk A 2006 Determination of optimum insulation thickness for building walls with respect to various fuels and climate zones in Turkey *Appl Therm Eng* **26** pp1301–1309

[8] Dombayci OA, Golcu M and Pancar Y 2006 Optimization of insulation thickness for external walls using different energy-sources *Appl Energy* **83** pp 921–928.

[9] Lollini, Barozzi, Fasano, Meroni and Zinzi 2006 Optimisation of opaque components of the building envelope. Energy, economic and environmental issues *Build Environ* **41** pp 1001–13

[10] Dombayci OA 2007 The environmental impact of optimum insulation thickness for external walls of buildings *Build Environ* **42** pp 3855–3859

[11] Mahlia TMI, Taufiq BN, Ismail and Masjuki HH 2007 Correlation between thermal conductivity and the thickness of selected insulation materials for building wall *Energy Build* **39** pp 182–187

[12] Bolatturk A 2008 Optimum insulation thicknesses for building walls with respect to cooling and heating degree-hours in the warmest zone of Turkey *Build Environ* **43** pp 1055–64

[13] Yildiz A, Gurlek G, Erkek M and Ozbalta N 2008 Economical. environmental analyses of thermal insulation thickness in buildings *J Therm Sci Technol* **28** pp 25–34

[14] Kaynakli O 2008 A study on residential heating energy requirement and optimum insulation thickness *Renew Energy* **33** pp 1164–1172

[15] Ucar A and Balo F 2009 Effect of fuel type on the optimum thickness of selected insulation materials for the four different climatic regions of Turkey *Appl Energy* **86** pp 730–736

[16] Yu J, Yang C, Tian L and Liao D 2009 A study on optimum insulation thicknesses of external walls in hot summer and cold winter zone of China *Appl Energy* **86** pp 2520–2529

[17] Ucar A and Balo F 2010 Determination of the energy savings and the optimum insulation thickness in the four different insulated exterior walls *Renew Energy* **35** pp 88–94

IOP Conf. Series: Materials Science and Engineering **404** (2018) 012050 doi:10.1088/1757-899X/404/1/012050

[18] Ozkan DB and Onan C 2011Optimization of insulation thickness for different glazing areas in buildings for various climatic regions in Turkey *Appl Energy* **88** pp1331–1342

[19] Daouas N 2011 A study on optimum insulation thickness in walls and energy savings in Tunisian buildings based on analytical calculation of cooling and heating transmission loads *Appl Energy* **88** pp 156–164

[20] Yu J, Tian L, Yang C, Xu X. and Wang J 2011 Optimum insulation thickness of residential roof with respect to solar-air degree-hours in hot summer and cold winter zone of China *Energy Build* **43** pp 2304–2313

[21] Mishra S, Usmani JA and Varshney S 2012 Optimum insulation thickness of the external walls and roof for different degree-days region *IJERT* **1** Issue 7 pp 1-8

[22] Mishra S, Usmani JA and Varshney S 2012 Energy saving analysis in building walls through thermal insulation system *IJERA* **2** Issue 5 pp 128-135

[23] Kayfeci M, Kecebas A and Gedik E 2013 Determination of optimum insulation thickness of external walls with two different methods in cooling applications *Appl Therm Eng* **50** pp 217–224

[24] Oktay Z, Coskun C and Erturk M 2014 A new point of view for insulation calculations *IJEE* **4** Issue 3 pp 111-117

[25] Ozkan DB and Onan C 2011 Optimization of insulation thickness for different glazing areas in buildings for various climatic regions in Turkey *Appl Energy* **88** pp 1331-1342

[26] Kaynakli O. 2012 A review of the economical and optimum thermal insulation for building applications study on residential heating energy requirement and optimum insulation thickness *Renewable and Sustainable Energy Reviews* **16** pp 415–425

[27] Jafri SAH, Bharti PK and Ahmad MJ 2015 Optimum insulation thickness for building envelope- A review *IJRET 4 pp* 214-218

[28] Al-Homoud, MS 2005 Performance characteristics and practical applications of common building thermal insulation materials *Building and environment* **40** pp 353-366

[29] Papadopoulos AM 2005 State of the art in thermal insulation materials and aims for future developments *Energy and Bulidings* **37** pp 77-86

IOP Conf. Series: Materials Science and Engineering **404** (2018) 012051 doi:10.1088/1757-899X/404/1/012051

Mathematical modelling and performance analysis of single pass flat plate solar collector

Mohd Irshad, Anshul Yadav, Rajnish Singh, Anil Kumar

Department of Mechanical Engineering, Kamla Nehru Institute of Technology, Sultanpur 228118, India.

E-mail: anshulyadav.iitk@gmail.com

Abstract. Various type of solar collectors has already been designed and developed. A solar flat plate collector is used to raise the temperature of the passing fluid (air/water) from it. In the present work, a single cover flat plate solar collector was fabricated using the standard methodology proposed by other researches for the construction of conventional solar air heaters. The variation of the solar radiation, the air outlet temperature and the variation of absorber plate temperature throughout the day were recorded. The efficiency, effective optical efficiency, and the effective heat loss coefficient of single cover flat plate solar collector were calculated. The effect of wind velocity on collector's performance was also studied. The effective optical efficiency of the fabricated collector, effective heat loss coefficient, and the efficiency of the collector was found to be 72.7 %, 8.422 W m^{-2} K^{-1}, 36.73 % respectively.

1. Introduction

Solar collectors are devices that transform solar energy into some sort of heat energy using a fluid medium. Solar collectors absorb solar radiation converting it into thermal energy and transfers this energy into a fluid flowing through the solar collector. Now, this heated fluid is used in space conditioning and many other application fields. The solar collector is also equipped with storage tanks so that it can be used when solar radiation is not present. Solar collectors are used in thermo-syphon, space heating and cooling, desalination, solar furnaces, refrigeration and heat generation.

Many researchers have reviewed various types of flat plate solar collectors with various applications [1]–[4]. The basic types of the solar collector are: concentrating and non-concentrating. For the non-concentrating collector, intercepting and absorbing radiation area are same, whereas a concentrating solar collector uses the concave surface to intercept and focus the radiation beam to a smaller receiving area which increases radiation flux. These two types of collectors are further divided into large numbers of collectors depending upon their uses and shapes.

Thermal performance of the flat plate solar collector depends upon the transmittance, absorption and conduction of solar energy and working fluid conductivity. The absorber plate is an important component of solar air heater to decide its performance. In designing the solar collector plate, efficiency factor and heat removal factor are important parameters. Sekha et al. [5] worked on the solar collector and reduced empiricism associated in the design of solar collector. Many researchers have analyzed the maximum energy conversion by the use of coatings applied to the absorber plate [6]–[9].

It has been found that the thermal conductivity of the collector's absorber plate depends upon fin efficiency factor and heat removal factor and cost factor is the most critical factor in designing a collector [10], [11].

As can be seen from the aforementioned arguments, solar energy systems can be used for a wide range of applications and provide significant benefits, therefore, they should be used whenever possible. Flat plate solar collectors are characterizing with the shape of the solar collector and thermal reflectivity of plate and absorptivity of the plate. The more dominant parameters are reflectivity and absorbing coefficient. The time for which solar energy is incident on the plate and tilt angles of the plate is also dependent parameters for performance analysis of flat plate solar collector. Choudhury et al. [12] proposed construction curves for conventional air heaters for economical designing of flat plate solar collector.

In the present work, a single cover flat plate solar collector is fabricated using the methodology used in the construction of conventional solar air heaters [13]–[15]. The variation of the solar radiation, the air outlet temperature and the variation of absorber plate temperature throughout the day were recorded. The efficiency, effective optical efficiency, and the effective heat loss coefficient of single cover flat plate solar collector are calculated and the effect of wind velocity on collector's performance is also studied.

2. Proposed methodology

For the construction of the collectors, the methodology proposed by Choudhury et al. [12] was selected. They proposed construction curves for the conventional solar air heater. In these curves, they proposed different curves for construction for three types of solar air heaters. In this work, a single cover flat plate solar collector using the curve data provide by Choudhury et al. [16].

In this methodology, for the construction of the collector's acceptable values following parameters as inputs have to be predetermined like length of the absorber (decided on the basis of the floor space available), pressure drop (decided by the capacity of the air blower), the mass flow rate of air (decided by the flow of heat energy required for the job to be done by the collector), the air temperature increment for unit incident flux (decided by the metrological data of the place where this collector was being used).

With the help of the design curves, the absorber's area, absorbers width and depth of the duct were calculated with the following steps:

1. First, select the length of the collector.
2. Then, select the pressure drop and the mass flow rate of air on the basis of where the collector is going to be used.
3. Collect meteorological data of the place where the collector is going to be used.
4. Then, calculate the temperature increment of the air per unit incident flux.
5. Now, with the use of graph between temperature increment of the air per unit incident flux and length for various fixed values of mass flow rate per unit area, find the mass flow rate per unit area.
6. Then, calculate the required air flow from the collector.
7. With the use of required air flow from the collector and mass flow rate per unit area, calculate the area of the collector.
8. Now, with the use of graph between length and duct depth for various fixed values of the mass flow rate per unit area, find the duct depth.

The solar flat plate collector fabricated by the above procedure is shown in Figure 1.

Figure 1. The fabricated solar collector

2.1. The positioning of Solar Collector

Flat plate solar collector's efficiency depends upon positioning of the solar collector. In the fabricated solar collector, the tilting mechanism is provided to obtain the maximum efficiency. The period during which 80 % of the collectable solar energy falls on the collector is known as a solar window. In general, the time period from 9 am to 3 pm is considered as "solar window". The sun's position determines optimal angles and direction of the solar panel of the collector at different times of the year.

In the northern hemisphere of the earth, the sun is highest in the sky summer season and also in the sky for more hours than it is in the winter. The opposite is true for winter. The solar panels are placed at an angle, 15^0 less than the latitude where they are installed in the summer and 15^0 more than the latitude of the place where the collector is installed. The collector faces south, towards the sun for maximum flux absorption. The effect of sun position in the northern hemisphere of the globe is also considered. Sun's radiation is tilted to 23.5^0 about its mean position which affects the incident energy and conversion energy of solar collector. Beam radiation has unique direction and can be converged by reflection and refraction technique but diffused radiation cannot be concentrated and does not follow the optical principles. The flat plate collector absorbs direct and diffused both types of radiations. Therefore, the flat plate collector does not require sun tracking system and also has a very little maintenance cost. Though, optical efficiency is less due to large surface area heat loss.

The concentrating ratio is defined as the ratio of the area of the aperture to the area of the receiver. The concentrating ratio for the study is found to be 1. The metallic tubes were soldered/welded to absorber plate. Hence, insulation prevented heat loss from the rear surface and sides of the collector. The glass cover permits short wavelengths solar radiations but opaque to larger infrared radiations. For operation from January to December the tilt can be taken as 0.9 times the latitude for Indian conditions.

2.2. Designing and Fabrication

As stated in the previous sections, some parameter's values were predetermined in this designing methodology. The length *(L)* was taken as 150 cm and pressure drop was taken as 30 Pa. From the metrological data in the month of March, April and May, the average ambient temperature (T_a) was found to be 35 °C and the average solar radiation (I) was found to be 875 Wm^{-2}. The solar collector was used for drying Chilies. Hence, air outlet temperature (T_o) should not be more than 65 °C. From above data temperature increment per solar radiation, $\frac{To-Ti}{I}$ = 0.034. For the experiments, 6 kg of chillies were taken for dying. The chillies water content before (m_i) and after drying (m_f) was found as 80 % and 10 % respectively. From the aforementioned data, the mass of water to be evaporated was calculated as 5 Kg. The required flow rate of the air was calculated as 50 Kg h^{-1} m^{-2}, the width of the collector was calculated as 65 cm and depth of duct was calculated as 10 mm with the help of construction curve between length and temperature increment per solar radiation.

3. Mathematical Modelling

In the present design, the transparent cover is placed directly above the absorber plate and there is a layer of air between cover and absorber. The fluid to be heated runs between the layer of insulation and the inner surface of the absorber. Energy flow in this design was studied from [7], [12], [17]–[21]. The theoretical and analytical results were compared from the benchmark results directly above the rear plate of the collector. Following equations can be obtained as the steady-state energy balance equations. The heat transfers that take place in single cover solar air heater are shown in Figure 2.

For the absorber plate:

$$\alpha_p \tau_g S = h_{cpf}(T_p - T_f) + h_{cpg}(T_p - T_g) + h_{rpb}(T_p - T_b) + h_{rpg}(T_p - T_g) \quad (1)$$

For the flowing air:

$$h_{cpf}(T_p - T_f) = h_{cbf}(T_f - T_b) + \frac{mC_f}{W}\frac{dT_f}{dx} \quad (2)$$

Bottom plate:

$$h_{cpf}(T_f - T_b) + h_{rpb}(T_p - T_b) = U_b(T_b - T_a) \quad (3)$$

For cover:

$$\alpha_g S + h_{rpg}(T_p - T_g) + h_{cpg}(T_p - T_g) = (h_w + h_{rs})(T_g - T_a) \quad (4)$$

The boundary conditions for the above sets of equations:

$$T_f(x = 0) = T_{fi}, T_f(x = L) = T_{fo} \quad (5)$$

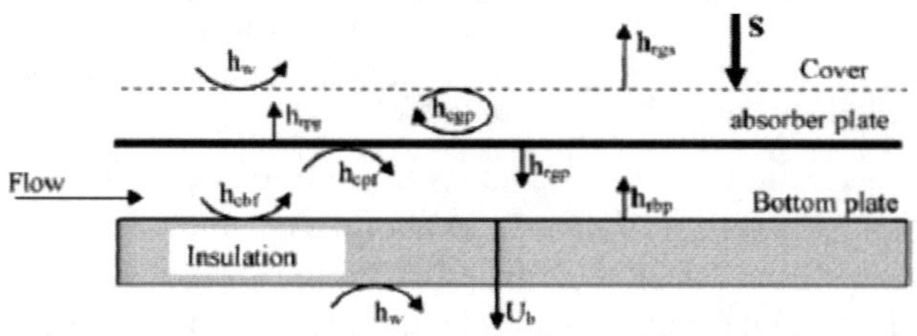

Figure 2. Schematic showing heat transfer in single cover solar air heater [12]

In the mathematical model of the flat plate solar collector proposed by Ong [13], there is no heat transfer in between the system and the top cover and there is negligible heat transfer between the cover and stagnant air cavity. Only free convection takes place in an inclined wall cavity. Hence the balance equations obtained for this design of solar collector are the same as proposed by Choudhury and Garg [12] instead of the equation for the flow of air which was replaced by the following equation:

$$h_{cpf}(T_p - T_f) = h_{cbf}(T_f - T_b) + \frac{2mC_f}{WL}(T_f - T_{fi}) \quad (6)$$

The equation for the flow of the air is replaced by

$$\frac{mC_f}{W}\frac{dT_f}{dx} = h_{pf}(T_p - T_f) + h_{bf}(T_b - T_f) \quad (7)$$

4. Performance of the solar collector

The performance of the collector is determined by the distribution of incident solar energy into useful energy and various thermal as well as leakage losses. To measure collector performance, it was exposed to solar radiation and we measure the fluid inlet T_i and outlet T_o temperatures and the flow rate of the fluid. The useful gain is calculated as

$$Q = m.C_p(T_o - T_i) \quad (8)$$

Under steady state condition, the useful heat delivered by the solar collector is equal to the energy absorbed in the metal surface minus the heat losses from the surface to the ambient environment. The useful heat output of the flat plate collector is given by

$$Q_c = A[I_c(\tau\alpha)_e - U_c\,(T_{in} - T_{out})] \quad (9)$$

Introducing the heat removal factor (FR) in the above equation,

$$Qc = A[I_c FR\,(\tau\alpha)e - FRUc\,(Tin - Ta)] \quad (10)$$

The efficiency of a general solar collector is stated as the ratio of the useful heat energy output of the collector to the solar energy flux incident on the collector.

$$Efficiency = \frac{useful\ heat\ output\ of\ the\ collector}{solar\ enery\ flux\ incident\ on\ the\ collector} \quad (11)$$

$$\eta = Qc/AIc \quad (12)$$

$$\eta = FR(\tau\alpha)e - FR * U_c((T_{in} - T_a)/I_c) \quad (13)$$

T_{in} is replaced by T_p in case of air heaters.

$FR(\tau\alpha)_e$ (can be calculated by determining the *y*-axis intercept of the efficiency line equation) is known as the effective optical efficiency and *FRU* (this can be calculated by determining the slope of the efficiency line equation) is known as effective heat loss coefficient. The outlet temperature of the collector heat transfer fluid is given by

$$T_{out} = T_{in} + Q_c/(m * C_p) \quad (14)$$

The stagnation temperature T_s, of the collector, is defined as the temperature of the absorber which is achieved when there is no flow of heat transfer fluid in the collector and therefore, its useful heat output and efficiency both are equal to zero.

$$T_s = T_{in} - T_a + I_c \times (FR(\tau\alpha)e)/(FR\,U_c) \quad (15)$$

5. Results and discussion

A single cover flat plate solar collector/air heater was constructed with the help of construction curve for conventional air heaters. Pyranometer was used for the measurement of the solar radiation, K-type thermocouples were used for measuring collector's absorber plate temperature, digital thermometer was used in measuring air outlet temperature and Anemometer was used in measuring the velocity of air and wind. The collector was mounted on the stand and the collector was made to face south in the sunshine. Three thermocouples were attached to the absorber plate at 15 cm, 75 cm, and 135 cm across the length of the collector. The observations from the flat plate collector were recorded at the regular interval of 30 minute from 9:00 hours to 16:00 hours and tabulated in Table 1. It can be seen that absorber plate temperature and air temperature at outlet increases as solar radiation increases but wind velocity plays important role in deciding the final value of these temperatures. With the increase in wind velocity both the temperature decreases but the extent by which outlet temperature decreases is more than the absorber plate temperature. The outlet temperature reaches its maximum value 66.9 °C at 1:30 pm and absorber plate temperature reaches its maximum value 82.1 °C at 2:00 pm.

Table 1. Experimental data obtained from the solar collector

Time (hh:mm)	V_a (m s^{-1})	I (W/m^2)	T_i (°C)	T_o (°C)	T_p (°C)	$\frac{Tp-Ti}{I}$	η (%)
09:00	0.1	340	28	37.9	40.6	0.037	42.07
09:30	0.3	380	30	39.1	46.7	0.044	34.60
10:00	0.3	560	32	45.7	57	0.0446	35.35
10:30	0.2	620	43	58.5	69.7	0.0431	36.13
11:00	0.3	640	42.8	58.6	70.9	0.0439	35.68
11:30	0.5	660	47.9	62.9	77.6	0.0450	32.84
12:00	0.1	680	48	66.3	75.8	0.0410	38.89
12:30	0.5	680	48	64.1	79.8	0.0468	34.21
13:00	0.5	680	47.8	64.2	78.8	0.0456	34.85
13:30	0.5	660	47.8	66.9	72.9	0.038	41.82
14:00	1.5	660	47.8	61.7	82.12	0.052	30.43
14:30	0.8	620	47.5	63.5	74	0.0428	37.29
15:00	0.8	620	42	60.9	64.3	0.036	44.05
15:30	1.8	580	40.8	55.6	64	0.040	36.88
16:00	0.8	540	40.5	53.9	63.2	0.042	35.86

To calculate the effective optical efficiency and the effective heat loss coefficient, a linear straight line was fitted as shown in Figure 3. The eefficiency of the collector was calculated as 36.73 %. The effective optical efficiency was calculated as 72.7 % and the effective heat loss coefficient was calculated as 8.422 W m^{-2} K^{-1}.

Table 2. Line fitting method for the plot between $(T_b - T_a)/I$ and efficiency

$x\left(\frac{Tp-Ti}{I}\right)$	y (η)	$x \times y$	x^2
0.037	42.07	0.0156	0.001369
0.044	34.60	0.0152	0.001936
0.0446	35.35	0.0158	0.001989
0.0431	36.13	0.0156	0.001858
0.0439	35.68	0.0157	0.001927
0.0450	32.84	0.0148	0.002025
0.0410	38.89	0.0159	0.001681
0.0468	34.21	0.0160	0.002190
0.0456	34.84	0.0159	0.002079
0.038	41.82	0.0159	0.001444
0.052	30.43	0.0158	0.002704
0.0428	37.29	0.0158	0.001832
0.036	44.05	0.0158	0.001296
0.040	36.88	0.0148	0.001600
0.0402	35.86	0.0151	0.001764
$\Sigma x = 0.6418$	$\Sigma y = 550.95$	$\Sigma xy = 0.2337$	$\Sigma x^2 = 0.0277$

IOP Conf. Series: Materials Science and Engineering **404** (2018) 012051 doi:10.1088/1757-899X/404/1/012051

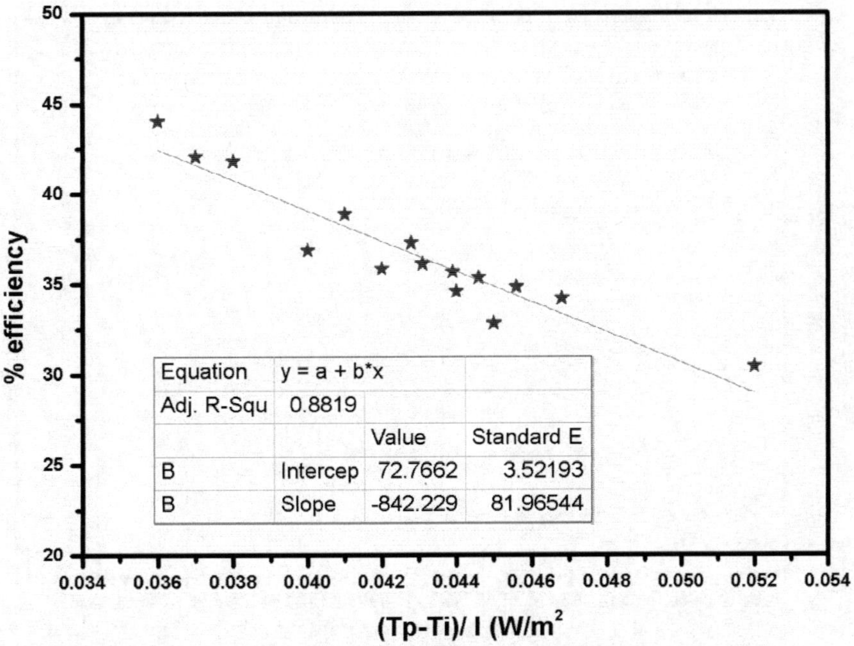

Figure 3. Plot of efficiency with $\frac{Tp-Ti}{I}$ for the designed solar collector

The variation of mean solar radiation on absorber plate throughout the day (dated 8/4/2016) (from 9:00 hours to 16:00 hours) is plotted in Figure 4. From the figure, it can be seen that the maximum solar radiation increased till noon and then decreased till 16:00 hours.

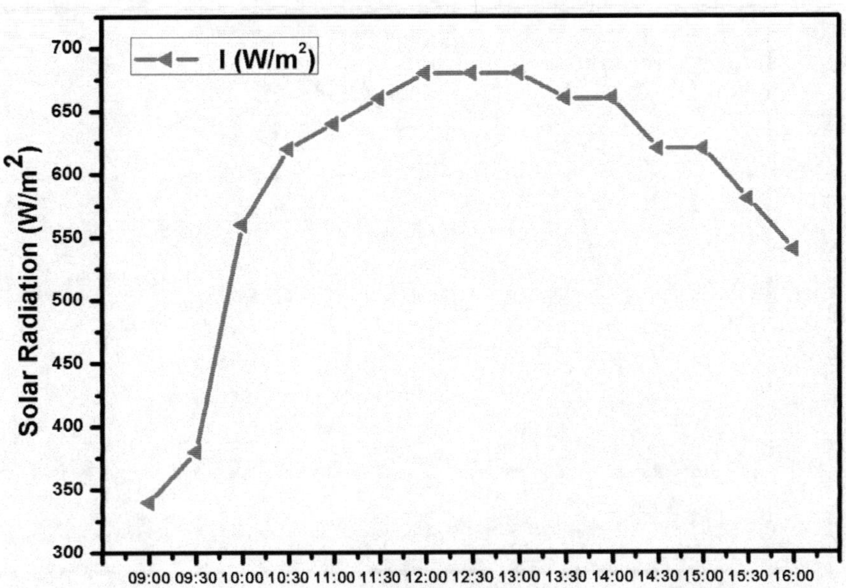

Figure 4. Plot showing variation of solar radiation with time

The temperature profile of the flat plate collector throughout the day (dated 8/4/2016) (from 9:00 hours. to 16:00 hours) is plotted in Figure 5. From the figure, it can be seen that the temperature of the collector plate increases as the sun's radiation increases.

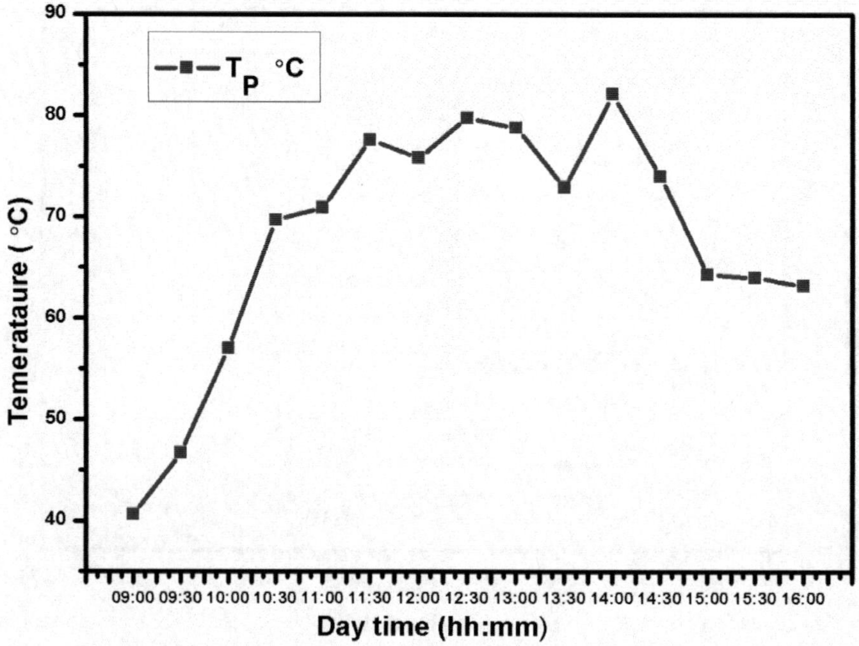

Figure 5. Plot showing variation of mean absorber plate temperature with time

The variation of the temperature of the outlet air throughout the day (dated 8/4/2016) (from 9:00 hours. to 16:00 hours) is plotted in Figure 6. From the figure, it can be seen that the outlet air temperature increases as the mean solar radiation on the flat plate collector increases.

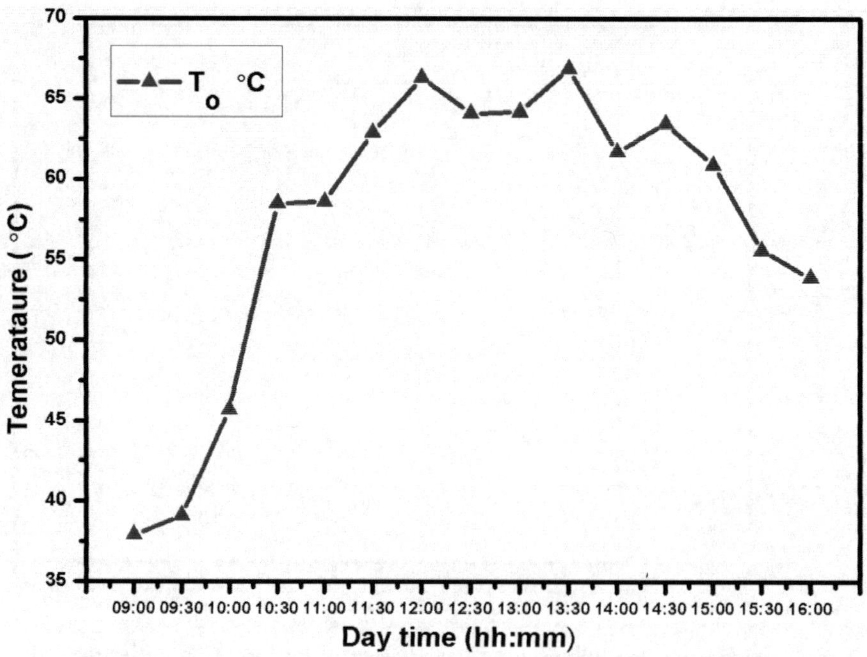

Figure 6. Plot showing variation of outlet temperature with time

Figure 7 shows the variation in wind velocity, mean temperature and thermal efficiency of flat plate collector. The thermal efficiency and flat plate temperature are not only affected by the intensity of solar radiation but this parameter widely affected wind velocity because high wind velocity may extract heat from flat plate collector by means of convection. From Figure 8, it may be concluded that the efficiency of the collector at low solar radiation and low wind speed is higher than as compared to the condition of higher radiation and high wind speed. Hence, it can be concluded that the wind velocity plays important role in the performance of collector. With the increase in wind velocity efficiency of the collector reduces this may be due to the reason that, with the increase in wind velocity the convective heat loss to environment increases.

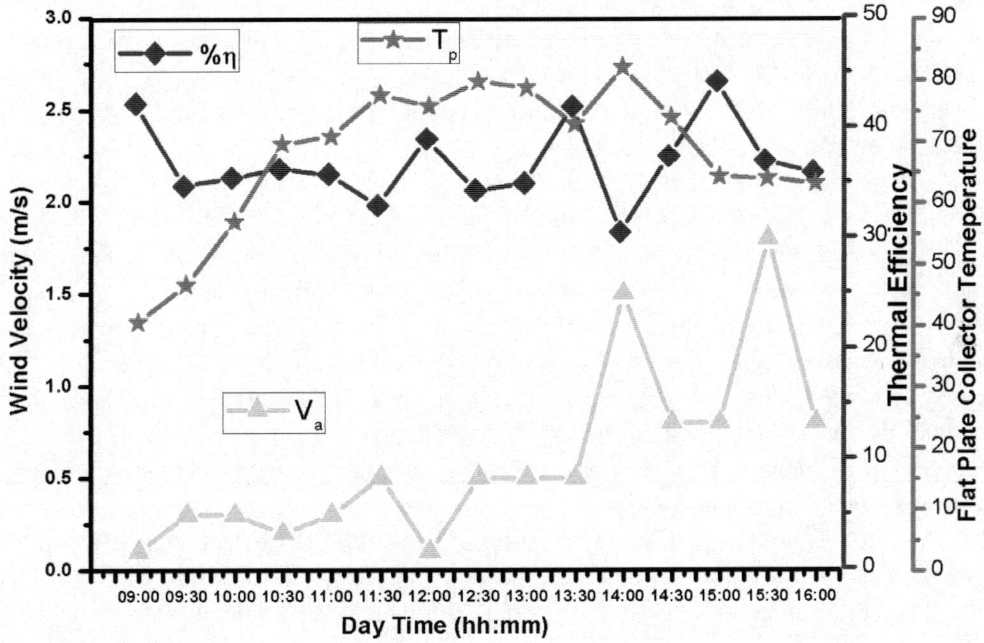

Figure 7. Plot showing variation of efficiency with wind velocity

6. Conclusions

In the present work, a single cover flat plate solar collector is fabricated and the efficiency, effective optical efficiency, and the effective heat loss coefficient of single cover flat plate solar collector are calculated and the effect of wind velocity on collector's performance is also studied. A linear straight line was fitted to calculate the effective optical efficiency and the effective heat loss coefficient. Following conclusions were drawn from the study:

1. The efficiency of the collector was calculated as 36.73 %. The effective optical efficiency was calculated as 72.7 % and the effective heat loss coefficient was calculated as 8.422 W m^{-2} K^{-1}.
2. The maximum solar radiation increased till noon and then decreased till 16:00 hours and the temperature of the collector plate increases as the sun's radiation increases.
3. The thermal efficiency and flat plate temperature are not only affected by the intensity of solar radiation but this parameter widely affected wind velocity because high wind velocity may extract heat from flat plate collector by means of convection.
4. The efficiency of the collector at low solar radiation and low wind speed is higher than as compared to the condition of higher radiation and high wind speed. Hence, it can be concluded that the wind velocity plays important role in the performance of collector.
5. With the increase in wind velocity efficiency of the collector reduces this may be due to the reason that, with the increase in wind velocity the convective heat loss to environment increases.

References

[1] Pandey K M, Chaurasiya R 2017 A review on analysis and development of solar flat plate collector. *Renewable and Sustainable Energy Reviews* **67** 641–650.

[2] Baskar P, Edison G 2013 A review of mathematical models for performance analysis of hybrid solar photovoltaic-thermal (PV/T) air heating systems. *Adv. Mater. Res.* **768** 29–39.

[3] Tchinda R 2009 A review of the mathematical models for predicting solar air heaters systems. *Renew. Sustain. Energy Rev.* **13** 1734–1759.

[4] Saxena A, Varun, El-Sebaii A A 2015 A thermodynamic review of solar air heaters. *Renew. Sustain. Energy Rev.* **43** 863–890.

[5] Sekhar Y R, Sharma K V, Rao M B 2009 Evaluation of heat loss coefficients in solar flat plate collectors. *ARPN J. Eng. Appl. Sci.* **4** 15–19.

[6] Zhou F, Ji J, Yuan W, Cai J, Tang W, Modjinou M 2018 Numerical study and experimental validation on the optimization of the large size solar collector. *Appl. Therm. Eng.* **133** 8–20.

[7] Wang N, Zeng S, Zhou M, Wang S 2015 Numerical study of flat plate solar collector with novel heat collecting components. *Int. Commun. Heat Mass Transf.* **69** 18–22.

[8] Hegazy A A 2000 Thermohydraulic performance of air heating solar collectors with variable width, flat absorber plates. *Energy Convers. Manag.* **41** 1361–1378.

[9] Hegazy A A 1999 Optimum channel geometry for solar air heaters of conventional design and constant flow operation. *Energy Convers. Manag.* **40** 757–774.

[10] Choi S U S, Eastman J A 1995 Enhancing thermal conductivity of fluids with nanoparticles. *ASME Int. Mech. Eng. Congr. Expo.* **66** 99–105.

[11] Shariah A M, Rousan A, Rousan K K, Ahmad A A 1999 Effect of thermal conductivity of absorber plate on the performance of a solar water heater. *Appl. Therm. Eng.* **19** 733–741.

[12] Choudhury C, Chauhan P M, Garg H P 1995 Performance and cost analysis of two-pass solar air heaters. *Heat Recover. Syst. CHP* **15** 755–773.

[13] Ong K S 1995 Thermal performance of solar air heaters: Mathematical model and solution procedure. *Sol. Energy* **55** 93–109.

[14] Yousef B A A, Adam N M 2012 Performance and cost analysis of double duct solar air heaters. *Int. J. Sci. Technol. Res.* **1** 108–116.

[15] Jannot Y, Coulibaly Y 1997 Radiative heat transfer in a solar air heater covered with a plastic film. *Sol. Energy* **60** 35–40.

[16] Choudhury C, Chauhan P M, Garg H P 1995 Design curves for conventional solar air heaters. *Renew. Energy* **6** 739–749.

[17] Saedodin S, Zamzamian S A H, Nimvari M E, Wongwises S, Jouybari H J 2017 Performance evaluation of a flatplate solar collector filled with porous metal foam : Experimental and numerical analysis. *Energy Convers. Manag.* **153** 278–287.

[18] Nikolić N, Lukić N 2015Theoretical and experimental investigation of the thermal performance of a double exposure flat-plate solar collector. *Sol. Energy* **119** 100–113.

[19] Dara J E, Ikebudu K O, Ubani N O, Chinwuko C E, Ubachukwu O A, 2013 Evaluation of a passive flat-plate solar collector. *Int. J. Adv. Res. Technol.* **2**.

[20] Ladas D I, Stathopoulos T, Rounis E D 2017Wind effects on the performance of solar collectors on rectangular flat roofs: A wind tunnel study. *J. Wind Eng. Ind. Aerodyn.* **161** 27–41.

[21] Kalogirou S A 2004 Solar thermal collectors and applications. *Progress in Energy and Combustion Science* **30** 231–295.

IOP Conf. Series: Materials Science and Engineering **404** (2018) 012052 doi:10.1088/1757-899X/404/1/012052

Mathematical modelling to predict mechanical properties of Copper (C101) feedstock in continuous extrusion

Devendra Kumar Sinha[1], Santosh Kumar[2], Anil Kumar[3], Anshul Yadav[4]

[1]School of Mechanical Engineering, Adama Science & Technology University, Adama, Post Box No.1888, Ethiopia.

[2]Department of Mechanical Engineering, Indian Institute of Technology BHU, Varanasi 221005, India.

[3,4]Department of Mechanical Engineering, Kamla Nehru Institute of Technology, Sultanpur 228118, India.

E-mails: [1]sinhadevendra25@rediffmail.com, [2]santosh.kumar.mec@iitbhu.ac.in, [3]anilap.knit@gmail.com, [4]anshulyadav.iitk@gmail.com

Abstract. The continuous extrusion process is a process for producing endless profiles with a high degree of accuracy. The two important and vital process parameters are extrusion wheel velocity and product diameter which affect the mechanical and metallurgical properties of the material. In this work, continuous extrusion of pure copper (C101) rod has been carried out on commercial setup TBJ 350 for feedstock diameter of 9.5 mm at different extrusion wheel velocities and product diameters using different die sizes. The yield strength and percent elongation of extruded feedstock at different wheel velocities and product diameters has been estimated by carrying out the tensile test. The continuous extrusion process has been carried out based on two factor three-level central composite rotatable design using response surface methodology. A mathematical model has been developed to predict the mechanical properties of the extruded feedstock through continuous extrusion process and its validation has been carried out using ANOVA technique.

1. Introduction

The continuous extrusion forming process uses the frictional force between a circular driving wheel and material and can produce significantly long continuous products of a variety of sectional shapes which are hard in the classical forming processes owing to their methodological limitation. Because its superiority and impact on the current forming technology are evident, demand from industry has been growing rapidly. Basic experimental and analytical studies have been underway continuously since the early 1970s [1, 2]. However, theoretical and numerical studies are still insufficient to analyze accurately the complicated process characteristics. Furthermore, a study on the effects of major process parameters such as wheel velocity, relative die opening width, and flash gap size is essential for the optimal design of continuous extrusion process.

Recently, [3] investigated the use of extrusion-cutting as a material test method operating at severe conditions of strain, strain-rate and temperature, such as in machining. Many researchers have tried to study the mechanical properties of pure copper in the extrusion process [4]–[6]. The extrusion wheel angular velocity of CONFORM process for Aluminium alloy has been studied by Kim et al. [7] and Cho et al. [8] by using the two-dimensional finite element technology. Wu et al. [9] carried out by three-dimensional finite element technology, the detailed distributions of the velocity field, stress field, strain field, thermal field and damage field. Zhao et al. [10] carried out the effect of deformation speed on the microstructure and mechanical properties of AA6063 during the continuous extrusion process. Their results indicated that the variations in extrusion wheel velocity directly influence material deformation and significantly influence the maximum extrusion temperature. They concluded that there exists an optimum extrusion wheel velocity which generates products with good mechanical and metallurgical properties. Palanivel et al. [11] carried out a systematic approach to develop the mathematical model through response surface methodology (RSM) for predicting the yield strength (YS), and percent elongation of AA6351 aluminium alloy which is widely used in automotive, aircraft and defence industries by incorporating friction stir welding (FSW) process parameters such as tool rotational speed, welding speed and axial force. Balasubramaniam et al. [12] developed a mathematical model using RSM to predict tensile properties of pulsed current gas tungsten arc welded Ti-6Al-4V alloy. Niranjan et al. [13] carried out numerical modelling of aluminium sheets formability using response surface methodology. Sinha et al. [14] fabrication of actual set-up and did virtual design using computer-aided design (CAD) software.

In the present work, continuous extrusion forming of the pure copper (C101) rod has been carried out on commercially available setup TBJ 350 at different extrusion wheel velocities and product diameters. A mathematical model through RSM is developed to analyse the influence of wheel velocity and product diameter on mechanical properties such as YS and percent elongation. An optimum value of the extrusion wheel velocity and product diameter has been determined to predict the best mechanical properties of the continuous extrusion forming products at different wheel velocities and product diameters. The adequacy of the model has also been tested by the analysis of variance test (ANOVA).

2. Experimental

For the continuous extrusion forming, pure copper (C101) rod available commercially of 9.5 mm diameter feedstock has been used as the raw material. The commercially available continuous extrusion set up TBJ350 has been used to carry out experimentation at different wheel velocity and Product Diameter. The pure copper (C101) rod has been subjected to continuous extrusion under the extrusion wheel velocities and different product diameters. The continuous extrusion products have not been subjected to any artificial ageing and treatments. The tensile testing of the samples has been performed on the INSTRON machine. The YS and percent elongation of the broken sample after tensile testing has been obtained and recorded for mathematical modelling.

3. Mathematical model

For finding out the relationship between the process parameters and the mechanical properties, second order polynomial response surface mathematical models can be considered [12, 13]

$$Y_u = b_0 + \sum b_i x_{iu} + \sum b_{ii} x^2_{iu} + \sum b_{ij} x_{iu} x_{ju} \quad (1)$$

where Y_u is the corresponding response; x_{iu} is the coded values of the ith continuous extrusion parameters for the uth experiments; and b_i, b_{ii}, b_{ij} are the second order regression coefficients. The second term under the summation sign of this polynomial equation corresponds to linear effect, while the third term denotes to the higher order effect. The fourth term of the equation includes the interactive effects of the process parameters.

IOP Conf. Series: Materials Science and Engineering **404** (2018) 012052 doi:10.1088/1757-899X/404/1/012052

The two factors with three levels and a central composite rotatable design matrix has been chosen to optimize the experimental conditions. The main aim of the factorial experiments is to depict the relationship between the response as a dependent variable and the parameter levels. This approach helps to understand how the change in levels of parameters affects the response. The combination of different levels of the parameters leads to certain optimum response. The present investigation studied the effects of factors such as extrusion wheel velocity and product diameter. The response parameters are YS and percent elongation of the extruded C101 feedstock during the continuous extrusion process. A 2^k factorial with central composite second order design has been used (here $k = 2$).

The coded and actual values of the parameters used in the present work have been listed in Table 1. Experiments were conducted and the output values of YS and percent elongation along with the design matrix is tabulated in Table 2. For the analysis of the data, the fitness of the model is necessary and well required. For checking the accuracy of the model includes a test for the significance of the regression model, the test for significance of model coefficient and test for lack of fit. For this purpose, analysis of variance (ANOVA) is performed.

Table 1. Experimental parameter and levels

Factor	Level		
	-1	0	1
Wheel velocity	4	6	8
Product diameter	6	7	8

3.1 Mathematical model for yield strength

The test of significance of YS has been carried out using the quadratic model. The results of the quadratic model for YS given in Table 4. The value of R^2 and adjusted R^2 are 95.96 % and 93.07 %. This means that the regression model provides a complete relationship between the dependent and independent variables. The associated p value for the model is less than 0.05 (i.e., $\alpha = 0.05$ or 95 % confidence level), indicating that the model is considered to be statistically significant. It is also seen from Table 3 that from the p values, the main effect of X_2 and second-order effect of X_1 and X_2 is much more significant. The other model terms can be regarded as insignificant due to their probabilities values being more than 0.05. Figure 1 depicts the normal probability of residuals for YS. It is observed from the plot that residuals values are distributed normally and in a straight line and hence the model is adequate.

Using the results presented in Table 3, the derived model is shown as:

$$\sigma = 34.5255-0.9350X_1 -1.4483X_2 -3.6547X_1^2 - 8.0953X_2^2 + 1.5700X_1X_2 \quad (2)$$

3.2 Mathematical model for percent elongation

The test of significance of percent elongation has been carried out using the quadratic model. The results of the quadratic model for percent elongation are given in Table 4. The value of R^2 and adjusted R^2 are 97.76 % and 96.15 %. This means that the regression model provides a complete relationship between the dependent and independent variables. The associated p value for the model is less than 0.05 (i.e., $\alpha = 0.05$ or 95 % confidence level), indicating that the model is considered to be statistically significant. It is also seen from Table 3 that from the p values, the main effect of X_2 and second-order effect of X_1 and X_2 is much more significant. The other model terms can be regarded as insignificant due to their probabilities values being more than 0.05. Figure 2 depicts the normal probability of residuals for percent elongation. It is observed from the plot that residuals values are distributed normally and in a straight line and hence the model is adequate.

Using the results presented in Table 3, the derived model is shown as:

$$L= 46.3973- 0.2500X_1 + 1.0000\, X_2- 5.8200\, X_1^2- 3.2575\, X_2^2+ 3.0000\, X_1X_2 \quad (3)$$

The adequacy of both YS and percent elongation models has also been tested through analysis of variance (ANOVA). The results of the analysis justify the closeness of the fit of the derived mathematical model. It has been concluded that the evolved mathematical models given by Eqs 2 and 3 are quite adequate at 95 % confidence level.

Table 2. Experimental plan and result based on central composite second order rotatable design

Expt. No.	Wheel velocity (RPM)		Product diameter (mm)		Yield strength (M Pa)	Percent elongation
	Coded	Actual	Coded	Actual		
1	-2	1	0	7	18.000	25.500
2	1	10	1	8	51.500	43.000
3	-1	4	1	8	42.750	34.000
4	0	7	0	7	33.000	46.500
5	0	7	-2	5	65.000	32.000
6	0	7	0	7	33.000	46.500
7	2	13	0	7	18.000	21.000
8	-1	4	-1	6	37.200	37.000
9	0	7	0	7	33.000	46.500
10	1	10	-1	6	39.670	34.000
11	0	7	2	9	65.000	35.000
12	0	7	0	7	33.000	46.500
13	0	7	0	7	33.000	46.500

Table 3. Test result of ANOVA for yield strength and percent elongation

Source	DF	Yield strength				Percent elongation			
		Sum of squares	Mean sum of squares	*F value*	*P value*	Sum of squares	Mean sum of squares	*F* value	*P* value
Regression	5	2430.48	486.10	21.86	0.000	889.097	177.819	61.01	0.000
Linear	2	35.66	35.66	0.80	0.486	12.750	6.375	2.19	0.183
Square	2	2384.96	1192.48	53.62	0.000	840.347	420.173	144.16	0.000
Interaction	1	9.86	9.86	0.44	0.527	36.000	36.000	12.35	0.010
Residual error	7	155.69	22.24			20.403	2.915		
Lack of fit	3	155.69	51.90			20.403	6.801		
Pure error	4	0.00	0.00	0.00		0.000	0.000		
Total	12	2586.17				909.500			

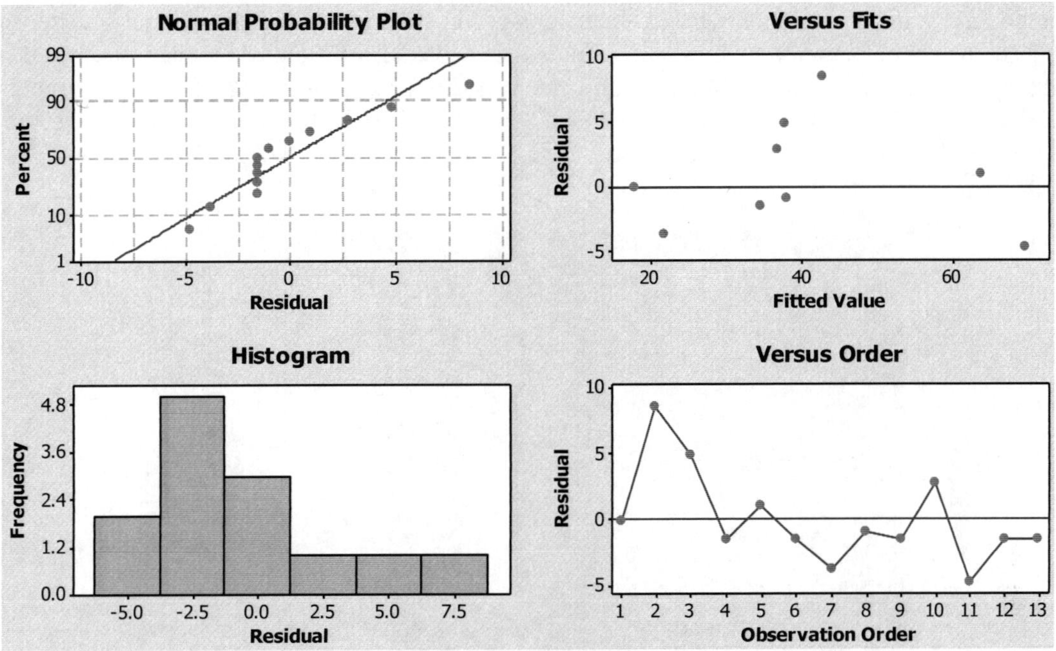

Figure 1. Residual plots for yield strength

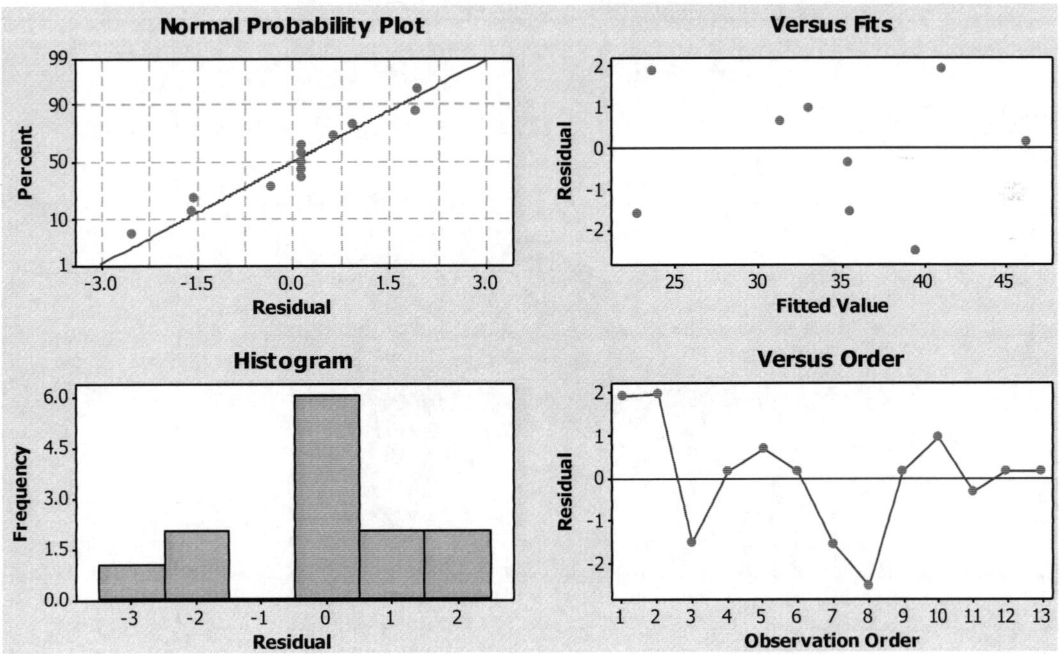

Figure 2. Residual plots for percent elongation

4. Results and Discussion

The effects of extrusion wheel velocity and product diameter on YS and percent elongation are shown in Figures 3 to 6 by contour and surface plots respectively. It can be inferred from Figures 3 and 4 that as the extrusion wheel velocity and product diameter increases, the YS of the extruded feedstock through continuous extrusion process increases and reaches to a maximum value and then further increase in the value of extrusion wheel velocity and product diameter results in decrease in the value of YS. The percent elongation value of the extruded feedstock also follows the same trend like that of Yield Strength with the variation in extrusion wheel velocity and product diameter. It can be inferred from Figures 5 and 6 With the increase in extrusion wheel velocity and product diameter increases the percent elongation of the extruded product to a certain maximum value and then decreases with further increase in the value of extrusion wheel velocity and product diameter.

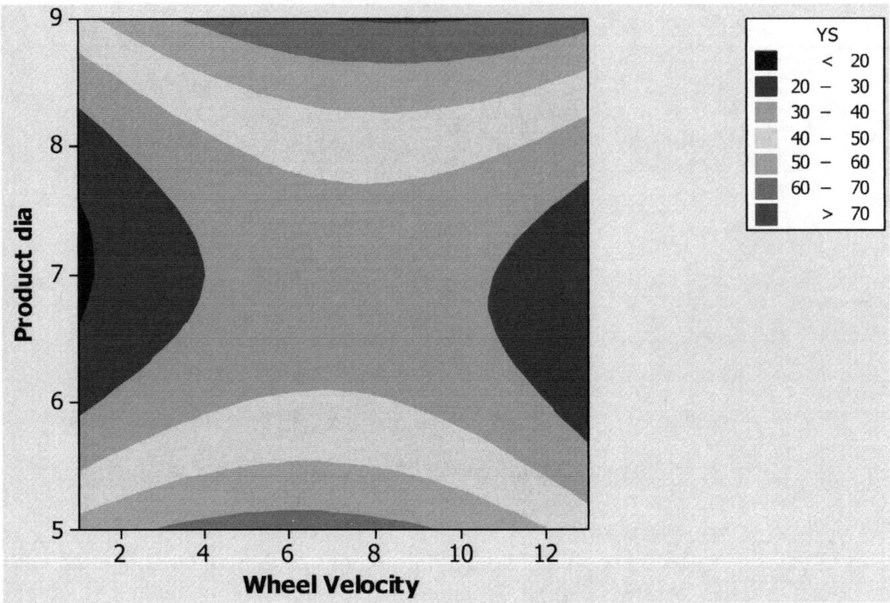

Figure 3. Contour plot of yield strength with variation in wheel velocity and product diameter

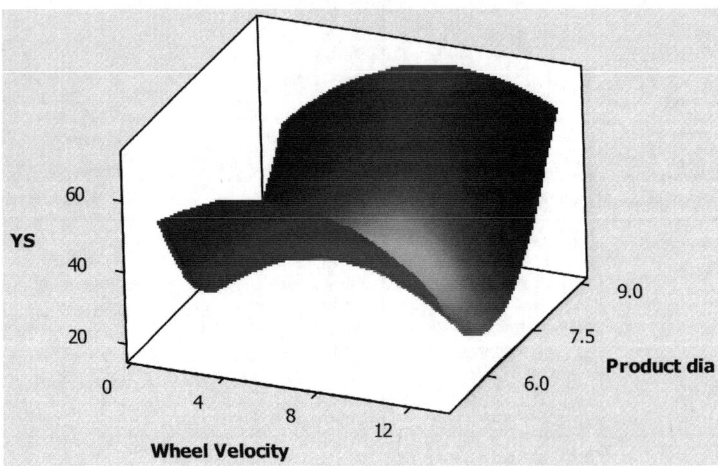

Figure 4. Surface plot of yield strength with variation in wheel velocity and product diameter

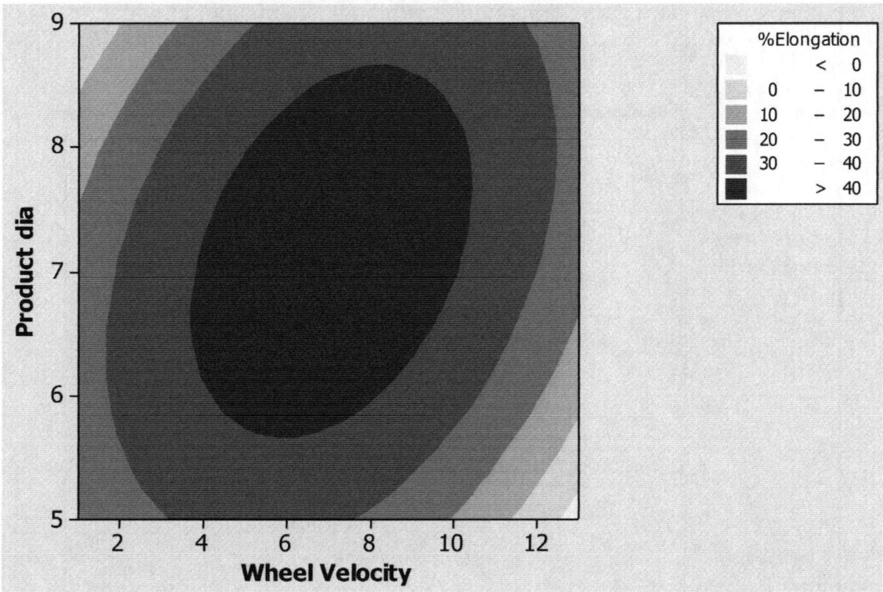

Figure 5. Contour plot of percent elongation with variation in wheel velocity and product diameter

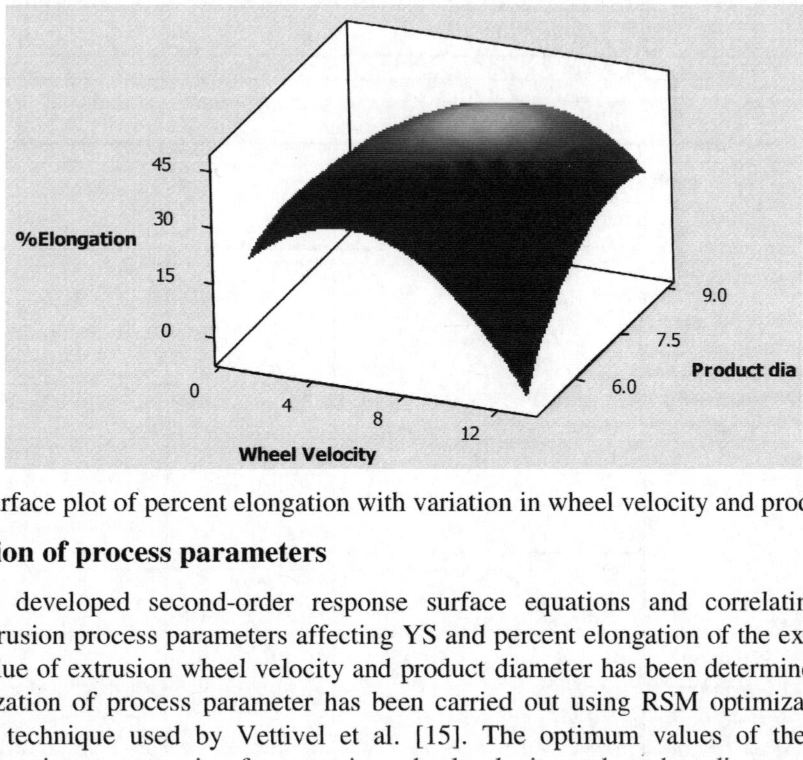

Figure 6. Surface plot of percent elongation with variation in wheel velocity and product diameter

5. Optimization of process parameters

Based on the developed second-order response surface equations and correlating the various continuous extrusion process parameters affecting YS and percent elongation of the extruded product, the optimal value of extrusion wheel velocity and product diameter has been determined. An analysis for the optimization of process parameter has been carried out using RSM optimization technique, similar to the technique used by Vettivel et al. [15]. The optimum values of the input process parameters in continuous extrusion for extrusion wheel velocity and product diameter are 8.63 RPM and 9.0 mm respectively for the optimal value of YS 70.93 MPa as shown in Figure 7. The optimum values of the input process parameters in continuous extrusion for extrusion wheel velocity and product diameter are 7.06 RPM and 7.18 mm respectively for the optimal value of percent elongation 46.46 as shown in Figure 8.

IOP Conf. Series: Materials Science and Engineering **404** (2018) 012052 doi:10.1088/1757-899X/404/1/012052

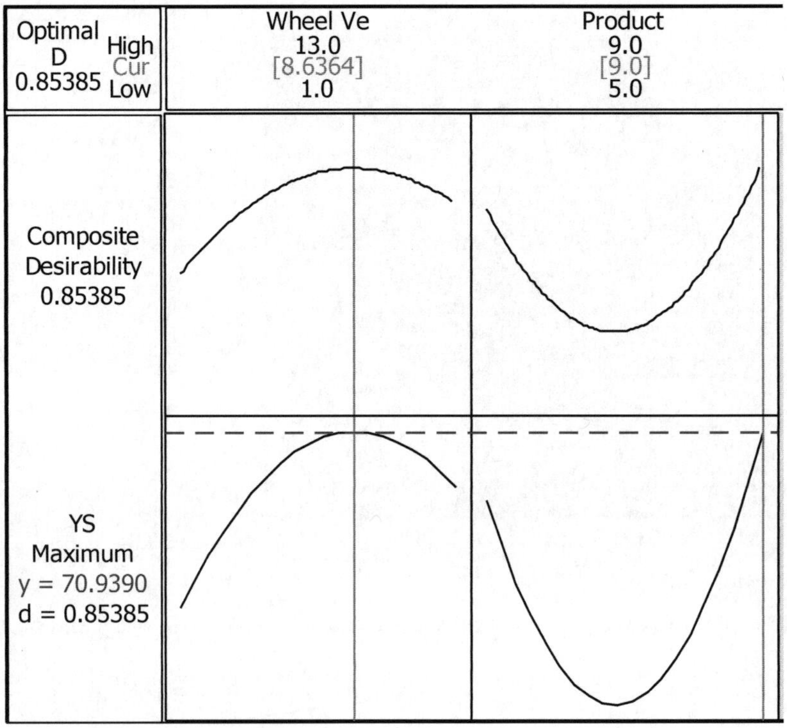

Figure 7. Optimization plot for yield strength with respect to wheel velocity and product diameter

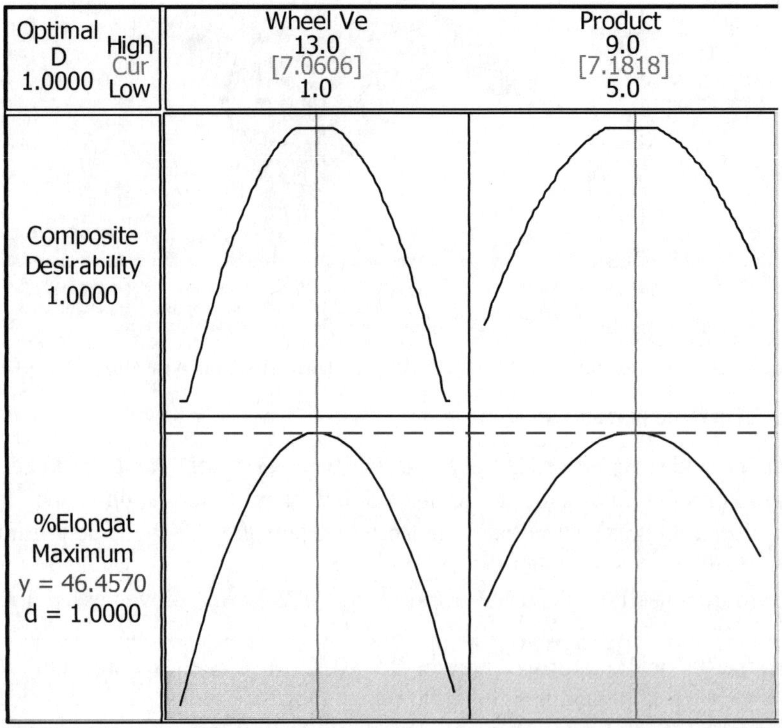

Figure 8. Optimization plot for percent elongation with respect to wheel velocity and product diameter

Table 4. Test for significance of yield strength and percent elongation

Term	Coefficient	*t*	*P*	Coefficient	*t*	*P*
Constant	34.5255	17.631	0.000	46.3793	65.424	0.000
Wheel Velocity	0.9350	0.687	0.514	-0.2500	-0.507	0.628
Product diameter	1.4483	1.064	0.323	1.0000	2.029	0.082
Wheel velocity × Wheel velocity	-3.6547	-3.710	0.008	-5.8200	-16.32	0.000
Product diameter × Product diameter	8.0953	8.217	0.000	-3.2575	-9.134	0.000
Wheel velocity × Product diameter	1.5700	0.666	0.527	3.0000	3.514	0.010

6. Conclusions

Continuous extrusion of pure copper (C101) feedstock has been successfully done. A mathematical model has been developed using the RSM approach. The developed RSM based mathematical modelling has the potential to evaluate YS and percent elongation under various process parameter settings. The adequacy of the developed mathematical model has also been tested through the analysis of variance (ANOVA). Following conclusions can be drawn from this study:

1. The YS increases with the increase of extrusion wheel velocity and is maximum at a particular value of extrusion wheel velocity and then decrease with further increase in the value of extrusion wheel velocity.
2. The percent elongation increases with the increase of wheel velocity and is maximum at a particular value of extrusion wheel velocity and then decrease with further increase in the value of extrusion wheel velocity.
3. The optimal input factor for YS was found as 8.63 RPM and 9.0 mm respectively for the optimal value of Yield Strength 70.93 MPa.
4. The optimal input parameter for percent elongation was found as 7.06 RPM extrusion wheel velocity and 7.18 mm to be product diameter for the maximum percent elongation value of 46.46.

References

[1] Green D 1972 Continuous extrusion of wire sections. *J. Inst. Met.* **100** 296–300.

[2] Green D 1974 Conform – A new concept for continuous extrusion forming of metals. *J. Manuf. Sci. Eng.* **96** 893–900.

[3] Segalina F, De Chiffre L 2017 Material Testing of Copper by Extrusion-cutting. *Procedia CIRP* **58** 375–380.

[4] Kral P, Kvapilova M, Dvorak J, Ponizil P, Sedivy O, Helisova K. *et al.* 2014 Quantitative characteristics of inhomogeneous microstructure in UFG copper. in *IOP Conf. Ser.: Mater. Sci. Eng.* **63** 012137.

[5] Sheppard T, Yiu H L 1984 Microstructure and properties of some extruded copper-phosphorus alloys. Metal Sci. **18** 439–447.

[6] Bazaz B, Zarei-Hanzaki A, Fatemi-Varzaneh S M 2013 Hardness and microstructure homogeneity of pure copper processed by accumulative back extrusion. *Mater. Sci. Eng. A* **559** 595–600.

[7] Kim Y H, Cho J, Jeong H S, Kim K S, Yoon S 1998 A study on optimal design for CONFORM process. *J. Mater. Process. Technol.* **81** 671–675.

[8] Cho J R, Jeong H S 2001 Parametric investigation on the curling phenomenon in CONFORM process by three-dimensional finite element analysis. *J. Mater. Process. Technol.* **110** 53–60.

[9] Wu P, Xie S, Li H, Yan M, Huan G, Cheng L 2007 Effect of extrusion wheel angular velocity on continuous extrusion forming process of copper concave bus bar. *Trans. Nonferrous Met. Soc. China* **17** 280–286.

[10] Zhao Y, Song B, Pei J, Jia C, Li B, Linlin G 2013 Effect of deformation speed on the microstructure and mechanical properties of AA6063 during continuous extrusion process. *J. Mater. Process. Technol.* **213** 1855–1863.

[11] Palanivel R, Mathews P K, Murugan N 2012 Development of mathematical model to predict the ultimate tensile strength of friction stir welded dissimilar aluminum alloy. *Mechanika* **18** 517–523.

[12] Balasubramanian M 2016 Statistical analysis of tensile strength and elongation of pulse TIG welded titanium alloy joints using Weibull distribution. *Cogent Eng.* **3** 1–9.

[13] Niranjan K, Lakshminarayanan P R 2013 Optimization of process parameters for in situ casting of Al/TiB2composites through response surface methodology. *Trans. Nonferrous Met. Soc. China* **23** 1269–1274.

[14] Sinha D K, Kumar S, Kumar A 2012 Virtual Design And Fabrication Of A Continuous Extrusion Setup With Process Analysis. *Int. J. Eng. Res. Technol.* **1** 1–7.

[15] Vettivel S C, Selvakumar N, Leema N, Lenin A H 2014 Electrical resistivity, wear map and modeling of extruded tungsten reinforced copper composite. *Mater. Des.* **56** 791–806.

[16] Seeman M, Ganesan G, Karthikeyan R, Velayudham A 2010 Study on tool wear and surface roughness in machining of particulate aluminum metal matrix composite-response surface methodology approach. *Int. J. Adv. Manuf. Technol.* **48** 613–624.

IOP Conf. Series: Materials Science and Engineering **404** (2018) 012053 doi:10.1088/1757-899X/404/1/012053

Development of mathematical model for material removal and surface roughness in electrolytic magnetic abrasive finishing process

Nitesh Pandey[1], Rajnish Singh[2], Piyush Pant[3], Anshul Yadav[4]

[1]Department of Mechanical Engineering, Shri Ramswaroop Memorial University, Lucknow, India.
[2,4]Department of mechanical engineering Kamla Nehru institute of Technology, Sultanpur, India.
[3]Department of Mechanical Engineering, Jadavpur University, Kolkata, India.

[1]n.pandey86@gmail.com, [2]rajnish.singh.knit@gmail.com, [4]anshulyadav.iitk@gmail.com

Abstract: Electrolytic magnetic abrasive finishing is a hybrid machining process which takes into account both electrolytic & magnetic abrasive finishing process. It is a high efficiency finishing process which is used for advanced engineering materials like super alloys, metal matrix composites and ceramics. In this study mathematical modeling of material removal has been performed by considering the effect of electrode gap on material removal and then surface roughness model has been developed by assuming the surface profile as triangular. This research also investigates the interaction between various input process and their effect on process output characteristics like MRR and surface finish. The simulated data has been validated with the previous literature result which were found to be in good agreement.

1. Introduction

The need for high accuracy and high efficiency machining of difficult to cut materials is making the applications of abrasive finishing technologies increasingly important. The most labor intensive, uncontrollable area is the manufacture of precision parts involves final finishing operation because the cost of surface finish increases sharply for a roughness value of less than one micron. In the field of manufacturing industry it is not only the demand to obtain the surfaces with high quality but it is also the urgent demand to fulfill this demand along with high efficiency also. However it is not possible to obtain these present demands by using single traditional machining processes only for both high quality and high efficiency. Thus, there need a compound finishing process which consist of several process and meets the current demand in the field of finishing having high quality and high finishing. Electrolytic magnetic abrasive finishing is a hybrid machining process which is used for high efficiency and high quality surface. It consist of Magnetic abrasive finishing process and electrolytic process in which magnetic abrasive particles form magnetic abrasive brush in the presence of magnetic field and then perform the finishing process by abrasion. Material is removed in the form of small micro-chips as given by Jain et al

IOP Conf. Series: Materials Science and Engineering **404** (2018) 012053 doi:10.1088/1757-899X/404/1/012053

[1].In electrolytic process material is removed due to the formation of passive layer formed on the surface of workpiece which is kept at anodic potential.

Yan et al[2]have performed experiments on the electrolytic magnetic abrasive finishing process and describes the principles of the process, the finishing characteristics of the surface roughness and material removal, and the associated mechanism. Mischler et al[3]has found out the role of oxide (passive) layer formed due to electrochemical reaction on the steel surface which has significant effect and found out that the formation of passive layer has significant on the increased wear. Garcia et al [4] have introduced the concept of active wear track area to consider the corrosion wear process on passive material. Jayswal et al [5] deals with the theoretical investigation of MAF process and a finite element model of this process has been developed to evaluate the distribution of magnetic forces on the workpiece surface. Kim and Choi [6] had developed a mathematical model of surface roughness to find out surface roughness as a function of finishing time which is based on the amount of material removed in magnetic abrasive machining. Kim and choi [7]have carried out magneto-electrolytic-abrasive-polishing of newly developed Cr coated rollers and their finishing characteristics have been analyzed which were used in video-tape manufacturing industries and also they have theoretically studied the effect of magnetic field on electrolytic process and give the theoretical basis behind the modification of electrolytic ion motion caused by magnetic field. Kim et al. [8]in another research have developed a non woven abrasive pad, which is a kind of efficient machining polishing material which has been used in magneto-electrolytic-abrasive -polishing system. Judal and Yadava [9] has developed response surface methodology model for the prediction of material removal and surface roughness during the machining of non-magnetic stainless steel in cylindrical electrochemical magnetic abrasive machining process (C-EMAM). Though significant work has been reported to simulate the magnetic abrasive finishing process numerically but very little effort has been made toward the modeling of electrolytic magnetic abrasive finishing process.

In the present work a mathematical model of material removal and surface roughness of EMAF process has been done by considering abrasion, electrolytic action considering the effect of electrode gap on material removal. Derivational mathematical model of material removal has been derived then the final surface roughness has been derived by considering total volume of material removed by considering uniform triangular surface profile.

2. Modeling of material removal ratein EMAF process

The total amount of material removed during EMAF process is equal to the sum of material removed during MAF process and electrolytic process.

2.1 Modeling of magnetic abrasive finishing system

For the purpose of brevity the assumptions made during modelling are not listed here and can be referred from [1,2,3,5] . In magnetic abrasive finishing since abrasives are distributed randomly and the primary cutting action is performed by abrasives while ferromagnetic particles supplies abrasion pressure. The number of abrasives taking part in abrasion action beneath each FP may vary from one instance to another instance.It has been assumed that cross-sectional area of magnetic brush is same as that of air gap, therefore the number of magnetic abrasive that are in actual contact with workpiece are [3]:

$$N = \frac{A_0}{(\pi/4)D^2} \quad (1)$$

The force acting on a magnetic abrasive is

$$f = \frac{F}{N} \quad (2)$$

IOP Conf. Series: Materials Science and Engineering **404** (2018) 012053 doi:10.1088/1757-899X/404/1/012053

$$f = \frac{D^2}{4} p \tag{3}$$

The force acting on cutting edge is

$$\Delta f = \frac{f}{n} \tag{4}$$

where n = number of cutting edges participating in the machining for a magnetic abrasive

$$\Delta f = \frac{D^2}{4n} p \tag{5}$$

P= machining pressure applied on the surface of workpiece due to ferromagnetic particles [13]

$$p = \frac{\mu_0 H^2 (1 - \frac{1}{\mu_a})}{2} \tag{6}$$

μ_a = relative permeability of abrasive particles [1]

$$\mu_a = \alpha \mu_{FP} + (1-\alpha)\mu_{ABR} \tag{7}$$

The characteristic shape of magnetic abrasive particle which have been considered during modeling is shown in Fig.1

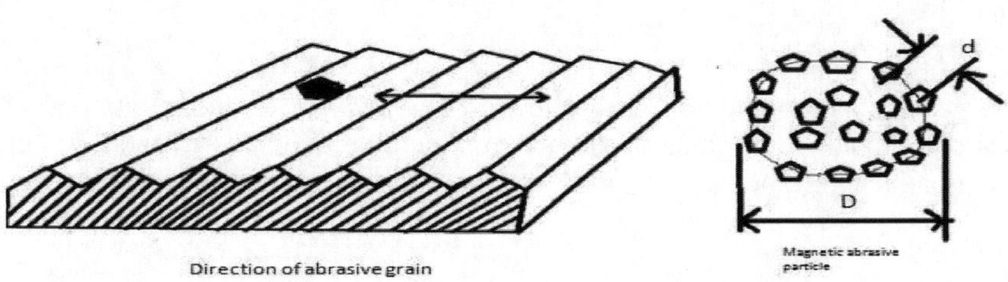

Fig 1 Characteristic shape of magnetic abrasive particle [3]

The total length of workpiece which has to be finished is $l_{tot} = v \times t$

Actual length of contact between workpiece and magnetic abrasive within total finishing length from the above Fig.2

$$l = (1 - \frac{R_a}{R_a^0}) vt \tag{8}$$

Stock of material removed due to MAM by cutting edge in time t =

actual length of contact X area of grain on which force is acting

actual length of contact X total force acting/force acting on grain on unit area

$$= vt(1-\frac{R_a}{R_a^0})[\frac{\Delta f}{H_m \pi \tan\theta_a}]$$

$$W_1 = \pi DNt(1-\frac{R_a}{R_a^0})[\frac{\Delta f}{H_m \pi \tan\theta_a}] \quad (9)$$

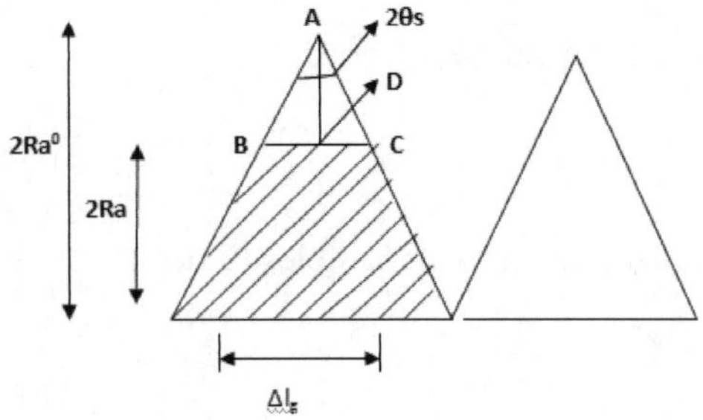

Fig.2.Shape of scratched machined surface

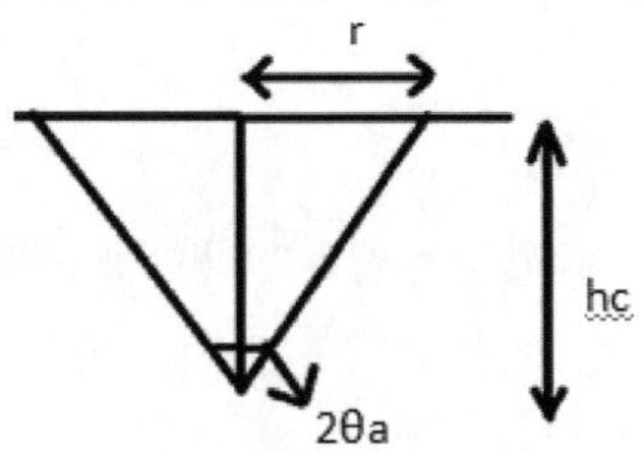

Fig .3 Wedge shaped abrasive

2.2 Calculation of material removal due to electrolytic process considering electrode gap

In electrochemical dissolution process the gap between workpiece and electrode is filled with electrolyte. The electrode is connected to cathode while workpiece is connected to anode of the DC power source. The material is removed atom by atom from the workpiece kept at anodic potential. Material removal in term of volume can be calculated as [10]

$$W_2 = \frac{\eta i t E}{F\rho} \quad (10)$$

In ECM process generally tool (cathode) is fed toward the workpiece (anode) at constant rate. During equilibrium, feed rate of cathode is equal to the rate at which the thickness of anode is being reduced i.e material removal along length**.** The inter-electrode gap between workpiece and electrode is given by[11]

$$y_e = \frac{(V-\Delta V)\eta k E}{\rho f F} \quad (11)$$

So, material removal by considering electrode gap is

$$W_2 = \frac{i t y_e f}{(V-\Delta V)k} \quad (12)$$

Thus, total material removed during EMAF process considering material removal due to magnetic abrasive finishing and electrolytic process considering electrode gap is

$$W = W_1 + W_2 \quad (13)$$

3. Modeling of surface roughness during EMAF process

The quality of surface can be improved by reducing unevenness from the surface profile of workpiece. It is generally reduced by the normal magnetic force which applies machining pressure on the surface of workpiece through magnetic abrasive particles which results in the penetration of cutting edges of abrasive grains in the workpiece. Due to rotation of MAPs, grooves are formed on the workpiece surface which decides the surface profile after the EMAF. Surface roughness is determined on the basis of the

IOP Conf. Series: Materials Science and Engineering **404** (2018) 012053 doi:10.1088/1757-899X/404/1/012053

surface profile achieved by equating the volume of the material removed to the volume of groove produced. It has been assumed that the initial surface of the workpiece have uniform triangular profile as shown in Fig.2. It has been assumed that initial arithmetic mean surface roughness of workpiece is $2R_a^0$. After processing time of T during EMAF process the final surface roughness achieved is $2R_a$ It has been assumed that the valleys of the surface irregularities are passivated initially, which prevents further electrolytic process. The passive film formed at valleys is inaccessible to magnetic abrasive brush during MAM, hence the major amount of material is removed from the peaks of the surface profile and the peaks get truncated as shown in Fig.2. The final surface roughness can be obtained from the volume of material removed during the EMAF process.

Thus total volume of material removed in a cell is

$$W = \frac{\left(R_a^0 - R_a\right)^2}{R_a^0} \times l_w \times \pi \times d_w \qquad (17)$$

By re-arranging equation 17 we can found out final surface roughness of the workpiece. Thus the final surface roughness of workpiece is

$$R_a = R_a^0 - \sqrt{\frac{R_a^0 \times W}{\pi d_w \times l_w}} \qquad (18)$$

From the above equation the final surface roughness of the workpiece can be calculated if the total volume of material removed from the machining zone is known. A critical surface roughness may exist in the machining region due to indentation of the cutting edge on the workpiece surface and due to the presence of this critical surface roughness the surface roughness of workpiece will not improve further. The shape of wedge shaped abrasive is shown in Fig.3.

Radius of conical shaped abrasive can be calculated as

$$r = h_c \tan\theta_a \qquad (19)$$

Force acting on single cutting abrasive is [3]

$$\Delta f = H_{mt} \pi r^2 \qquad (20)$$

$$\Delta f = H_{mt} \pi (h_c \tan\theta_a)^2 \qquad (21)$$

Thus the value of critical depth of penetration is

$$h_c = \frac{1}{\tan\theta_a} \sqrt{\frac{\Delta f}{H_{mt}\pi}} \qquad (22)$$

Since the critical surface roughness is the mean of depth of penetration

$$(R_a)_{crt} = {h_c}/{2} \qquad (23)$$

Higher the value obtained from both the (18) and (23) will be taken as final surface roughness.

IOP Conf. Series: Materials Science and Engineering **404** (2018) 012053 doi:10.1088/1757-899X/404/1/012053

4. Results and Discussion

In the present work, a MATLAB code has been written for the material removal and surface roughness model of EMAF process. The simulation result of EMAF has been compared with the experimental result [9]of material removal and surface roughness with respect to machining time. The process parameters which has beenused for the validation of EMAF model has been given in the table 1.Fig.4 shows the simulation result of material removal at 0.5A electrolytic current at different workpiece rotational speed of 200 and 500 rpm as shown in Fig.4 (a) and (b) respectively. It can be observed that material removed at 200 rpm is 22 mg while it is 35 mg at 500 rpm for a finishing time of 5 min for a given electrolytic current of 0.5A. Higher material removal at higher rotational speed is due to the reason that higher rotational speed can cause the easy removal of passive film formed on the workpiece surface. The simulation result is close to the experimental result [9] and also follow nearly similar trend. Fig.5shows the simulation result of surface roughness 0.5A electrolytic current for 200 and 500 rpm of workpiece rotational speed. It can be observed from Fig.5 (a) that at 200 rpm the surface roughness can be improved to 0.03μm Ra after the finishing time of 5 min while it can be seen from Fig.5(b) that at 500 rpm surface roughness can be improved to 0.017 μm Ra. This is due to the reason that at smaller rate of rotation less amount of material is removed. Thus higher rate of rotation is associated with better surface roughness. The trend of simulated result of surface roughness is same as the experimental result [9].

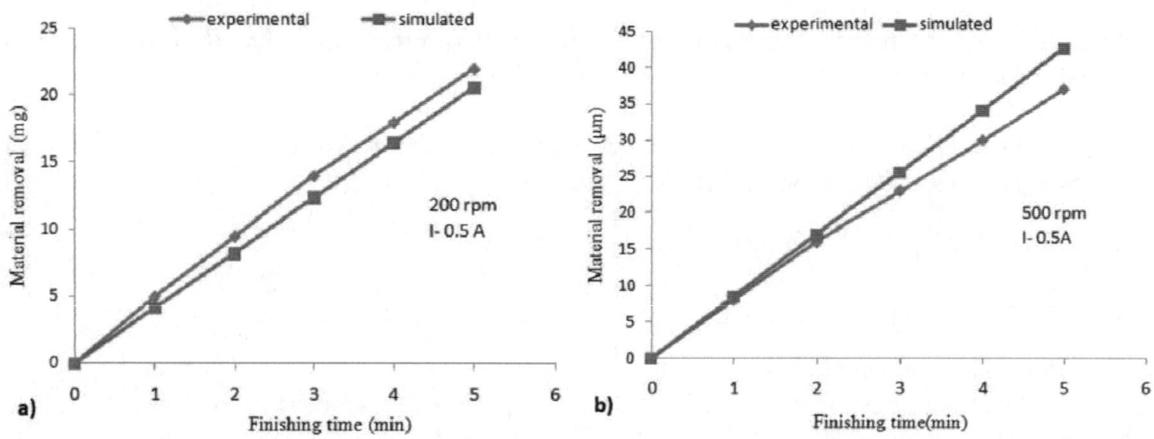

Fig.4. Variation of material removal with finishing time for various rate of workpiece rotation (a) for 200 rpm (b) for 500 rpm at electrolytic current of 0.5A

Fig.6. (a) and (b) shows the simulation result of material removal at an electrode gap of 3mm and 5mm respectively, it can be observed that smaller the electrode gap is more will be the material removal however the difference is not much more. During the processing rate of workpiece rotation is kept at 500 rpm while electrolytic current is kept at 0.5 A. Due to more material removed at smaller electrode gap surface roughness obtained at smaller electrode gap is better than that obtained at larger electrode gap Fig.7 (a) and (b) shows the surface roughness obtained after the finishing time of 5 min at an electrode gap of 3 mm and 5 mm respectively, the surface roughness obtained at the electrode gap of 3 mm is 0.017μm and 0.021μm at an electrode gap of 5 mm. The reason of less material removal and poor surface roughness at higher electrode gap is when the electrode gap increase, ions are acted upon by Lorentz force and under this effect the anion cannot reach the electrode surface to participate in the reaction and thus electrolytic reaction decreases and thus material removal also decreases. It can also be seen that surface roughness decreases rapidly at the start of the process and then the decrease in roughness is quite slow, this is due to the reason that the peak of the surface roughness before the start of the process is much higher so that they can be removed

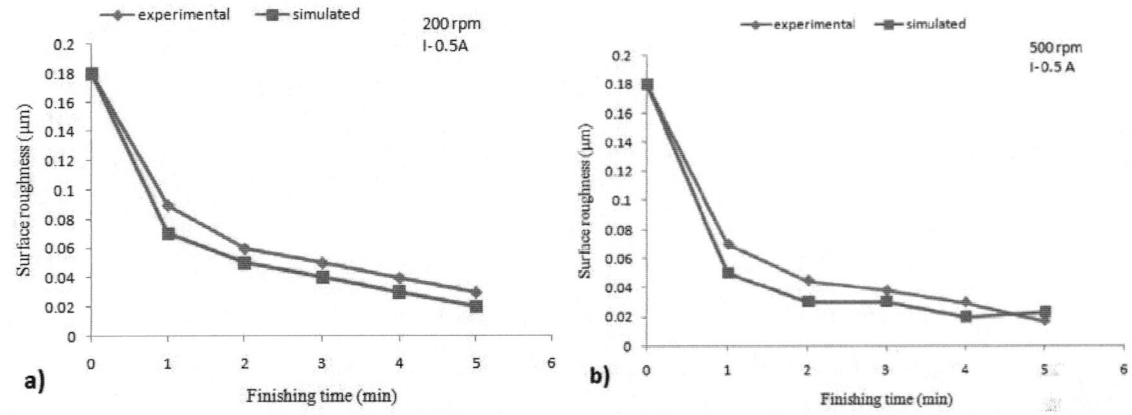

Fig.5. Variation of surface roughness with finishing time for various rate of workpiece rotation (a) for 200 rpm (b) for 500 rpm at an electrolytic current of 0.5A

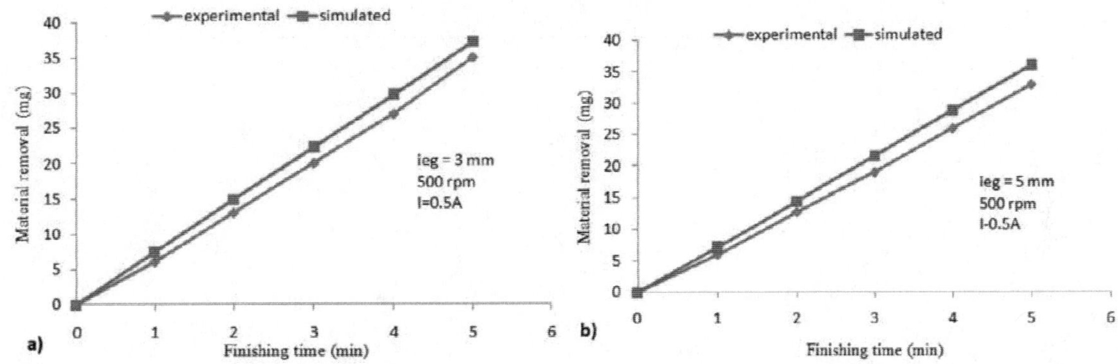

Fig.6. Variation of material removal with finishing time at different electrode gap (a) for 3mm of electrode gap (b) with 5mm of electrode gap at 500 rpm and an electrolytic current of 0.5A

easily, but once the peaks get removed then the height of roughness peaks get smaller so it becomes quite difficult to remove the roughness peaks latter. Thus, surface roughness improves rapidly at beginning and less at the end of the process. The simulation result is close to experimental result [9] and follow nearly similar trend. Fig 8(a) and 8(b) shows the simulated result of material removal at an electrolytic current of 0.5A and 1.5A respectively at 200 rpm and 500 rpm respectively. It can be observed that higher the electrolytic current more will be the material removal. From Fig. 8 (a) when the workpiece revolution is 200 rpm, EMAF with electrolytic current of 1.5 A will cause 55 mg of material removal while it is only 35 mg at 0.5 A of electrolytic current. This is due to the fact that higher the electrolytic current, more will be the formation of passive film at the workpiece surface and easier will be the material removal. From Fig.8 (b) it can be observed that when the workpiece revolution is 500 rpm, material removal at 0.5A current and 1.5 A current is more as comparison to that at 200 rpm. Reason behind the removal of more material at higher workpiece rotation is, at higher speed it's much easier to remove the passive film which

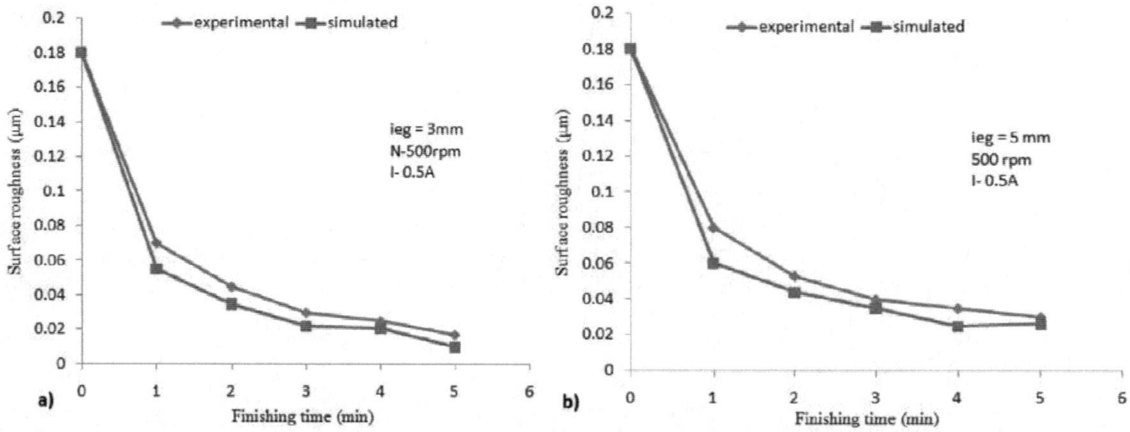

Fig.7. Variation of surface roughness for various inter electrode gap (ieg) (a) for 3mm (b) for 5mm at N=500 rpm ,0.5A of electrolytic current.

is formed on the surface in comparison to lower rate of rotation. The computed result of material removal follows the same trend and agrees with the experimental result.

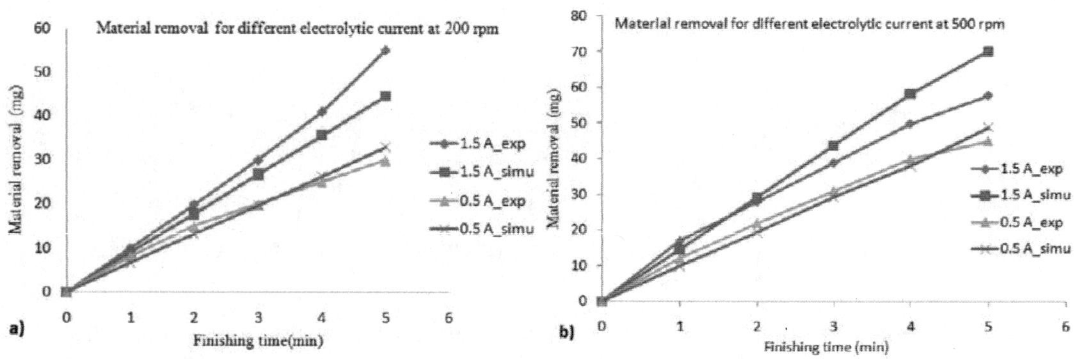

Fig .8 Effect of electrolytic current a) at 200 rpm, b) at 500 rpm on material removal

5. Conclusions

A model for material removal by considering the electrode gap has been developed and its effect on the surface roughness has been considered by developing a surface roughness model considering triangular profile. The effect of electrode gap has been captured in the model of material removal. From result and discussion following points have been drawn. .

1. EMAF is a hybrid machining process which improves material removal and surface roughness drastically in comparison to the constituent process when they are applied individually.

2. Material removal becomes more when higher rate of workpiece revolution is used in comparison to lower rate of revolution.

3. Using higher electrolytic current more material removal and better surface finish is obtained in comparison to lower electrolytic current because of more formation of passive film at higher electrolytic current.

4. Lower inter-electrode gap will cause more material removal while higher inter-electrode gap will cause less material removal, however the difference is not much more.

5. Simulation result of the model being used agreed with the previous literature result and follows the same trend.

References

[1] Singh. D.K, Jain, V.K , Raghuram ,V. Experimental study into forces acting during Magnetic abrasive finishing process. *International journal of Advanced Manufacturing Technology 30 (2006), 652-662.*

[2] Yan,B.H.,Chang,G.W., Cheng,T.J, Hsu,R.T.Electrolytic Magnetic abrasive finishing. *International Journal of Machine Tool and Manufacture. 2003; 43,1355-1366.*

[3] Mischler,S.,Debaud,S.,Landolt,D.Wear accelerated corrosion of passive metal in tribocorrosion system. *Journal of the Electrochemical society 1998; 145, 750-758*

[4] Garcia, I., Drees, D., Celis, J.P. Corrosion-wear of passivating materials in sliding contacts based on a concept of active wear track area. *Wear2001; 249,452–460.*

[5] Jayswal, S.C., Jain, V.K., Dixit, P.M. Modeling and simulation of magnetic abrasive finishing process. *International journal of Advanced Manufacturing Technology, 2006; 26,477-490.*

[6] Kim, J.D, Choi, M.S.Simulation for the prediction of surface accuracy in magnetic abrasive machining, *Journal of material processing technology 1995; 53,630-642.*

[7] Kim, J.D, Choi M.S, Development of the magneto–electrolytic–abrasive polishing system (MEAPS) and finishing characteristics of a Cr-coated roller, *Int. J. Mach. Tools Manuf. 1997; 37 (7) 997-1006.*

[8] Kim, J.D, Xu Y,M, Kang Y.H, Study on the characteristics of magneto–electrolytic–abrasive polishing by using the newly developed nonwoven-abrasive pads *Int. J. Mach. Tools Manufacturing. 1998; 38, 1031-1043.*

[9] Judal, K.B.,Yadava,V.Cylindrical electrochemical magnetic abrasive machining of AISI-304 stainless steel. *Material and Manufacturing Processes 2013; 28,449-456.*

[10] Shinmura,T, Takazawa,K. Hatano, E, Study on Magnetic Abrasive finishing on process principle and a few finishing characterstic, 1986*JSPE 52(5),851-856.*

[11] Jain VK (2002) Advanced machining processes. *Allied Publishers, Delhi.*